Evolution of Peculiar Red Giant Stars

International Astronomical Union
Union Astronomique International

Evolution of Peculiar Red Giant Stars

*Proceedings of the 106th colloquium of the
International Astronomical Union
Held in Bloomington, Indiana, USA
27–29 July 1988*

Edited by

HOLLIS R. JOHNSON
Indiana University, Indiana, USA

BEN ZUCKERMAN
University of California, California, USA

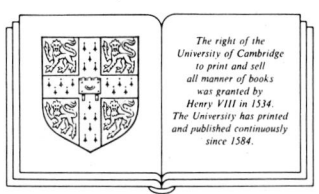

CAMBRIDGE UNIVERSITY PRESS

Cambridge

New York   Port Chester

Melbourne   Sydney

Published by the Press Syndicate of the University of Cambridge
The Pitt Building, Trumpington Street, Cambridge CB2 1RP
32 East 57th Street, New York, NY 10022, USA
10 Stamford Road, Oakleigh, Melbourne 3166, Australia

© Cambridge University Press 1989

First published 1989

Printed in Great Britain at the University Press, Cambridge

British Library cataloguing in publication data available

Library of Congress cataloguing in publication data available

ISBN 0 521 36617 8

CONTENTS

Foreword	ix
Organizing Committees	xiii
List of Participants	xvi

1. Properties and Kinematics of the Peculiar Red Giant Stars — 1

Spectral Classification of Peculiar Red Giants (Invited paper) *P. C. Keenan*	3
The Spatial, Temporal, and Photometric Properties of AGB Stars (Invited paper) *S. G. Kleinmann*	13
The Kinematics of Peculiar Red Giants (Invited paper) *M. W. Feast*	26
Peculiar Red Giants in External Galaxies (Invited paper) *H. B. Richer*	35
The Ratio of the Numbers of Carbon Stars to M Stars in Galaxies *Yu. L. Frantsman*	51
Identification of the (+2) Sequence of the CN Red System ($A^2\Pi$–$X^2\Sigma$) in Carbon Stars *Y. Fujita*	52
Spectroscopy and Photometry of the Southern R Stars *T. Lloyd Evans*	53
Classification of C-Rich Stars According to Their Mid-IR Signatures *R. Papoular*	54
Multi-Wavelength Observations of the Peculiar Red Giant HR 3126 *J. E. Pesce, R. E. Stencel, J. Doggett, F. M. Walter, P. A. Whitelock and J. Dachs*	55
Kinematics of Short-Period Mira Variables – A Progress Report *J. Hron*	56
Photometric Observations of RCB Stars in the LMC *W. A. Lawson and P. L. Cottrell*	57
The Infrared HR Diagram – The IRAS Two-Colour Diagram of AGB Stars and Its Evolutionary Explanations *Jiang Shi-Yang*	58
A Photometric Monitoring of Barium Stars *A. Jorissen and J. Manrfroid*	59
Are the Barium and Am Stars Related? *J. Hakkila*	60
Molecular Bands in the 1.1–1.4 μm Spectra of M-S-C Stars *K. H. Hinkle, D. L. Lambert and R. F. Wing*	61
Photometric Classification of Carbon-Rich Stars *V. Straizys*	62
A Deep Visual-Red and Near Infrared Objective Prism Spectral Survey of the Milky Way *O. M. Kurtanidze and M. G. Nikolashvili*	63

2. Model Photospheres and Chemical Compositions 65

Photospheric Models for Cool Giant Stars (Invited paper) 67
M. S. Bessel and M. Scholz
Testing the Models Against Observations (Invited paper) 81
T. Tsuji
The Chemical Composition of Asymptotic Giant Branch Stars (Invited paper) 101
D. L. Lambert
The Role of Technetium in the Evolution of Red Giants (Invited paper) 131
I. R. Little-Marenin
Long-Period Radial-Velocity Variations of Arcturus 144
A. W. Irwin, B. Campbell, C. L. Morbey, G. A. H. Walker and S. Yang
S-Process Dispersion in G and K Field Giants 145
A. McWilliam
Heavy Element Abundances in FG Satittae 146
T. Kipper and M. Kipper
HD 39853: A High Velocity K5III Star With an Exceptionally Large Li Content 150
R. G. Gratton and F. D'Antona
Hydrogen Deficiency in Peculiar Red Giants 151
R. F. Wing and P. Saizar
Relative CNO Abundances in Upper AGB Stars of the Magellanic Clouds: A Search for
    Envelope Burning 152
J. M. Brett and M. S. Bessell
Fluxes in M Giants With Improved Water Vapor Opacity 153
D. R. Alexander, G. C. Augason, J. A. Brown and H. R. Johnson
S-Process Deficiencies in Low-Mass Supergiant Variables 154
H. E. Bond and R. E. Luck
The Spectrum of TX Psc 155
U. G. Jørgensen
BD-21.3873: An Heavy Element-Rich Symbiotic? 156
A. Jorissen
Comparison of Blanketing in a 3000 K, Oxygen Rich, Spherically Symmetric Model with the
    Blanketing in Non-Mira, M Giant Stars 157
G. C. Augason, J. A. Brown and D. R. Alexander

3. Evolution of Peculiar Red Giant Stars 159

Evolution and Mixing on the AGB (Invited paper) 161
J. Lattanzio
s-Process Enrichment in Low-Mass AGB Stars (Invited paper) 176
R. Gallino
The Role of Binarity in the Evolution of Peculiar Red Giants (Invited paper) 196
R. D. McClure
Peculiar Red Giants – What Kind of White Dwarfs Do They Become? (Invited paper) 205
I. Iben, Jr.
A Statistical Study of Spectroscopic Binaries Containing A Late-Type Giant Star 222
H. M. J. Boffin
Theoretical and Observational Tests for the Mass Transfer Scenario of Ba II Stars 223
H. M. J. Boffin
The Evolution of Stars on the AGB: The Mass Loss Intensity and the Formation of Carbon Stars 224
Yu. L. Frantsman
Evolution of a Star of 7 $M_\odot$ With Mass Loss 225
H. Runqian and J. Suyun
Red Giant Mass Loss and Planetary Nebula Formation 226
P. J. Huggins and A. P. Healy
Neutron Nucleosynthesis in a Low-Mass, Low-Metallicity AGB Star 227
D. Hollowell and I. Iben Jr.

Contents

The Effects of Main-Sequence Mass Loss on Surface C/N Abundance Ratios During the Ascent
  of the First Giant Branch 228
J. A. Guzik and T. E. Beach
The Production of Low Mass Carbon Stars: Carbon-Rich Dredge Up or Oxygen-Rich Mass Loss? .. 229
R. E. Stencel, J. E. Pesce and K. M. MacGregor
Early Shaping of Asymmetric Planetary Nebulae 232
N. Soker
IRAS 21282+5050: A Transitional Planetary Nebula 233
L. Likkel, M. Morris, A. Omont and T. Forveille
5 Ceti: A Long-Period Binary Evolving Through Mass Exchange 234
J. A. Eaton
The N/O–Core Mass Relation in Planetary Nebulae 235
J. B. Kaler, R. A. Shaw and K. B. Kwitter
Models of AGB Stars Envelopes and Atmospheres 236
M. Forestini
Comparison of s-Processing Occurring in a Low Mass AGB Star of Low Metallicity and the
  Results of s-Classical Analysis 237
M. Busso, G. Picchio, R. Gallino, F. Kappeler and C. M. Raiteri
Evolution of a 10 $M_\odot$ Star With LMC Metallicities 238
N. Rathna Sree and A. Ray

4. The Variability–Evolution Connection   239

Observations of Long Period Variable Stars (Invited paper) 241
T. Lloyd Evans
Irregular Red-Giant Variable Stars (Invited paper) 258
M. Querci and F. Querci
Model Atmospheres with Periodic Shocks (Invited paper) 269
G. H. Bowen
Evolution of Oxygen-rich and Carbon Stars on the Asymptotic Giant Branch (Invited paper) .. 284
S. Kwok, K. M. Volk and S. J. Chan
IRAS LRS Spectral Class and Light Curve of M & S Miras 292
M. S. Vardya
Pulsational Periodicities in R CrB 293
J. D. Fernie
IRAS Infrared Fluxes of RV Tauri Stars 294
W. P. Bidelman
Another Look at the RV Tauri Period–Luminosity Relation 295
G. M. Wahlgren
Spectroscopic and Polarimetric Observations of AC Her and UU Her 296
M. A. Nook, J. A. Cardelli and K. H. Nordsieck
Shock-Induced Behavior of Atomic Species in LPV Atmospheres 297
J. N. Pierce
Investigation of a Sample of Long-Period Variable Stars Possessing Maser Emission 298
I. L. Andronov, L. S. Kudashkina and G. M. Rudnitskij
Line Brightness Variations in Alpha Orionis – Phase Differences and Radial Velocities 299
P. Joras

5. Chromospheres, Winds and Mass Loss   301

Chromospheres of Chemically Peculiar Giant Stars (Invited paper) 303
P. G. Judge
Molecular Radio Line Observations of Mass Loss From Red Giants (Invited paper) 321
H. Olofsson
Mass-Losing Peculiar Red Giants: The Comparison Between Theory and Observations (Invited paper) 339
M. Jura
Pre-Planetary Nebulae and R Corona Borealis Stars (Invited Paper) 348
D. Schönberner

Carbon Stars with Oxygen-Rich Circumstellar Chemistry — 359
M. S. Vardya

The Violet Flux Deficiency of Cool Carbon Stars — 366
D. G. Luttermoser and H. R. Johnson

Observations and Models for Red Giants with Unusual Dust — 367
I. Griffin, C. J. Skinner and B. R. Whitmore

CO in OH/IR-Stars – on Excitation and Mass Loss — 368
A. Heske, H. J. Habing, W. E. C. J. Van der Veen, A. Omont and T. Forveille

Pulsation and its Role for Circumstellar Features — 369
A. Heske

IRAS09371+1212: A Unique Red Giant With Strong Emission in the 40–70 Micron Bands of Ice — 370
A. Omont, S. H. Moseley, T. Forveille, P. M. Harvey and L. Likkel

ER Del: A True Symbiotic Star? — 371
H. R. Johnson and T. B. Ake

Fluorescence in the Outer Atmospheres of Red Giant Stars — 372
K. G. Carpenter

Summary of Some Observations of Peculiar Red Giants With the IRAM 30M Radiotelescope — 373
A. Omont, S. Guilloteau, R. Lucas, J. J. Benayoun, J. Cernicharo, E. Nercessian, M. Jura, M. Morris, P. F. Goldsmith and D. C. Lis

Grains of Meteorites, Originating in Cool Carbon Stars — 374
U. G. Jørgensen

He I λ10830 in the S Star HR 1105 — 378
J. A. Brown, H. R. Johnson and K. H. Hinkle

Some Condensation Calculations Using Chemical Equilibria — 379
C. M. Sharp

OH Maser Survey of Very Cool IRAS Sources — 380
P. Te Lintel Hekkert, A. Heske and A. M. Le Squeren

Molecular Cooling in the Outer Atmospheres of Red Giants — 381
D. Muchmore

A Unified Formula for Mass Loss Rate of O to M Stars and its Effect on Stellar Evolution — 382
S. P. Tarafdar

On the Nature of Radio Emission of Late-Type Giants — 383
G. M. Rudnitskij

Possible Proto-Planetary Nebulae — 384
M. Parthasarathy

IRAS Observations of Symbiotic Stars — 391
M. Parthasarathy and H. C. Bhatt

Formation of Helium Lines and Continua in a Late-Type Giant Star — 401
R. De La Reza and C. Batalha

The 8–22 μm Excess in Carbon Stars From IRAS LRS Spectra — 402
S. J. Little and I. R. Little-Marenin

Absorption Spectrum of R CrB During the Light Minimum — 403
N. Kameswara Rao, S. Giridhar and B. N. Ashoka

The Variability of Water Maser Emission Associated with Long Period Variable Stars — 404
I. R. Little-Marenin, P. J. Benson and D. Goodwin

6. Outstanding Problems in Research on Peculiar Red Giant Stars — 405

Panel Discussion: — 407
A. E. Glassgold, Chair, B. Gustafsson, A. Renzini, G. Wallerstein, P. R. Wood and B. M. Zuckerman

Index — 435

FOREWORD

Red-giant stars display fascinating physical processes and regimes -- among them double-shell nuclear burning, complex mechanisms of convection and mixing, such peculiarities in surface chemical composition as enhancement of carbon and s-process elements, extended photospheres, pulsations, light variability, shocks, chromospheres, grain formation, circumstellar envelopes, and mass loss -- a wonderful variety of unusual natural experiments. At the same time the most evolved of these stars stand poised to shed their remaining envelopes as planetary nebulae -- indeed, some have already done so -- and enter a long cooling phase as white dwarfs.

This book reports the proceedings of Colloquium 106 of the International Astronomical Union, which considered red giant stars, focusing specifically on the chemically peculiar red giants. It was held July 27-29, 1988 at Indiana University, with its own long tradition of research in red-giant stars. The spacious Indiana Memorial Union Building provided a comfortable and stimulating setting for the conference.

Arising from a suggestion of the Working Group on Peculiar Red Giant Stars of IAU Commission 29, the conference was endorsed by the presidents of IAU Commissions 27, 29, 35, 36, and 45. We are indebted to the Chairpersons of these commissions -- Bela Szeidl, Guisa Cayrel de Strobel, Daiichiro Sugimoto, Keiichi Kodaira, and Robert F. Garrison, respectively -- for their foresight and timely support. Originally conceived as a sequel to the 1984 Strasbourg meeting on "Cool Stars with Excesses of Heavy Elements", the conference assumed a broader perspective and became in effect a continuation to several recent significant conferences. Interest in the late stages of stellar evolution continues at such a pace that frequent meetings are important to maintain community among workers.

Our goal was to trace and explain the evolutionary status and history of the many groups of peculiar red-giant stars, such as M, MS, S, SC, C, Ba, R, and CH stars. In particular, we were interested in discussing the properties -- luminosity, temperature, composition, binarity, and galactic distribution -- of each group of stars and then relating these to initial mass and composition.

Generous financial support by the International Astronomical Union helped provide travel assistance for several scientists. That support was a necessary and deeply appreciated part of the meeting, and the resulting international flavor of the conference was evident. Derek McNally, Assistant General Secretary of the IAU, was encouraging and helpful. Indiana University's office of Research and Graduate Development contributed significantly to the conference by a large grant, for which we thank its director, Morton Lowengrub (now Dean of the College of Arts and Sciences), who also officially opened the conference.

Meetings do not plan themselves. The Scientific Organizing Committee consisted of Michael Feast, Bengt Gustafsson, Carlos Jaschek, Hollis Johnson (Chairman), James Kaler, Robert McClure, Alvio Renzini, Vitautas Straizys, Allen Sweigert, Takashi Tsuji, Peter Wood, and Ben Zuckerman. Many suggestions regarding topics and speakers came from this group, and they are in large measure responsible for the excellent scientific content of the meeting. The hard-working Local Organizing Committee, headed by Kent Honeycutt, consisted of Haldan Cohn, Richard Durisen, Hollis Johnson, Paula Jentgens, Phyllis Lugger, and Karyl Robb. All conference participants benefited from the excellent arrangements.

The Conference Bureau of Indiana University and its representative, Karyl Robb, managed smoothly many details of the meeting. Much of the typing and mailing was handled by the secretaries of the Astronomy Department, Paula Jentgens and Brenda Records. Many graduate students pitched in to help with transportation, audio-visual equipment, and many other small and large tasks.

Foreword

Richard Durisen, Chairman of Astronomy, was strongly supportive. Icko Iben entertained the participants with his after-dinner speech. Ann Honeycutt, Barbara Burkhead, and Grete Johnson helped spouses enjoy the event. Thanks to all!

These proceedings follow the format of the meeting; each (half-day) session begins with four invited papers and concludes with poster papers. Chairpersons of the five sessions of invited papers and discussion were R.F. Wing, M.S. Vardya, C. Chiose, V. Straizys, and J.B. Kaler, respectively. The full text of the invited papers and an abstract of all contributed papers are printed. A few short papers selected by the editors have also been included. Finally we present the full transcript of the final lively panel discussion. We hope this volume will be of substantial value in furthering interest and research on the evolution of the peculiar red-giant stars.

December 1988

Hollis R. Johnson
Ben M. Zuckerman

Organizing Committee

IAU COLLOQUIUM NO. 106

EVOLUTION OF PECULIAR RED GIANT STARS

July 27-29, 1988

Bloomington, Indiana, U. S. A.

Sponsored by Indiana University and
IAU Commission 27: Variable Stars
IAU Commission 29: Stellar Spectra
IAU Commission 35: Stellar Constitution
IAU Commission 36: Theory of Stellar Atmospheres
IAU Commission 45: Stellar Classification

Scientific Organizing Committee

M.W. Feast (South Africa) (IAU Comm. 27)
B. Gustafsson (Sweden) (IAU Comm. 36)
C.O. Jaschek (France) (IAU Comm. 29)
H.R. Johnson (USA) (Chairman)
J.B. Kaler (USA)
R.D. McClure (Canada) (IAU Comm. 45)

A. Renzini (Italy)
V. Straizys (Lithuania, USSR)
A.V. Sweigart (USA) (IAU Comm. 35)
T. Tsuji (Japan)
P.R. Wood (Australia)
B.M. Zuckerman (USA)

Local Organizing Committee

H.N. Cohn
R.H. Durisen
R.K. Honeycutt (Chairman)
H.R. Johnson

P.P. Jentgens
P.M. Lugger
K. Robb

IAU Colloquium No. 106
Evolution of Peculiar Red Giant Stars
July 27-29, 1988
Bloomington, Indiana, USA

Participants on Picture

ROW 1 (Bottom) Left to Right

Haldan Cohn, Phyllis Lugger, Kent Honeycutt, Hollis R. Johnson, Glenn M. Spiczak, Benjamin F. Peery, John P. Petrakis, Richard H. Durisen (with hat), Shelby Yang, Irene Little-Marenin, Uffe Grae Jorgensen

ROW 2 Left to Right

Lauren Likkel, Robert McClure, Mark Nook, M.S. Vardya, James White, Danny R. Faulkner, Stephen J. Shawl, Stephen J. Little, James Pierce, Yoshio Fujita, Rao Kameswara, George Wallerstein, Al Glassgold

ROW 3 Left to Right

Icko Iben, Jr., Thomas Beach, Jeffery Brown, Brian Murphy, Run-Qian Huang, Shi-yang Jiang, Peter Wood, Per Joras, Renaus Papoular

ROW 4 Left to Right

Amos Harpaz, Joyce A. Guzik, T. Lloyd Evans, Phil Lockett, Ben Zuckerman, Ian Griffen, David Lambert, P.J. Huggins, Raffaele Gratton, Manuel Forestini, David Muchmore, Michael Bessell, John Brett

ROW 5 Left to Right

David Hollowell, Alessandro Chieffi, Kenneth Carpenter, John Lattanzio, Stephen Becker, Howard E. Bond, Romas Mitalas, M. Parthasarathy, Robert F. Garrison (with cap), Steve Cederbloom, Monique Querci, Andrew McWilliam

ROW 6 Left to Right

Roberto Gallino, Vittoria Caloi, Glenn Wahlgren, Alak Ray, Robert Stencel, John E. Littleton, Philip Judge, Inge Thiering, Ingrid Wing, Robert Wing, Sun Kwok, Susan Kleinmann, Josef Hron

ROW 7 Left to Right

Noam Soker, Christopher Sharp, Tae Seog Yoon, Detlef Schonberner, Don Fernie, Ernst Dorfi, Gordon Augason, Shankar Prosad Tarafdar, David R. Alexander, Frank K. Edmondson, W.P. Bidelman, John Hakkila, Donald G. Luttermoser, Lee Anne Willson, Astrid Heske, Alain Jorissen

ROW 8 Left to Right

Oscar Straniero, Dorothy D. Locanthi, Alan W. Irwin, Bengt Gustafsson, Philip Keenan, Michael Feast (behind Keenan), Harvey Richer, Michael Jura, Cesare Chiosi, Paul Bode, Takashi Tsuji, Henri Boffin, Bob Grabhorn, Hans Olofsson, George Bowen, Yitzchak Tuchman

EVOLUTION OF PECULIAR RED GIANT STARS

(IAU COLLOQUIUM NO. 106)

LIST OF PARTICIPANTS

Name	Affiliation
Alexander, David R.	Wichita State University
Augason, Gordon	NASA–Ames Research Center
Beach, Thomas E.	Iowa State University
Becker, Stephen A.	Los Alamos National Laboratory
Bessell, Michael S.	Mt. Stromlo Observatory
Bidelman, William P.	Warner & Swasey Observatory
Bode, Paul	Indiana University
Boffin, Henri	Universite de Bruxelles
Bond, Howard E.	Space Telescope Science Institute (STScI)
Bowen, George H.	Iowa State University
Brett, John M.	University of Sussex
Brown, Jeffery A.	Indiana University
Burkhead, Martin S.	Indiana University
Busso, Maurizio	Osservatorio Astronomice di Torino
Caloi, Victoria	Instituto Astrofisica Spaziale
Carpenter, Kenneth G.	CASA – University of Colorado
Cederbloom, Steven E.	Indiana University
Chieffi, Alessandro	Istituto de Astrofisica Spaziale CNR
Chiosi, Cesare	University of Padova
Cohn, Haldan N.	Indiana University
Cutright, Lori C.	Indiana University
Dorfi, Ernst	Universitat Wein
Durisen, Richard H.	Indiana University
Eaton, Joel A.	Indiana University
Edmondson, Frank K.	Indiana University
Faulkner, Danny	University South Carolina Lancaster
Feast, Michael W.	South African Astronomical Observatory
Fernie, J. Donald	David Dunlap Observatory
Forestini, Manuel	Universite Libre de Bruxelles
Frantsman, Yu L.	Radioastrophysical Observatory, Riga
Fujita, Yoshio	University of Tokyo
Gallino, Roberto	Ist. di Fisica Generale Dell'Universita Torin
Garrison, Robert F.	University of Toronto
Glassgold, A.E.	New York University

List of Participants

Grabhorn, Robert P.	Indiana University
Gratton, Raffaele	Observatorio Astronomico Di Roha
Greggio, Laura	Universita di Bologna
Griffin, Ian	University College London
Gustafsson, Bengt	Uppsala Astronomical Observatory
Guzik, Joyce Ann	Iowa State University
Hakkila, Jon	Mankato State University, Minnesota
Harpaz, Amos	Technion, Haifa
Heske, Astrid	Sterrewacht Leiden
Hollowell, David	University of Illinois
Honeycutt, Kent	Indiana University
Hron, Josef	Universitat Wein
Huang, Run Qian	Yunnan Observatory Academia Sinica
Huggins, Patrick J.	New York University
Iben, Jr., Icko	University of Illinois
Irwin, Alan W.	University of Victoria
Jiang, Shi–yang	Beijing Astronomical Observatory
Johnson, Hollis R.	Indiana University
Joras, Per	University of Oslo
Jorgensen, Uffe G.	Niels Bohr Institute
Jorissen, Alain	Universite de Bruxelles
Judge, Philip	Joint Institute for Lab. Astrophysics (JILA)
Jura, Michael	University of California (UCLA)
Kaler, James B.	University of Illinois
Keenan, Philip C.	Ohio State University
Kipper, Tonu	Tartu Astrophysical Observatory
Kleinmann, Susan	Five College Radio Astronomy, Massachusetts
Koszut, Keith A.	Indiana University
Kwok, Sun	University of Calgary
Lambert, David L.	University of Texas
Lattanzio, John C.	Institute of Geophysics & Planetary Phy.
Likkel, Lauren	University of California (UCLA)
Little, Stephen J.	Bertley College
Little–Marenin, Irene R.	Wellesley College
Littleton, John E.	West Virginia University
Lloyd Evans, T.	South African Astronomical Observatory
Locanthi, Dorothy D.	Jet Propulsion Laboratory
Lockett, Philip	University of Kentucky
Lugger, Phyllis	Indiana University
Luttermoser, Donald G.	Indiana University
Maizels, Cecelia	University of California (UCLA)
McWilliam, Andrew	Cerro Tololo Inter–American Observatory
Miller, Lynn	Indiana University
Mitalas, Romas	University of Western Ontario
Muchmore, David	Universite de Montreal
Mufson, Stuart	Indiana University
Murphy, Brian W.	Indiana University
Nook, Mark A.	University of Wisconsin
Olofsson, Hans	Onsala Space Observatory
Omont, Alain	Grenoble Observatory

Papoular, R.	Astrophysique/Saclay
Parthasarathy, M.	Indian Institute of Astrophysics
Peery, Benjamin F.	National Science Foundation & Howard University
Petrakis, John P.	Indiana University
Pierce, James N.	Mankato State University, Minnesota
Querci, Monique	Observatoire Midi–Pyrenees, Toulouse
Rao, Kamesware	Indian Institute of Astrophysics
Ray, Alak	Tata Institute of Fundamental Research
Renzini, Alvio	Astronomy Department of Bologna
Richer, Harvey B.	University of British Columbia
Schonberner, Detlef	Institut fur Theoret. Physik & Sternwart, Kiel
Sharp, Christopher	Max–Planck Institute
Shawl, Stephen J.	University of Kansas
Smith, Verne	University of Texas
Soker, Noam	University of Virginia
Spiczak, Glenn M.	Indiana University
Stencel, Robert E.	University of Colorado
Straizys, Vitautas	Institute of Physics, Vilnius
Straniero, Oscar	C.N.R. (Italy)
Tarafdar, Shankar P.	Tata Institute of Fundamental Research
Thiering, Inge	Hamburg Observatory
Tsuji, Takashi	Tokyo Astronomical Observatory
Tuchman, Y.	The Hebrew University
Turner, George	Indiana University
Vardya, Mahendra S.	Tata Institute of Fundamental Research
Vesper, David N.	Indiana University
Wahlgren, Glenn M.	Space Telescope Science Institute (STScI)
Wallerstein, George	University of Washington
White, James C.	Indiana University
Williams, Reva K.	Indiana University
Willson, Lee Anne	Iowa State University
Wing, Robert F.	Ohio State University
Wood, Peter R.	Mt. Stromlo & Siding Spring Observatories
Yang, Shelby	Indiana University
Yoon, Tae Seog	Indiana University
Zuckerman, Ben	University of California (UCLA)

1. Properties and Kinematics of the Peculiar Red Giant Stars

SPECTRAL CLASSIFICATION OF PECULIAR
RED GIANTS

Philip C. Keenan
Perkins Observatory
Ohio State and Ohio Wesleyan Universities

Abstract

Among field stars recent work includes development of
classification notation of the strong-CN stars, and the
recognition that the barium star with weak CN (HD 130255)
may actually be a dwarf barium star. The now established
presence of weak-G-band stars on the AGB of globular clusters
suggests that such field stars as HD49500 may be in the AGB
stage.
Such progress as has been made in the application of
modern detectors to the classification of faint stars in the
nuclear bulge of our galaxy, and in the Magellanic Clouds, is
summarized.

By the time of the Strasbourg Colloquium of 1984, most of the so-called "peculiar" groups of the cooler stars had been recognized. Since then their spectral definition has been discussed in the book "The Classification of Stars" by Carlos and Mercedes Jaschek (1987), and in the review paper "Spectral Types and Their Uses" (Keenan, 1987).

I shall mention now only one group that has resisted interpretation in terms of the surface composition of its members.

1. The Strong-CN Stars.

These are G- and K-type giants with the blue $\lambda 4216$ band of CN unusually strong for their temperature and luminosity. The notation developed last year, using positive CN abundance indices (Keenan, Yorka, Wilson, 1987), should serve well enough until analyses of their atmospheres finally answer the question of whether the abundances of the common metals are markedly greater in them than in the sun. They were termed "super-metal-rich" (SMR) by Spinrad and Taylor (1969) but the lack of agreement between published values of their metal abundances led Taylor in 1982 to be more cautious and refer to them merely as Very-Strong-Line stars. There is no doubt about the excess of CN in their atmospheres, however.

The best-known member of the group is μ Leo, which can be considered the prototype. Recently, Harris, Lambert and Smith (1987) have made careful determinations of the abundances of C, N, and O in μ Leo, using infrared bands of CO and CN, along with published data for Ca and [O I] lines. They accepted the metal excess of [Fe/H]=0.48 from the work as Branch, Bonnell and Tomkin (1978). Their results were [C/H]=0.22 [N/H]=0.93 [O/H]=0.32 for the excess abundances compared to the solar values. Alternative analyses with solar metal abundance gave lower concentrations of C, N, and O, but did not remove the excesses. The authors concluded that these abundances were to be expected if μ Leo has undergone the first red-giant dredge-up, and that there was no necessity to invoke the binary coalescence model of Campbell (1986) to explain them, although such a process is not ruled out by the new abundances.

2. HD 130255; A Peculiar Barium Star.

There are a few stars which seem to be mavericks - they refuse to satisfy the tidy definitions that we drew up to specify the members of any of these groups. These truly peculiar stars often turn out to be of the greatest interest from the standpoint of stellar evolution.

One such star is HD 130255 which was listed as a probable barium star by MacConnell, Frye and Upgren (1972), and included in

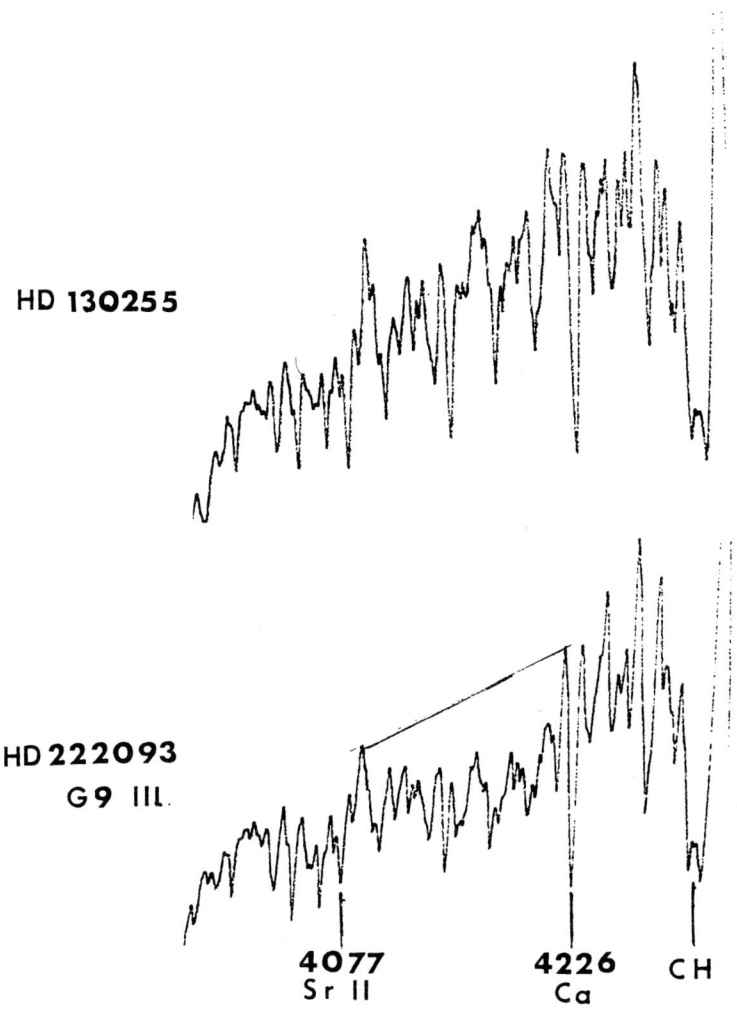

Fig. 1 - Region of the blue CN ΔV=-1 band system. In the typical G9 giant (HD 222093) the extent of the CN absorption can be judged by comparing the local continuum with the interpolated continuum shown by the straight line. The head of the shortward-shaded 0-1 band is at λ 4216.0 A. In HD 130255 the sharp feature at that position is due chiefly to Fe and SrII lines.

the catalogs of barium stars published by Yamashita and Norimoto (1981) and Lü et al. (1983). These catalogs gave the intensity of Ba 4554 in its spectrum as 1. Three classification spectrograms of HD 130255 were taken by Dr. Pitts a few years ago, but there is an ambiguity in its classification. The SrII and BaII lines would suggest a giant with a mild excess of S-process elements - a type of G9 III Ba1. Fig. 1, however, shows the extreme weakness of the blue CN band, and the 0, 0 band near 3883A is also weak, though in barium stars it is usually so strong as to appear saturated. The weakening would require a CN index of -3 if the star is giant. Since I have never seen another late-G barium star that did not have strengthened CN bands in its spectrum, the alternative possibility is that HD 130255 is one of the long-sought dwarf barium stars. Then the relative strengths of CN and CH would be explained but we should have to classify it as a strong barium star, perhaps G9V Ba2 in order to account for the strength of the SrII and BaII lines. One difficulty with its identification as a dwarf is that the ratio $\lambda 4376/\lambda 4383$ suggests a higher luminosity. This may not be a fatal objection, for one contributor to the $\lambda 4376$ blend is a line of Y II, and the abundance of yttrium sometimes is surprisingly high in stars in which even a small neutron flux has occurred.

For HD 130255, $m_V \cong 8.4$ and the components of its proper motion are given in the S.A.O. Catalogue as:

$\mu\alpha = 0^s.0022$, $\mu\delta = -0".049$.

The motion in declination would be large for a giant, but not surprising for a dwarf. Clearly, there is need for not only a good analysis to give log g and atmospheric abundances for HD 130255, but also for monitoring of its radial velocity. As soon as the luminosity of one such star has been determined, we shall easily be able to recognize and classify any others that may exist.

3. AGB in Globular Clusters and Field Stars.

The most important development in spectral classification, now and in the foreseeable future, is the application of the sensitive new detectors, particularly the CCD, to extend the schemes of classification to distant part of our galaxy and to outlying systems.

Before turning to the really faint stars, however, let us consider classification in the asymptotic giant branches (AGB) of globular clusters. Not all globular clusters have

AGB's so well defined that stars on at least their lower parts (sometimes termed early AGB stars) can be selected just from their positions on luminosity - color diagrams, but this is possible for several of the best-known clusters. As long ago as 1973, Zinn noticed stars with an unusually weak G-band in the AGB of M92. The work was extended to several other clusters with well-defined AGB's (M13, M15 and NGC 6397) by Norris and Zinn (1977) and others. The results have been brought together by G.H. Smith (1987) in his review of the properties of globular clusters.

In all of these clusters CH-weak stars were found in the AGB, and Norris and Zinn concluded that 67% of all the lower AGB stars had weak G-bands, the deficiency in total absorption amounting to as much as 2A. Only a few of the stars that were not clearly on the AGB showed the effect, and in several of those the weakening was doubtful. The clusters were all quite poor in metals, and it was soon noticed (e.g. Mallia, 1978) that some clusters (NGC6752, M22 and 47 Tuc) possess few, if any, weak G-band stars. Later, Smith (1987) pointed out that such stars were generally less common in clusters of intermediate luminosity.

This evidence suggests that when a weak G-band is observed in a G- or K-giant, that star has a high probability of being in the AGB stage - at least for stars of the halo and thick-disk population. It is interesting to look for AGB members among field stars by means of this diagnostic. It is somewhat embarrassing, however, to find that practically all the well-known weak-CH stars are low-velocity population-one objects. This may be partly because the rapid decrease in G-band strength as metal deficiency becomes more extreme makes it difficult to recognize a spectrum in which the band behaves abnormally.

There is one star, however, HD 49500, which is a possible AGB candidate. This star has tended to fall between several stools of classification, the best type that I can give it being K2-III Fe-2 CH-1 $H_\delta 1$.

Thus it is slightly metal-deficient, and slightly deficient in the G-band. Since V=7.19, the S.A.O. values for the proper motion, $\mu\alpha=0".030$, and Heard's Vr=+61, would not be inconsistent with its belonging to the thick-disk population. Here again a good analysis of the atmosphere of HD 49500 would help in the classification of similar stars.

There is the obvious possibility that all weak G-band

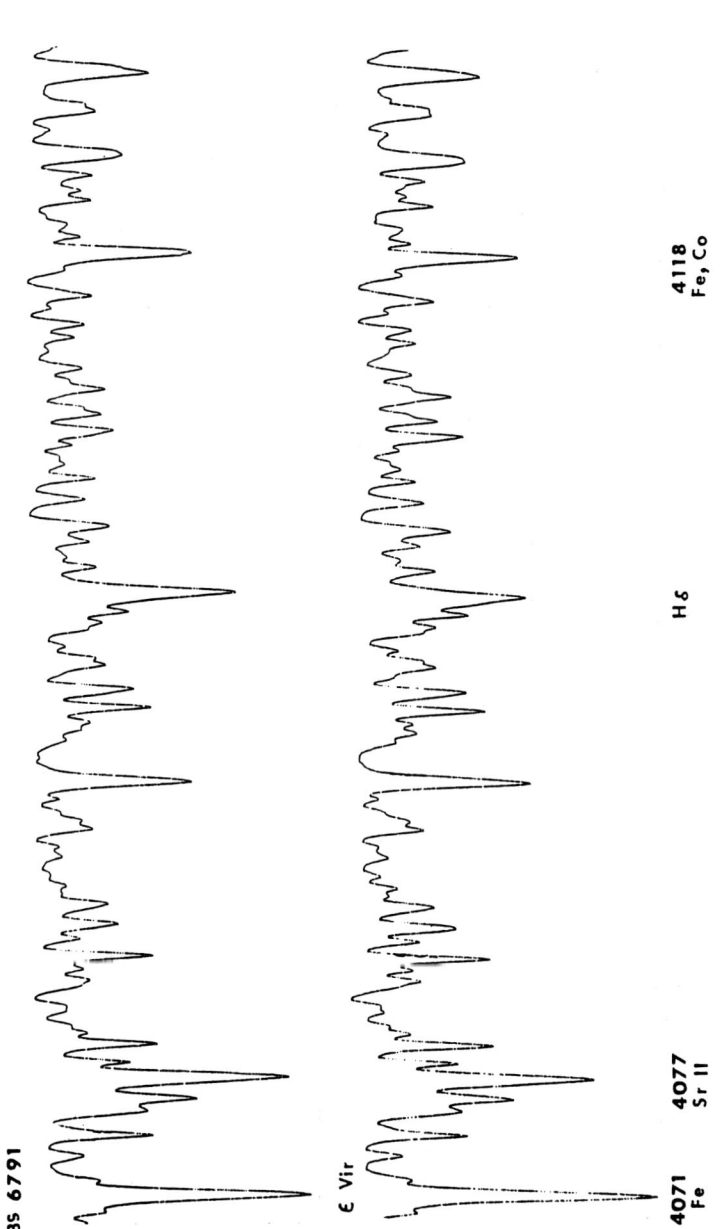

Fig. 2 – Comparison of the weak-G-band star HD6791 (G8 III CN-1 CH-3) with the standard G8 giant ε Vir. Resolution of the Palomar spectrograms was about 0.3Å.

giants can be assigned to the AGB. One constraint on their
interpretation is that only the CH band appears to have
abnormal intensity in these spectra. Figure 2 shows a short
section of two Palomar coude' spectrograms. BS 6791 is one of
the well-known weak G-band stars, while ε Vir has often been
cited as a standard G8 giant of nearly solar composition.
Actually, ε Vir is a rather late G8 star, and BS 6791 is a
trifle hotter, as can be seen from the greater strength as
H_δ in its spectrum. Although SrII 4077 seems slightly
stronger in BS 6791, Ba II 4554 shows no strengthening on
the same spectrograms, and since BS 6791 may be slightly the more
luminous of the two, there is no convincing evidence of a marked
excess of S-process elements in BS 6791. The close agreement
of both the relative and the absolute strengths of the lines
in these two spectra suggests further that the opacities also
in these stars are the same. Thus spectra of fairly high
resolution appear to confirm the evidence from low-dispersion
classification that the material dredged up into the atmospheres
of weak-G-band stars differs from that in normal G or K giants
essentially in the conversion of some of the carbon to nitrogen,
with little changes in metal abundance.

If next we consider open clusters, even the oldest do not
contain enough giants to allow AGB stars to be recognized from
their color-magnitude diagrams. The program of Garrison at
Toronto to determine a consistent set of spectral types for
giants and subgiants in open clusters offers the possibility of
detecting even slightly peculiar members.

4. New Programs of Classification of Remote Stars.
This survey must be limited to the extension to faint stars of the
more accurate classification that can be carried out with
resolutions of a few angstroms, and barely mention the immensely
fruitful classifications of the low-dispersion surveys. The
success of the surveys is due to the fortunate circumstance
that the stronger bands of such molecules as TiO, ZrO and CN
can be distinguished at scales even below 2000 A/mm. This
has made possible the extensive discovery lists of S, SC and C
stars, most of which, it is generally agreed, are late AGB
stars (cf. Smith and Lambert, 1986). Examples of on-going
objective-prism programs are the Abastumani survey of stars
down to V=16 within 5° of the galactic plane (Nicolashvili,
1987), and the Case survey which reaches to R=17 and extends

to high galactic latitudes (Sanduleak and Pesch, 1988). The powerful technique of GRISM spectroscopy, developed by McCarthy, and Blanco, 1978 has been extended by them (Blanco and McCarthy, 1983) and also Westerlund et al. (1986) to the Magellanic Clouds and other galaxies of the local system.

We have now become accustomed to the dominance of M stars in the nuclear bulge of our galaxy (cf. Blanco et al., 1984) in contrast to the much greater frequency of carbon stars in the Magellanic Clouds. Such differences between galactic populations make it the more important to refine the classifications with higher resolution. As early as 1983 Whitford and Rich had examined a sample of K giants in the nuclear bulge observed through Baade's window, using Reticon detectors with resolutions of 2.5 to 4.5 A. The sample has since been extended to 112 stars (Rich, 1988), for which the mean abundance of metals was 1.96 times the solar value. For M giants in the bulge, Whitford (1986) had combined the results of GRISM surveys with photometric data to conclude that they also are metal rich. All of these results are in accord with the evidence for increasing metal abundance towards the nuclear regions of spiral galaxies.

Since Blanco and McCarthy had found no carbon stars in Baade's window, a search for possible ones in that vicinity was carried out by Azzopardi et al. (1985), who found three that seemed fairly definitely to be early carbon stars of rather low luminosity. Lloyd Evans (1985) obtained spectra of one of them, BW-C6, at resolutions of 6A in the red and 2A in the blue, and found it to be a metal-rich R star with strong ^{13}C.

If we turn now to recent work on the Magellanic Clouds, we find copies of objective-prism spectrograms taken with the United Kingdom 1.2m Schmidt telescope used by Kontizas et al. (1987) to classify members of globular clusters in the clouds. At their scales of 830 or 2440 A/mm it was possible only to assign whole types: K, M, C, etc. These data were then used to find the ratios of carbon to M stars and thus to estimate the ages of the clusters.

The value of higher resolution is seen in the work of Lundgren (1987) on M giants and supergiants in the LMC. With the IDS spectrograph on the 3.6m and 1.5m telescopes of ESO at La Silla he was able to identify a number of MS stars. This was the first time that the presence in the Clouds of stars at this stage of AGB evolution was confirmed.

5. I want to finish by considering a question that is important to the planning of future programs. Is it worthwhile to observe spectra at classification resolutions ($\cong$ 2A) when the new detectors make it possible to carry out atmospheric analyses of the same stars at higher resolutions?

I believe that spectral types of these stars are essential, for three reasons:

[1] When one sees an abundance analysis of an unclassified star it is difficult to know how to characterize that star for comparison with others without writing a whole descriptive paragraph. A spectral type is designed to meet this need by giving the most compact summary possible of the luminosity, effective temperature, and atmospheric composition of a star.

[2] As soon as a few good analyses of classified stars in any population become available, the classification indices allow quite accurate interpolation between the physical variables measured in the analyses. Also, the spectral types need less frequent revision than the analyses, which are strongly model dependent.

[3] Since the spectral resolution required for good analysis is usually about five times higher than that required for classification, a classification program with a given telescope and detector generally reaches a limit one or two magnitudes fainter. With the data usually in digital form, either operation will presumably be carried out automatically or semi-automatically, and the saving of telescope time by the classification program is considerable.

References

Azzopardi, M., Lequeux, J., Robeirot, E. 1985, Astr. Astroph., 145, L4.
Blanco, V.M., McCarthy, M.F. 1983, A.J., 88, 1442.
Blanco, V.M., McCarthy, M.F., Blanco, B.M. 1984, A.J., 89, 636.
Branch, D., Bonnell, J., Tomkin, J. 1978. Ap.J., 225, 902.
Campbell, B. 1986, Ap.J., 307, 750.
Harris, M.J., Lambert, D.L., Smith, V.V. 1987, Pub. Astr. Soc. Pacific, 99, 1003.
Jaschek, C., Jaschek, M., 1987. The Classification of Stars. Cambridge University Press.
Keenan, P.C., 1987, Pub. Astr. Soc. Pacific, 99, 1003.
Keenan, P.C., Yorka, S., Wilson, O.C., 1987, Pub. Astr. Soc. Pacific, 99, 629.
Kontizas, E., Kontizas M., Xiradaki, E., 1987, Astr. Astroph. Suppl., 71, 575.
Lloyd-Evans, R., 1985, Monthly Not. R.A.S., 216, 29D.
Lü, P.K., Dawson, D.W., Upgren, A.B. Weis, E.W., 1983, Ap. J. Suppl., 52, 169.
Luck, R.E., Bond, H.E. 1982, Ap. J., 259, 792.
Lundgren, K. 1987, ESO Messenger, No. 49, 4.
MacConnell, D.J., Frye, R.L., Upgren, A.R. 1972, A.J., 77, 384.
McCarthy, M.E., Blanco, V.M. 1978, Mem. Soc. Astron. Ital., 49, 281.
Mallia, E.A. 1978, Astr. Ap., 70, 115.
Nikolashvili, M.G. 1987, Astrofisika, 26, 209, 1987 (trans. in Astrophysics n. 125.)
Norris, J., Zinn, R. 1977. Ap.J., 215, 74.
Rich, R.M. 1988, A.J., 95, 828.
Sanduleak, N., Pesch, P. 1988, Ap.J. Suppl., 66, 387.
Smith, G.H. 1987, Pub. Astr. Soc. Pac., 99, 67.
Smith, V.V., Lambert, D.L. 1986, Ap. J., 311, 843.
Spinrad, H., Taylor, B.J. 1969, Ap.J., 157, 1279.
Westerland, B.F., Azzopardi, M., Breysacher, J. 1986, Astr. Ap. Suppl. 65, 79.
Whitford, A.E., Rich, R.M., 1983, Ap.J., 274, 723.
Whitford, A.E. 1986, in Spectral Evolution of Galaxies, ed. C. Chiosi and A. Renzini, Reidel, Dordrecht, p. 157.
Yamashita, Y., Norimooto, Y. 1981. Ann. Tokyo Obs., (2), 18, 125.
Zinn, R. 1973, Ap. J., 182, 183.

THE SPATIAL, TEMPORAL, and PHOTOMETRIC PROPERTIES OF AGB STARS

S. G. Kleinmann
University of Massachusetts

The Two Micron Sky Survey (Neugebauer & Leighton 1969;TMSS) provides a census of AGB stars which is relatively insensitive to interstellar or circumstellar reddening, temporal variations, or differences in photospheric temperature. This paper summarizes results from recent analyses of all carbon, S type, and mass-losing M stars in the TMSS, including local surface densities, scale heights, and mass loss rates. All three groups are concentrated toward the plane; the mass-losing M stars appear least concentrated toward the plane but most strongly concentrated toward the galactic center. Results from the IRAS survey were used to determine the range of infrared colors of stars in each class, and to estimate their mass loss rates. Carbon stars have relatively higher 60 μm flux densities than oxygen-rich stars, and have relatively higher mass loss rates. The total mass loss rate is dominated by a small fraction of the stars in this sample. IRAS photometry and IRAS Low Resolution Spectometer data do not unambiguously distinguish carbon-rich and oxygen-rich stars in this sample. Future searches for stars with the greatest mass loss rates might concentrate on sources found to be variable in the IRAS survey, since a large fraction of the TMSS stars with the most massive envelopes are known Miras or infrared variables.

INTRODUCTION

The Asymptotic Giant Branch (AGB) of the Hertzsprung-Russell diagram (B-V > 0.5 mag., 0<M(V)<-3) is associated with the final phases of nuclear burning in stars having masses between 1 and 6 $M_\odot$. Though stars in this branch are quite luminous and can be seen to large distances, they are rare because their lifetime in this phase is strongly limited by their rapid mass loss rates (up to 10^{-4} $M_\odot$ yr^{-1}). Unfortunately, since stars in this mass range are concentrated toward the plane of the Milky Way, interstellar obscuration limits the range over which they can be detected in optical surveys. Interest in obtaining an accurate census of these stars is stimulated by the goals of measuring their range of lifetimes in the AGB phase, and the extent to which they dominate the chemical evolution of the Galaxy.

Infrared surveys carried out in the past two decades are far less sensitive to interstellar dust obscuration and far more sensitive to dust emission from warm circumstellar envelopes than previous surveys of the sky at shorter wavelengths. The infrared surveys will therefore provide a far more accurate picture of the types and distribution of mass-losing stars than was previously available.

In order to exploit infrared sky surveys, it is necessary first to have some minimal information on the classification of stars found in the survey, since stars in different classes have different luminosities, and may have very different spatial distributions. For example, it is necessary to distinguish the evolved mass-losing stars from those which are bright in the infrared simply because they are nearby or because they are reddened by interstellar dust. It is also necessary to distinguish the evolved mass-losing stars from very young stars whose infrared-bright circumstellar envelopes are a remnant of their birth rather than a product of their old age. Finally, it is important to distinguish stars belonging to specific sub-classes, such as Carbon stars (O/C < 1), S stars (O/C ~ 1), and M stars (O/C > 1), in order to test and measure the standard scenario for chemical evolution among these stars (M -> MS -> S -> SC -> C).

The necessary classification information is only just now becoming available for the stars found twenty years ago in the Two Micron Sky Survey (Neugebauer & Leighton 1969). This survey covered the sky between $-33° < \delta < +82°$ to a limit of m(K) = +3.0 mag. (39 Jy) at 2.2 μm, and obtained simultaneous measurements of each star in a second broad band with an effective wavelength near 0.9 μm. The total number of stars found in this survey was 5612, and the average surface density is about the same as the surface density of stars listed in the Bright Star Catalog, which contains sources brighter than V = 6.5 mag.

This review summarizes recent studies of carbon stars, S stars, and mass-losing M stars in the TMSS presented by Claussen et al. (1987), Jura (1988), and Kleinmann et al. (1988) respectively. These authors combined data from the Bidelman's (1980) catalog of spectral classes of sources in the TMSS with photometry presented in the TMSS and the IRAS Point Source Catalog (1985) and temporal information given in the General Catalog of Variable Stars (Kholopov 1985; hereafter GCVS) to learn the space distribution and mass loss rates of the various classes of AGB stars detected in the TMSS. The results of these studies are summarized in Table 1. One of the major conclusions of these analyses, and of a more recent study of very rapidly mass-losing stars in the solar neighborhood (Jura & Kleinmann 1988), is that only a few of the stars found in the TMSS make a significant contribution to the total rate of mass return to the interstellar medium.

The IRAS Sky Survey provides a far more sensitive probe of stars losing mass rapidly. Unfortunately, classifications based on high resolution spectra are unavailable for many IRAS sources at this time, which has led to the use of low-resolution spectra and broad-band photometric data to identify these stars. The adequacy of these techniques is addressed here, by reviewing the IRAS colors and spectral characterizations of stars found in the TMSS. Infrared source variability is discussed as an alternative means of identifying the stars in the IRAS survey that are losing mass most rapidly.

CARBON STARS IN THE TWO MICRON SKY SURVEY

Using spectral classifications given in Bidelman's (1980) catalog and in a compilation made by C. Payne Gaposchkin, Claussen et al. (1987) obtained a list of all known carbon stars in the TMSS. (Dr. Bidelman subsequently pointed out that at least one of the objects included in that list -- TMSS -10433 -- is a K-type supergiant, and not a carbon star.) The total number of carbon stars in this flux-limited sample (m(K) < +3.0 mag.) is 214. The Yale Bright Star Catalog lists fewer than one-tenth as many carbon stars.

The low median galactic latitude of carbon stars ($6.6°$) implies that, even at the relatively high flux levels of the TMSS, they must be being viewed at distances that are large compared to their scale height. Distances to individual carbon stars could be deduced by adopting the mean absolute K magnitude for carbons stars in the Magellanic Clouds, (M(K) = -8.1 mag.), since Frogel et al. (1980) found the dispersion in in this value is relatively small, ~0.5 mag. (It remains to be seen, however, whether the same mean and dispersion apply to carbon stars in the neighborhood of the sun, where stars are relatively richer in metals than in the Magellanic Clouds.) In most cases, the total fluxes and derived distances indicated total luminosities $~10^4$ Lo. For a few objects, namely those where $F(12 \mu m) > F(2.2 \mu m)$, circumstellar extinction depresses the observed 2.2 μm continuum, so that distances derived by assuming a constant K-band luminosity are uncertain for these objects. However even with the assumption of a constant K-band luminosity, only one object -- TMSS +10216 -- would have appeared to have a total luminosity far exceeding an AGB limit of $~3 \times 10^4$ Lo. This object is the most heavily obscured carbon star in the sample. No stars as red as TMSS +10216 have been detected in the nearby dwarf ellipticals in Draco and Sculptor (Jura 1986).

With the derived distances, and making allowance for interstellar dust obscuration amounting to ~ 0.2 K mag./kpc (Jura et al. 1989), we deduced that the carbon stars seen in the TMSS must be being observed out to a distance of ~1.5 kpc. This value is, as expected, much larger than their scale height. Thus, we can accurately estimate the surface density of carbon stars projected onto the galactic plane: it is N = ~40 kpc^{-2}. (In deriving this number, a 25% correction was made for the incomplete sky coverage of the TMSS.) The distribution of stars perpendicular to the disk is well fit with an exponential, and the derived scale height is ~ 200 pc. This implies a local number density of ~100 kpc^{-3}.

A comparison of the TMSS and the Case survey (Stephenson 1973) showed that the TMSS missed a large fraction of the carbon stars found in the optical survey. Analyses of these stars suggests that they must belong to another class of carbon stars which is at least 10 times less luminous in the near-infrared than the average carbon star sampled in the TMSS.

On the other hand, the Case survey missed ~7% of the carbon stars that were found in the TMSS survey. The missing stars have extremely red colors; over half of them are brighter at 12 μm than at 2.2 μm. It seems plausible to assume that this redness is due to obscuration by circumstellar dust in massive envelopes; this obscuration hindered their detection in the shorter-wavelength Case survey.

The Revised Air Force Four-Color Infrared Sky Survey (Price & Murdock 1983) lists nearly 50 infrared-bright carbon stars that were not detected in the TMSS (Kleinmann et al. 1981). Kinematic distances to these objects, derived from the velocities of the line centers of their strong CO J = 1-0 emission lines, suggest that about a third of them lie within the 1.5 kpc radius sampled by the TMSS. As indicated below, these objects play a major role in the rate of mass return by carbon stars in the solar neighborhood.

A two-color diagram of the IRAS colors of carbon stars in the TMSS is shown in Figure 1a. The stars are seen to exhibit strong long-wavelength excesses due to emission by dust in their circumstellar envelopes. The median mass loss rate for carbon stars, computed from their 60 μm IRAS fluxes according to a prescription given by Jura (1987), and assuming an average outflow velocity of 15 km sec^{-1} (Knapp & Morris 1985; Zuckerman et al. 1986) is ~2×10^{-7} M$_\odot$ yr^{-1}. Only ~10% of the carbon stars in the TMSS have mass loss rates higher than ~2×10^{-6} M$_\odot$ yr^{-1}; yet these stars contribute 85% of the mass returned by carbon stars to the interstellar medium. This fraction is really a lower limit, since heavily dust-enshrouded carbon stars detected in the Air Force survey and the IRAS survey make a contribution to the rate of mass return that is twice the total rate lost by all of the carbon stars in the TMSS (Jura & Kleinmann 1988).

S STARS IN THE TWO MICRON SKY SURVEY

Jura (1988) studied all 65 of the stars in the TMSS that were classified as type S by Wing & Yorka (1977). The completeness of this list is not well known. Stars having transitional spectral classes of MS and SC were excluded.

The space distribution of the S stars studied by Jura (1988) is similar to that of the carbon stars in the TMSS. The median galactic latitude of the S stars is 6.0°, essentially the same as the carbon stars in the TMSS. If the S type stars seen in the TMSS have a scale height that is roughly comparable to that of the carbon stars, then they are being viewed at comparable distances; i.e., they must have the same average K-band luminosity as the TMSS carbon stars. Thus, as Jura (1988) pointed out, the TMSS samples more luminous S stars than those found in globular clusters (Lloyd Evans 1984).

The IRAS two-color diagram for S stars in the TMSS is shown in Figure 1b. In the range 12 μm to 60 μm, these stars appear to have continua that more closely resemble those of the carbon stars, than of the mass-losing M stars in the TMSS (see below). Relatively fewer S stars than carbon stars have strong far-infrared excesses. Also, most of the S stars in this sample have mass loss rates lower than the median mass loss rate of carbon stars in the TMSS (Jura 1988). Only two S stars in this sample have mass-loss rates exceeding 2×10^{-6} M$_\odot$ yr^{-1}. No heavily dust-enshrouded S stars have been identified from the IRAS

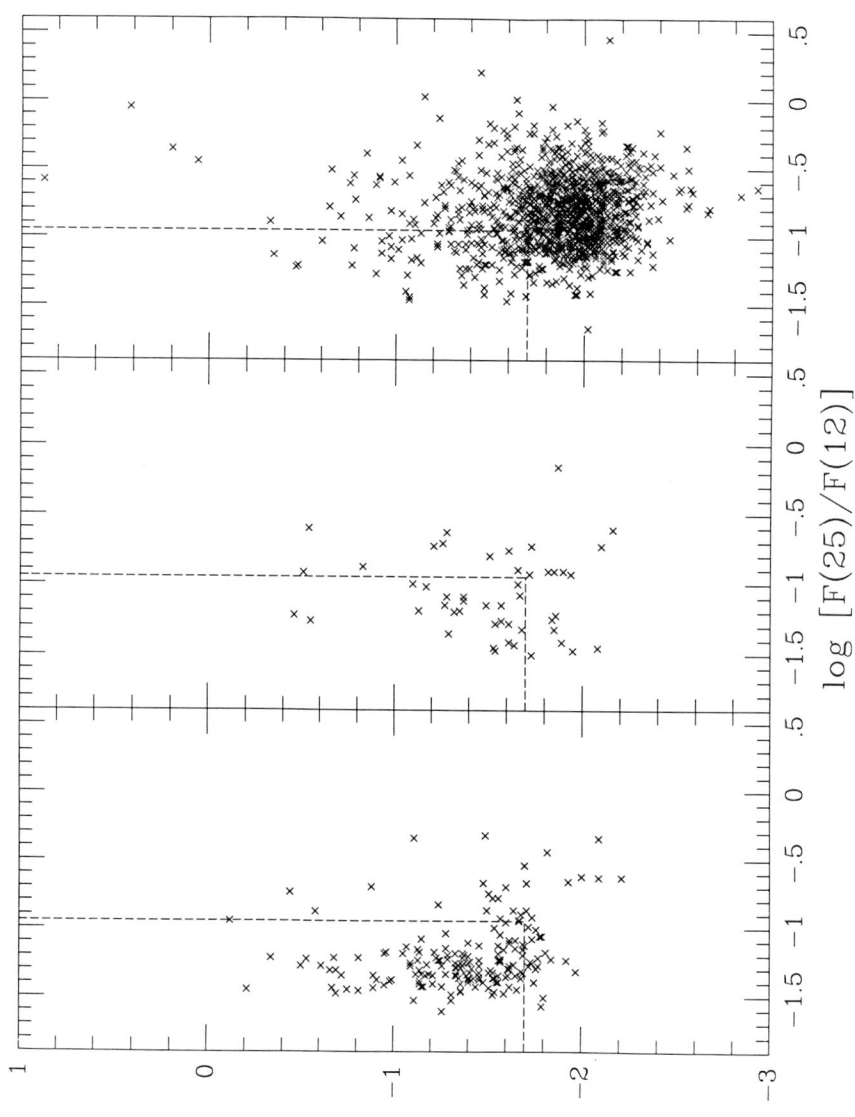

Figure 1. A two-color diagram of stars in the TMSS. Flux densities at 12, 25, and 60 μm were taken from the PSC, Version 2; they were not color-corrected. A 4000 K blackbody would lie at -0.61, -0.77 in these plots. Left: Carbon stars; Middle: S Stars; Right: Mass-losing M stars.

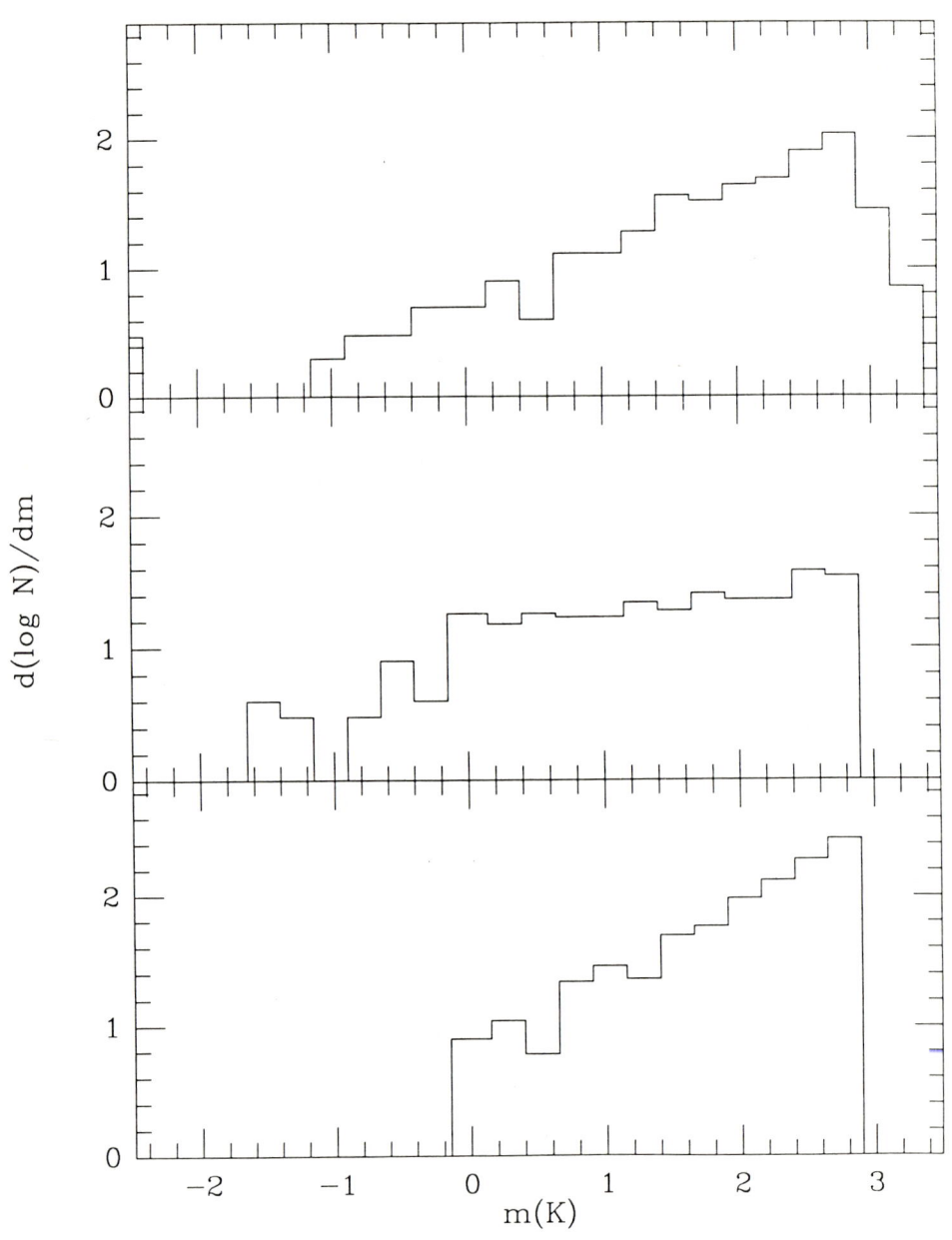

Figure 2. Distribution of mass-losing M stars in the TMSS vs. m(K). Top: Known Mira variables and infrared variables not identified with stars in the GCVS; Middle: Known Semi-Regular variables; Bottom: Irregular variables and stars not identified with known variables.

survey. It is not clear at what level this is a selection effect, i.e.,
that the spectroscopic criteria distinguishing an S star from a carbon
or M star are masked by strong circumstellar absorption.

M STARS IN THE TWO MICRON SKY SURVEY

Kleinmann et al. (1988) have reviewed the properties of AGB M stars
detected in the TMSS. In this study, the assumption was made that the
AGB stars could be distinguished from the first ascent giant branch M
stars by their high mass loss rates. A significant infrared excess at
$\lambda > 10$ μm was taken as an indicator of the presence of a massive
circumstellar envelope.

The distribution of colors among carbon stars were used as a guide
in selecting the mass-losing M stars in the TMSS. The carbon stars have
uncorrected colors (K - m[12]) as low as 0.9 mag., which is the expected
uncorrected color for a 4000 K blackbody. All stars having colors
redder than this, but not known to be carbon or S stars, were taken to
be mass-losing M stars. Stars with peculiar spectral types (e.g.,
Wolf-Rayet stars, early-type emission line stars), or peculiar variable
types (e.g., symbiotic stars) were eliminated from this list.
Supergiants listed in Bidelman's (1980) catalog were also eliminated.
Finally, 8 objects listed in Bidelman's (1980) catalog with spectral
classes of M0 or earlier were eliminated. The remaining list numbered
1705 stars.

Studies of M stars in the Magellanic Clouds indicate that they are
not characterized by a constant luminosity; rather, the Miras appear to
follow a period-luminosity relationship. Thus, a knowledge of the
temporal properties of the sample is critical to an understanding of
their spatial distribution. Comparison of the sample of mass-losing M
stars with the GCVS shows that ~52% of the sample are known variables.
The TMSS and PSC provide an independent measure of source variability;
those sources which appeared to vary in the infrared during the course
of multiple observations made in these surveys are flagged in the
catalogs. There are 523 such infrared variables among the mass-losing M
stars. Of these, 413 are identified with variables in the GCVS; most of
these previously known variables are Miras. Since Miras have larger
amplitudes than Semi-Regulars or Irregulars (the only other variable
classes in the sample), and since Miras have smaller amplitudes at
infrared than at optical wavelengths, it seems reasonable to assume that
most of the remaining 110 infrared variables are also Miras; these stars
might not have been recognized as optical variables, since they are
heavily enshrouded in thick circumstellar envelopes (as indicated by
their extremely red colors).

The lack of temporal information for all of the M stars in our
sample leads to significant selection effects. The brightness
distribution of the Miras and the infrared variables (shown in Figure
2a) is consistent with that expected for stars confined to a disk whose
thickness is small compared to the maximum distance of stars in the
sample. On the contrary, the brightness distribution of the
Semi-Regular stars (Figure 2b) increases less rapidly at faint flux
levels than would be expected for such a distribution. The Irregular
variables (Figure 2c) (including the unidentified stars which are not
infrared variables) increase more rapidly than expected even for stars

being viewed at distances small compared to their scale height. The deficiency of faint Semi-Regular stars, and the excess of faint Irregular stars and stars not identified with variables, suggest that many of the Irregulars and stars not identified with variables may be unrecognized Semi-Regular variables. The brightness distribution derived by combining the Semi-Regulars with the Irregulars and the stars not identified with variables is the same as the distribution of the Miras, i.e., consistent with a flux-limited sample of stars being seen at distances that are large compared to their scale height.

The distribution of periods among known variables in the sample is bimodal; 80% of the stars with P $\geq$ 250 days are Miras, while 89% of the stars with P < 250 days are Semi-Regulars. The median period for all identified Miras in the sample is 360 days, while the median period for all identified Semi-Regulars is 150 days. If the period-luminosity relationship given by Glass et al. (1987) were applied both to Miras and Semi-Regulars, then the Miras would have a median absolute K magnitude of -8.1, similar to the carbon stars, and the Semi-Regulars would have a median absolute K magnitude of ~ -6.7.

This expectation is in marked contrast to the observed distribution of galactic latitudes among the Miras (including the infrared variables) and the Semi-Regulars (including Irregulars and stars not identified with variables). The median galactic latitude of the former class is 14.5°, while the median galactic latitude of the latter class is actually less, only 11.4°. If the Semi-Regulars had the same scale height as Miras, their median galactic latitude should be nearly twice as large, since the TMSS should detect them at distances only about half as great as the faintest Miras in the sample. Lacking any obvious reason why the Semi-Regulars should be more confined to the galactic plane than the Miras, we deduce from their proximate median latitudes that they must have a comparable K-band luminosities. In this case, the TMSS samples mass-losing M stars and carbon stars to comparable distances, ~1.5 kpc. The surface density of all mass-losing M stars is then 320 kpc^{-2}.

Table 1 shows that the carbon stars appear more closely confined to the galactic plane than the mass-losing M stars (either the Miras or the Semi-Regulars). If the K-band luminosities of these stars are comparable, then the scale height of M-type Miras must be about twice that of the carbon stars, i.e., ~ 400 pc. This result implies that the progenitors of most of the mass-losing M stars in this sample may be somewhat less massive than the progenitors of the carbon stars.

A check on the luminosities of M stars in our sample can be obtained by considering their longitude distribution. If the TMSS views these stars to distances ~1.5 kpc, and if these stars are distributed in an exponential disk with a scale length of ~3.5 kpc (like the integrated light from other well-studied spirals), then there should be an apparent concentration of these sources toward the galactic center. Figure 3 shows the longitude distribution for the all TMSS carbon stars, Mira-type M stars (including infrared variables), and Semi-Regular M stars (including Irregulars and stars not identified with variables), in the brightness range 3 $\leq$ K < 2. (The "dip" between 220° < l < 360° is due to the fact that the TMSS was limited to δ > -33°, and therefore did not sample all galactic latitudes at these longitudes.) With the

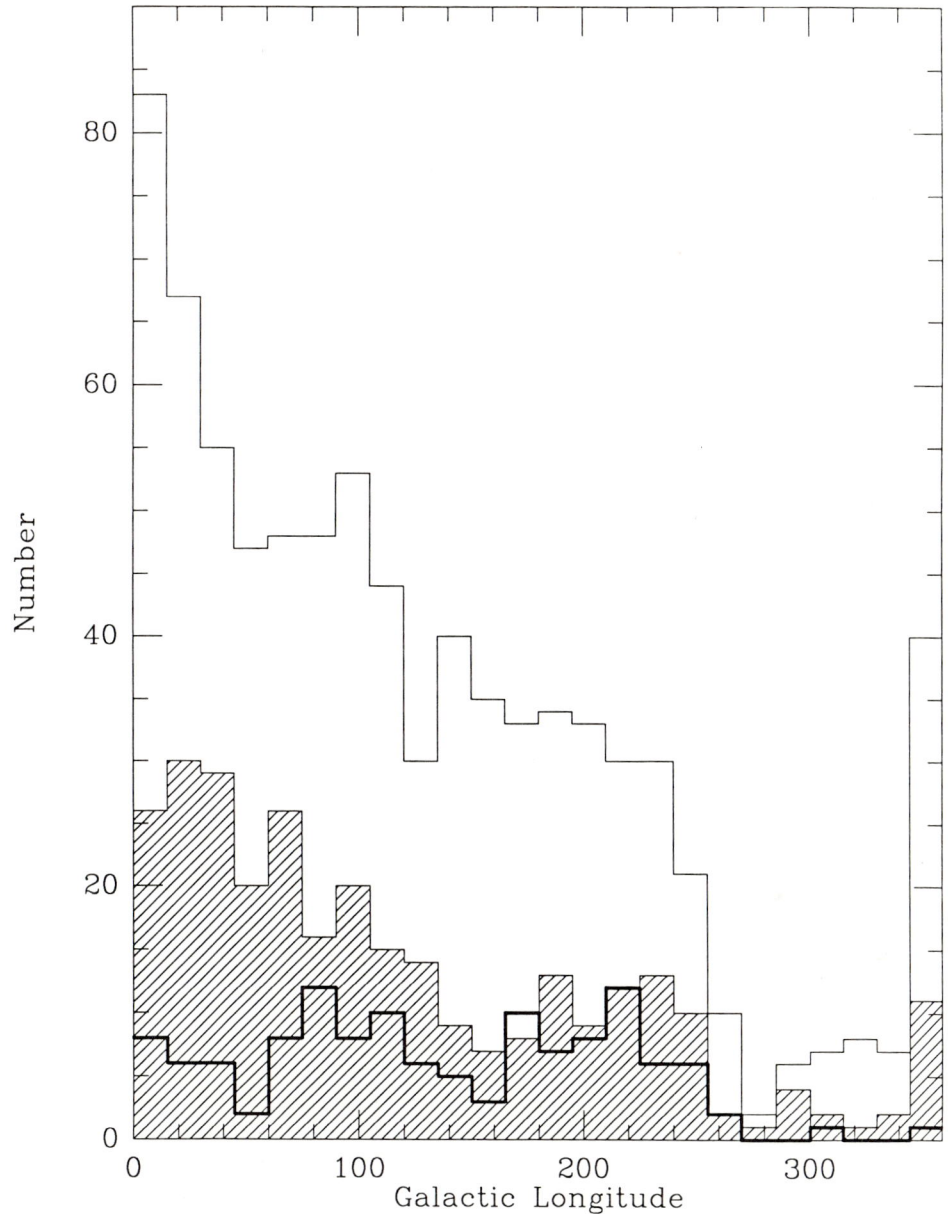

Figure 3. The distribution of certain classes of AGB stars in the TMSS vs. galactic longitude. The heavy line denotes carbon stars. The hatched area denotes Semi-Regular variable, mass-losing M stars (including Irregular variables and stars not identified with known variables). The thin line denotes Mira variables and infrared variables not identified with stars in the GCVS among the mass-losing M stars. Only stars having $3 \leq m(K) < 2$ are included in the Figure. The TMSS covered only a fraction of the sky for longitudes between $235° < l < 15°$.

TABLE 1
AGB Stars in the TMSS

	C	S	M
Number	214	65	1705
Miras	38	22	354
IR Variables	13	2	110
Semiregulars	79	14	331
Irregulars	53	14	210
Others	31	13	700
Median Galactic Latitude	$6.6°$	$6.0°$	$12.4°$
Fraction Within $330° < l < 30°$	0.13	0.21	0.19
Surface Density (kpc^{-2})	40	12	320
Scale Height (pc)	200	200	400
Median Mass-Loss Rate (10^{-7} $M_\odot$ yr^{-1})	2.0	0.1	1.5
Total Projected Mass-Loss Rate (10^{-4} $M_\odot$ yr^{-1} kpc^{-2})	0.3	0.03	1.0

TABLE 2
LRS Characterizations for AGB Stars in the TMSS

Index Range	Description	C	S	M
10 - 19	Featureless Blue Continuum	33	17	572
20 - 29	Blue Continuum with 10 μm Emission	14	7	409
30 - 39	Blue Continuum with 10 μm Absorption	2	1	12
40 - 49	Blue Continuum with 11 μm Emission	88	2	21
50 - 59	Featurless Red Continuum	1	0	4
60 - 69	Red Continuum with 10 μm Emission	0	0	2
70 - 79	Red Continuum with 10 μm Absorption	0	0	1

assumptions stated above, the number of stars should be greater, roughly by a factor of 2, between longitudes centered at $l = 0°$ and the anti-center direction. Just such an increase is observed for both Miras and Semi-Regulars. The sharp increase in the number of Semi-Regulars at $l < 30°$ may be due to unrecognized supergiants in the 5 kpc ring.

Interestingly, the carbon stars show no such variation with longitude, although their numbers in the TMSS are too small to make a definitive assessment. Jura et al. (1989) have, however, studied the carbon stars found in Fuenmayor's (1981) deep I-band survey near the galactic anti-center. They found no difference between the local density of carbon stars and the density of carbon stars at the deepest levels of Fuenmayor's survey (which samples distances out to ~ 4 kpc beyond the solar circle). These results imply that the density of carbon stars falls much less steeply with galacto-centric radius than the density of other stars, an effect which might be attributed to the higher incidence of carbon stars among low-metallicity populations.

The fractions of all carbon, S, and mass-losing M stars in the TMSS that are concentrated in the direction of the galactic center, $330° < l < 30°$ are summarized in Table 1. The overall average ratio of mass-losing M stars to carbon stars within the volume sampled by the TMSS is 8. Since the M stars appear to be more concentrated toward the galactic center than do the carbon stars, and since the TMSS missed a large fraction of the galactic plane near the galactic center, this ratio is a lower limit.

As shown in Figure 1, the median infrared colors of mass-losing M stars in the TMSS differ significantly from those of the carbon stars. This difference was first noted by Hacking et al. (1985), in their analysis of the brightest high-latitude 12 µm sources. The average color, $F(60)/F(K)$, of mass-losing M stars in the TMSS is lower than that of the carbon stars. Thus, if their outflow velocities and K-band luminosities are similar, then their inferred mass loss rates must be lower. The median, $\sim 1.5 \times 10^{-7}$ M☉ yr^{-1}, is ~25% lower than the median for carbon stars in the TMSS; some of this discrepancy may be due to differences in the properties of the dust grains (Zuckerman and Dyck 1986). Only 5% of the mass-losing M stars in the TMSS have mass loss rates exceeding $\sim 2 \times 10^{-6}$ M☉ yr^{-1}, about half the fraction of carbon stars that are losing mass rapidly. Though the number of M stars is much greater than that of carbon stars, longer-wavelength infrared surveys (e.g., the RAFGL) failed to show that the number of heavily dust-enshrouded M stars significantly exceeds that of heavily dust-enshrouded carbon stars. This result is in substantial agreement with the recent analysis by Jura & Kleinmann (1988) showing that carbon and M stars contribute about equally to the local rate of mass return to the interstellar medium.

DEEPER INFRARED SURVEYS

The IRAS survey offers an opportunity to obtain a census of heavily mass-losing stars at distances much greater than those sampled by the RAFGL or TMSS. In analyzing the IRAS data, it will be necessary to distinguish between carbon-rich and oxygen-rich stars because of their different spatial distributions. Unfortunately, the IRAS data do not provide a robust means of making this distinction. Though there is a significant difference in the median colors of carbon stars and M stars, as seen in Figure 1, there is also a large degree of overlap in the colors of stars belonging to different spectral subclasses. Thronson et al. (1987) attempted to find the carbon stars in the PSC by selecting those within the region bounded by the dotted lines in Figure 1. More than 25% of the TMSS carbon stars, including +10216, lie outside the region used by Thronson et al. (1987) to select carbon stars; it seems reasonable to imagine that a survey at a longer wavelength would detect a relatively higher fraction of extremely red stars like those not in the Thronson et al. sample. More than 50% of the TMSS sources lying within the bounded region are, in fact, not carbon stars. This fact emphasizes the difficulty in using IRAS broad-band colors to deduce spectral classifications.

It is instructive to determine whether the classifications provided by the Low Resolution Spectrometer (LRS) experiment in IRAS might usefully distinguish between carbon-rich and oxygen-rich stars. Unfortunately, LRS classifications are only available for the brightest stars in the IRAS survey, and therefore samples only a small fraction of the volume represented in the PSC. For example, the LRS provides classifications for only 60% of the AGB stars in the TMSS. These results are summarized in Table 2: the number of sources in each of 8 ranges of classification are given for carbon stars, S stars, and M stars. Fewer than half of the stars classified by the LRS exhibited any infrared spectral feature. The LRS classes assigned to the remaining objects are sometimes discordant from the optical spectral classifications. For example, 20% of the sources with 11 μm emission, often attributed to SiC dust and commonly seen in carbon stars, have been classified M or S on the basis of optical spectra. Of the sources with 10 μm emission or absorption, often attributed to silicate dust and commonly seen in M stars, 4% have been classifed as carbon stars on the basis of optical spectra.

Will it take decades of optical and near-infrared spectroscopy to exploit the IRAS data to learn the distribution of mass-losing stars? Maybe not. Analyses of the carbon stars and mass-losing M stars in the TMSS shows that the stars with high infrared variability typically have the highest mass loss rates. Of the 214 carbon stars in the TMSS, 24 were found to vary (probability of variability $\geq$ 90%) during the IRAS survey. These stars (11% of the all the TMSS carbon stars) contribute nearly half of the mass loss rate by carbon stars in the TMSS. Similary, 10% of the mass-losing M stars in the TMSS were found to be variables by IRAS, and these stars account for nearly 40% of the mass loss rate by AGB M stars in the TMSS. Thus, studies focussed just on the infrared variable sources in the PSC might provide good estimates of the numbers and distribution of mass-losing stars in the Galaxy to distances of at least 5 kpc.

ACKNOWLEDGEMENTS

This work benefitted from helpful discussions with M. Jura. This work was supported by the AFOSR, under grant No. 88-0070, and by the National Aeronautics and Space Administration under JPL contract No. 958046.

REFERENCES

Bidelman, W. P. 1980, Pub. Warner and Swasey Obs., 2, No. 6.
Claussen, M. J., Kleinmann, S. G., Joyce, R. R., and Jura, M. 1987, Ap. J. Suppl., 65, 385.
Frogel, J. A., Persson, S. E., and Cohen J. G. 1980, Ap. J., 239, 495.
Fuenmayor, F. J. 1981, Rev. Mexicana Astr. Ap., 6, 83.
Glass, I. S., Catchpole, R. M., Feast, M. W., Whitelock, P. A., and Reid, I. N. 1987, in Late Stages of Stellar Evolution, ed. S. Kwok and S. R. Pottach (Dordrecht: Reidel), p. 51.
Hacking, P. et al. 1985, Pub. A.S.P., 97, 616.
IRAS Point Source Catalog. 1985, Joint IRAS Science Working Group (Washington, DC: GPO) (PSC).
Jura, M. 1986, Ap. J., 301, 624.
Jura, M. 1987, Ap. J. 313, 743.
Jura, M. 1988, Ap. J. Suppl., 66, 33.
Jura, M., Joyce, R. R., and Kleinmann, S. G. 1989, Ap. J., in press.
Jura, M., and Kleinmann, S. G. 1988, Ap. J., submitted.
Kholopov, P. N. 1985-1987, General Catalogue of Variable Stars, Vols. 1-3, (4th ed.; Moscow: Nauka Publishing House) (GCVS).
Kleinmann, S. G., Jura, M., Joyce, R. R., and Claussen, M.J. 1988, in preparation.
Kleinmann, S. G., Gillett, F. C., and Joyce, R. R. 1981, Ann. Rev. Astr. Ap., 19, 411.
Knapp, G. R. and Morris, M. 1985, Ap. J., 292, 640.
Lloyd Evans, T., 1984, M.N.R.A.S., 209, 825.
Neugebauer, G., and Leighton, R. B. 1969, Two Micron Sky Survey (NASA SP-3047) (TMSS).
Price, S. D., and Murdock, T. L. 1983, The Revised AFGL Infrared Sky Survey Catalog (AFGL-TR-83-0161).
Stephenson, C. B. 1973, Pub. Warner and Swasey Obs., 1, No. 4.
Thronson, H. A., Jr., Latter, W. B., Black, J. H., Bally, J., and Hacking, P. 1987, Ap. J., 322, 770.
Wing, R. F., and Yorka, S. B. 1977, M.N.R.A.S., 178, 383.
Zuckerman, B. and Dyck, H.M. 1986, Ap. J. 311, 345.
Zuckerman, B., Dyck, H. M., and Claussen, M. J. 1986, Ap. J., 304, 401.

THE KINEMATICS OF PECULIAR RED GIANTS

M.W. Feast
South African Astronomical Observatory
P O Box 9 Observatory 7935
SOUTH AFRICA

Abstract. The ages, masses and evolutionary state of peculiar red giants are discussed on the basis of kinematic and other data.

INTRODUCTION

In discussing the kinematics of the peculiar red giants, and the contribution that the kinematics make to our understanding of the evolution of these stars it is convenient to deal first with those objects whose peculiarities are believed to arise from evolutionary processes in single stars and then those (e.g. BaII stars, CH stars) whose peculiarities seem related to their presence in binary systems. This division into single and double systems is as yet only tentative (see the recent discussion of the probable binary nature of some S type stars (Jorissen & Mayor 1988). In addition, some peculiarities may simply reflect unusual abundances in the material from which the stars formed, as has been suggested as one possible explanation for some of the peculiar red giants in the globular cluster Omega Centauri (cf. Lloyd Evans 1983).

An earlier review of this general topic was published by Catchpole & Feast (1985).

S STARS, C STARS, MIRAS

Overall, mean, results for the kinematics of C, S and SC stars in the general field can be obtained in a straightforward way from radial velocities. Table 1 shows estimates of the total velocity dispersion, $\sigma_T = (\sigma_u^2+\sigma_v^2+\sigma_w^2)^{1/2}$, for various groups of objects. For the S, Se and SC stars these values have been obtained from the unpublished data of Catchpole (cf. Catchpole & Feast 1985). The C star data is from Dean (1976). Note that the stars with emission lines are generally Mira type variables. The relation between σ_T and age deduced by Wielen (1977) from stars of known age in the solar neighbourhood can then be used to estimate rough ages and these are shown for some of the groups in the table. Estimates of the corresponding initial masses are also given. These were derived from Buzzoni's interpolation formula for the Yale isochrones assuming that the initial abundances are approximately solar (Iben & Renzini 1983, equation 2). The values of σ_T for the Me type Mira variables, which are listed for comparison with the S and C stars, were obtained from the data in Feast et al. (1972).

The results in Table 1 indicate that the bulk of the S and C type stars are low mass objects of intermediate age. There is some indication from the kinematics that the C stars as a group may be younger and (initially) more massive than the S and SC stars. It would be desirable to confirm this result from larger samples of stars with each sample analysed using identical methods. Dean (1976) omitted from his sample of C stars all CH stars, and all stars with large radial velocities ($|\rho|$ >100 kms^{-1}) on the grounds that the latter stars were CH-star candidates. This exclusion may be reasonable in view of the fact that CH stars are likely to be a quite different type of object from other C stars. The possible exclusion of some high velocity non-CH stars in this process might be feared to have artificially reduced the velocity dispersion of "true" C stars. However it can hardly have affected the comparison with the S and SC stars in Table 1 since in fact none of the S stars and only one of the SC stars, in the samples used, had $|\rho|$ >100 kms^{-1} (the SC star was Case 598 with a velocity, corrected for local solar motion, of -101 kms^{-1}). An uncertainty in the comparison is introduced by the fact that some of the C, S and SC stars used are probably sufficiently distant for the effects of galactic rotation to be important, so that the results depend somewhat on the adopted absolute magnitudes.

TABLE 1

The velocity dispersions of peculiar red giants and Mira variables

Type	Number of stars	σ_T kms^{-1}	Age	$M_i/M_\odot$
1(a) S and C Stars				
S	124	55 ⎫		
Se*	29	61 ⎬ ~5 Gyr		~1.3
SC	30	66 ⎭		
C	427	37 ⎫ ~2 Gyr		~1.6
C type Miras	36	<41 ⎭		
1(b) M-type Miras				
P < 145 days	22	81		
145 < P < 200	46	180		
200 < P < 250	71	101		
250 < P < 300	77	88		
300 < P < 350	83	69		
350 < P < 410	54	58		
P > 410 days	35	50		

* These will be mainly S type Miras.

A fuller interpretation of these results is complicated by the fact that there may be a spread in ages and initial masses within the groups of S, C and SC stars in Table 1. This seems likely for C stars since such stars are found in Magellanic Cloud clusters with ages ranging from $\sim 10^9$ to $\sim 10^{10}$ years (cf. Frogel & Blanco 1984; Cohen 1982; Lloyd Evans 1984).

Another group of objects which indicates a spread of ages amongst the PRGs is the Mira variables. It is now clear that we must include Mira variables, or at least the more luminous ones, amongst the PRGs. This was long suspected to be the case, not only from the occurrence of S and C type Miras but also from the frequent occurrence of MS stars amongst Miras. These conclusions have been strongly reinforced by the recent work of Little et al. (1987) who find that most (M type) Miras with periods greater than about 300 days (M_{Bol} = -4.6 on the Mira period - luminosity relation) have TcI lines which show that there has been a recent dredge-up phase. The M type Miras are not a kinematically homogeneous group. They show a dependence of both velocity dispersion (cf. Table 1) and asymmetrical drift on period (Feast 1963; Feast et al. 1972). These kinematic results indicate that the longer period (~ 400 day) M type Miras are of intermediate age (~ 5 Gyr) and of somewhat over one solar (initial) mass, the shorter period (~ 200 day) M type Miras belong to an older population (>10 Gyr) (cf. Feast & Whitelock 1987).

It is clear from the above discussion that in order to make full use of the kinematic information one requires some parameter which will enable one to divide the spectroscopic groups (C,S,SC etc) into subgroups which are homogeneous as regards age and initial mass. The evidence indicates that in the case of the M type Miras, the period can be used as this parameter. The period has a further importance for Miras because of the existence of an infrared and bolometric, period-luminosity relation. This relation has been best demonstrated for the Miras in the Large Magellanic Cloud (cf. Glass et al. 1987) but is also shown by the relatively few known SMC Miras (cf. Lloyd Evans et al. 1988; Feast 1988a), by Miras in globular clusters (Menzies & Whitelock 1985; Feast 1987) and by Miras in the galactic bulge (Glass & Feast 1982; Feast 1986). At K (2.2μm), the same relation fits Miras of spectral types M, C and S rather well. In M_{Bol} the M and S type Miras in the LMC fall on the same relation whereas the C type Miras apparently fall ~ 0.25 mag fainter than the other types at a given period (See Glass et al. 1987). It is not yet entirely clear whether this displacement is real or an artifact of the very different energy distributions of the C and M type stars which could lead to spurious systematic differences when deriving M_{Bol} by integrating multicolour, broad-band photometry. In the following we refer mainly to the PL relation shown by M and S type Miras.

By combining the ages (or initial masses) of M type Miras as a function of period, derived from the kinematics, with the corresponding PL luminosities we can compare the results with predictions for the top part of the AGB which can be obtained, e.g. from figure 2 of Iben and Renzini (1983) (cf. Feast & Whitelock 1987). With an age of ~5 Gyr and an M_{Bol} of -4.7 at a period of 400 days (and corresponding pairs of values at other periods) one finds that the Miras lie ~0.5 to 1.0 magnitudes brighter, at a given age, than the predicted values for the start of thermal pulsing on the AGB and that in fact they are close to the locus of the highest luminosity predicted on the AGB if the mass loss parameter $\eta \sim 2/3$.

The narrowness of the PL relation (σ = 0.16 mag in M_{Bol}) taken together with the fact that Miras in any one globular cluster are closely clustered at the top of the AGB, and the results just discussed, suggest that M type Miras are confined to a very narrow luminosity range at each age/initial-mass. Thus one concludes that they lie distinctly above the luminosity at which thermal pulsing begins. Evidence on the luminosities of small amplitude variables in globular clusters (Whitelock 1986; see also Feast 1988b) tentatively suggests that these stars may populate the region between the start of thermal pulsing and the AGB tip occupied by the Miras.

As already indicated the work of Little et al. (1987) has shown that technetium is not present in the short period Miras, indicating that dredge-up of s-process elements is not taking place in these stars. This is in broad agreement with theoretical predictions that such dredge-up will be absent in low mass objects (cf. Iben and Renzini 1983).

There is a general tendency for the C type Miras to have longer mean periods in the LMC (cf. Glass et al. 1987) and in the general solar neighbourhood (cf. Merrill 1960) than the M type Miras. However, the most striking feature of these distributions is the substantial overlap in periods between the two classes. This overlap in periods and the absence of C type Miras in the galactic bulge (Blanco et al. 1984) seem potentially important clues for our understanding of PRG's. There are a number of possible alternative explanations of these results, some of which are discussed below.

Suppose the similarity of the PL relations for M and C type Miras is taken to indicate similar current and initial masses at a given period for stars in these two groups. In that case there are at least two possibilities. (1) At a given initial mass there is a range in [Fe/H] values. The more metal-rich stars do not evolve into C stars (cf. Iben & Renzini 1983). This explanation requires that there is a significant spread in metal abundance at a given age in the solar neighbourhood and in the Magellanic Clouds. The relatively metal poor component would be missing in the galactic bulge. (2) An alternative is suggested by the result, discussed above, that Miras (at least M type Miras) occupy a narrow range of luminosities (at a given age) well above the luminosity at which thermal pulsing on the AGB begins. Theory predicts (cf. Iben & Renzini 1983) that a star will decrease in luminosity by 0.5 - 1.0 magnitudes (bolometric) between thermal pulses (the exact amount depending on the mass). The possibility then exists that a star will only enter the Mira stage at a bright part in the thermal pulse cycle, dropping back to lower luminosities and small amplitude pulsation (or constancy) between thermal pulses. A star would then enter the Mira phase several times but each time with a higher C/O ratio and a higher abundance of s-process elements due to dredge-up. The attraction of this scheme over one in which a star evolves from an M to a C type Mira entirely within the Mira phase, is that the occurrence of low amplitude (or constant) C (and S) stars can be naturally accommodated into the scheme. The second of these two schemes does not require a spread in metal abundance in any one environment.

An entirely different possibility is tentatively suggested by the data in Table 1. These show that the velocity dispersion of C type Miras is smaller than that of even the longer period groups of (optical) M type Miras. It would be particularly important to confirm this result since it suggests that the C type Miras belong to a younger, higher initial mass, population than even the longer period M type Miras. The current masses are not necessarily different at a given period since the C type Miras may have evolved from stars which underwent heavy mass loss. There is evidence (Feast 1985) that the very long period (600-2000 day) OH/IR stars extend the M type Mira sequence to higher luminosities and presumably higher initial masses. Thus the C type Miras could, in this scheme, evolve from the OH/IR stars or from objects of somewhat larger initial mass. Further kinematic work should enable us to clarify the evolutionary connection between these different groups of stars.

Whilst the S type stars fit, at least qualitatively, into the picture of AGB dredge-up, there are some apparent anomalies. The outstanding kinematic puzzle is probably NT Tel an Se Mira at $\ell = 347°.6$ b = $-25°.4$ with a radial velocity of $+325$ kms^{-1} and a period of 252 days (see Andrews 1975 and also Catchpole & Feast 1985). No other S star is known with a velocity, $|\rho| > 100$ kms^{-1}. Most observational and theoretical evidence leads us to expect that single PRGs will be absent from the halo population to which this star presumably belongs. One possibility is that it is in a binary system in which mass exchange has taken place (e.g. it is an evolved BaII star). Alternatively it could perhaps have formed from s-process rich material (as mentioned earlier this is one possible explanation for some PRGs in Omega Centauri).

Further work is required on the four S type stars in the region of the Eta Carinae complex to decide definitely on their membership of the complex. This grouping is very young (it contains early O type stars) and S type stars would not be expected theoretically. (See the discussion in Catchpole & Feast 1985).

It has been recognized for sometime that because carbon stars can be detected out to large distances they are of considerable potential importance for the study of the kinematics not only of our own galaxy but also of the Magellanic Clouds and of other galaxies (c.f. Richer 1988). Many carbon stars are periodic or quasi-periodic variables. However we can only expect the PL relation discussed above to hold if we restrict the sample studied to Mira type variables (note that the M type low amplitude variables in globular clusters show a PL relation of much lower slope (Whitelock 1986 see also Feast 1988b). It is not surprising therefore that Claussen et al. (1987) find no evidence for a period-luminosity relation for a sample of galactic C type variables, many of which are non-Miras.

Schechter et al. (1988) have recently carried out an extensive radial velocity study of distant C stars in our galaxy. A considerable range in absolute magnitudes is expected amongst these stars since in the Magellanic Clouds the C stars range in M_K (the absolute magnitude at 2.2μm) from -7 to -9. Thus in analysing the data Schechter et al. have to allow for significant Malmquist-type bias. Using their analysis and adopting M_K = -7.9 (± 0.6), a mean value derived from the Magellanic Clouds, one can derive a value for the distance to the galactic centre R_0 = 8.4 ± 0.6 kpc and an Oort constant A = 14.8 ± 1.4 kms^{-1} kpc^{-1}. These values agree quite well with values recently derived from a study of the kinematics of Cepheids (R_0 = 7.8 ± 0.7 kpc, A = 14.6 ± 1.8 kms^{-1} kpc^{-1}) (Caldwell & Coulson 1987).

BaII STARS, CH STARS

The evidence is now strong that BaII and CH stars are members of binary systems (cf. McClure 1988). The BaII stars can be divided into a number of spectroscopic subgroups (BaII weak, BaII strong, metal weak BaII star). As discussed by, e.g. Catchpole et al. (1977) and Catchpole & Feast (1985) these subgroups have different kinematic properties (cf. Table 2) indicating that the group as a whole is a heterogeneous one with a range of ages and absolute magnitudes. The CH stars are particularly interesting from the kinematic point of view because they seem to define a rather pure halo population. Several of the CH stars, including some towards the galactic poles, have radial velocities (without regard to sign) greater than 200 kms^{-1} and one, the Fehrenbach-Duflot star in the direction of the LMC, has a radial velocity of +440 kms^{-1} (Fehrenbach & Duflot 1981). Table 2 lists the galactic components of the velocity dispersions as well as the total dispersions for CH stars and some other groups. Hartwick & Cowley (1985) have drawn attention to the fact that whilst the total (or one co-ordinate mean) velocity dispersions of CH stars, metal poor giants and globular clusters are about the same, the ratio of the axes of the velocity ellipsoids are possibly different. Thus σ_u/σ_v may be different for the CH stars and for the metal-poor giants (cf. Table 2). Hartwick & Cowley suggest that this may indicate different kinematic histories for the various groups. Improved statistics are necessary to test this possibility. Mould et al. (1985) note that their results are suggestive of a velocity ellipsoid whose major axis points always to the galactic centre rather than being parallel to the galactic plane.

The CH stars obviously have an important role to play in the study of the galactic halo. There is a drawback in that they cover a rather wide range in absolute magnitudes ($M_V \sim 0$ to -3, even excluding the subdwarf CH stars) (cf. Catchpole & Feast 1985; Hartwick & Cowley 1985; Bond 1974 and references therein). Though this can possibly be overcome using a colour-luminosity relation (cf. Hartwick & Cowley 1985). There is also some difficulty in defining the group precisely. For instance the carbon star V CrB ($\rho = -115$ kms^{-1} b = $+51°$) is sometimes considered a CH star and sometimes not. It seems possible that the CH stars could in fact form a continuous sequence of kinematic groups with the various subgroups of BaII stars (cf. Table 2). Indeed a star such as HD 36598 which has been classified both as a BaII (with strong CN and CH (MacConnell et al. 1972, Catchpole et al. 1977) and as a carbon star (Stephenson 1973; Sanduleak & Davis Philip 1977) might perhaps be as well classified with the CH stars, and the star HD 115444, which is classified as a BaII, has [Fe/H] = -2.95 (Griffin et al. 1982) and is presumably a member of the halo.

Hartwick & Cowley (1985) find the remarkable result that the local space density of CH stars must be high (5.6×10^{-9} pc^{-3}). They point out that this is only one third of the local space density of metal

TABLE 2

The velocity dispersions of CH stars, BaII and some other objects

Type	Ref.	Number	σ_U	σ_V kms^{-1}	σ_W	σ_T
CH stars	(1)	51	$\sqrt{-2966 \pm 48}$	161±24	125±18	196
CH stars	(2)	9			101±24	
Metal Poor Giants	(3)	52	141±16	106±23	56±30	185
Globular Clusters*	(4)					204
BaII weak stars*	(5)	25				29
BaII strong stars*	(5)	40				43
Metal weak BaII stars*	(5)	27				57

* $\sigma_T = \sqrt{3}$ (dispersion in radial velocity)

(1) Hartwick & Cowley 1985 (2) Mould et al. 1985 (3) Hartwick 1983
(4) Frenk & White 1980 (5) Catchpole et al. 1977.

poor giants. It is not clear that this is consistent with the relatively few CH stars in globular clusters. Perhaps this is an indication that only a limited proportion of the CH stars belong to a globular cluster-like population. If Harwick & Cowley's conclusions are even approximately correct they show that large numbers of CH stars remain to be discovered. They predict approximately one CH star per square degree brighter than V = 15 at the galactic poles. In that case one might expect a significant number of galactic CH stars to occur in the objective-prism surveys of the Magellanic Clouds for carbon stars, especially ones such as the LMC survey by Sanduleak & Davis Philip (1977) which was carried out in the blue region and therefore found preferentially the hotter carbon stars. This survey in which 474 carbon stars were found, covered about 72 square degrees and went down to V = 16 with all except three stars of V > 14 or fainter. The two brightest are the BaII/CH star HD 36598 mentioned above (V = 8.5, ρ = +48 kms^{-1}) and HD 269343 (V = 12) which has a very similar spectrum and ρ = +226 kms^{-1} (Feast & Spencer Jones, unpublished). This velocity is within 10 kms^{-1} of that of the LMC in this direction, but it seems unlikely that it is a member since its brightness would then be M_v = -6.5. Presumably it is a member of the halo of our galaxy. At M_v = 16 an object with a CH-like spectrum could conceivably be a low luminosity CH star in the outer halo of our galaxy or a high luminosity CH star in the LMC. A colour-absolute magnitude relation might be used to distinguish between these possibilities. It should be noticed that even a survey like the Sanduleak-Davis Philip one will not uncover all the CH stars. Their survey missed the high velocity CH star found by Fehrenbach & Duflot (1981), presumably because it has very weak C_2 bands. The Case survey of the Northern Galactic Cap (cf. Report of Warner and Swasey Obs. Bull AAS 20, 136, 1988) should contain many new CH stars.

CONCLUSIONS

Kinematic studies have made major contributions to our understanding of peculiar red giant stars. However a number of the results discussed in this review are still quite tentative. A major extention of kinematic studies of these stars in our own and other galaxies would greatly help in defining more precisely the ages, masses and evolutionary state of these objects.

REFERENCES

Andrews, P.J. (1975). M.N.R.A.S., 173, 701.
Blanco, V.M., McCarthy, M.F. & Blanco, B.M. (1984). A.J., 89, 636.
Bond, H.E. (1974). Ap. J., 194, 95.
Caldwell, J.A.R. & Coulson, I.M. (1987). A.J., 93, 1090.
Catchpole, R.M. & Feast, M.W. (1985). In Cool Stars With Excesses of
    Heavy Elements, ed. M. Jaschek and P.C. Keenan, p.113.
    Dordrecht: Reidel.
Catchpole, R.M., Robertson, B.S.C. & Warren, P.R. (1977). M.N.R.A.S.,
    181, 391.
Claussen, M.J., Kleinmann, S.G., Joyce, R.R. & Jura, M. (1987). Ap. J.
    Suppl., 65, 385.

Cohen, J.G. (1982). Ap. J., 258, 143.
Dean, C.A. (1976). A.J., 81, 364.
Feast, M.W. (1963). M.N.R.A.S., 125, 367.
Feast, M.W. (1985). The Observatory, 105, 85.
Feast, M.W. (1986). In Light on Dark Matter, ed. F.P. Israel, p.339. Dordrecht: Reidel.
Feast, M.W. (1987). In The Galaxy, ed. G. Gilmore & B. Carswell, p.1. Dordrecht: Reidel.
Feast, M.W. (1988a). In The Extragalactic Distance Scale, ed. S. van den Bergh (Victoria Symposium). In press.
Feast, M.W. (1988b). In IAU Colloquium 111 (Lincoln, Nebraska). In press.
Feast, M.W. & Whitelock, P.A. (1987). In Late Stages of Stellar Evolution, ed. S. Kwok & S.R. Pottasch, p.33. Dordrecht: Reidel.
Feast, M.W., Woolley, R. & Yilmaz, N. (1972). M.N.R.A.S., 158, 23.
Fehrenbach, C. & Duflot, M. (1981). Astr. Astrophys., 101, 226.
Frenk, C.S. & White, S.D.M. (1980). M.N.R.A.S., 193, 295.
Frogel, J.A. & Blanco, V.M. (1984). In Observational Tests of the Stellar Evolution Theory, IAU Symp. 105, ed. A. Maeder and A. Renzini, p.175. Dordrecht: Reidel.
Glass, I.S., Catchpole, R.M., Feast, M.W., Whitelock, P.A. & Reid I.N. (1987). In Late Stages of Stellar Evolution, ed. S. Kwok & S.R. Pottasch, p.51. Dordrecht: Reidel.
Glass, I.S. & Feast, M.W. (1982). M.N.R.A.S., 198, 199.
Griffin, R. & R., Gustafsson, B. & Vieira, T. (1982). M.N.R.A.S., 198, 637.
Hartwick, F.D.A. (1983). Mem. Soc. Astron. Ital., 54, 51.
Hartwick, F.D.A. & Cowley, A.P. (1985). A.J., 90, 2244.
Iben, I. & Renzini, A. (1983). Ann. Rev. Ast. Astrophys., 21, 271.
Jorissen, A. & Mayor, M. (1988). Astr. Astrophys., 198, 187.
Little, S.J., Little-Marenin, I.R. & Bauer, W.H. (1987). A.J., 94, 981.
Lloyd Evans, T. (1983). M.N.R.A.S., 204, 975.
Lloyd Evans, T. (1984). M.N.R.A.S., 208, 447.
Lloyd Evans, T., Glass, I.S. & Catchpole, R.M. (1988). M.N.R.A.S., 231, 773.
McClure, R.D. (1988). This volume.
MacConnell, D.J., Frye, R.L. & Upgren, A.R. (1972). A.J., 77, 384.
Menzies, J.W. & Whitelock, P.A. (1985). M.N.R.A.S., 212, 783.
Merrill, P.W. (1960). In Stellar Atmospheres, ed. J.L. Greenstein, p.509. Chicago: University of Chicago Press.
Mould, J.R., Schneider, D.P., Gordon, G.A., Aaronson, M. & Liebert, J.W. (1985). P.A.S.P., 97, 130.
Richer, H.B. (1988). This volume.
Sanduleak, N. & Davis Philip, A.G. (1977). Pub. Warner Swasey Obs., 2, No.5.
Schechter, P.L., Aaronson, M., Cook, K.H. & Blanco, V.M. (1988). In The Outer Galaxy, ed. L. Blitz and F.J. Lockman. In press.
Stephenson, C.B. (1973). Pub. Warner Swasey Obs., 1, No.4.
Whitelock, P.A. (1986). M.N.R.A.S., 219, 525.
Wielen, R. (1977). Astr. Astrophys., 60, 263.

Harvey B. Richer
University of British Columbia
Vancouver, B.C. V6T 1W5 Canada

ABSTRACT. Study of the late-type stellar content in external galaxies provides numerous clues for the theory of stellar evolution, for star-formation scenarios in galaxies, and for proper models of the luminosity evolution of galaxies which are then used in cosmological studies. In addition, these late-type stars can be used as distance indicators themselves and yield a local value of the Hubble constant consistent with recent Cepheid determinations.

1. SURVEYS OF PECULIAR RED GIANT STARS

For the purposes of this review we will define a peculiar red giant (PRG) star observationally as a star with $(V-I)_o > 1.5$ and $M_{bol} < -4$. This will then include only objects brighter and redder than the tip of the M92 giant branch and will eliminate most first-ascent giant stars. A few supergiants pass the above criteria, but these are so rare (and anyways can be easily eliminated in the surveys via their brightness) that their statistical contribution is entirely negligible. These observational selection criteria will then include virtually all late-type stars that exhibit peculiarities in their spectra including M, MS, S, SC, and C stars. Of course, a pure photometric selection criterion as described above provides no spectral information. Even so, Reid and Mould (1984) were able to use only these photometric criteria to provide important information on the LMC AGB luminosity function. Some of the surveys that have been carried out in external galaxies do provide some spectral information from low dispersion objective prism observations (Westerlund 1960; Sanduleak and Philip 1977; Westerlund et al 1978), or transmission grating spectroscopy (Blanco, McCarthy, and Blanco 1980; Azzopardi, Lequeux, and Westerlund 1985), or from narrow band imagery (Richer, Crabtree, and Pritchet 1984; Cook, Aaronson, and Norris 1986; Pritchet et al 1987).

2. WHY ARE PRG'S IN EXTERNAL GALAXIES IMPORTANT?

Over the past decade the study of PRG's in external galaxies has provided important constraints on a wide range of important astrophysical problems. We highlight 4 of these below.

(1) In the field of stellar evolution theory of late type stars (see review by Iben and Renzini 1983, as well as more recent contributions of Lattanzio 1986a, 1986b, 1987) observations, particularly of carbon stars in the Magellanic Clouds allowed, for the first time, isolation of a complete sample of such objects. This was made possible by the grism surveys of Blanco and his collaborators (Blanco, Blanco, and McCarthy 1978; Blanco, McCarthy, and Blanco 1980; Blanco and McCarthy 1983) as well as Westerlund and his group (Westerlund 1964, 1965; Westerlund et al 1978). Luminosity functions constructed from these data and compared with theory (Richer 1981a, 1981b, Frogel et al 1981) showed conclusively that the existing theory predicted too few low luminosity carbon stars and too many high luminosity ones. This was critical because at the time it was felt that luminous AGB stars were the sole source of s-process elements in the solar system distribution. While some progress has been made in resolving the difference between the theoretical and observed luminosity functions (Lattanzio 1987), the problem of the origin of the s-process elements remains, and must be considered an outstanding problem in stellar nucleosynthesis. The current view regarding the lack of luminous AGB stars seems to have converged on the idea that mass loss rates are much higher than that obtained with a Reimer's mass-loss coefficient between a third and a half (Frogel and Richer 1983, Reid and Mould 1984).

(2) The question of the star formation history in a galaxy is important for understanding galactic evolution. In turn, a model of galactic evolution is critical in interpreting the colors and magnitudes of galaxies at cosmologically interesting red shifts. In NGC 205, Richer, Crabtree, and Pritchet (1984) showed that this galaxy must currently be experiencing a burst of star formation as the number of PRG stars presently observed was much smaller than expected given a constant star formation rate and the number of luminous blue stars found. Effectively, the PRG's and the blue luminous stars allowed the star formation rate to be sampled at two different epochs. An equally important result was found by Reid and Mould (1984), who detected a spatial variation in the AGB luminosity function in several fields of the LMC. More luminous AGB stars were found in regions of more active star formation. Reid and Mould conclude that the AGB luminosity function is a sensitive probe of the star formation history of a galaxy. What is currently required, however, before the AGB luminosity function can be used to model quantitatively this history, is an empirical determination of the relation between stellar age and the height to which a star rises on the AGB.

(3) The evidence is now rather compelling that galaxy colors and luminosities evolve (Djorgovski, Spinrad, and Marr 1985). Thus, in order to use galaxies as probes of the cosmos to determine cosmologically interesting parameters such as the deceleration parameter q_o, their evolution must be understood. For example

to take a simple case, stellar evolution will make a distant galaxy appear brighter, and unless this is taken into account its distance will be underestimated.

Chokshi and Wright (1987) have shown, in a theoretical paper, that a population of AGB stars can make significant changes to the red and infrared color evolution of a galaxy. Lilly (1987) quantified these results by showing that adding the effects of the AGB to galaxy evolution models provides much better fits to the observed (U-V) versus (V-H) plots of distant (z>0.5) galaxies.

(4) Theory predicts, and observations confirm (Richer, Pritchet, and Crabtree 1985), that some PRG's (in particular carbon stars) are useful standard candles for the determination of distances to nearby galaxies. Carbon stars have a small dispersion in their absolute I magnitude (+/-0.47 magnitudes for a complete LMC sample, Richer 1981a), are luminous ($<M_I>=-4.75$), are easily detected in galaxies with distances out to 5 Mpc, and should not suffer from undue foreground contamination. While M supergiants are easily confused with Galactic M dwarfs, extragalactic carbon stars suffer virtually no Galactic contamination. Empirical knowledge of the carbon star luminosity function in a wide variety of galaxies is thus desirable to test their utility as distance indictors.

3. DETECTION OF PRG'S IN EXTERNAL GALAXIES

Within the immediate environs of the Milky Way Galaxy, PRG's can be located from low dispersion objective prism or from transmission grating spectroscopy. A summary of applications of this method to the dwarf spheroidal companions of our Galaxy can be found in Aaronson and Mould (1985). However, in more distant systems, or in more crowded regions of nearby galaxies, the smearing out of a stellar image even in a very low resolution spectrum can produce image crowding to such an extent that all information is effectively lost. Figure 1 provides a dramatic example of this. The upper panel of this Figure contains a direct image of a region of the plane of the Milky Way in the direction of the constellation Circinus. The lower panel is a low resolution prism plate of the same region. The crowding is so severe in this latter frame that classification of the spectra is impossible. However, it certainly does appear possible to carry out stellar photometry on the images in the upper panel.

This was the motivation that led us to develop a purely photometric system that would be capable of providing spectral information for PRG's in extremely crowded systems. The advent of CCD cameras, with their high quantum efficiency and good red response provided the ideal detector for such a system. The photometric system developed uses two narrow band filters which closely mimic bandpasses in Wing's (1971) eight-color system. These two filters are nominally centered at 7800 and 8100Å, both with $\Delta\lambda$=140Å. The 8100 filter measures CN ($\Delta\nu$=+2), while the 7800 filter is sensitive to TiO ($\Delta\nu$=-1) in M stars and serves as a continuum filter for carbon stars. The

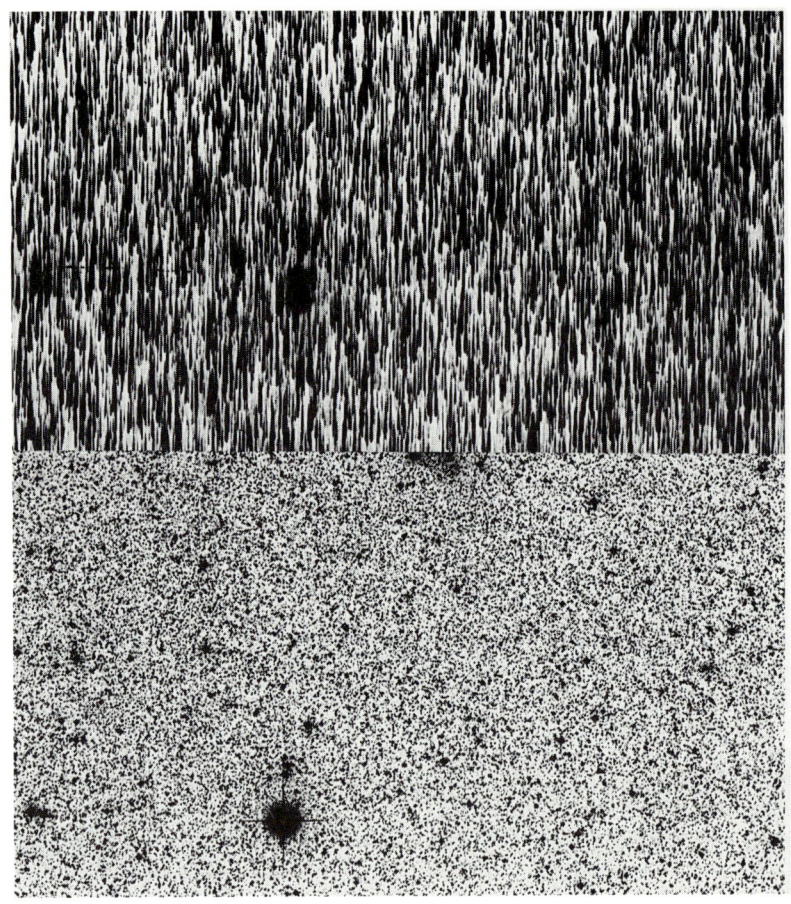

Figure 1: A direct V image of a region of the plane of the Milky Way is seen in the upper panel. In the lower panel an objective prism photograph of the same field is displayed. Crowding in this field makes it almost impossible to classify the spectra.

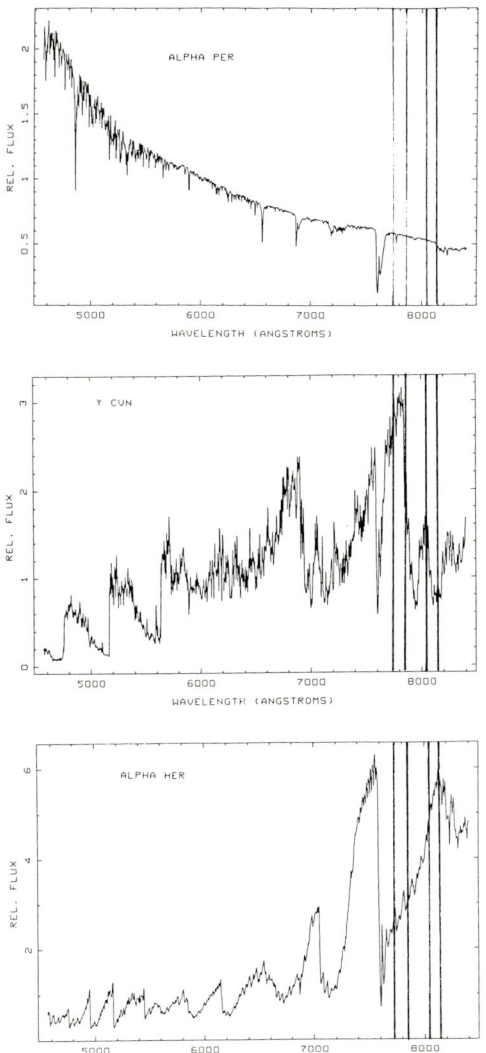

Figure 2: Operation of the photometric system developed to locate PRG's in external galaxies. The 7800 Å filter measures continuum in carbon stars and TiO in M stars. The 8100 Å filter measures CN in carbon stars and continuum in M stars. Very hot stars have (81-78) colors similar to that of carbon stars.

operation of this filter system is illustrated in Figure 2 where the filter bandpasses are overlaid on the spectra of three types of stars; an F star, a carbon star, and an M star. Carbon stars have positive (81-78) colors, while M stars have a negative index in this color. However, measurement of the colors of very hot stars yields colors similar to that of the carbon stars because of the decreasing flux with wavelength through the region of the spectrum defined by the filter system. Hence, with only the single (81-78) color it is not possible to distinguish carbon stars from O and B stars The solution to this is to include in the system two broad band filters (usually V and I). Note also from the spectrum of Alpha Per that the narrow band color index is insensitive to temperature in stars of intermediate spectral classes.

This photometric system was tested on PRG's of known spectral type in the LMC as well as on several Galactic standards of earlier spectral type. The resulting color-color diagram is shown in Figure 3 (taken from Richer, Pritchet, and Crabtree 1985). The main point to note is the insensitivity of early type stars to the (81-78) color (these stars have little or no CN or TiO), and the bifurcation of the diagram for (V-I) greater than 2.0 cleanly into two regions. One of these regions contains exclusively oxygen-rich stars, while the other is occupied only by carbon-rich ones. These two areas are separated by about 1 magnitude in the (81-78) color.

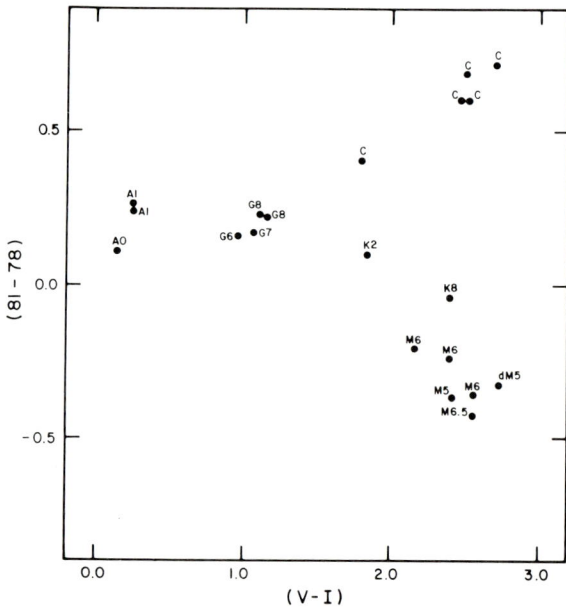

Figure 3:   The (81-78) narrow band color plotted against broad band (V-I) for a selection of PRG's in the LMC and some Galactic standards. Note the separation of oxygen-rich and carbon-rich stars for (V-I) >2.0.

We have applied this photometric system to about half a dozen galaxies, while Aaronson and his collaborators (Cook, Aaronson, and Norris 1985) have used a similar photometric system on 5 extragalactic systems. One of the galaxies that both groups have investigated is M31. Figure 4 illustrates this galaxy together with the location and approximate size of the single field which we observed in it in the left panel. The right panel shows the CCD field observed through the two narrow band and two broad band filters. The color-color diagram constructed from this data is shown in Figure 5 (Richer and Crabtree 1985). From a comparison with Figure 3, five carbon stars and numerous M stars are clearly present in this field.

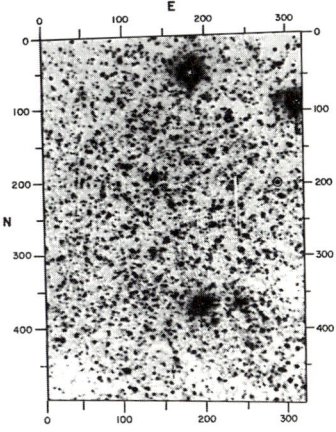

Figure 4: Left panel: The Andromeda galaxy indicating the location and approximate size of the CCD field surveyed in it to locate PRG's. Right panel: The I CCD frame of the field observed in M31. These latter data were obtained at the Cassegrain focus of CFHT.

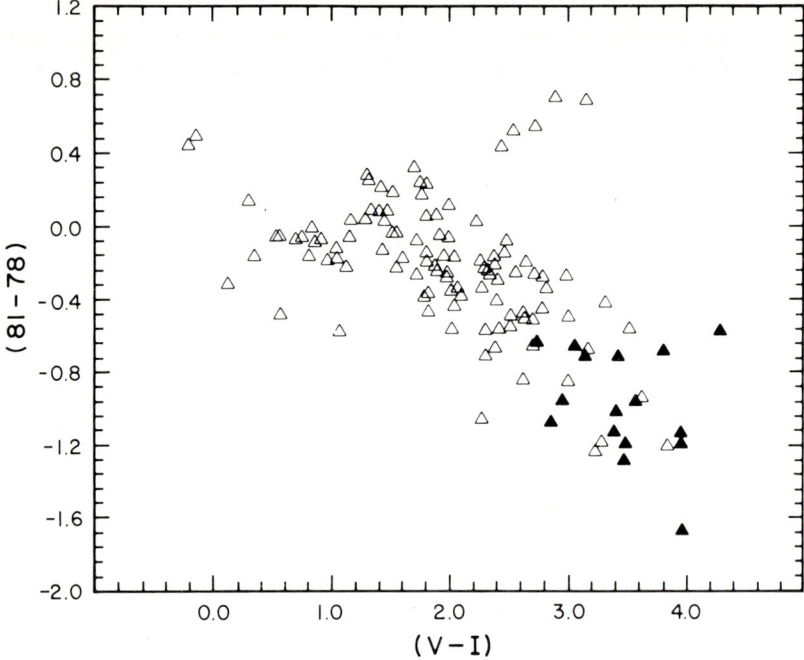

Figure 5: Color-color diagram of the M31 data. Five carbon-rich stars are seen together with 41 M stars with spectral types later than M5.

4. RESULTS OF SURVEYS FOR PRG'S IN EXTERNAL GALAXIES

The surveys carried out by our group, by Aaronson and his collaborators, by Reid and Mould, and by Blanco and his collaborators have resulted in new insight into the uses of PRG's in understanding the late phases of stellar evolution, but especially they have been extremely useful in providing new ideas into the star formation history in nearby galaxies and as distance indicators. In the following three subsections I indicate what, in my view, have been the three most important results from these surveys, and suggest future directions.

4.1 THE CARBON TO M (C/M) STAR RATIO

One of the first results to come out of the surveys for PRG's in external galaxies was the observation by Blanco, Blanco, and McCarthy (1978) that the ratio of the number of carbon to late M type stars (M5 or later) in the SMC, LMC, and Galactic bulge was

correlated with the metal abundance of that system. The sense of
the correlation is that more metal poor systems possess more
carbon stars per late M type star (C/M larger as [Fe/H] lower).
The most current version of this correlation involves 9 galactic
systems that are spirals or irregulars including two that are
outside the Local Group. This correlation is shown in Figure 6
and is taken from Pritchet et al (1987). The correlation is
quite remarkable covering more than two dex in C/M and more than
one dex in metallicity.

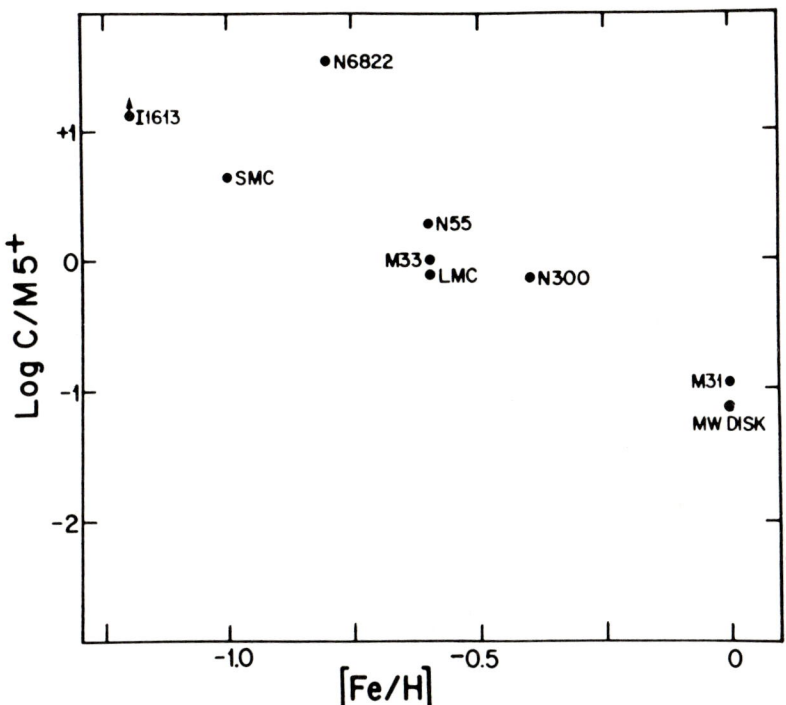

Figure 6: Plot of the C/M ratio versus [Fe/H] for 9 spiral and irregular type galaxies that have been surveyed for PRG's.

There are two major competing explanations for the correlation seen in Figure 6. The first suggests that it occurs because decreasing metallicity drives the giant branch to higher temperatures, thus decreasing the number of oxygen-rich giants with spectral types later than M5. The second explanation suggests that metal-poor stars are more easily turned into carbon stars because they begin with a low oxygen content, and thus a smaller amount of carbon has to be convected to the surface in order to produce C/O>1. It may be possible to distinguish between these two explanations by considering the total population of carbon stars in these different systems. If it is true that the high C/M ratio seen in low [Fe/H] galaxies is due only to the color of the giant branch, then the number of carbon stars in a galaxy should scale simply as the luminosity of that galaxy, and be independent of the metal abundance of that system. However, if the correct explanation to the correlation lies in the ease of producing carbon-rich stars in low metal abundance systems, then one would expect that the number of C stars per unit of galaxy luminosity should exhibit a strong dependance on [Fe/H]. The relation between the number of carbon stars per unit galaxy luminosity, and metallicity is shown in Figure 7 with the data taken from Pritchet et al (1987). The open circles are data points for the dwarf spheroidals surrounding the Milky Way (Aaronson and Mould 1985) while the open circle with a cross in the center is a single point for the Galactic globular cluster system (Aaronson and Mould 1985). Because the number of carbon stars per unit galaxy luminosity varies strongly with [Fe/H], we can exclude the variation in the temperature of the giant branch as the sole cause of the correlation between C/M and [Fe/H].

While the data in Figure 7 yield a correlation between the number of carbon stars per unit of galaxy luminosity and metal abundance, it is important to note that the correlation exhibits much scatter. In particular, the Magellanic Clouds contain an anomalously large population of carbon stars for their luminosities and abundances, while NGC 55 and 300 contain few such objects. This strongly suggests that some parameter other than metallicity strongly affects the carbon star production rate in galaxies. The most likely candidate is star formation history.

At the moment, we lack a complete explanation of the correlation between the C/M ratio and [Fe/H} observed in extragalactic systems. However, it appears that the time is ripe for a detailed theoretical attack on the problem. The observational data with which we can constrain the theory is well in hand. The new carbon star models (Lattanzio 1987) seem capable of producing the required low luminosity carbon stars seen in the Magellanic Clouds, and these coupled to star formation scenarios should be capable of yielding C/M ratios as a function both of metal abundance of the system and its star forming history.

4.2 THE PRG LUMINOSITY FUNCTIONS IN EXTERNAL GALAXIES

Reid and Mould (1984) first pointed out the importance of

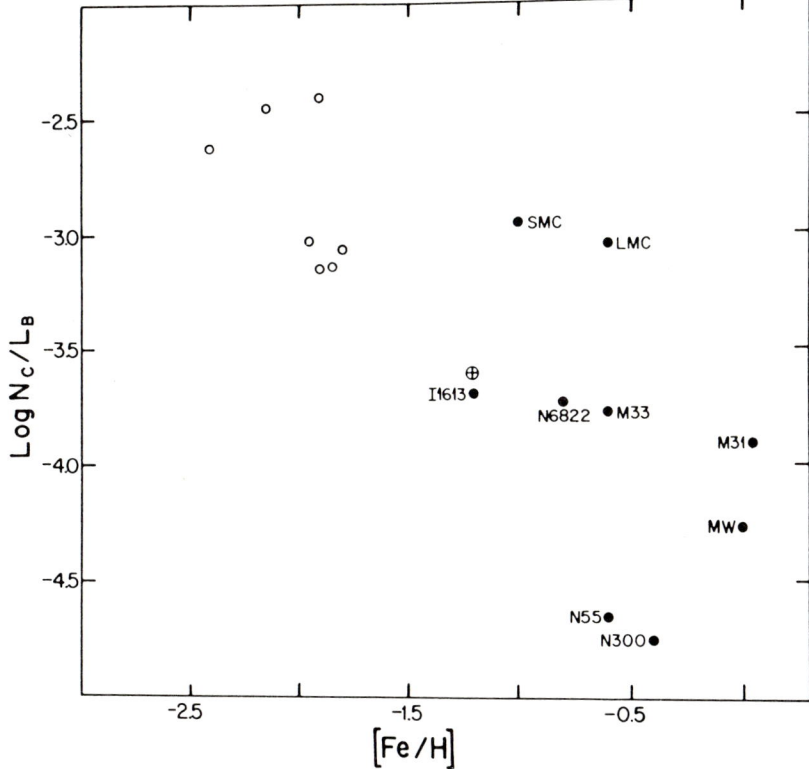

Figure 7: Plot of the number of carbon stars per unit of galaxy luminosity versus metallicity. Open circles-dwarf spheroidals. Closed circles-spiral and irregular galaxies. Cross inside open circle-Galactic globular clusters.

studying the AGB luminosity function in external galaxies. Earlier, several authors (Richer 1981a, 1981b; Frogel et al 1981) had shown that the carbon star luminosity function in the LMC did not agree with theoretical luminosity functions constructed from the models of Renzini and Voli (1981) which were the state-of-the-art models for these stars at that time. Reid and Mould extended this idea to all AGB stars thus ruling out the then fashionable idea that the observed carbon star deficiency at high luminosity was due to nuclear processing of their carbon-rich atmospheres back to more normal composition.

In a series of papers on nearby galaxies (Richer, Pritchet, and Crabtree 1985; Richer and Crabtree 1985; Pritchet et al 1987) we were able to show that the AGB luminosity functions in NGC 300, M31, and NGC 55 were all similar to that in the Magellanic Clouds. These data are collected in Figure 8 wherein we also include the Reid and Mould (1984) LMC AGB luminosity function which is compared to a theoretical function incorporating a constant star formation rate. No reasonable star formation history or IMF slope is capable of producing a model that agrees well with the observations. The agreement between the LMC AGB luminosity function and that in the other galaxies only strengthens the conclusion that star formation history is not the major reason why too few luminous AGB stars are seen. It must be due to the evolution of the stars themselves, and the current idea is that the mass loss rates incorporated into the models have been badly underestimated.

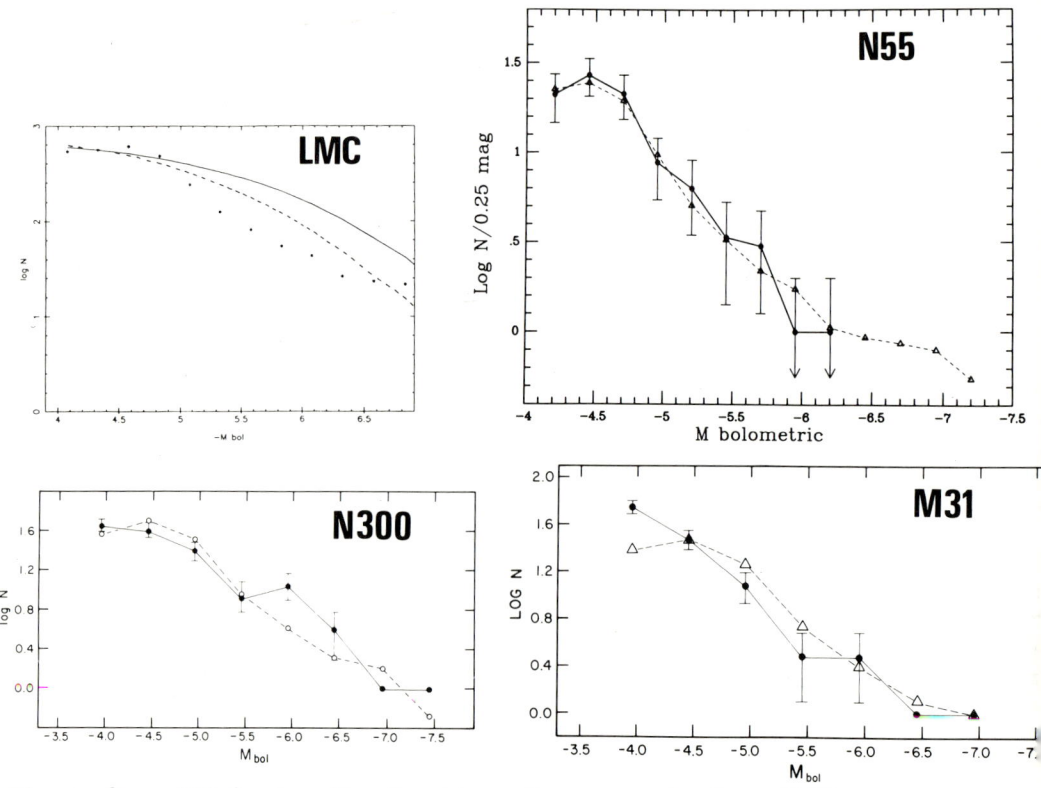

Figure 8: AGB luminosity functions for stars in four different galaxies. The upper left panel is for the LMC. In this panel the points are the data and the solid line is the model AGB assuming a Salpeter mass function and a constant star formation rate. The dashed line is for an IMF with a slope of 3.35. In the remaining three panels the LMC data are replotted for comparison (broken line). In all systems, there is a deficiency of luminous AGB stars compared with the theory.

In a new study, Hudon and Richer (1989) have determined the luminosity function for the AGB in a field in the galaxy NGC 2403. This galaxy, a member of the M81 group, probably represents the limit for ground based photometric studies of PRG's in external galaxies. The apparent distance modulus in V to this system is about 27.4. Thus typical carbon stars are found at about V=26. The luminosity function constructed for this galaxy is shown in Figure 9. Even in this remote system, far removed from the Local Group, we see that, within the errors, the AGB luminosity function remains similar to that of the LMC.

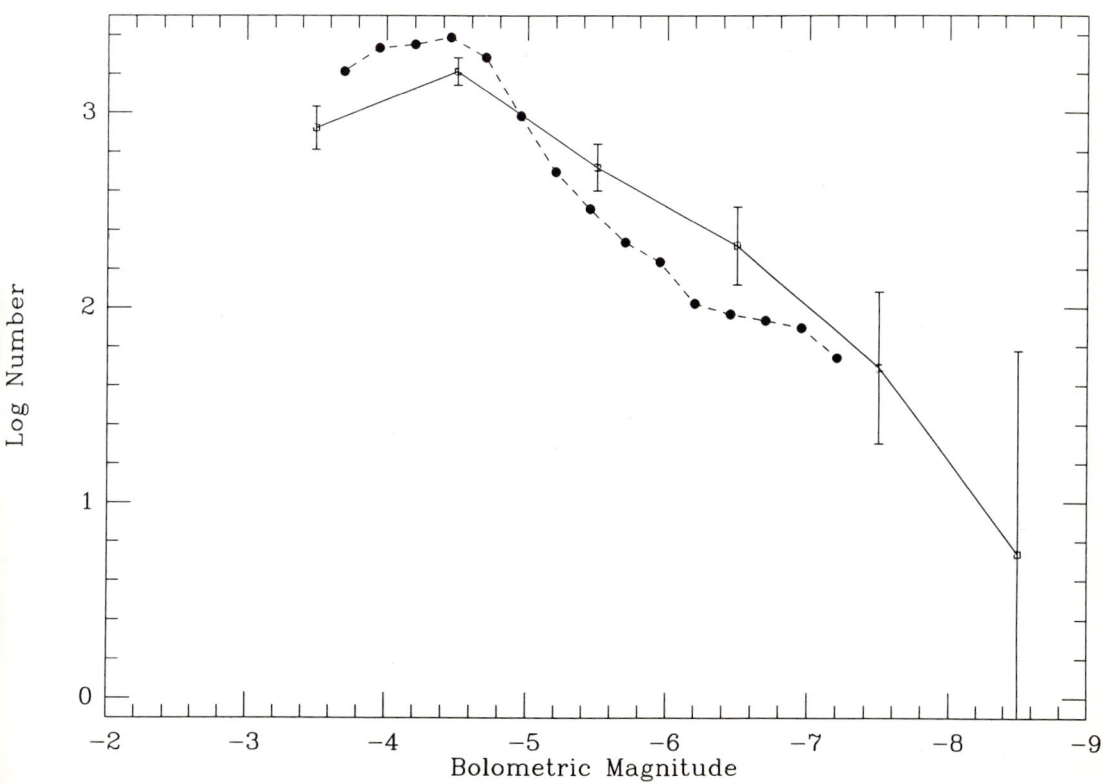

Figure 9: AGB luminosity function for a single CCD field in NGC 2403. The dots refer to the LMC luminosity function.

It should not be concluded, however, that star formation history plays no part in controlling the AGB luminosity function. Reid and Mould (1984) did find some variation in the LMC AGB luminosity functions with position in the galaxy in the sense that more luminous AGB stars were found in regions of more active star formation. An important study that could easily be carried out on several nearby galaxies is to observe their AGB luminosity functions in fields with widely differing star formation rates. This may provide the much needed observational input into the role that star formation history plays in controlling the AGB.

5.3 CARBON STARS AS EXTRAGALACTIC DISTANCE INDICATORS

Richer, Pritchet, and Crabtree (1985) have outlined the reasons why carbon stars should be useful extragalactic distance indicators. Briefly, these reasons are as follows. Carbon stars are luminous, have a small dispersion in their absolute luminosity, have the same luminosity function (within the errors) in diverse extragalactic systems, are fairly easy to locate in an external galaxy, are relatively numerous, and are unlikely to be contaminated by foreground stars in our own Galaxy. A summary of the distances derived to nearby galaxies using carbon stars exclusively is contained in Table 1. These distances are actually obtained differentially with respect to the LMC whose true distance modulus is taken to be 18.45 (Welch et al 1984).

Table 1

Distances to Nearby Galaxies Using Carbon Stars

Galaxy	D(Mpc)	H	M_H	ΔV (km s^{-1})
M 31	0.77	0.91	-23.52	548
M 33	0.78	4.38	-20.08	236
NGC 55	1.34	....	......	...
NGC 300	1.50	6.87	-19.01	222

Also included in Table 1 is the apparent H magnitude of the galaxy (Aaronson, Mould, and Huchra 1980) its absolute H magnitude using the distance derived from the carbon stars, and, in the last column, the observed 21-cm velocity width (van den Bergh 1984).

We can use the data in Table 1 to provide an infrared calibration of the Tulley-Fisher relation (Tulley and Fisher 1977). The linear least squares fit between H and log ΔV yields

$$M_H = 5.36 - 10.55 \log \Delta V \tag{1}$$

using the data for the three calibrating galaxies. This relation can be used to estimate a value of the Hubble constant using galaxies in the Virgo cluster. The infrared Tulley-Fisher relation defined by the Virgo galaxies yields

$$H = 36.69 - 10.77 \log \Delta V \tag{2}$$

(van den Bergh 1984), an equation with a slope remarkably similar to that of the calibrating galaxies. Ignoring the small difference in the slope of the two equations, we can then derive a Virgo distance modulus of 31.33. This corresponds to a distance of 18.5 Mpc. From this value and the cosmological redshift of 1322 km s^{-1} for the Virgo cluster, one obtains a Hubble constant H=71.5 km s^{-1} Mpc^{-1}.

This preliminary result appears promising. What is currently needed are several more calibrating galaxies. Good candidates here are NGC 2403, M81, NGC 247, and NGC 253. Studies of the PRG content of these systems should provide rich rewards.

References

Aaronson, M., and Mould, J.R. 1985, Ap.J., 290, 191.
Aaronson, M., Mould, J.R., and Huchra, J. 1980, Ap.J., 237, 655.
Azzopardi, M., Lequeux, J., and Westerlund, B.E. 1985, Astr. Ap., 144, 388.
Blanco, V.M., Blanco, B.M., and McCarthy, M.F. 1978, Nature, 271, 638.
Blanco, V.M., McCarthy, M.F., and Blanco, B.M. 1980, Ap.J., 242, 938.
Blanco, V.M., and McCarthy, M.F. 1983, A.J., 88, 1442.
Chokshi, A., and Wright, E.L. 1987, Ap.J., 319, 44.
Cook, K.H., Aaronson, M., and Norris, J. 1986, Ap.J., 305, 634.
Djorgovski, S., Spinrad, H., and Marr, J. 1985, in New Aspects of Galaxy Photometry, ed. J.L. Nieto (Lecture Notes in Physics, 232), p. 193.
Frogel, J.A., Cohen, J.G., Persson, S.E., and Elias, J.H. 1981, in Physical Processes in Red Giants, eds I. Iben Jr., and A. Renzini (Dordrecht:Reidel), p. 159.
Frogel, J.A., and Richer, H.B. 1983, Ap.J., 275, 84.
Hudon, D., and Richer, H.B. 1989, in preparation.
Iben I. Jr. and Renzini, A. 1983, Ann. Rev. Astr. Ap. 21, 271.
Lattanzio, J.C. 1986a, Ap.J., 311, 708.
Lattanzio, J.C. 1986b, in Calgary Workshop on Late Stages of Stellar Evolution, ed. S. Kwok and S.R. Pottasch, (Dordrecht:Reidel), p. 235.
Lattanzio, J.C. 1987, Ap.J. (Letters), 313, 15.
Lilly, S.J. 1987, M.N.R.A.S., 229, 573.
Pritchet, C.J., Richer, H.B., Schade, D., Crabtree, D., and Yee, H.K.C. 1987, Ap.J., 323, 79.
Reid, N., and Mould, J.R. 1984, Ap.J., 284, 98.
Renzini, A., and Voli, M. 1981, Astr. Ap., 94, 175.
Richer, H.B. 1981a, Ap.J., 243, 744.
Richer, H.B. 1981b, in Physical Processes in Red Giants, ed. I. Iben Jr., and A. Renzini (Dordrecht:Reidel), p. 153.
Richer, H.B., and Crabtree, D.R. 1985, Ap.J. (Letters), 298, L13.
Richer, H.B., Crabtree, D.R., and Pritchet, C.J. 1984, Ap.J., 287, 138.
Richer, H.B., Pritchet, C.J., and Crabtree, D.R. 1985, Ap.J., 298, 240.
Sanduleak, N., and Phillip, A.G.D. 1977, Pub. Warner and Swasey Observatory, Vol. 2, No. 5, p. 105.
Tulley, R.B., and Fisher, J.R. 1977, Astr. Ap., 54, 661.
van den Bergh, S. 1984, Q. Jour. R. Ast. Soc., 25, 137.
Welch, D.L., McAlary, C.W., McLaren, R.A., and Madore, B.F. 1984, in IAU Colloquium 82, Cepeids: Observations and Theory, ed. B.F. Madore (New York: Cambridge University Press), p. 219.

Westerlund, B.E. 1960, Uppsala Astron. Obs. Ann., 4, No. 7.
Westerlund, B.E. 1964, M.N.R.A.S., 127, 429.
Westerlund, B.E. 1965, M.N.R.A.S., 130, 45.
Westerlund, B.E., Olander, N., Richer, H.B., and Crabtree, D.R. 1978, Astr. Ap. (Suppl.), 31, 61.
Wing, R.F. 1971, in Proceedings of the Conference on Late-Type Stars, eds. G.W. Lockwood and H.M. Dyck (KPNO Contr. No. 554), p.145.

The Ratio of the Numbers of Carbon Stars to M Stars in Galaxies

Yu. L. Frantsman
Radioastrophysical Observatory, Latvian Academy of Sciences, Riga

Simulated populations of the AGB stars were calculated with different assumptions about mass loss, initial chemical composition and dredge-up efficiency. The early-AGB (E-AGB) phase was taken into account. The numbers of carbon and oxygen stars per 10^6 generated stars and the ratio (N_C/N_M) of these numbers were obtained. It is possible to match theoretically obtained N_C/N_M with the observations only if the luminosity of observed stars $M_{bol} < -3.5$; otherwise it is necessary to take into account the E-AGB phase. The data in the Table are for all AGB stars in the Galaxy and for stars with $M_{bol} < -1.80$ in the LMC.

Table

Object	$\dot{M}$ law	N_C	N_M	N_C/N_M
Solar neighborhood	$\alpha = 3.0$	1	650	0.002
	I	6	664	0.009
	II	8	661	0.01
LMC $(m-M)_o = 18.6$	$\alpha = 3.0$	11	424	0.03
	II	31	426	0.07

Here α is a coefficient in Reimers's mass-loss law; I: $\alpha = \alpha_o + \alpha_1 \exp(M_c)$, where ($M_c$) is the mass of the C-O core, and α_o and α_1 are chosen such that $\alpha = 0.33$ if $M_c = 0.5\ M_\odot$ while $\alpha = 10$ if $M_c = 1.0\ M_\odot$; II: $\alpha = 1$ if $\log(L/L_\odot) \leq 4.1$, $\alpha = 10$ if $\log(L/L_\odot) > 4.1$.

Identification of the (+2) Sequence of the CN Red System ($A^2\Pi - X^2\Sigma$) in Carbon Stars

Yoshio Fujita
Department of Astronomy, University of Tokyo
Bunkyo-ku, Tokyo 113, Japan

Identification of spectral lines from $\lambda 7700$ to $\lambda 8800$ has been carried out in twenty-four carbon stars: AQ And, UU Aur, W CMa, Y CVn, X Cnc, ST Cas, WZ Cas, AX Cyg, RS Cyg, U Cyg, V 460 Cyg, RY Dra, UX Dra, BL Ori, W Ori, RX Peg, Z Psc, 19 Psc, S Sct, Y Tau, VY UMa, HD 137613, HD 156074, and HD 182040. The spectrograms used for this purpose were obtained at the coude foci of spectrographs attached to 100-inch reflector of Mt. Wilson, 200-inch reflector of Mt. Palomar, and 74-inch reflector of Okayama. From the microphotometric tracings of each spectrogram, it was found that the main contributors to the spectral features of the region investigated are lines of the CN Red System ($A^2\Pi - X^2\Sigma$), and lines of the C_2 Phillips system ($b^1\Pi u - x^1\Sigma g$) are minor ones. The wavelengths of the band heads of the vibrational transitions of these diatomic molecules are in the CN System (+2) sequence: (2,0) $\lambda 7873$; (3,1) $\lambda 8067$; (4,2) $\lambda 8271$; (5,3) $\lambda 8485$; and (6,4) $\lambda 8709$; and in C_2 Phillips System (+3) sequence (3,0) $\lambda 7715$; (4,1) $\lambda 7908$; (5,2) $\lambda 8108$; and (6,3) $\lambda 8316$. In this note, a comparative survey of the spectral features around each band head of the (+2) Sequence of CN Red System was carried out. Generally speaking, except for U Cyg almost all subclasses from early-type carbon stars to late-type show similar and well developed features of this sequence. Identification of each spectral feature in the microphotometric tracings is graphically illustrated.

Spectroscopy and Photometry of the Southern R Stars

T. Lloyd Evans
South African Astronomical Observatory
Observatory 7935, South Africa

UBVRI photometry and 1-2Å resolution spectra have been used to classify 85 stars listed as type R by various authors. Some are N stars, including a few characterized by strong NaD, weak C_2 and enhanced ^{13}C which are distinct from the cooler counterparts of the strong-banded ^{13}C rich stars which are the most numerous of the R stars. The latter are also distinguished from non-^{13}C rich stars by having redder V-I at given B-V in the bifurcated distribution in the B-V, V-I diagram.

Classification of C-Rich Stars According to Their Mid-IR Signatures

R. Papoular
Service d'Astrophysique, CEN-Saclay, 91191 Gif-sur-Yvette Cedex, France

The spectra of the IRAS low-resolution-spectrometer in tape form have been submitted to a systematic morphological analysis, using classical quantitative discriminants (O.Gal et al. 1987, A & A **183**, 29; Y. Baron et al. 1987, A & A **186**, 271; R. Papoular 1988, A & A , in press). Spectra which display the 11.5µ feature of SiC fall into 4 classes of average spectral excesses. They differ by the width of the SiC feature and by the presence or absence of secondary features at ~8.6, ~11.7 and ~12.8µm.

A majority of these spectra have a lower 12-25µm colour temperature than do most optically selected C-stars, presumably because of thicker dust envelopes. While most spectra belong to LRS class 4n, 20% of the total were found among the brightest 20% of the much larger class 1n, suggesting that the relative abundance of C-stars is much higher than previously assumed.

Besides SiC, the dust appears to include graphitic and amorphous hydrogenated carbon (HAC), the latter being responsible for the secondary features. Here are the distinctive features of the 4 classes and their associations with optical spectral types in as much as these were identified.

SiC(a), fig.1. Mostly in LRS 4n, also in 1n. Late C types of the disc component. The SiC feature nicely matches the extinction curve for the purest and finest grains of laboratory α-SiC. The relative intensities of the secondary bands with respect to the SiC band increase as the colour temperature decreases, but remain constant with respect to the CS continuum. $F_{100\mu}/F_{60\mu}$ is often > 1.

SiC(b), fig.2. Only in LRS 4n. Late C types of the flat component. Weaker secondary features.

SiC(c), fig.3. Mostly in LRS 1n, also in 4n. Optical types S and M of the spheroidal component. The SiC feature is wider than (a) and (b) and similar to that of polluted, coarser grains of α-SiC.

SiC(d), fig.4. Only in 4n. Irregulars of the thin disc. 8.6-µm feature shifted to the blue, 11.7 µm absent. Graphitic component dominant. High value of $F_{100\mu}/F_{60\mu}$.

Multi-Wavelength Observations of the Peculiar Red Giant HR 3126

Joseph E. Pesce, Robert E. Stencel, Jesse Doggett, Frederick M. Walter
(Center for Astrophysics and Space Astronomy, U. of Colorado),
Patricia A. Whitelock (South African Astronomical Observatory) and
Joachim Dachs (Astronomisches Institut, Ruhr Univ., Bochum FRG)

The M2 II star HR 3126 (HD 65750) is remarkable because it sits near the center of a $0.7 M_\odot$ butterfly-shaped reflection nebula of several **arc minutes** extent (Dachs *et al.* 1978, *A & A* **63**, 353.). If it is a member of the nearby open cluster NGC 2516, the distance (375 pc), main sequence turnoff age ($\sim 10^8$ years) and implied mass ($\sim 5 M_\odot$) suggest that HR 3126 is a red bright giant past the initiation of helium burning in its core.

Ultraviolet observations were obtained 1987 Sept. 28 by Stencel and Pesce [IUE program MLJRS]: SWP 31915 (30 minute exposure) and LWP 11744 (159 minute exposure). Only Mg II emission and a photospheric continuum longward of about 2700Å were detected. The Mg II emission suggests that a chromosphere is present.

Doggett obtained blue-visual spectra of HR 3126 on 1987 September 25th and 26th at CTIO. They appear to be typical red giant spectra. The Ba II feature at 4554Å is present and is possibly enhanced compared to other M giants. In addition, a 5640-7040Å region spectrum was obtained by Walter at Cerro Tololo (1987 April 4) showing an overly strong Li I line at 6707Å. High lithium abundances in evolved stars have been taken as evidence that the stars have undergone helium shell flashes and are second ascent giants.

Whitelock observed HR 3126 on 1987 September 15/16 and comparison of (J-H), (H-K) and (K-L) colors with a standard star show longwave excesses. A low resolution ($\Delta\lambda/\lambda \sim 0.01$) spectrum from 1.2 - 4.0 μm obtained in December 1981 at South Africa is typical of an M giant, the CO strength indicates a type of M3 and there is no sign of Mira-like H_2O absorption. In addition, lack of large amplitude variability indicates that the star probably is not a Mira.

Finally, IRAS observed this object as well, detecting it in all four bands, 12 to 100 μm and with the Low Resolution Spectrograph (LRS). The infrared colors suggest a far infrared excess remains after a 3250K blackbody is subtracted. Further analysis suggest the source was spatially resolved by the IRAS detectors at all frequencies.

We surmise that HR 3126 passed through a red giant phase several million years ago and experienced comparatively high mass loss and dust production. The expansion of that dust shell is the present day IC 2220 nebula. The star subsequently may have undergone a blueward excursion in the HR diagram, onto the horizontal branch. At present, it appears to be evolving redward again and may soon begin its ascent of the asymptotic giant branch (while helium shell burning) on its way to carbon core ignition. The multiwavelength data appears to offer the possibility to test atmospheric response to evolutionary changes.

We are pleased to acknowledge NASA support for some of this work under grants NAG5-816 and JPL 957632, to the University of Colorado. This paper will appear in the proceedings of "A Decade of UV Astronomy with the *IUE* Satellite" conference, held in April 1988.

Kinematics of Short-Period Mira Variables - A Progress Report

J. Hron
Institut fur Astronomie der Universitat Wien

First results of an observing program recently started at the Figl Observatory for Astrophysics, Austria, are presented. Radial velocities derived by a correlation technique from the TiO bandheads near 7050 Å will be used to identify the stellar populations present among the short period ($P \leq 200d$) Mira variables. From the first observations with the Reticon-equipped Echelle Spectrograph of the FOA (resolution 0.3 Å at 7000 Å) we conclude that for Mira stars with a limiting visual magnitude of about 10m velocities can be determined with an external accuracy of 6 km/s.

Photometric observations of RCB stars in the LMC

W.A. Lawson and P.L. Cottrell
Mount John University Observatory,
Department of Physics,
University of Canterbury,
Christchurch, New Zealand.

We present UBV photometric observations of two R Coronae Borealis (RCB) stars, W Mensae and HV12842, which are members of the Large Magallenic Cloud (LMC). These data have been obtained over the last two years using single-channel photometers on both the 0.6m and 1m telescopes at Mount John. They form part of an ongoing long-term program to investigate photometric and spectroscopic variations in the hydrogen-deficient carbon (HdC) stars both in the LMC and in our Galaxy.

These two stars show quite different properties, indicative of the nature of these objects. HV12842 (Fig.1) shows semi-regular pulsations not dissimilar to the bright southern RCB star, RY Sgr (see Lawson, Cottrell and Bateson 1988), whereas W Mensae has a V magnitude which seems to have no obvious periodic variations, similar to the variations in R Coronae Borealis itself. In addition, the V magnitude curve of HV12842 appears to show beating due to multiple mode pulsations, as the amplitude of the pulsations has changed from ≈0.5 mag. to <0.05 mag. over a period of about 450 days. Further confirmation of this effect is shown by the recent observations (JD 2447300 onwards) which have a V amplitude of ≈0.1 mag.

Lawson,W.A.,Cottrell,P.L.&Bateson,F.M.1988,*Publ.Var.StarSect.R.A.S.N.Z.*,**14**, 38.

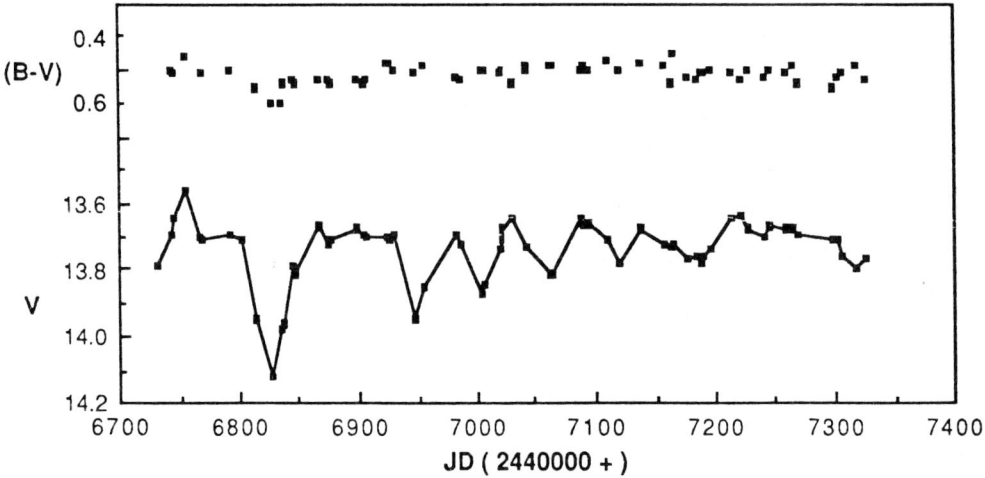

Figure 1. (B-V) and V curves for HV12842.

The Infrared HR Diagram - The IRAS Two-Colour Diagram of AGB Stars and Its Evolutionary Explanation

Jiang Shi-yang
Beijing Astronomical Observatory, Chinese Academy of Sciences
Beijing, China

In recent years several people have used the IRAS data to produce ([25] - [60]) / ([12] - [25]) two-colour diagrams of late type stars (Van der Veen and Habing 1988; Qing-Quan 1988). The distribution of all these AGB stars are concentrated in a limited area, and any special series of stars are limited on a definite curve which can be represented by an exponential equation like

$$[25] - [60] = A + B \exp([12] - [25]).$$

Here $[X] = -2.5 \log$ flux and A, B, C are constant coefficients. On the other hand, we can make a model calculation to fit all these distributions. There are two kinds of models. One is a temporal model for different kinds of objects; another is an evolutionary model for stars with different initial mass. All these models are determined by the initial depth of the dust-gas envelope, the emission and absorption coefficients of the particles within the envelope, etc. So by comparing the real two-colour diagrams with these theoretical models we can determine many basic properties of late-type stars, especially those of AGB stars. Our results follow. (1) The initial masses of AGB stars are between 1 to 8 solar mass. (2) The mass loss rate of AGB stars varies from 10^{-7} to 10^{-5} solar mass per year. (3) The temperature of the photosphere of AGB stars are between 2000 to 3000 K. (4) The optical depths at the 12 micron band are between 10^{-3} and 10^{-2}.

Finally, the evolutionary scenario for AGB stars are as follows. At first the mass loss rates are still very low so the IR excess are small; at a certain moment in the AGB life the mass-loss rate increases steeply so that the IR excess increases and the position on the two-colour diagram moves from lower left to the upper right. At the same time the 9.7 and/or 11.3 micron feature changes from strong emission to strong absorption gradually, and the star changes from a mira variable to an OH/IR source.

References

Van der Veen, W.E.C.J., Habing, H.J., 1988, Astron. Ap. 194, 125.
Tang Qing-Quan, Thesis, 1988, Beijing Astronomical Observatory.

A Photometric Monitoring of Barium Stars*

A. Jorissen
Institut d'Astronomie,
d'Astrophysique et de Geophysique
Universite Libre de Bruxelles

J. Manrfroid+
Institut d'Astrophysique
Universite de Liege

An uvby photometric monitoring of a sample of 19 Ba II stars has been performed since July 1984 at ESO on the Danish and ESO 50 cm telescopes, as part of the "Long Term Program for Variable Stars" (Sterken, 1983). The sample contains southern Ba II stars with variable radial velocities (from McClure, 1983 and Jorissen and Mayor, 1988) as well as Ba II stars already monitored by Landolt (1983). The observations are done differentially, using two comparison giants of the same spectral type located within a few degrees of the Ba II star. The reduction to the standard uvby system is performed using the algorithm described in Manfroid and Heck (1983), taking into account <u>all</u> the measurements performed on the considered telescope since the beginning of the monitoring. With such a reduction procedure, the scatter in the magnitude difference between two comparison stars (of magnitude 5 to 7) can be as low as 0.003-0.004 mag rms.

There is only one star (HD 46407) in our sample that seems to show an irregular long term variation of small amplitude in y ($\approx$0.01 mag), but of larger amplitude in u ($\approx$0.05 mag). A few stars, however, display a jitter of 0.005 to 0.01 mag amplitude, which is greater than the photon noise for the considered stars. That jitter seems to be correlated to the position of these stars in the (b-y,c1) diagram, since they are all located in the region b-y>1.0 and c1<-0.1. It is not yet clear whether such a jitter is an artifact, or whether it could be attributed to something like wind accretion in the binary system (as it is observed for Mira B; Warner, 1972).

References

Jorissen, A., Mayor, M. 1988, <u>Astron. Astrophys.</u> **198**, 187.
Landolt, A.U. 1983, <u>Publ. Astron. Soc. Pacific</u> **95**, 644.
Manfroid, J., Heck, A. 1983, <u>Astron. Astrophys.</u> **120**, 302.
McClure, R.D. 1983, <u>Astrophys. J.</u> **268**, 264.
Sterken, C. 1983, <u>Messenger</u> **33**, 10.
Warner, B. 1972, <u>Monthly Notices Roy. Astron. Soc.</u> **159**, 95.

*Based on observations carried on at the European Southern Observatory (ESO, La Silla, Chile).
+Senior Research Associate, Belgian National Fund for Scientific Research.

Are the Barium and Am Stars Related?

J. Hakkila (Mankato State University)

A study of barium star kinematics from 112 stars with known radial velocities indicates a solar motion of

$u_\odot$(km/s)	$v'_\odot$(km/s)	$w_\odot$(km/s)	$S_\odot$(km/s)
-15.6	16.3	10.5	24.9

All values have errors of roughly ±2.6 km/s.

Parameters of the velocity ellipsoid from 32 barium stars with independently-determined absolute magnitudes are:

$\langle u^2 \rangle^{1/2}$(km/s)	$\langle v^2 \rangle^{1/2}$(km/s)	$\langle w^2 \rangle^{1/2}$(km/s)
24	16	15

These values, a scale height of barium stars perpendicular to the galactic plane of 230 pc (Lü 1988, private communication), and an estimate of the spatial distribution of stars in the Galactic plane all indicate that barium stars belong to an intermediate disk population with masses between 1.5 and 3.0 solar masses and ages of 5×10^8 to 2×10^9 years.

Given the masses of barium stars and their tendency for binaricity, one is tempted to look for likely progenitors. One class of binary main-sequence star in this mass range is the Am (or metallic-lined A) star.

Mass transfer in a binary system is often thought to lead to the formation of a barium star, and excess s-process elements are conveniently present in Am atmospheres (probably due to diffusion resulting from tidal braking and subsequent slow rotation). Am stars have closer orbital separations and larger mass ratios than barium stars, which provide reasonable constraints to any mass transfer/loss model of barium star formation. This relationship between stellar types could explain why barium stars exhibit (1) peculiar atmospheric abundances of heavy elements with no sign of recent s-process production, and (2) white dwarf companions. Numerically, Am stars are roughly ten times more common than barium stars, indicating that such an evolutionary relationship is reasonable. Mass transfer/loss also introduces a random element which might explain statistical barium star peculiarities such as their huge range of absolute magnitudes.

A few evolutionary models are presented, even though the combination of diffusion, mass transfer, and normal stellar evolution make such models far from unique.

Molecular Bands in the 1.1-1.4 μm Spectra of M-S-C stars

Kenneth H. Hinkle	David L. Lambert	Robert F. Wing
K.P.N.O., N.O.A.O.[1]	University of Texas	Ohio State University

Spectra are presented in the J band (7400 to 9700 cm^{-1}) for four Miras ranging in spectral type from M through C. All the program stars have been observed near minimum light. The program stars cover a considerable range in C/O and the spectral features exhibit a progression as a function of C/O. The S-type stars contain strong bands not previously reported. Especially striking are two sets of triple-headed bands in the J-band spectrum of the S-type Mira R And. The bandheads, which are degraded to longer wavelengths, are at 7877, 7957, 8030 cm^{-1} and 8379, 8459, 8530 cm^{-1}. The former triplet, which is the stronger of the two, also is present in the mild S star χ Cyg but not in the M star R Cas. Additional heads are found in R And at 7477 cm^{-1}, near the short wavelength edge of strong telluric absorption, and at 8968, 9031, 9063 cm^{-1}. The bands are identified as the $\Delta v = -1$, 0, 1, and 2 sequences of a predicted ($^3\Pi-^3\Delta$) transition of ZrS. Additional conspicuous features in the spectra of χ Cyg and R Cas are identified with VO, TiO, and H$_2$O bands. These observations provide additional evidence that ZrS is responsible for the majority of the Keenan-Wing bands in the near infrared (0.7-1.1 μm). With additional laboratory work, the ZrS bands will provide an opportunity to measure sulfur abundances in late-type stellar photospheres.

[1]Operated by the Association of Universities for Research in Astronomy, Inc. under contract with the National Science Foundation.

Photometric Classification of Carbon-Rich Stars

V. Straizys
Astrophysical Department, Institute of Physics
Vilnius, Lithuania, USSR

The Vilnius seven-color photometric system with the mean wavelengths of bandpasses at 345, 374, 405, 466, 516, 544, and 656 nm is intended for photometric determination of spectral classes, absolute magnitudes, and metallicities of stars and the amount of interstellar reddening. At the same time, different reddening-free diagrams of the system make it possible to recognize stars with different peculiarities. Carbon-rich stars are among these types of objects. For separation of carbon, barium, and CH stars from normal stars, the diagram Q (345, 374, 466, 544), Q (405, 516, 656) is the best. Its merit is based on sensitivity of the 405 magnitude to the violet depression and of the 374 magnitude to the absorption of the C_3, SiC, and CN bands. This diagram can be used to estimate the C/O abundance ratio with corresponding calibration at hand.

A method for simultaneous determination of metallicities and O/C ratios for late-type giants is developed. This method may be used after exclusion of the interstellar reddening effect.

A Deep Visual-Red and Near Infrared Objective Prism Spectral Survey of the Milky Way

O.M. Kurtanidze   M.G. Nikolashvili
Abastumani Astrophysical Observatory

The deep visual-red and near infrared low dispersion (1250 Å/mm at Hγ and 7000 Å/mm at Å band) objective prism spectral survey of the Milky Way equatorial ten degree belt has been done by 70 cm meniscus telescope equipped with $2°$ prism ($30° \leq 1 \leq 115°$ infrared).

The Kodak IIIa-J and IIIa-F spectroscopic plates were hypersensitized by baking in dry air or in dry nitrogen. The silver-nitrate ($AgNO_3$) treatment was used for Kodak IV-N plate hypersensitization. The limiting visual magnitudes of the survey performed are $16\overset{m}{.}0$ (J, F) and $18\overset{m}{.}0$ (N, V-I = 3.0).

Carbon stars are identified by the presence of the C_2 and CN molecules absorption band system at $\lambda\lambda 5165$, 5635 A and $\lambda\lambda 7945$, 8025, 8320 A respectively. More than eleven hundred new C stars are revealed.

The study of the latitude and longitude distribution of all C stars discovered in the region $90° \leq L \leq 165°$ shows that the first one is uniform and the other one is nonuniform. The later comes from significant increasing in the surface density of Carbon stars from anticenter to Cas-Cyg directions. The mean surface density of C stars in the studied region is 1.0 per sq. degree i.e., on average, it has increased 2.5 times, meanwhile in the near infrared survey 5.0 times.

By the "nearest-neighbour" method it is shown that statistically significant number of pairs and members of the open clusters are not observed. The connection of Carbon stars with dark clouds were also studied.

2. Model Photospheres and Chemical Compositions

PHOTOSPHERIC MODELS FOR COOL GIANT STARS

M.S. Bessell[1] and M. Scholz[2]

[1] Mount Stromlo and Siding Spring Observatory
Canberra, Australia
[2] Institut für Theoretische Astrophysik
der Universität Heidelberg, Heidelberg, F.R.G.

Abstract. Models for M and C giant stars differ from those for hotter stars by having complicated state equations and opacities dominated by lines from diatomic and polyatomic molecules. In addition many cool giants have atmospheres which are extended, and which cannot therefore be adequately modelled using a plane-parallel approximation. Mira variable stars have atmospheres which are even more extended due to the regular passage of shock waves through their atmosphere. In this review we will discuss recent modelling of such atmospheres and show comparisons with observations of variable and non-variable M stars.

I. HISTORICAL OVERVIEW

Early attempts at constructing model atmospheres for cool stars were undertaken by Auman (1967,1969), Tsuji (1966), Alexander et al(1972), and Johnson (1974) for M stars (oxygen-rich, O/C >1), and by Querci et al (1974,1976), and Sneden et al (1976) for C stars (carbon-rich, O/C <1). Temperatures from models by Tsuji (1978) were shown to be in good agreement with the lunar occultation temperature scale for M stars (Ridgway et al, 1980), a significant success. Various methods of handling the complex line absorption opacities have been considered, mean opacities (eg Golden 1969; Tsuji 1966; Zeidler et al, 1982), Elsasser Band model (Tsuji 1968,1978), opacity distribution functions (ODF)(eg Querci et al 1971,1974; Gustaffson et al, 1975; Saxner et al, 1984) and opacity sampling (OS) method (eg Peytremann,1974; Sneden et al,1976). All this modelling was done under the assumption of plane-parallel (PP) geometry. More recent PP models for M stars have been published by Tsuji (1981), and Johnson et al (1980,OS). Steiman-Cameron et al (1986) compared these latter models with observations.

Schmid-Burgk et al (1975) surveyed the location of extended-photosphere stars in the HR diagram, and the properties of extended M-type photospheres were outlined by Watanabe et al (1978,1979); Schmid-Burgk et al (1981); Wehrse (1981), Scholz et al (1982), Kipper (1982) and Scholz (1985); and extended C-type photospheres by Scholz et al (1984). Scholz et al (1984) and Lambert et al (1986) found that extension was much less in C-type photospheres than in M-type photospheres of the same atmospheric parameters, but extension of C-type photospheres depended very sensitively upon the low-temperature opacities of carbon-rich element mixes, namely the hitherto little investigated polyatomic molecules such as HCN, C_2H and C_3 (Jørgensen et al 1985, 1987). Detailed discussions of extended model photospheres for M stars and comparison with observations have been given Scholz (1985), Brett (1988), and Bessell et al (1988a), hereinafter called BBSW1. Limb darkening and monochromatic radii of extended models (including miras) were discussed by Scholz et al (1987). Extension of atmospheres of pulsating stars caused by periodic shocks were investigated by Fedorova (1973), Wood (1979) and Bowen (1988). Exploratory extended atmospheres for mira-type variable M stars based on Wood's dynamical models were discussed by Bessell et al (1988b), hereinafter called BBSW2. More extensive analysis of model photospheres for miras of different period and composition, and at various phases, is underway.

In this review we will outline some of the complications associated with such modelling and show comparisons between recent model computations and observations.

II. MODELLING PROBLEMS CONCERNING COOL PHOTOSPHERES

1. Obtaining opacities.

(a) Cool models involve complicated state equations with large number of composed particles (including dust at very low temperatures).
(b) Missing particles (in particular at very low temperatures) in the state equation may lead to spurious particle pressures of the absorbing particles.
(c) Poorly known physical input data of the state equation (cf. revisions of TiO and CN dissociation energy in the past) may lead to spurious partial pressures of the absorbing particles.
(d) Absorption is dominated by very large numbers of lines, in particular molecular-band lines (plus dust at very low temperatures). Missing absorbers (in particular at very low temperatures) may lead to grossly incorrect opacities. This is especially a problem for element mixes with $C/O \geq 1$ or ≈ 1.
(e) Oscillator strengths of important bands may be unknown or poorly known (cf. several TiO bands, VO, HCN, C_3 etc).

2. Handling opacities.

(a) Continuum (H^-, H_2^-) is fading away towards low temperatures and opacities become line-dominated.
(b) Treatment of molecular lines by the mean-opacity technique reduces the required computer capacity (storage, time) but may be seriously inadequate, in particular for saturated lines.
(c) Treatment of lines by the ODF or the OS techniques requires large computer capacity. OS is more flexible, including the treatment of macroscopic velocity fields.
(d) Complicated molecules (e.g. H_2O, HCN, C_3) with complicated line patterns are hardly treatable with line-by-line techniques like ODF or OS.

3. Modelling and analyzing the photosphere.

(a) Strong wavelength dependence of opacities may lead to difficulties in finding the stratification which fulfills the energy equation, in particular via temperature correction methods which use weighted mean opacities (e.g. Lucy).
(b) Very accurate modelling is required if an important absorber which has great influence upon the stratification reacts sensitively to stratification changes (e.g. H_2O, HCN, C_3).
(c) Gravity sensitive features are hard to find at very low temperatures.
(d) Since all elements are coupled with each other via the composed particles in the state equation, a consistent set of element abundances has to be determined instead of carrying out an element by element analysis.
(e) Since absorption is dominated by lines, the abundance input for model construction and the abundances deduced from analysis have to be checked carefully for consistency. Not even the continuous absorption is a safe pre-given quantity because the free e^- forming H^- and H_2^- depend on the abundances of the alkali-earth and the alkali metals (which are less easily accessible than the Fe-type metals supplying the free e^- in hotter stars).
(f) An element or molecule which is not accessible by observation may still be important concerning (d) and (e).

III. MODELLING PROBLEMS CONCERNING GIANT PHOTOSPHERES

1. Modelling the photosphere.

(a) Non-variable (static) stars: Density or gas pressure scale height (scaled with the radius)

$$\frac{H_\rho(r)}{r} \approx \frac{H_{Pg}(r)}{r} = \frac{1}{r}\frac{\Re T}{\mu g_{eff}} \propto \frac{T_{eff}}{Rg_s} \propto \frac{\sqrt{L}}{T_{eff}M}$$

is large for high-luminosity, low-mass giants (not as large for supergiants) meaning extended photospheres.

(b) Mira (pulsating) stars: $H_\rho(r)/r$ is large because outgoing shock fronts produce flat density gradients between fronts; dynamical models imply more extended photospheres.
(c) Extension effects are important for extensions > 5%. (i) Dilution of radiative energy is proportional to $1/r^2$. Quasi plane modelling i.e. plane stratification with $1/r^2$ modification of radiative flux, is possible up to ??%. Needs to be tested. (ii) Peaking of radiation. Cannot be approximated in any modified PP approximation.
(d) Extension requires radius definition. Flat density gradient leads to noticeable photospheric mass fractions. Both reasons demand careful fitting of interior and photospheric models.
(e) Dynamical models should treat consistently the interior and photospheric layers.
(f) Grids of extended static model photospheres with given element mixture must be three-dimensional (L,M,R) or (T_{eff}, g_s, extension) rather than two-dimensional (T_{eff}, g_s).
(g) Dynamical models should take into account the possibly slow relaxation of the heated post-shock material towards local thermodynamic dissociation and radiative equilibria.
(h) Dynamical models should take into account the radial macroscopic velocity fields modifying the line absorption coefficients used for model construction.

2. Analysing the photosphere.
(a) Because of low densities, deviations from LTE may be important (possibly also relevant for modelling).
(b) Because of possibly slow relaxation of the heated post-shock material of a dynamical model, deviations from LTE may be very strong (possibly also relevant for modelling).
(c) In the case of large extension, analysis of a static photosphere may yield three parameters (L,M,R) or (T_{eff}, g_s, extension) rather than only two (T_{eff}, g_s).
(d) Analysis of lines of a mira spectrum must take into account the radial macroscopic velocity fields affecting the line absorption coefficients.
Remark : up to some extension limit ??% (see 1c.), analysis of a line profile by means of a plane code is possible if the density-temperature-stratification of a quasi-plane ($1/r^2$) or extended model is used. (We intend to publish our T, Pg, r, rhox stratifications from an extended code and believe that they could be so used to good approximation.)

3. Monochromatic radii.
(a) Extended atmospheres have wavelength-dependent radii. Radius measurements must be carefully evaluated by using models for proper interpretation and for deriving a suitable $R(T_{eff})$ of the star.
(b) Monochromatic radii may be used as probes of the photospheric stratification.

IV. DETAILS OF EXTENDED MODEL CONSTRUCTION

The following discussion concerns mainly the spherical models of Scholz (1985) and BBSW1,2.
1. The static models are in radiative and hydrostatic equilibrium:

$$4\pi r^2 \pi F_{rad} = L$$

$$\frac{dP_g}{dr} = \frac{dP}{dr} + \rho k_F \cdot \frac{\pi F_{rad}}{c} = \rho \frac{GM}{r^2}\left(1 - k_F \frac{G}{4\pi c}\frac{L}{M}\right)$$

where r is the distance from the star's center, k_F is the flux-weighted mean opacity, other symbols have usual meaning. Convective energy transport does occur in red giant photospheres but is negligible. The radiative acceleration term is usually not important and turbulent acceleration is here neglected. The radiation transport equation is

$$\cos\theta \frac{\partial I_\nu}{\partial r} + \frac{\sin^2\theta}{r}\frac{\partial I_\nu}{\partial \cos\theta} = -\rho k_\nu (I_\nu - S_\nu)$$

where θ is the polar angle in spherical coordinates, I_ν is the intensity in frequency ν and the source function S_ν is expressed in LTE by the Planck function B_ν and the mean intensity J_ν: $k_\nu S_\nu = \kappa_\nu B_\nu + \sigma_\nu J_\nu$. The monochromatic opacity k_ν is the sum of the LTE absorption and the scattering coefficients: $k_\nu = \kappa_\nu + \sigma_\nu$. The first equation of radiative equilibrium above is equivalent to the condition of vanishing radiative flux gradient,

$$\int_0^\infty k_\nu (J_\nu - S_\nu) \, d\nu = 0,$$

which is very sensitive to violations of the radiative equilibrium condition in the upper model layers.

The models discussed by BBSW1,2 were computed with a code based on the Schmid-Burgk method (Schmid-Burgk et al, 1984) of solving the spherical radiation transport equation. They were iterated for a flux constancy of 0.01% and radiative flux gradient of 1% (usually much better than these values). The LTE approximation is used to solve the equation of state and to calculate absorption and continuum scattering coefficients which are averaged over 72 wavelength meshes for model construction (Scholz et al 1984; Tsuji 1978). The absorption coefficients used were mean absorption coefficients, smeared over finite wavelength intervals.

The stellar radius R, which is the third basic photospheric parameter besides L and M, enters the system of equations through the outer boundary conditions:
$P(r_0) = P_0$ and $I_\nu(r_0, \theta) = 0$ for $\theta < 0$, where r_0 denotes the transition layer from the photosphere to any other "non-photospheric" outer atmospheric region. This r_0 is a poorly defined quantity and in extended photospheres can be considerably larger than the observed radius or the outer boundary of an interior model. For many models it is adequate to define the radius R as the distance from the center of the star to the point where the standard optical depth (Rosseland mean opacity) is 1, and to define r_0 to be the distance to the point where the optical depth is 10^{-5}. See Scholz (1985) for details. The extension d is defined as:
$$d = (r_0 - R)/R.$$
(When $d \ll 1$, the atmosphere is called compact. For $d < 0.05$ to 0.1 the atmosphere can often be treated in the plane-parallel approximation). The effective temperature T_{eff} can defined by the relation
$$T_{eff}^4 = L/[4\pi a R^2(\tau_{Ros} = 1)].$$
There is an alternative definition of radius and efective temperature that is useful for interior models esp. with $T_{eff} < 3000K$. From the $T \sim r$ stratification of the model, if one evaluates the quantity $L/(4\pi a r^2)$ at each value of r and compares it with the quantity T^4 (T is the temperature at r), then at some radius $r = \tilde{R}$, T will equal $(L/(4\pi a r^2))^{1/4}$. This temperature can be defined as the effective temperature $\tilde{T}_{eff}$ and $\tilde{R}$ can be defined as the radius of the star. That is
$$\tilde{T}_{eff} = T(\tilde{R}) = (L/(4\pi \tilde{R}^2))^{1/4}$$
In this definition, the stellar radius R does not depend on the special choice of reference τ scale, but it does depend sensitively on the photospheric temperature profile and on the model and opacity details. Some of our coolest models near 2200K produced differences of almost 200K from the two definitions, $\tilde{T}_{eff}$ being less than Teff. These aspects are more fully discussed by BBSW2.

2. In the mira models that we have constructed, the condition of hydrostatic equilibrium is not imposed, instead, $\rho(r)$ or $P_g(r)$ is adopted from a dynamic envelope configuration. Wood (1979, 1989) computes non-adiabatic pulsating envelope-atmosphere models using grey opacities. The configuration of exploratory models (BBSW2) were based on typical density profiles, at specific phases, of older pulsation models, with the density steps modified to agree with measured shock velocities (Hinkle et al, 1984). Recently we have started to use the actual structure from more realistic models, computed with specially computed Rosseland opacities for various metal mixes and temperatures between 10000K

and 750K. Scholz and Tsuji plan to compute similar carbon-rich opacities tables. The same technique as noted above is used to solve the spherical radiation transport and energy equations. LTE is assumed for the equation of state and opacity calculations.

3. Synthetic spectra computations.

The wavelength averaged opacity mesh used for model construction was too coarse to permit reasonable comparison with observed spectra, therefore an additional spectrum was computed for each model using a table of opacities at a much higher wavelength resolution. Several hundred wavelength points were computed down to 10Å resolution. Brett (1988) discussed the computation of these opacities and the comparison of the resultant spectra with observations of early M stars. Examples will be given below.

IV RESULTANT MODEL STRATIFICATIONS

1. Comparison with different techniques.

Scholz et al (1984) showed in the plane-parallel case, that the use of the smeared mean averaged opacities gave close to the same temperature stratification for models of early M stars as the OS technique of Johnson et al (1980). For the carbon-rich material the results were very different and they suspected that the carbon-rich line list used for the OS must be incomplete. Details of the Lambert et al (1986) carbon-rich model stratifications made using ODFs are not yet published, so a similar comparison cannot be made, however, Lambert et al note that IR fluxes from their models give a similar temperature calibration to that derived by Tsuji (1981) from his models based on the Elsasser Band model of mean opacities. These results support the proposition that mean-opacities can be used with cautious confidence for model building. The Elsasser Band model treatment is similar to construction of ODFs.

2. Comparison of different extended models.

Scholz (1985) found that usually (i) the temperature in the upper layers of the models decreases with increasing photospheric extension due to dilution of radiation and higher photon escape probability (in Fig.1a is shown the $T(\log P_g)$ stratification of three 3000K models with increasing photospheric extension): (ii) higher surface gravity shifts the $T(P_g)$ curves towards higher pressures by $\Delta \log Pg \approx \Delta \log g_s$; and (iii) lower metal abundances shift the $T(P_g)$ curves towards higher pressures due to the higher average transparency of matter. The main absorbers in O-rich atmospheres H^-, TiO, CO and H_2O react differently to abundance changes and are efficient at different temperatures in different layers, consequently the T(Pg) curves can have quite different shapes for models with different chemical compositions, different effective temperatures and different extensions. Water vapor plays an important role in this behaviour. When H_2O begins to form it acts as an efficient cooling mechanism, causing a steepened temperature gradient in these layers. Polyatomic molecules including C such as HCN and C_3 very likely produce the same effects in C stars.

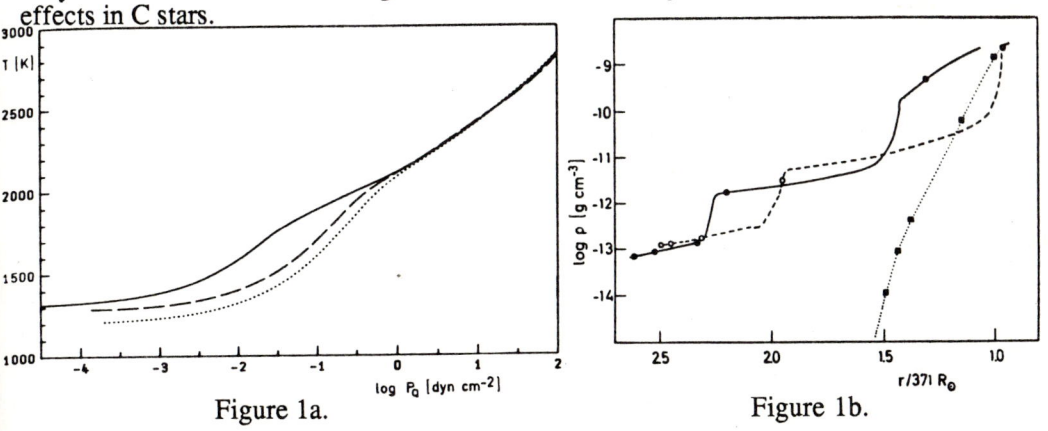

Figure 1a. Figure 1b.

In Fig.1b the T(r) stratification of two mira models is shown in comparison with a static model. The lowest temperature points marked on the curves indicate mean Rosseland optical depths of 10^{-4} so that one can see that the mira models have more than 3 times the extension of the static model which has d = 0.54. Scholz et al (1987) show that in such models the radius measured at the wavelength of a strong TiO band will be more than twice that measured in a continuum window. These model predictions agree nicely with the monochromatic radius observations for miras reported by Labeyrie et al (1977) and Bonneau et al (1982). Some of the models investigated were so extended that, owing to special conditions of density and opacity stratification, evidence of faint emission by H_2O was seen in the spectra. In such cases, slight limb-brightening is also possible (Scholz et al 1987).

The formation of H_2O is enhanced by low temperatures and high pressures, therefore H_2O opacity becomes important in high gravity models at a higher effective temperature than in a low gravity model, and is stronger in an extended model compared to a less extended model of the same T_{eff} and g. Scargle et al (1979) noted that the IR spectra of Tsuji's dwarf models resembled their observed mira spectra more than the giant and supergiant spectra. We can understand this now on the basis of the cool extended atmospheres of the mira models.

In the hotter models of M stars the extension d varied approximately linearly with the pressure scale height, however, in those models where H_2O formation became important, the extension doubled for little change in the pressure scale height, reflecting the increased opacity in the cool upper layers due to the polyatomic absorber. We anticipate that polyatomic absorbers will be similarly extremely important in extending the atmospheres of cool carbon-rich models.

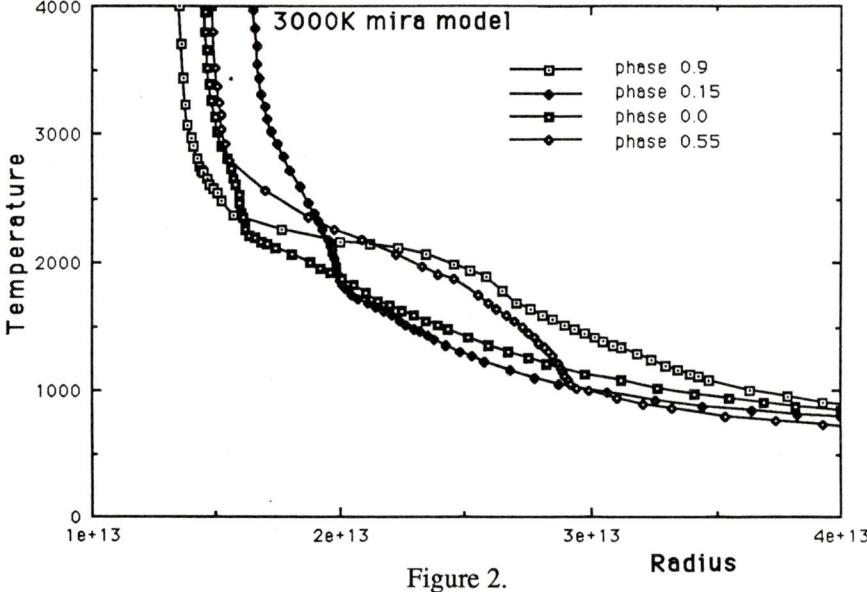

Figure 2.

In Fig.2 are shown the the T and r stratification at 4 phases, of some very preliminary self-consistent 3000K mira models. These show nicely the change in the temperature and density profiles as the shock-wave traverses the atmosphere. Wood's modelling assumes that the material behind the shock rapidly reaches radiative and thermodynamical equilibrium, whereas other workers, eg. Fedorova (1978) and Bowen (1987) use very approximate approaches for treating the de-excitation of the heated gas behind the shock-front. We show below that the Bowen (1988) T and ρ stratification results in unrealistic photospheric spectra compared to observations and that Wood's assumptions therefore are more likely to be correct.

V. RESULTANT PHOTOSPHERIC SPECTRA

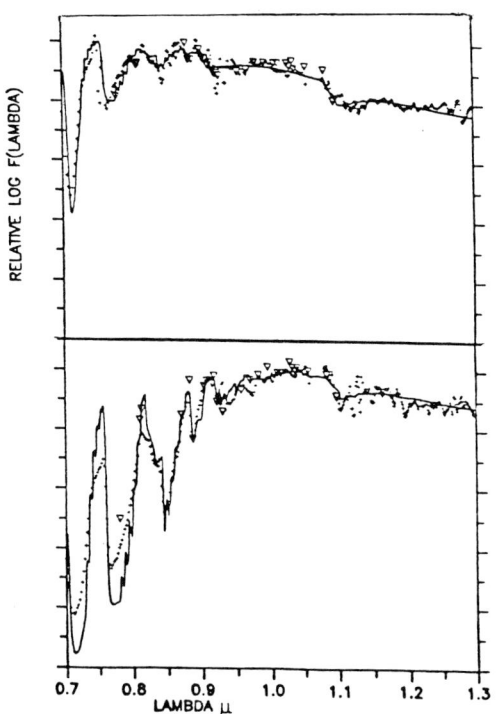

1. In Fig.3 are shown the near-IR spectra of some of our static models in comparison with observations. The top panel shows M3III observations and the 3650K model flux. The bottom panel shows M5III observations and the 3350K model flux. The observations are from Gunn and Stryker (1983) and Wing (1967). These comparisons are from Brett (1988). Spectral types were adopted from Wing (1978). The extension and gravity of the models were relevant for class III giants. The temperature scale thus indicated is in good agreement with the scale of Ridgway et al (1980), and the model spectral features are similar to observations. The grid of static spectra are discussed in detail in BBSW1.

Figure 3.

Fig.4 shows the calibration of two colors which have been used for temperature derivation in M stars. Both colors are reasonably insensitive to extension and gravity, as long as H_2O absorption is weak. Amongst the many features calculated, the TiO and H_2O bands were found to be most sensitive to the extension, while the CN and CO bands were very sensitive to the gravity, however, both effects must be accounted for in fitting data.

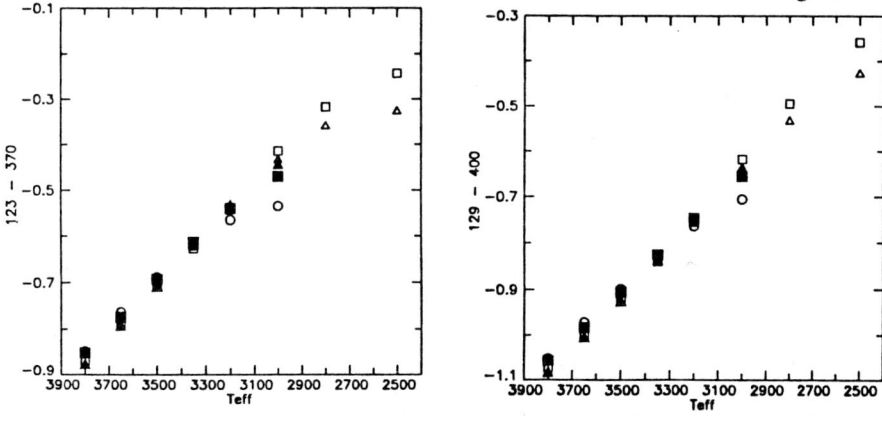

Figure 4.

Fig.5 shows the variation of an IR color measuring the H_2O for giant models with different gravity and extension. The solid squares and open circles are higher gravity, low extension models; the open squares are more extended models than the open triangles, but have the same low gravity. Fig.6 shows the variation of a color measuring the CO second-overtone bands. The difference in gravity between the open square-triangle sequence and the open circle (higher gravity) sequence is 1.3 in log g. Details are given in BBSW1.

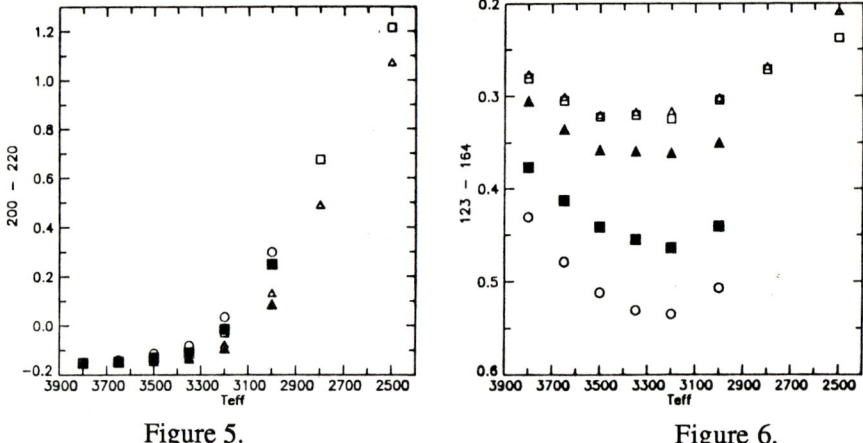

Figure 5.   Figure 6.

In cool M stars, the IR broad-band magnitudes measure spectral regions which contain a variety of spectral features. In Fig.7 the photometric bands are shown with respect to the spectrum of a T_{eff} = 3000K, log g = -0.70, d = 0.54 model. Strong CO absorption is seen in the H and K bands, while H_2O is beginning to be seen in both wings of the H and K bands and the blue wing of the L band. In the J band TiO and VO are beginning to strengthen. The subsequent behaviour of the broad-band colors can be appreciated when one considers the modification of the continuum colors by the band absorption at cool temperatures.

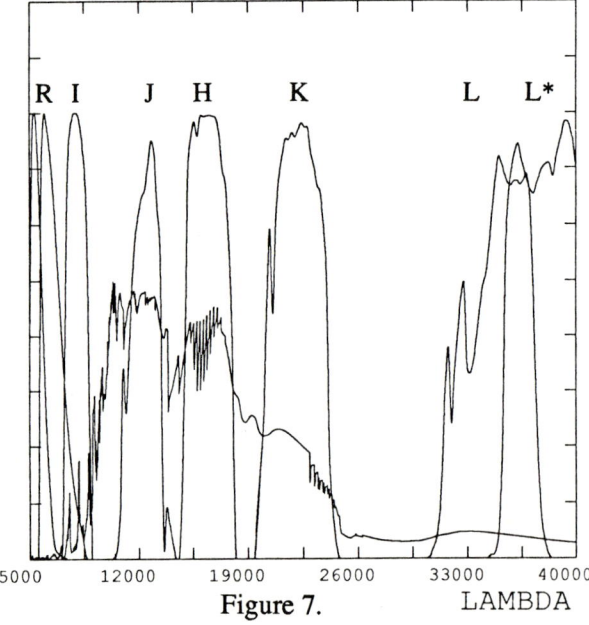

Figure 7.

Brett (1988) has investigated the CN and Zr composition of peculiar redgiants in the Magellanic Clouds using modified mean opacities for bands. He demonstrated that such an approach is capable of determining abundances in very cool giants, providing specific observations are made. In the late-M and S stars, CN bands and ZrO bands are often masked by strong TiO bands. This requires that the 1.08μ CN and the 9300Å bands be used. The 1.08μ region is impossible to reach with a CCD detector but will be achievable for faint objects with the new IR arrays. Brett (1988) used data from Smith et al (1985) and Wing (1967) to fit the red ZrO band. He found that for T ≥ 3000K no significant bands of ZrO appear in models with unenhanced Zr abundance, irrespective of the C/O ratio, consistent with the column density results of Piccirillo (1980). He intends to extend this investigation to the cooler mira models in future as many ZrO strong objects have been found amongst long period miras in the Clouds.

With regard to the relevance of the coolest static models, we note that the spectra and IR colors of a $T_{eff} \approx 3000K$ model corresponds to a spectral type of ~ M7III. BBSW2 showed that semi-regular M-type variables were observed to have such IR colors but that most mira variables in the solar-neighbourhood had redder, or more extreme colors. It is likely therefore that mira models will be more relevant for temperatures below 3000K.

2. In Fig.8 we show the near-IR CCD spectra of some late-type stars, mainly miras.

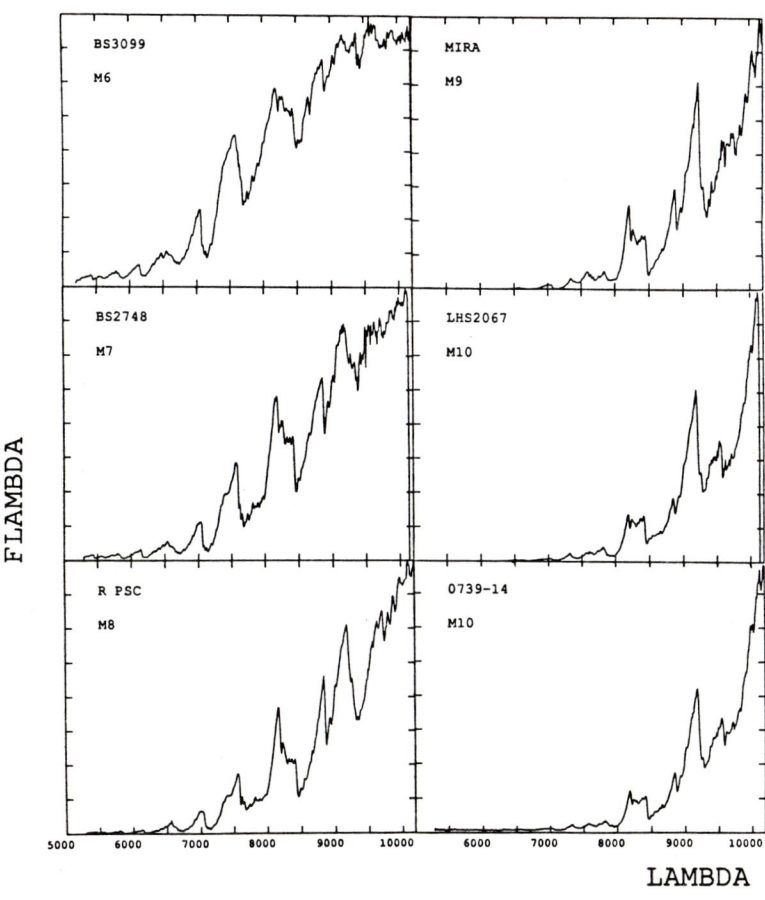

Figure 8.

For comparison in Fig.9 are shown some model spectra. Although some of the strong bands are too strong in the models due to saturation not being taken into account in the mean opacities, the overall appearance of the model spectra, in particular the relative VO and TiO strengths, resembles that of the observations and indicates the relative effective temperatures corresponding to the later spectral types.

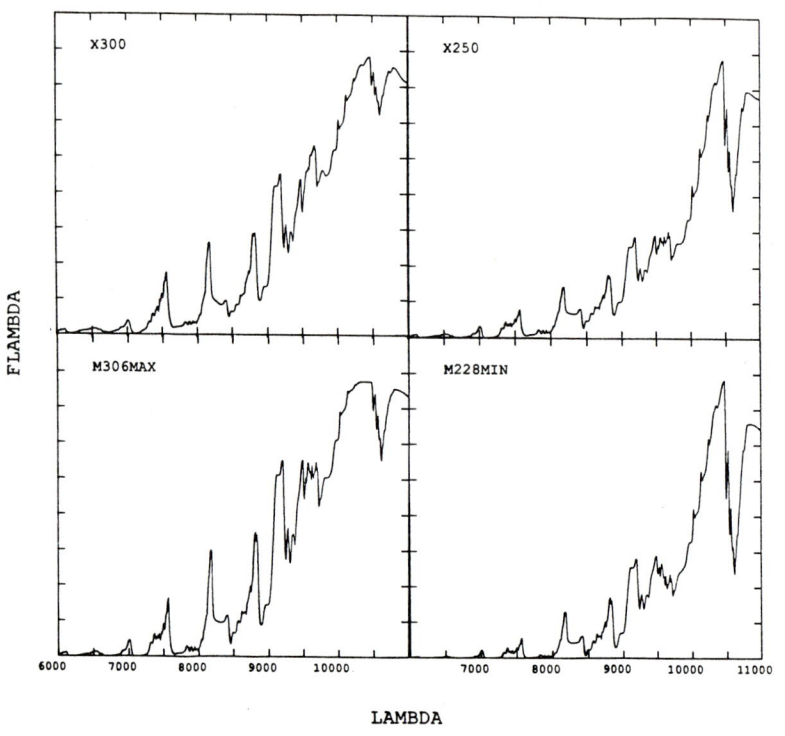

Figure 9.

BBSW2 investigated the spectral differences between static models and exploratory mira models with the same T_{eff}. They found that mira models had comparable continuum, or color temperatures to the static models but that the strength of the TiO γ 0,0 band near 0.7μ was similar to that of a static model 200K cooler. The redder TiO and VO bands also showed a difference, but not as large. As a result of this, different spectral types would be assigned to a mira model, depending on which TiO band was used. Such an observed effect had been long noted in mira variables (eg. Wing 1967). The phase of maximum (luminosity) mira models also have H_2O absorption similar to a static model 200K cooler, although the absorption in the minimum phase models were comparable to that in the static models. The greater strength of the H_2O absorption in the extended mira models results in the H, K and L fluxes, in particular H, being depressed.

One interesting aspect of the IR fluxes of both cool static and mira models is that black-body fitting to broad-band JHKL fluxes is not a sensitive way of determining effective temperatures. In Fig.10 are plotted the IR color temperature, so derived, and the model T_{eff}. It is clear that there is little change in the IR color temperature as T_{eff} changes from 3200K to 2500K. Also plotted are the data for some miras whose effective temperatures have been derived from lunar-occultation. In this aspect at least, the behaviour of the models clearly resemble that of real stars.

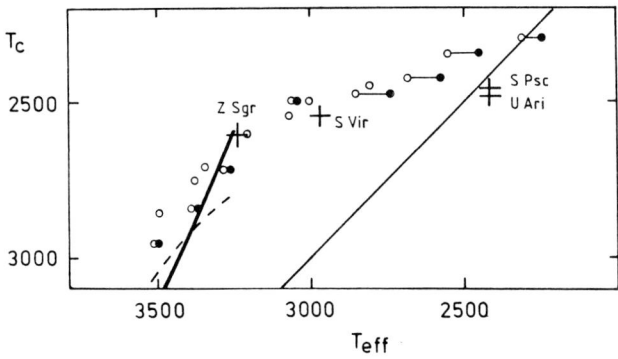

Figure 10.

In Fig.11 and 12 the IR spectra of two miras (one near maximum the other near minimum) with periods near 1 year are shown in comparison with the exploratory 3000K mira spectra of BBSW2. Although the H_2O bands in the models are deeper than in the stars, possibly another instance of saturation not being properly considered, the IR model spectra again showed encouraging similarity to the observations. Black-body fits to the integrated fluxes are also shown for instruction.

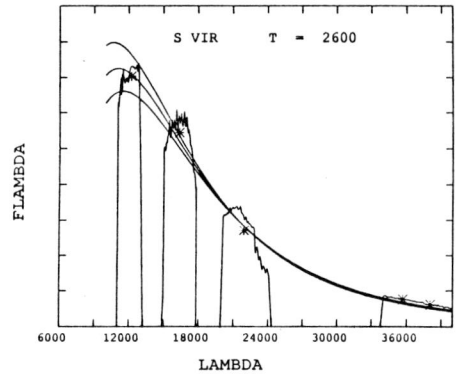
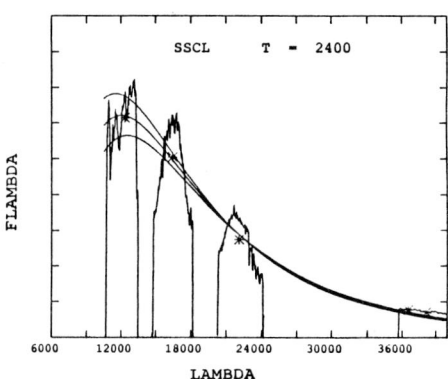

Figure 11.

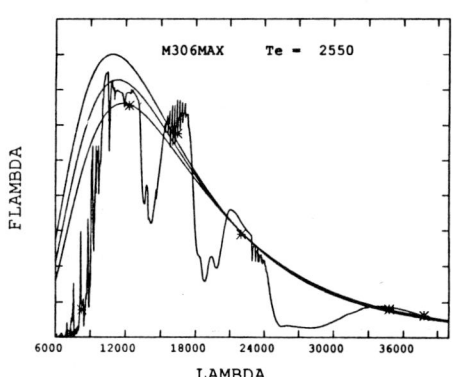
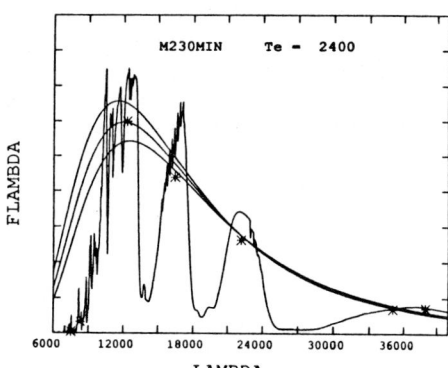

Figure 12.

Very recently, Wood (1988) calculated a more realistic model for a 1.5M$_\odot$, 300 day, 3000K mira pulsating in the fundamental mode. Several hundred models are computed for each pulsation cycle, and after a periodic behaviour was well established with a bolometric amplitude of 0.6mags, we selected models at particular phases in order to produce non-grey spherically symmetric model atmospheres based on the P$_g$, r stratification of the dynamical grey-models. In a preliminary examination for this conference we have computed 4 models at phases 0.9, 0.0, 0.15 and 0.55. The T and ρ stratifications were shown in Fig.2.

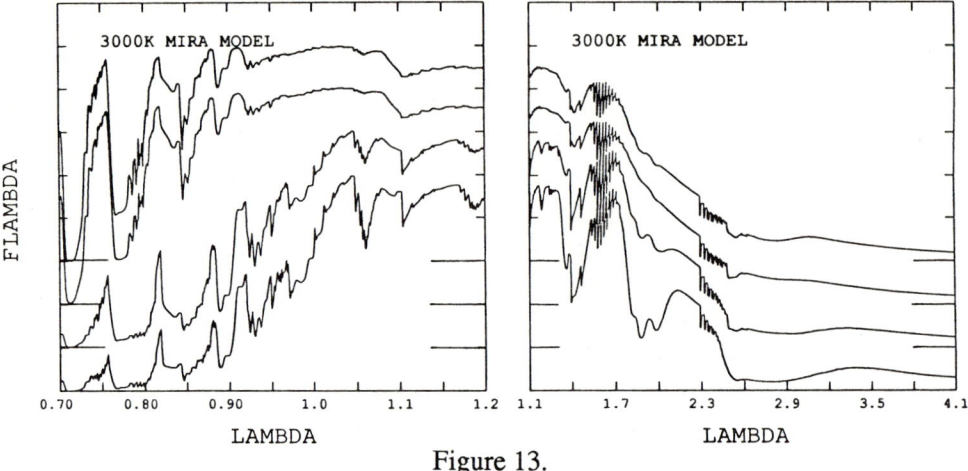

Figure 13.

In Fig.13 is shown the near-IR and the IR spectra of these models. The zero levels of each spectrum is offset as indicated on the ordinate. The T$_{eff}$ for these models were about 3390K, 3330K, 2970K and 2850K respectively, and the general appearance of the near-IR spectra suggests that the spectral-type varies from M4-5 to M8, similar to observed for 300 day miras. The great strength of the TiO γ 0,0 band in the models near maximum luminosity is very spectacular and resembles that shown in published spectra of some Magellanic Cloud miras (Bessell 1983). Although such self-consistent modelling is in the earliest stages we are encouraged by the preliminary spectra and plan to investigate a range of periods, masses, luminosities and compositions.

An alternative approach to the modelling of mira atmospheres by Bowen (1988) and Beach et al (1988) was mentioned above. We have taken their published T and ρ stratification based on non-equilibrium assumptions, and computed a spectrum using ion and molecular abundances calculated from the LTE equation of state and the LTE absorption and scattering coefficients. This is of course, an inconsistent treatment but it should still give some clue whether or not this stratification is suited for interpreting the observed spectra of mira variables. In Fig.14 is shown the resultant spectrum.

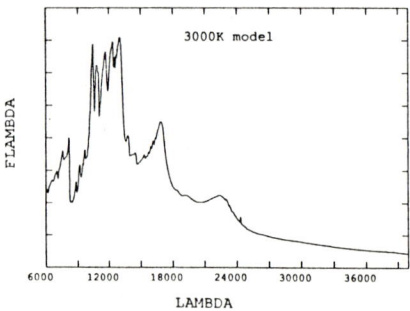

Figure 14.

This is to be compared to the observed spectra of Figs.8, 11 and 12, and the model spectra of Fig. 13. Clearly, the spectra bear little resemblence to observations and we believe that this is good evidence that the Bowen stratification and hence their present approximations concerning rates of cooling behind the shock front are less realistic than the assumption of radiative equilibrium used by Wood. We note that our models completely neglect the hot region in the vicinity of the shock front which produces the hydrogen emission lines and UV MgII emission lines seen at some phases in mira variables, but we suggest that the region of excited material is confined to such a small volume that its effect on the bulk of the photospheric material is negligible at most phases.

VI. SUMMARY

The current modelling of the photospheres of M stars including mira variables is producing good agreement with many of the observed properties of such stars. Some of this work uses mean opacities which does not handle saturated bands so well as other techniques; however, adoption of Tsuji's use of the Elsasser Band model, which is equivalent to the ODF technique, will provide improvement. It would be of great interest to use the OS technique, together with the velocity gradient information, to produce more realistic spectra from our stratifications. Generally, the modelling of M stars has produced good insight into the interpretation of observations and we anticipate a very productive phase of M-S star analyses to follow.

Lambert et al (1986) discuss their grid of C-rich models and use the models to interpret high resolution spectra of C stars. These models appear to be excellent, but the authors comment on a few molecular features that do not fit the observations as well as they would like. They have not yet published the models or the fluxes or colors. Jørgensen et al (1988) is continuing the investigation of additional polyatomic molecules relevant for C stars. We intend to construct extended static and mira models for C stars using Tsuji's opacities, but Tsuji continually reminds us of the uncertainties associated with the polyatomic carbon molecules so important at cool temperatures. Nevertheless, we anticipate advances in the interpretation of C-rich stars also, following the publication of the Lambert et al model grid and of the Scholz-Tsuji study.

VII. REFERENCES

Alexander, D.R., Johnson, H.R. 1972, **Astrophys.J.** 176, 629
Auman, J.R. 1967, **Astrophys.J.Suppl.** 14, 171
Auman, J.R. 1969, **Astrophys.J.** 157, 799
Beach, T.E., Willson, L.A., Bowen, G.H. 1988, **Astrophys.J.** 329, 241
Bessell, M.S. 1983, in **IAU Symposium No.108**, ed.Van Den Bergh, S., De Boer, K.p171
Bessell, M.S., Brett, J.M., Scholz, M., Wood, P.R. 1988a, **Astron.Astrophys.** in press
Bessell, M.S., Brett, J.M., Scholz, M., Wood, P.R. 1988b, **Astron.Astrophys.** in press
Bonneau, D., Foy, R., Blazit, A., Labeyrie, A. 1982, **Astron.Astrophys.** 106, 235
Bowen, G.H. 1988, **Astrophys.J.** 329, 299
Brett, J.M. 1988 **Astron.Astrophys.** submitted
Brett, J.M. 1988 Thesis ANU
Fedorova, O.V.1973, **Astrofizika** 14, 239
Golden, S.A. 1969, **Journal Quantit.Spectrosc.Radiat.Transfer** 9, 1067
Gunn, J.E., Stryker, L.L. 1983, **Astrophys.J.Suppl.** 52, 121
Gustafsson, B., Bell, R.A., Eriksson, K., Nordlund, A. 1975, **Astron.Astrophys.** 42, 407
Johnson, H.R. 1974, NCAR-TN/STR-95
Johnson, H.R., Bernat, A.P., Krupp, B.M. 1980, **Astrophys.J.Suppl.** 42, 581
Jørgensen, U.G., Almhof, J., Gustafsson, B., Larsson, M., Siegbahn, P. 1985, **J.Chem.Phys.** 83, 3034
Jørgensen, U.G., Almhof, J., Siegbahn, P. 1988, **J.Chem.Phys.** submitted
Kipper, T. 1982, W.Struve nimeline Tartu Astrofüsika Observatoorium 66, 3

Labeyrie, A., Koechlin, L., Bonneau, D., Blazit, A., Foy, R. 1977, **Astrophys.J.** 218, L75
Lambert, D.L., Gustafsson, B., Eriksson, K., Hinkle, K.H. 1986
        Astrophys.J.Suppl. 62, 581
Petreymann, E. 1974, **Astron.Astrophys.** 33, 203
Piccirillo, J. 1980, **Mon.Not.Roy.astr.Soc.** 190, 441
Querci, F., Querci, M., Kunde, V.G. 1971, **Astron.Astrophys.** 15, 256
Querci, F., Querci, M., Tsuji, T. 1974, **Astron.Astrophys.** 31, 265
Querci, F., Querci, M.1976, **Astron.Astrophys.** 49, 443
Ridgway, S.T., Joyce, R.R., White, N.M., Wing, R.F. 1980, **Astrophys.J.** 235, 126
Saxner, M., Gustafsson, B. 1984, **Astron.Astrophys.** 140, 334
Scargle, J.D., Strecker, D.W. 1979, **Astrophys.J.** 228, 838
Scholz, M. 1985, **Astron.Astrophys.** 145, 251
Scholz, M., Takeda, Y. 1987, **Astron.Astrophys.** 186, 200
Scholz, M., Tsuji, T. 1984, **Astron.Astrophys.** 130, 11
Scholz, M., Wehrse, R. 1982, **Mon.Not.Roy.astr.Soc.** 200, 41
Schmidt-Burgk, J., Scholz, M. 1975, **Astron.Astrophys.** 41, 41
Schmidt-Burgk, J., Scholz, M., Wehrse, R. 1981, **Mon.Not.Roy.astr.Soc.** 194, 383
Smith, V.V., Lambert, D.L. 1985, **Astrophys.J.** 294, 326
Sneden, C., Johnson, H.R., Krupp, B.M. 1976, **Astrophys.J.** 204, 281
Steiman-Cameron, T.Y., Johnson, H.R. 1986, **Astrophys.J.** 301, 868
Tsuji, T. 1966, **Pub.Astr.Soc.Japan** 18, 127
Tsuji, T. 1968, in **Low Luminosity Stars**, ed. S.Kumar, Gordon and Breach, N.Y., p.457
Tsuji, T. 1978, **Astron.Astrophys.** 62, 29
Tsuji, T. 1981, **J. Astrophys.Astron.** 2, 95
Watanabe, T., Kodaira, K. 1978, **Pub.Astr.Soc.Japan** 30, 21
Watanabe, T., Kodaira, K. 1979, **Pub.Astr.Soc.Japan** 31, 61
Wehrse, R. 1981, **Mon.Not.Roy.astr.Soc.** 195, 553
Wing, R.F. 1967, Thesis University of California, Berkeley
Wood, P.R. 1979, **Astrophys.J.** 227, 220
Wood, P.R. 1988, work in progress
Zeidler-K.T., E.-M., Koester, D. 1982, **Astron.Astrophys.** 113, 173

TESTING THE MODELS AGAINST OBSERVATIONS

Takashi Tsuji
Institute of Astronomy
The University of Tokyo
Mitaka, Tokyo, 181 Japan

Abstract: Tests of the model atmospheres of red giant stars by photometric observations and angular diameter measurements (including limb-darkening) revealed that the present models reasonably represent the thermal structure of the continuum-forming region of red giant atmospheres. On the other hand, tests by line spectra (including line-blanketed fluxes) revealed that classical model atmospheres of red giant stars cannot yet satisfactorily represent the line-forming region. Recent progress in observations also provide some tests of improved models incorporating non-LTE or sphericity effects. More importantly, recent observations offer the means to test some new approaches such as: hydrodynamical simulation of photospheric convection, modeling atmospheres with thermal bifurcation, and incorporating the effects of outer atmosphere having inhomogeneous structure consisting of warm chromosphere and cool molecular dissociation zone.

1 INTRODUCTION

It is now more than 20 years since initial attempts to study atmospheric structure of red giant stars by means of the model-atmosphere method (e.g., Auman 1967; Tsuji 1967). In these 20 years, computations of model atmospheres for red giant stars have made considerable progress with increasing sophistication in treating numerous molecular lines both in oxygen-rich giants (e.g., Gustafsson et al. 1975; Tsuji 1978a; Johnson et al. 1980) as well as in carbon-rich stars (e.g., Querci et al. 1974; Johnson 1982; Eriksson et al. 1984; Johnson & Yorka 1986). Further, some attempts to relax the assumption of LTE (Auman & Woodrow 1975) or of plane-parallel geometry (e.g., Watanabe & Kodaira 1978; Schmid-Burgk et al. 1981; Kipper 1982; Scholz 1985) have been undertaken. As for details of these developments, see the reviews on model atmospheres of cool stars (Carbon 1979; Johnson 1985,1986).

Now, the more important problem should be how to assess the validity of theoretical model atmospheres by observations. However, what happened actually was not so simple as constructing the models and then making the observational tests. For example, initial attempts to construct model atmospheres of cool stars were largely motivated by the progress of infrared astronomy in the early 1960's. Also, in later developments of our studies of red giant stars, what actually happened more often was that observations preceded and perhaps motivated the models. Thus, testing the models against observations is not a simple one-way process; rather, it is a complicated process of successive iteration between observations and models, by which our understanding of red giant stars could be successively improved.

Major observational tools for testing theoretical model atmospheres of red giant stars are (1) photometry, especially in the infrared, and (2) angular diameter measurements. Initial tests of model atmospheres on these points provided a rather optimistic view of the validity of the classical model atmospheres of red giant stars (Sects. 2 & 3). On the other hand, the situation appeared to be not so optimistic once line spectra were examined, either by line-blanketed fluxes (Sect. 4) or by high resolution spectra (Sect.5). Thus, it is still difficult to have a unified understanding of the line-forming region on the basis of the present models that were so successful in understanding the continuum-forming region. Inspection on such tests, however, reveals that observations and theories consistently indicate new pictures of atmospheres of red giant stars (Sect.6).

The subject of testing models has already been discussed as a part of more general reviews on model atmospheres of red giant stars (e.g., Carbon 1979; Johnson 1985, 1986), or in connection with problems of determining fundamental stellar parameters (Gustafsson & Jorgensen 1985) or stellar abundances (Lambert et al. 1986; Gustafsson 1989). Thus, we emphasize more recent results in this review. Also, we concentrate on models of photospheres of cool giant stars (M, S, and C types), except when problems related to chromospheres, circumstellar envelopes, or G-K giants strongly impact our main subject. Also, problems related to Mira variables may be discussed by other reviews in this conference.

2 SPECTRAL ENERGY DISTRIBUTION AND ANGULAR DIAMETER

A classical model atmosphere is characterized by 4 parameters; effective temperature (T_{eff}), surface gravity (g), turbulent velocity (v_{tur}), and chemical composition. If the effect of spherical geometry is important, T_{eff} and g should be replaced by more fundamental parameters - mass (M), radius (R), and luminosity (L). For testing a model by observations, however, it is still convenient to define T_{eff} for spherical models as well. As the thermal structure of a stellar atmosphere is primarily determined by T_{eff}, it is natural that the correct assignment of T_{eff} is the first step in testing a model atmosphere by observations. Also, a comparison between T_{eff} determined by direct methods based on measured angular diameters and that determined by indirect methods based on analysis of photometric observations can be regarded as a test of a model atmosphere, since indirect methods depend largely, but indirect methods depend little, on model atmospheres.

2.1 Oxygen-Rich Giants

A test of model atmospheres by the method noted above can best be illustrated by a well-studied bright giant star α Boo. While recent analyses of flux distribution by model atmospheres (FDM) provided rather high temperatures of 4410 $\pm$88 K (Blackwell & Shallis 1977), 4462$\pm$33 K (Augason et al. 1980), or 4375 $\pm$ 50 K (Frisk et al. 1982), some angular diameter measurements suggested very low values near 4000 K (see e.g., Trimble & Bell 1981). This confused situation seems to be well resolved by two recent papers. First, Di Benedetto & Foy(1986), after careful analysis of the bias affecting the visibility curve obtained by Michelson interferometry with two telescopes (I2T) in the near infrared, showed that direct determination of effective temperature with accuracy better than 1% is now feasible and that T_{eff} = 4294 ± 30 K for α Boo. Second, Blackwell et al. (1986) showed that an indirect method based on the infrared flux method (IFM) also provides a value of the effective temperature with an accuracy of about 2% after careful evaluations of photometric data, infrared calibration, and line-blocking effect; the result is T_{eff} = 4230 ± 80 K or 4307 ± 80 K, depending on whether absolute flux calibration of the standard star Vega is by direct comparison with a standard furnace or by a model atmosphere, respectively.

Similar tests of model atmospheres by comparisons of direct and indirect effective temperatures have also been done for cooler stars extending to late M giant stars. An initial attempt in this direction revealed that T_{eff} by FDM only showed reasonable agreement with the upper estimates of T_{eff} based on angular diameters then available, and this fact was interpreted as due to some biases not well recognized in the angular diameter measurements (Tsuji 1978a). This fact also suggested a possibility that the effective temperature scale of M giants should be revised upward against the one used before (e.g., Johnson 1965). Extensive analysis on angular diameters from lunar occultation by Ridgway et al. (1980) provided conclusive evidence of the general increase in effective temperatures of M giant stars between M0 and M6. Meanwhile, T_{eff} by indirect method based on FDM (Manduca et al. 1981) as well as on

Fig.1. Effective temperature scale for K and M giants. Dotted lines recall two historical scales: a (Kuiper 1938) and b (Johnson 1965). Solid lines are those based on direct measurements of angular diameters: c (Ridgway et al. 1980) and d (Di Benedetto & Rabbia 1987). Dashed lines are those based on model-atmosphere analysis (the infrared flux method): e (Tsuji 1981a) and f (Leggett et al. 1986).

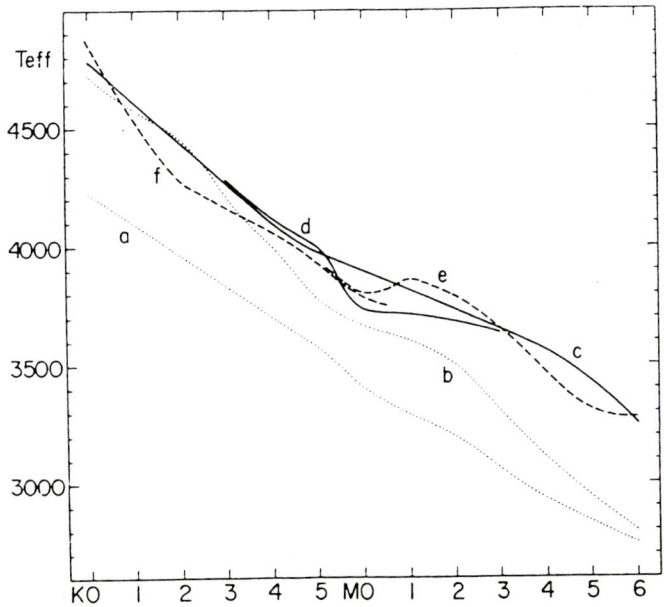

IFM (Tsuji 1981a) showed agreement within 100 K with the scale of Ridgway et al. (Fig.1). More recently, the direct method based on I2T in the near infrared provided effective temperatures with internal accuracy of 61 K for a large number of red giant stars cooler than K0 (Di Benedetto & Rabbia 1987), and this demonstrated a new possibility of the interferometric technique for measurements of stellar angular diameters; with such a high accuracy in angular diameter measurement, previous determinations of T_{eff} are confirmed in general (Fig.1), and, further, any possible bias in model atmospheres can now be investigated. For example, systematic deviation of the results by FDM or by IFM from those by interferometric measurements suggests that the cross-section of H^- opacity should still be improved. For such a purpose, however, higher accuracy in photometric data would be required.

2.2 Carbon Stars

For cool carbon stars, T_{eff}'s by the direct method (Ridgway et al. 1977) and by IFM (Tsuji 1981b) show reasonable agreement, and carbon stars turn out to be cooler than M giant stars. Such a recent temperature scale of cool carbon stars has been shown to be fairly consistent with various temperature indicators including infrared colours, band strengths of polyatomic molecules, and molecular excitation temperatures by Lambert et al. (1986), and it is expected that a more detailed analysis of the temperature scale based on their new models will be carried out. Especially, the effective temperature scale of cool carbon stars showed inverse correlation with the temperature class of the C-classification by Keenan & Morgan (1941), which is essentially related to excitation temperature. Such a contradiction between the effective temperature scale and the excitation temperature scale may reflect some serious model effect due to heavy line-blanketing by molecular bands (Tsuji 1985). To resolve the issue, empirical determination of excitation temperatures in cool carbon stars may also be useful.

3 STELLAR SURFACE BRIGHTNESS DISTRIBUTION

For testing of stellar structure, observations of limb-darkening or, more generally, brightness distribution over the stellar disk should be more informative than the single angular diameter measurement. In fact, recent progress in observations with high spatial resolution provides some interesting possibilities.

3.1 K-M Giant Stars

For the K5 giant star α Tau, Ridgway et al. (1982b) have obtained multiple wavelength measurements between 0.4 and 4.0 μm by the lunar occultation technique, and showed that the apparent diameter which is determined by a uniform-disk-model fit to the observed occultation diffraction pattern increases toward longer wavelength. This result showed reasonable agreement with the angular diameter variation based on predicted limb darkening by Manduca et al. (1977) for models of red giant stars by Gustafsson et al. (1975), although the observed variation with wavelength may be stronger than the predicted one. Thus, the limb-darkening effect as predicted by model atmospheres, or possibly somewhat greater, is confirmed for a red giant star. Also, angular diameters measured in the optical region by lunar occultations for α Tau (White & Kreidl 1984) and for μ Gem (Ridgway et al. 1977) show excellent agreements with those by Michelson interferometry in the near infrared, if limb-darkening effects are corrected (Di Benedetto & Rabbia 1987). This fact in turn confirmed the limb-darkening effect predicted by model atmospheres.

For other M giant stars, multichannel measurements by the lunar occultation technique also suggested wavelength dependence of angular diameters: an 20% decrease from 0.9 to 1.7 μm, followed by a slight rise at 2.2 μm (Ridgway et al. 1982a). This initial result was confirmed by more extensive observations, but the ratio of angular diameters at 2.2 μm to that at 1.7 μm is not a monotonically increasing function of spectral type (Schmidtke et al. 1986). Such an observation is difficult to understand by an analysis based on spherically extended models (Scholz & Takeda 1987), which showed that the angular diameter at 1.7 μm is a good measure of the photospheric radius defined by the continuum but that interpretation of monochromatic angular diameters are difficult when molecular bands appear in the higher level of photosphere so that the brightness distribution shows an extended halo in the outer disk.

Based on lunar occultation data, Bogdanov & Cherepashchuk (1984) tried to derive brightness distribution over the disk of μ Gem. It is shown that the effect of brightness distribution on the lunar occultation diffraction pattern is rather subtle for this star, and requirements for resolution and S/N ratio are quite severe. The conclusions so far reached by them are: the plane-parallel model with limb-darkening coefficient u > 0.7 cannot be excluded, and atmospheric extension of this star is rather small, if any, consistent with the extended spherical model atmospheres such as by Watanabe & Kodaira(1979).

3.2 Carbon Stars

Radial brightness distribution over the disk of the carbon star TX Psc has been estimated from the lunar occulation data by the same method as used for μ Gem (Bogdanov,1979): TX Psc may consist of central stellar disk and an extended halo which is more prominent than in μ Gem.

3.3 M Supergiant Stars

The M supergiant star α Ori has been well observed by interferometric techniques since its stellar angular diameter was first measured by Michelson & Pease (1921). However, the increase of observational data at first increased the discrepancy with model-atmosphere predictions; the observed angular diameters tended to be larger than the photospheric angular diameter based on model analysis of photometric data and also to be larger at shorter wavelength, in contradiction to the expectation from limb-darkening. Also, angular diameters measured at TiO bands tended to be larger than those at continuum bands, and the lower effective temperature was consistent not only with a larger diameter but also with the TiO-related variations of the measured diameters. (Belega et al. 1982). Further, a possibility of time variation in the photospheric radius has been suggested (White 1980).

More recent observations based on speckle and other interferometric techniques finally overcame these contradictions by resolving the central stellar disk against the extended halo in α Ori. For example, Ricort et al. (1981) have proposed a model consisting of core (diameter 35 ± 10 mas) and halo (diameter 100 ± 40 mas) for α Ori from their speckle data. Also, Roddier & Roddier (1983, 1985) showed evidence, from their two-dimensional interferometric image reconstruction, for a relatively small stellar disk with angular size of 37 mas and a highly asymmetric circumstellar envelope. Further, Cheng et al. (1986) showed that multi-wavelength speckle data can be interpreted as indicating a central stellar disk of 42.1 ±1.1 mas, if a bias possibly induced by the extended halo, between 1 and 5 R_*, is accounted for. Also, these authors showed that wavelength-dependent limb-darkening, whose coefficients turned out to range from u= 0.9 in strongly line-blanketed spectral regions to u=0.4 in less blanketed regions, are consistent with model prediction of plane-parallel atmospheres (Tsuji 1976). The same speckle data have been reanalyzed by a refined algorithm of high spatial resolution imaging by Christou et al. (1988), who showed that the recovered images of α Ori can be resolved into two Gaussian components which represent the disk and extended components. A mean value for the widths of the Gaussian fits to the disk component at 6 spectral band-passes was found to be 19.8 ± 0.7 mas. Observed apparent radii, however, decreased towards longer wavelengths for the 4 measurements of continuous fluxes, and this is again opposite to what can be expected from the limb-darkening effect. On the other hand, the remaining 2 apparent diameters measured at chromospheric lines around 6563 and 8542 A turned out to be substantially larger than the continuum diameters. Another component, the extended low power halo, could be found at Hα light, as is already known from the detailed analyses of the extended chromosphere by Hα imaging (Hebden et al. 1986, 1987). Furthermore, the extended halo has been found at all the continuum wavelengths out to a distance of 4 R_* in the recovered images.

The anaylses reviewed above consistently showed that the angular diameter of the stellar disk of α Ori is relatively small – in the range between 35 and 41 mas – and this confirms the rather high effective temperature of about 3900 K based on model atmosphere analyses (Tsuji 1976; Vieira 1986). At the same time, the presence of an extended halo around α Ori is now well established, and the problem is to decide what kind of model could explain this finding. A possible model that relies upon dust to provide the scattering to explain the rather large apparent angular diameter of α Ori has been proposed before (Tsuji 1978b; White 1980). Recently some observational evidences for dust condensation near the stellar photosphere have been found (e.g., Roddier & Roddier 1985; Roddier et al. 1986), and the possibility of dust formation in the stellar chromosphere by the so-called condensation instability has been proposed (Stencel et al. 1986). On the other hand, the main objection to such a model is recent observations on the possible presence of a dust-free region out to 10 R_* in α Ori by infrared interferometry or by infrared imaging (e.g., Bloemhof et al. 1984). If the dust model for extended halo cannot be applied, the only possibility remaining appears to be gaseous particle that may scatter or re-emit stellar radiation. Generally, models of the outer atmosphere of α Ori assume a rather low density of $n_H \simeq 3\ 10^{+9} cm^{-3}$ at $r \simeq 1.1 R_*$ (e.g.,

Hartmann & Avrett 1984; Skinner & Whitmore 1987), and the hydrogen column density in the extended halo is roughly $N(H) \simeq 10^{+23} cm^{-2}$ if $R_* \simeq 600 R_\odot$. Then, with a cross-section of $\sigma \simeq 7.4\ 10^{-29} \lambda (\mu m)^{-4}$ for Rayleigh scattering by hydrogen atoms, the scattering radial optical depth is roughly, $\tau = \sigma N(H) \simeq 10^{-4}$ at $\lambda = 0.5\ \mu m$. This is far short of explaining the observed halo. Also, if H^- b-f emission is considered instead of scattering, the radial optical depth of the halo is still smaller. Note that H^- b-f emission may have some effect on the integrated flux if a rather high density is assumed in the thin transition layer very close to the stellar surface (e.g., Skinner & Whitmore 1987), but this cannot explain the spatially resolved extended halo. Thus, if a small amount of dust does not exist in the inner halo of the outer atmosphere, a larger amount of gas must exist in the inner halo than that assumed in the present models for the outer atmosphere of α Ori.

4 LINE-BLANKETED FLUXES AND COLOURS

Once effective temperature – one of the major parameters that specify a model atmosphere – is well settled (Sects. 2 & 3), and global thermal structure of the photosphere is confirmed, one may expect that line-blanketed fluxes and colours could provide further tests of model atmospheres. While such tests provide positive results for models of G-K giant stars, many problems remain for those of cooler giants.

4.1 Oxygen-Rich Giants.

For G-K giant stars, line-blanketed colours predicted from model atmospheres have been tested against observed colours by various photometric systems for stars with well defined fundamental parameters, and the results showed fine success in general, except for minor discrepancies in the UV or in strong molecular bands (Gustafsson & Bell 1979). Also, the predicted surface brightness-colour relations such as F_V - (V-R) agree very well with those observed for $6000 > T_{eff} > 4000$ K (Bell & Gustafsson 1980).

Some attempts in the same direction have been made for cooler giant stars, including late K and M giants. Piccirillo et al. (1981) showed that the empirical relation between colour temperature and effective temperature can be well reproduced by considering the effect of weak TiO bands in predicting the near-infrared colours defined by Wing (1971). Recently, Steiman-Cameron & Johnson(1986) showed that agreements between predicted and observed colours are qualitatively good in general but not quantitatively, partly because of the difficulty of defining zero-points of photometric systems. Especially, these authors showed that the predicted TiO bands are systematically too weak relative to observations for models with T_{eff} below 3500K, and suggested it might be due to sphericity effect not accounted for in their models. In fact, it is known that the TiO absorption is stronger in models with larger atmospheric extension (Watanabe & Kodaira 1979) and, further, this fact can be used to measure the geometrical extension as a new classificaton parameter besides T_{eff} and g (Scholz & Wehrse 1982). Such a proposition is still to be tested by observations, for which it is essential to find stars with known fundamental parameters such as mass, radius, and luminosity, since TiO is already known to be sensitive to temperature. Also, cool giants having weak TiO bands for their color tend also to show weaker Mg II fluxes than stars of similar color (Steiman-Cameron et al. 1985), and this may indicate that other factors not yet considered above could be important in testing models by TiO.

Other examples of molecular bands that remains difficult to understand are the CO vibration-rotation bands in the spectrophotometric scans observed by Strecker et al. (1979); predicted strengths of the fundamental bands appear to be too weak compared to observed ones while predicted overtone bands appear to be stronger than those observed (Manduca et al. 1981). It is noted that this difficulty appears already in late K and early M giants. It is not clear if the same problem appears in cooler giants in the analysis of similar observational data by Scargle & Strecker(1979), since observed data are rather forced to fit predicted ones by adjusting several parameters that are not necessarily checked by other observations. However, we will see the same difficulty of understanding CO line strengths in an analysis of the high resolution spectra (Sect.5.3).

4.2 Carbon-Rich Stars

So far as diatomic opacity sources are concerned, predicted line-blanketed fluxes (Querci & Querci 1974) and observed infrared fluxes (Goebel et al. 1980) show relatively good agreement. However, the nature of opacity sources, especially of polyatomic molecules, in cool carbon stars is not yet fully understood. For example, atmospheric structure of cool carbon stars is drastically changed by polyatomic opacities, especially if the veil opacity due to many combination bands of HCN are included (Jorgensen et al. 1985), and a large reduction of gas pressure in the surface layers suggested a possibility of large extension of the outer layers (Eriksson et al. 1984). On the other hand, a preliminary computation of spherical models, including HCN opacity based on transitions confirmed in the laboratory and by observation (Ridgway et al. 1978) revealed that the atmospheric extension in carbon stars may be rather modest as compared with that in M giant stars (Scholz & Tsuji 1984). Detailed observational tests of model atmospheres based on polyatomic molecular opacity including the veil opacity were done by Lambert et al. (1986): the predicted depression due to HCN absorption as a whole is too strong compared with that observed, while predicted strengths of the individual lines of the fundamental bands of HCN identified in the spectrum of TX Psc show reasonable agreement with those observed. This fact may suggest that the veil opacity was over-estimated, but other possibilities suggested are: departure from LTE, heating of the surface layers by mechanical energy flux or by dust formation, errors in adopted effective temperature scale, etc. (see also Jorgensen, this conference).

Another problem associated with opacity is the origin of the violet depression in cool carbon stars, a classical interpretation of which was to assume a pseudo-continuum due to C_3 or grain opacity due to SiC. Recently, this problem has been reexamined in detail by new photometric data, including IUE data, and the two previously suggested opacity sources, C_3 and SiC, are both shown to be inadequate (Faulkner et al. 1988). Also, recent measurements of the Balmer decrement in Mira type carbon stars revealed that the the source of the violet opacity may be located deep in the photosphere (Orlati 1987). Finally, it is shown that the violet opacity in cool carbon stars is due to the cumulative effect of several sources, primarily b-b transitions (including the Mg I resonance line) and b-f opacities from low lying states of neutral metals, with partial contribution of CH photo-dissociation and C_3 pseudo-continuum (Johnson et al. 1988). These authors also noted that the violet flux deficiency in cool carbon stars could have been over-estimated and may largely be a temperature effect, if effective temperatures of carbon stars are lower than those of M giant stars (Sect.2).

5 LINE SPECTRA

Some of the difficulties encountered in testing model atmospheres of red giant stars by line-blanketed fluxes (or colours) may suggest the necessity of detailed analyses of line spectra. However, detailed analyses of line spectra for testing atmospheric structure of cool giant stars is rather meager, because most of these analyses have so far been done in connection with chromospheric and circumstellar problems. This may largely be due to the intrinsic difficulty of measuring line profiles accurately in the spectra of cool stars because of the intrinsic blending and of the uncertain continuum level. Such difficulties can partly be removed by the recent progress in high resolution and high quality spectroscopy, especially in the infrared.

5.1 Atomic Lines

One problem in using atomic lines as probes of atmospheric structure is that atomic line spectra may not always be interpreted under the assumption of LTE. For example, considerable departures from LTE in ionization equilibrium have been predicted from model analyses by Auman & Woodrow (1975). The predicted over-ionization may be consistent with observations for red supergiant stars (Ramsey 1981), but may be over-estimated for red giant stars if the result of Ramsey (1977) is reinterpreted by a more recent effective temperature scale (Sect.2) which is revised upward compared with that used by Ramsey (1977).

Observational evidence for pronounced departures from LTE in line formation was shown for β Gem (KOIII) by Ruland et al. (1980). Abundances of iron-peak elements determined by LTE analysis of ionized and high-excitation neutral lines appeared to be normal while those by similar analysis of low-excitation neutral lines appeared to be deficient by about 0.3 dex. A similar effect, indicating substantial departure from LTE, was noted in the Zr/Ti abundance ratio for a larger sample of G-K giant stars (Brown et al. 1983). A physical explanation of such observations was offered by Steenbock (1985), who has solved detailed statistical equilibrium equations of Fe I/Fe II in β Gem and showed that non-LTE abundance corrections can be as large as 0.2 dex. The observed effect in β Gem is consistent with such a model, even though the observed effect is somewhat larger than predicted.

Similar analyses, both theoretical as well as observational, on the possible departures from LTE would be highly desirable for cooler giant stars. At present, quantitative analyses of atomic lines (e.g., for abundance determination) in cooler giants are usually done differentially with respect to some standard stars such as α Tau to minimize the non-LTE effects (e.g., Smith & Lambert 1986). In this connection, a detailed quantitative analysis of the high resolution infrared spectrum of α Ori based on solar f-values by Vieira (1986) is quite interesting. The analysis is self-consistent within the framework of quantitative stellar spectroscopy based on classical model atmospheres. Yet the result shows a surprisingly low metal abundance for this typical population I supergiant star. Whether this is due to a real abundance anomaly or whether this is due to some serious model effect deserves further study.

5.2 H_2 Molecule

A possible importance of the quadrupole transitions of H_2 molecule for testing atmospheric structure of cool stars was well recognized at the infancy of infrared spectroscopy (Spinrad 1966; Lambert et al. 1973). So far, however, clear identification of H_2 quadrupole lines due to fundamental VR transition near 2.4 μm is limited to S type stars (Hall & Ridgway 1977) and cool carbon stars (Johnson et al. 1983). The H_2 1-0 S(0) line in a large sample of cool carbon stars has been used to test model atmospheres of carbon stars (Lambert et al. 1986). Models with polyatomic molecular opacity (including veil opacity) and with a certain range of the C/O ratio reproduce well the observed H_2 intensities for the entire range of effective temperatures of cool carbon stars, while models without polyatomic molecular opacities predict H_2 intensities much stronger than observations. The major reason for this difference is that the high opacity due to HCN lowers the gas pressure in the upper layers where H_2 lines are formed. However, one remaining problem is that the HCN veil opacity, which makes the gas pressure so low, has not yet been confirmed by observation (Sect.4.2). Although this problem should still be settled, the analysis of Lambert et al. (1986) showed a possibility of explaining the observed H_2 intensity without an assumption of hydrogen deficiency in cool carbon stars - a problem that was examined in detail by Johnson et al. (1985).

For M giant stars, the H_2 1-0 S(1) line is not clearly identified and the observed strengths are anyhow appreciably less than those predicted by the present model atmospheres (Tsuji 1983). This may suggest that temperature may be hotter (including the possibility of a chromosphere or inhomogeneity; see Sect.5.3) or gas pressure may be lower in the surface layers of actual stars than in present model atmospheres.

5.3 CO Molecule

Observed strengths of molecules other than H_2 depend not only on atmospheric structure but also on chemical abundance. Thus, a single line cannot be used for the testing of model atmospheres, but quantitative analysis of a large number of lines covering the range from weak to strong lines could provide a useful test of atmospheric structure from deep to outer layers. For this purpose, CO vibration-rotation bands, which consist of many measurable transitions, may be the best candidate, since the assumption of LTE is well warranted for the first and second overtones (Carbon et al. 1976), and molecular data such as oscillator strengths are relatively well known (e.g., Chackerian & Tipping 1983). Actually, abundance analysis based on CO VR lines can also be regarded as such a test, since abundance determination is based on the assumption that equivalent widths of different transitions with different strengths can consistently be understood by a model atmosphere with well defined parameters such as T_{eff}, g, v_{tur}, and chemical composition.

Such an analysis of CO has been done for oxygen-rich (Smith & Lambert 1985, 1986; Tsuji 1986) as well as for carbon-rich (Lambert et al. 1986) giants. Each of these results appear to be internally consistent, if low excitaton strong lines are excluded. For example, medium strong lines of the first overtone bands at 2.3μm can consistently be understood by a unique set of parameters to be interpreted as carbon abundance and turbulent velocity; an example of such an analysis is shown in Fig.2a, from which we obtain log A_C =7.68 (by standard notation with log A_H = 12.00) and ξ_{micro} = 3.20 km sec^{-1}(Tsuji 1986). However, for several M giant stars in common with the analysis by Smith & Lambert (1985,1986), the differences in the resulting carbon abundance are as large as 0.6 dex (e.g., 30 g Her). To trace the origin of this discrepancy, the second overtone bands are analyzed by the same method (Fig. 2b), and also the first and second overtone bands are analyzed together (Fig.2c), as was done by Smith & Lambert (1985). The results from the second overtone alone are log A_C = 8.01 and ξ_{micro} = 4.28 km sec^{-1} while

Fig.2. Abundance parameter y = log(C/H)+3.45 obtained from CO lines with reduced equivalent width of x = log(W/ν)+6.0 for several assumed values of the micro-turbulent parameter ξ_{micro}, by using a model atmosphere with T_{eff} = 3200K, log g=0.0 and v_{micro} = 3.0km sec^{-1} for 30 g Her (M6III). The analysis is based on high resolution infrared spectra by 4m FTS of Kitt Peak National Observatory (NOAO). (a) CO first overtone band (excluding low excitation strong lines); (b) CO second overtone band; and (c) CO first and second overtone bands together.

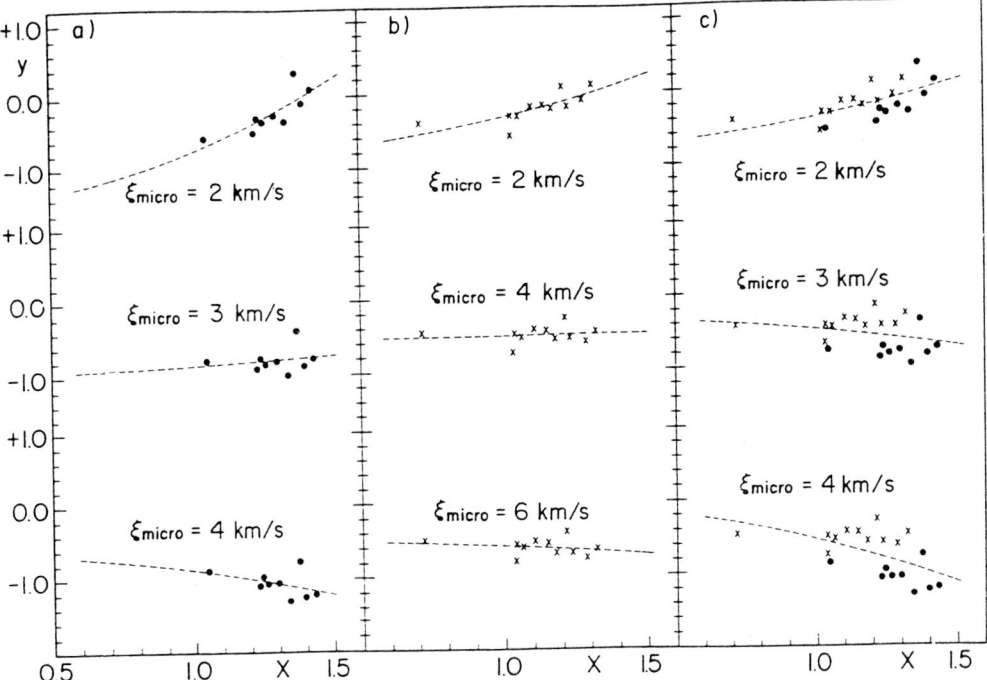

those by the first and second overtones are log A_C = 8.04 and ξ_{micro} = 2.46 km sec^{-1}. The resulting abundance parameters agree rather well with each other and with the result by Smith & Lambert (1985), but differ appreciably from the result by the first overtone alone (Fig.2a). Probably, micro-turbulent velocity may be more difficult to determine from weak lines of the second overtone alone than from saturated lines of the first overtone, and more detailed analysis should be needed before the difference of the turbulent velocities due to different bands can be confirmed. This fact also reveals that the nature of turbulence in M giant stars cannot yet be regarded as well known, whereas the distribution of turbulent velocities in G-K giant stars is rather well established (e.g., Gray 1982). The turbulent velocity determined from the combined analysis of the first and second overtone bands now shows good agreement with that by Smith & Lambert (1985, 1986), but this result is also open to question, since this is biased by the systematic effect that is revealed by the separate analyses of the first and second overtone bands.

Now, it is evident that the first and second overtone bands show drastically different results in abundance and turbulent velocity or, in other words, the first and second overtone bands of CO cannot be understood consistently by a unified model within the framework of classical atmosphere. More or less similar differences in the derived parameters from the first and second overtone bands have been found in some 20 M giant and supergiant stars while such an anomaly did not appear in K giant stars. This later result implies that the difficulty just found for M giant stars cannot be due to systematic error in the absolute scale of f-values of CO, which is accurate to within 10% (Tipping 1988). Other possible sources of such difference can be systematic differences in line blending, in continuum location, or in unrecognized opaci-

Fig.3. The result of the analysis by Fig.2 (notations have the same meanings as in Fig.2): CO second overtone (cross) gives log A_C =8.01 and ξ_{micro} =4.28 km sec^{-1} while CO first overtone (filled circle) gives log A_C =7.68 and ξ_{micro} = 3.20 km sec^{-1}. The low excitation strong lines of the CO first overtone (open circle) give an unreasonably large abundance parameter, indicating the presence of excess absorption that cannot be accounted for by the photospheric model.

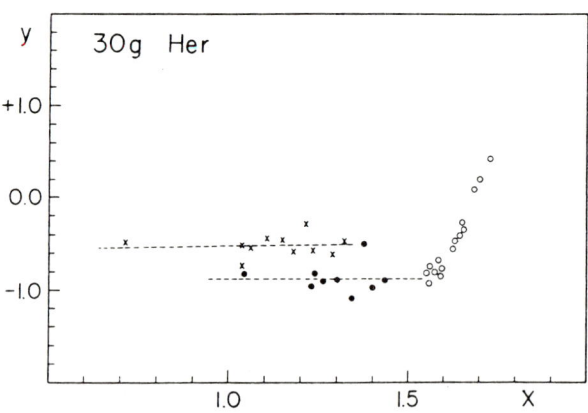

ty sources. Such a difference in CO abundances from the first and second overtones has also been noted in cool carbon stars with T_{eff} < 2900K (Lambert et al. 1986) and has been attributed to poorly determined continua in cooler stars. In M giant stars, such a difference already appears in early M giants, and a similar inconsistency also appears in photometrically calibrated line-blanketed fluxes which can be analyzed independently of continuum level (Sect.4.1). Thus, there is a possibility that the contradictory results from the first and second overtone bands may imply a serious difficulty for the classical model atmospheres in the line forming layers. Probably, the weak lines of the second overtone are fairly well represented by classical models, and may give the best estimate of the carbon abundance. The apparent decrease of carbon abundance, as revealed by the high excitation lines of the first overtone, reminds us of the case of the Sun in which CO lines gave carbon abundance about 0.3 dex less than that implied by other indicators (Tsuji 1977; Lambert 1978). This result was interpreted as due to inhomogeneity in the surface layers of the Sun, and recent observations of CO with high spatial resolution confirmed such a possibility (Ayres et al. 1986). Now, it is interesting to consider a similar possibility for M giant stars.

Further, strong low excitation lines of the CO first overtone bands, which have not been used in the analyses in Figs.2a-c, show excess absorption that cannot be explained by the photospheric models. As shown in Fig.3, carbon abundances derived from low excitation CO lines of the first overtone (open circle) turn out to be much larger than those from high excitation CO lines of the first overtone (filled circle; from Fig.2a) and those from CO lines of the second overtone bands (cross; from Fig.2b). Also, the low excitation CO lines show shifts and asymmetries that indicate excess absorption in the blue wing in some stars and in the red wing in other stars. Thus, it may be reasonable to assume that the excess absorption originates from an extra molecular layer in the outer atmosphere distinct from the stellar atmosphere, and the possible presence of a quasi-static molecular formation zone in the outer atmosphere of red giants has been suggested (Tsuji 1988a). Such a possibility can also be inferred for cool carbon stars from the presence of excess absorption in low excitation CO lines (Dominy et al. 1978) and from the distinct radial velocity difference of low excitation CO lines relative to high excitation lines (Lambert et al. 1986). Also, such an excess absorption cannot be the effect of such photospheric structure as a sphericity effect because of the different kinematical behaviours of the excess absoption relative to the photospheric spectrum. The effect of geometrical extension is not very clear in the predicted spectrum of CO lines based on a spherical model for α Her (Höflich et al. 1986) since other effects, such as those due to fundamental parameters, are not well separated.

At present there is no actual model to be tested by observations outlined above, but some interesting theoretical ideas have already been proposed. A possible origin of inhomogeneity or bifurcation in the solar upper atmosphere through efficient CO cooling in the presence of mechanical heating was suggested by Ayres (1981). Further, Kneer (1983) suggested that the atmospheres of cool stars might be destabilized by the radiative instability due to the high temperature sensitivity of CO and other molecular formation. Under such CO cooling, radiative equilibrium models are still obtained by a time-dependent numerical simulaton, but the resulting models have a bistable character in that they are distinctly divided into hot or cool models depending on effective temperature (Muchmore & Ulmschneider 1985), or they have an even more distinctively bistable character in that two different temperature structure can be possible in the same model for limited effective temperature range (Muchmore 1986). Further, it was suggested that the similar cooling instability due to SiO formation may induce autocatalytic molecular formation in cool envelope of red giant stars (Muchmore et al. 1987). These works provide a theoretical basis for the possible presence of inhomogeneities in stellar surface layers and a molecular formation zone in the outer atmospheres of red giant stars.

Now, an interesting possibility may be to assume that the high excitation lines of the CO first overtone in Fig.3 originate from upper layers with bifurcated structure. More recently, actual application of such an idea to Arcturus has been made by Cuntz & Muchmore(1988), who showed that an atmosphere with a hot chromospheric component and a cool molecular component is generated, depending on the strength of the shock due to acoustic waves. This result lent support to the suggestion based on high resolution spectra of the CO fundamental band that the upper atmosphere of Arcturus should be composed of hot and cool components (Heasley et al. 1978). Although CO first overtone bands revealed no marked anomaly in Arcturus, possibly because they are not yet as strong as in M giant stars, the first overtone bands in cooler M giants may already be strong enough to show some effects of a bifurcated atmosphere that appear only in CO fundamental bands in Arcturus.

Further, the identification of the quasi-static molecular dissociation zone in the outer atmosphere of red (super)giant stars, as revealed by the low excitation lines of CO first overtone bands, can be regarded as an observational support to the theory of molecular formation by a thermal instability (Kneer 1983; Muchmore et al. 1987) since otherwise it is difficult to explain the presence of a rather large abundance of CO in the outer atmospheres of red (super)giant stars. As the excitation temperatures of the molecular dissocation zone are relatively high - between 1000 and 2000 K (Tsuji 1988a), it may be located near the photosphere. Also, the fact that the molecular dissociation zone produces distinct absorption indicates that it may not be so extended as to produce emission. However, as emissivity at $2.3 \mu m$ decreases rapidly at about 1300 K, it is possible that the cooler part of the molecular dissociation zone is significantly extended. For Betelgeuse, the CO column density of the molecular dissociation zone is $10^{+20} cm^{-2}$ and, with C/H ratio of $2.5\ 10^{-4}$ (Lambert et al. 1984), the hydrogen column density is $4\ 10^{+23} cm^{-2}$ or the mass column density of the molecular component of the outer atmosphere is $0.4\ g\ cm^{-2}$. This is probably a lower-limit estimate, since possible emission from the extended part of the CO dissociation zone may have filled in the absorption. For comparison, the mass column density of the hot component of the chromosphere deduced from Ca II emission is $4.3\ g\ cm^{-2}$ (Basri et al. 1981).

Finally, the development of inhomogeneities in surface layers and in outer atmosphere may have its root in the convective zone, and granular convective cells themselves will appear as temperature inhomogeneities in red (super)giant stars (e.g., Schwarzschild 1975). An attempt to model the granular convection in the solar atmosphere has already been done (Nordlund 1985) and tested by detailed analysis of line asymmetries and shifts (Dravins et al. 1981). While detailed theoretical models are not yet available for red giant stars, empirical approaches to such a model have already been done by an analysis of infrared CO lines whose bisectors showed a systematic red shift in the line wings, in consistency with a granular convective model (Ridgway & Friel 1980), or by measuring differential shifts of CO lines relative to some Fe I lines, which show a dependence on excitation as expected for convective model (Nadeau & Maillard 1988).

5.4 Some Other Molecules

OH: Strong lines of VR bands show much larger strengths than those expected from model atmospheres in a curve-of-growth analysis for Betelgeuse, and this result was interpreted as due to anomalous atmospheric structure or of depth dependent turbulent velocity (Lambert et al. 1984). More recently, pure rotation lines of OH in the mid-infrared have been observed in Betelgeuse (Jennings et al. 1986); the measured equivalent widths show better agreement with predictions based on a semi-empirical model (Basri et al. 1981) than those based on a classical model atmosphere (Johnson et al. 1980).

TiO: This molecule is highly sensitive to temperature and could be used to test the surface layers of model atmospheres, once basic parameters such as T_{eff} are well established. Such a test on models of Arcturus and Aldebaran showed that the best agreement with observations can be obtained for models having boundary temperatures close to the empirical values (Hänni & Sitska 1986). To resolve the problems that appeared in colours blanketed by TiO (Sect.4.1), similar analyses by high resolution spectra are desirable.

MgH: Generally, hydrides are known as good luminosity criteria and hence can be indicators of surface gravity. This idea was applied to MgH in Arcturus (Bell et al. 1985); the surface gravity estimated from MgH showed good agreement with those by pressure broadening of strong metal lines and by FeI/FeII ionization equilibrium. Other problems related to spectroscopic determination of stellar mass have been reviewed by Trimble & Bell (1981) with special reference to Arcturus. Unfortunately, determination of surface gravity is more difficult in cooler giants than in Arcturus.

6 CONCLUDING REMARKS

A brief survey of major observational tests revealed that red giant atmospheres are far more complicated than those described by available models. In fact, positive confirmation of present model atmospheres of red giant stars is limited to the global thermal structure in the continuum-forming region, as can be tested by the flux distribution and angular diameter measurements(Sect.2) together with limited information on limb-darkening (Sect.3). On the other hand, it is difficult to find positive confirmation of the atmospheric model in the line-forming region in general, as is evidenced by the fact that many observations related to line spectra cannot be satisfactorily explained by classical model atmospheres (Sects.4 & 5). There is no doubt that model atmospheres of the first generation (classical model atmospheres) played a major role in interpretations and analyses of new observations in the 1960's and 1970's and served to provide basic knowledges on red giant stars. However, it is also clear that the classical models are now challenged by new observations of higher quality.

One problem in line spectra of cool stars is to obtain measurements of line intensity or profile of sufficient accuracy under the inherent difficulties in locating the continuum and in finding spectral regions free from intrinsic blending. Even with the recent progress in observational techniques, such difficulties could not entirely be resolved, and some of the difficulties encountered in testing models by line spectra may be due to difficulty in defining the true continuum, for example. In this regard, in addition to high resolution spectroscopy, well calibrated spectrophotometry in line-blanketed regions is very important. In fact, limited information by line-blanketed fluxes also reveals serious difficulties of classical models (Sect.4) and thus this problem cannot be attributed to the difficulty of continuum location alone. Some of these difficulties can be resolved by further improvements of the classical models (e.g., opacities) or in extending the available sophistications (e.g., non-LTE, spherical geometry). At the same time, observational tests already suggest that more radical improvements may be needed.

One major problem revealed by the recent analyses of line spectra is that inhomogeneity may prevail throughout the subphotospheric convective zone, through the surface layers, and on to the outer atmospheres. Theoretical modeling of the granular convection in red giant stars will certainly be very important in the future, and observational tests of such a model are already possible with available observational techniques (Sect.5.3). Also, an attempt of two-dimensional imaging of the stellar disk by temperature sensitive molecular bands, using the technique of speckle interferometry, will provide more direct tests of such inhomogeneities (Lynds et al. 1976). The temperature inhomogeneity due to the granular convection and the mechanical energy flux generated in the convective zone may further induce bifurcated structure in the surface layers of cool stars because of the characteristic molecular cooling function (Ayres 1981; Muchmore & Ulmschneider 1985). Observational tests of such a model are rather difficult by spatially unresolved spectroscopic observations, but some evidences for such a model can be found through systematic effects in temperature sensitive lines. Further, such a bistable structure in the upper atmosphere may develop into thermal instability due to molecular cooling (Kneer 1983) and may induce autocatalytic molecular (Muchmore et al. 1987) or dust (Stencel et al. 1986) formation in the outer atmosphere. Observational identification of a molecular dissociation zone in the outer atmosphere by low excitation CO lines can be regarded as the observational manifestation of such a possibility (Tsuji 1988a). Thus the presence of a cool molecular component, in addition to the warm chromosphere, in the outer atmospheres of red giant stars appears to be confirmed, and such a component may play an important role in determining the structure of the outer atmosphere and, eventually, in producing molecular outflow and mass-loss (Tsuji 1988b).

Now, the problem of model atmosphere of red giant stars may be closely related to an effort to have a unified model of the photosphere (including subphotospheric convective zone) and outer atmosphere (including a chromosphere with cool and warm components, dust envelope, and cool wind). For this purpose, observational as well as theoretical backgrounds are being developed, and there is no doubt that further efforts towards understanding the physical structure of the photosphere, together with the outer atmosphere, of red giant stars will be rewarding in promoting our basic understanding of red giant stars, including their evolution.

Acknowledgements
I thank Dr. Hollis R. Johnson for helpful suggestions and comments in improving the text.

References
Auman,J.R.: 1967, in Colloquium on Late-type Stars, ed. M.Hack, (Trieste: Astron. Obs. Trieste.), p.313
Auman,J.R.,Woodrow,J.E.J.: 1975, Ap.J. 197, 163
Augason,G.C.,Taylor,B.J.,Strecker,D.W.,Erickson,E.F.,Witteborn,F.C.: 1980, Ap.J. 235, 138
Ayres,T.R.: 1981, Ap.J. 244, 1064
Ayres,T.R.,Testerman,L.,Brault,J.W.: 1986, Ap.J. 304, 542
Basri,G.S.,Linsky,J.L.,Eriksson,K.: 1981, Ap.J. 251, 162
Belega,Y.,Blazit,A.,Bonneau,D.,Koechlin,L.,Foy,R.,Labeyrie,A.: 1982, Astron.Astrophys. 115, 253
Bell,R.A.,Edvardsson,B.,Gustafsson,B.: 1985, M.N.R.A.S. 212, 497
Bell,R.A.,Gustafsson,B.: 1980, M.N.R.A.S. 191, 435
Blackwell,D.E.,Booth,A.J.,Petford,A.D.,Leggett,S.K.,Mountain,C.M., Selby,M.J.: 1986, M.N.R.A.S. 221, 427
Blackwell,D.E.,Shallis,M.J.: 1977, M.N.R.A.S. 180, 177
Bloemhof,E.E.,Townes,C.H.,Vanderwyck,A.H.B.: 1984, Ap.J.Lett. 276, L21
Bogdanov,M.B.: 1979, Sov.Astron. 23, 577
Bogdanov,M.B.,Cherepashchuk,A.M.: 1984, Sov.Astron. 28, 549
Brown,J.A.,Tomkin,J.,Lambert,D.L.: 1983, Ap.J.Lett. 265, L93
Carbon,D.F.: 1979, Ann.Rev.Astron.Astrophys. 17, 513
Carbon,D.F.,Milkey,R.W.,Heasley,J.N.: 1976, Ap.J. 207, 253
Chackerian,C.,Jr.,Tipping,R.H.: 1983, J.Mol.Spectros. 99, 431
Cheng,A.Y.S.,Hege,E.K.,Hubbard,E.N.,Goldberg,L.,Strittmatter,P.A., Cocke,W.J.: 1986, Ap.J. 309, 737
Christou,J.C.,Hebden,J.C.,Hege,E.K.: 1988, Ap.J. 327, 894
Cuntz,M.,Muchmore,D.: 1988, Astron.Astrophys. submitted
Di Benedetto,G.P.,Foy,R.: 1986, Astron.Astrophys. 166, 204
Di Benedetto,G.P.,Rabbia,Y.: 1987, Astron.Astrophys. 188, 114
Dominy,J.F.,Hinkle,K.H.,Lambert,D.L.,Hall,D.N.B.,Ridgway,S.T.: 1978, Ap.J. 223, 949
Dravins,D.,Lindgren,L.,Nordlund,A.: 1981, Astron.Astrophys. 96, 345
Eriksson,K.,Gustafsson,B.,Jorgensen,U.G.,Nordlund,A.: 1984, Astron. Astrophys. 132, 37
Faulkner,D.R.,Honeycutt,R.K.,Johnson,H.R.: 1988, Ap.J. 324, 490

Frisk,U.,Bell,R.A.,Gustafsson,B.,Nordh,H.L.,Olofsson,S.G.: 1982,
    M.N.R.A.S. 199, 471
Goebel,J.H.,Bregman,J.D.,Goorvitch,D.,Strecker,D.W.,Puetter,R.C.,
    Russell,R.W.,Soifer,B.T.,Willner,S.P.,Forrest,W.J.,Hauck,J.R.,
    McCarthy,J.F.: 1980, Ap.J. 235, 104
Gray,D.F.: 1982, Ap.J. 262, 682
Gustafsson,B.: 1989, Ann. Rev. Astron. Astrophys. vol.27, in press
Gustafsson,B.,Bell,R.A.: 1979, Astron.Astrophys. 74, 313
Gustafsson,B.,Bell,R.A.,Eriksson,K.,Nordlund,A.: 1975, Astron.Astrophys.
    42, 407
Gustafsson,B.,Jorgensen,U.G.: 1985, in Calibration of Fundamental Stellar
    Quantities, ed. D.S.Hayes, (Dordrecht: Reidel), p.303
Hall,D.N.B.,Ridgway,S.T.: 1977, in Les Spectres des Molecules Simples au
    Laboratoire et en Astrophysique, (Liege: Univ.Liege), p.243
Hänni,L.,Sitska,J.: 1986, Contr. Tartu Astrophys. Obs. 51, 159
Hartmann,L.,Avrett,E.H.: 1985, Ap.J. 284, 238
Heasley,J.N.,Ridgway,S.T.,Carbon,D.F.,Milkey,R.W.,Hall,D.N.B.: 1978,
    Ap.J. 219, 970
Hebden,J.C.,Christou,J.C.,Cheng,A.Y.S.,Hege,E.K.,Strittmatter,P.A.:
    1986, Ap.J. 309, 745
Hebden,J.C.,Eckart,A.,Hege,E.K.: 1987, Ap.J. 314, 690
Höflich,R.,Lowe,R.P.,Moorhead,J.,Scholz,M.,Wehlau,W.,Wehrse,R.: 1986,
    M.N.R.A.S. 220, 377
Jennings,D.E.,Deming,D.,Weidemann,G.R.,Keady,J.J.: 1986, Ap.J.Lett.
    310, L39
Johnson,H.L.: 1965, Ann.Rev.Astron.Astrophys. 4, 193
Johnson,H.R.: 1982, Ap.J. 260, 254
Johnson,H.R.: 1985, in Cool Stars with Excesses of Heavy Elements,
    eds. M.Jaschek & P.C.Keenan, (Dordrecht: Reidel), p.271
Johnson,H.R.: 1986, in The M-Type Stars, eds. H.R.Johnson & F.Querci,
    NASA SP-492, p.323
Johnson,H.R.,Alexander,D.R.,Bower,C.D.,Lemke,D.A.,Luttermoser,D.G.,
    Petrakis,J.P.,Reinhart,M.D.,Welch,K.A.,Goebel,J.H.: 1985,
    Ap.J. 292, 228
Johnson,H.R.,Bernat,A.P.,Krupp,B.M.: 1980, Ap.J.Suppl. 42, 501
Johnson,H.R.,Goebel,J.H.,Goorvitch,D.,Ridgway,S.T.: 1983, Ap.J.Lett.
    270, L63
Johnson,H.R.,Luttermoser,D.G.,Faulkner,D.R.: 1988, Ap.J. 332, 421
Johnson,H.R.,Yorka,S.B.: 1986, Ap.J. 311, 299
Jorgensen,U.G.,Almlof,J.,Gustafsson,B.,Larsson,P.,Siegbahn,P.: 1985,
    J.Chem.Phys. 83, 3034
Keenan,P.C.,Morgan,W.W.: 1941, Ap.J. 94, 501
Kipper,T.: 1982, W.Struve Nimeline Tartu Astrouf.Obs. 66, 3
Kneer,F.: 1983, Astron.Astrophys. 128, 311
Kuiper,G.P.: 1938, Ap.J. 88, 429
Lambert,D.L.: 1978, M.N.R.A.S. 182, 249
Lambert,D.L.,Brooke,A.L.,Barnes,T.C.: 1973, Ap.J. 186, 573
Lambert,D.L.,Brown,J.A.,Hinkle,K.H.,Johnson,H.R.: 1984, Ap.J. 284, 223
Lambert,D.L.,Gustafsson,B.,Eriksson,K.,Hinkle,K.H.: 1986, Ap.J.Suppl.
    62, 373
Leggett,S.K.,Mountain,C.M.,Selby,M.J.,Blackwell,D.E.,Booth,A.J.,
    Haddock,D.J.,Petford,A.D.: 1986, Astron.Astrophys. 159, 217

Lynds,C.R.,Worden,S.P.,Harvey,S.W.: 1976, Ap.J. 207, 174
Manduca,A.,Bell,R.A.,Gustafsson,B.: 1977, Astron.Astrophys. 61, 809
Manduca,A.,Bell,R.A.,Gustafsson,B.: 1981, Ap.J. 243, 883
Michelson,A.A.,Pease,F.G.: 1921, Ap.J. 53, 249
Muchmore,D.: 1986, Astron.Astrophys. 155, 172
Muchmore,D.,Nuth III,J.A.,Stencel,R.E.: 1987, Ap.J.Lett. 315, L141
Muchmore,D.,Ulmschneider,P.: 1985. Astron.Astrophys. 142, 393
Nadeau,D.,Maillard,J.P.: 1988, Ap.J. 327, 321
Nordlund,A.: 1985, Solar Phys. 100, 209
Orlati,M.A.: 1987, Ap.J. 317, 819
Piccirillo,J.,Bernat,A.P.,Johnson,H.R.: 1981, Ap.J. 246, 246
Querci,F.,Querci,M.: 1974, Highlights Astron. 3, 341
Querci,F.,Querci,M.,Tsuji,T.: 1974, Astron.Astrophys. 31, 265
Ramsay,L.W.: 1977, Ap.J. 215, 827
Ramsay,L.W.: 1981, Ap.J. 245, 984
Ricort,G.,Aime,A.,Vernin,J.,Kadiri,S.: 1981, Astron.Astrophys. 99, 232
Ridgway,S.T.,Carbon,D.F.,Hall,D.N.B.: 1978, Ap.J. 225, 138
Ridgway,S.T.,Friel,E.D.: 1981, in Effects of Mass Loss on Stellar
    Evolution, eds. C.Chiosi & R.Stalio, (Dordrecht: Reidel),
    p.119
Ridgway,S.T.,Jacoby,G.H., Joyce,R.R., Siegel,M.J., Wells,D.C.: 1982a,
    Astron.J. 87, 808
Ridgway,S.T.,Jacoby,G.H., Joyce,R.R., Siegel,M.J., Wells,D.C.: 1982b,
    Astron.J. 87, 1044
Ridgway,S.T.,Joyce,R.R., White,N.M., Wing,R.F.: 1980, Ap.J. 235, 126
Ridgway,S.T.,Wells,D.C.,Joyce,R.R.: 1977, Astron.J. 82, 414
Roddier,C.,Roddier,F.: 1983, Ap.J.Lett. 270, L23
Roddier,F.,Roddier,C.: 1985, Ap.J.Lett. 295, L21
Roddier,F.,Roddier,C.,Petrov,R.,Martin,F.,Ricort,G.,Aime,C.: 1986,
    Ap.J.Lett. 305, L77
Ruland,F.,Holweger,H.,Griffin,R.,Griffin,R.,Biehl,D.:1980,
    Astron.Astrophys. 92, 70
Scargle,J.D.,Strecker,D.W.: 1979, Ap.J. 228, 838
Schmid-Burgk,J.,Scholz,M.,Wehrse,R.: 1981, M.N.R.A.S. 194, 383
Schmidtke,P.C.,Africano,J.L.,Jacoby,G.H.,Joyce,R.R.,Ridgway,S.T.: 1986,
    Astron.J. 91, 961
Scholz,M.: 1985, Astron.Astrophys. 145, 251
Scholz,M.,Takeda,Y.: 1987, Astr.Astrophys. 186, 200
Scholz,M.,Tsuji,T.: 1984, Astron.Astrophys. 130, 11
Scholz,M.,Wehrse,R.: 1982, M.N.R.A.S. 200, 41
Schwarzschild,M.: 1975, Ap.J. 195, 137
Skinner,C.J.,Whitmore,B.: 1987, M.N.R.A.S. 224, 335
Smith,V.V.,Lambert,D.L.: 1985, Ap.J. 294, 326
Smith,V.V.,Lambert,D.L.: 1986, Ap.J. 311, 843
Spinrad,H.: 1966, Ap.J. 145, 195
Stencel,R.,Carpenter,K.G.,Hagen.W.: 1986, Ap.J. 308, 859
Steenbock,W.: 1985, in Cool Stars with Excesses of Heavy Elements,
    eds. M.Jaschek & P.C.Keenan, (Dordrecht: Reidel), p. 231
Steiman-Cameron,T.Y.,Johnson,H.R.: 1986, Ap.J. 301, 868
Strecker,D.W., Erikson,E.F., Witteborn,F.C.: 1979, Ap.J.Suppl. 41, 501
Steiman-Cameron,T.Y.,Johnson,H.R.,Honeycutt,R.K.: 1985, Ap.J.Lett. 291,
    L51

Tipping,R.H.: 1988, private communication
Trimble,V., Bell,R.A.: 1981, Quart. J. Roy. Astron. Soc. 22, 361
Tsuji,T.: 1967, in Colloquium on Late-type Stars, ed. M.Hack, (Trieste: Astron. Obs. Trieste.), p.260
Tsuji,T.: 1976, Publ.Astron.Soc.Japan 28, 567
Tsuji,T.: 1977, Publ.Astron.Soc.Japan 29, 497
Tsuji,T.: 1978a, Astron.Astrophys. 62,29
Tsuji,T.: 1978b, Publ.Astron.Soc.Japan 30,435
Tsuji,T.: 1981a, Astron.Astrophys. 99, 48
Tsuji,T.: 1981b, J.Astrophys.Astron. 2, 95
Tsuji,T.: 1983, Astron.Astrophys. 122, 314
Tsuji,T.: 1985, in Cool Stars with Excesses of Heavy Elements, eds. M.Jaschek & P.C.Keenan, (Dordrecht: Reidel), p.93
Tsuji,T.: 1986, Astron.Astrophys. 156,8
Tsuji,T.: 1988a, Astron.Astrophys. 197,185
Tsuji,T.: 1988b, in Atmospheric Diagnostics of Stellar Evolution: Chemical Peculiarity, Mass-Loss, and Explosion, ed. K.Nomoto, (Berlin: Springer-Verlag), p.158
Vieira,T.: 1986, Ph.D.Thesis, Uppsala Astronomical Observatory
Watanabe,T.,Kodaira,K.: 1978, Publ.Astron.Soc.Japan 30, 21
Watanabe,T.,Kodaira,K.: 1979, Publ.Astron.Soc.Japan 31, 61
White,N.W.: 1980, Ap.J. 242, 646
White,N.W.,Kreidl,T.J.: 1984, Astron.J. 89, 424
Wing,R.F.: 1971, in Proc. Conf. on Late-type Stars, eds. G.W.Lockwood & H.M.Dyck, (Tucson: Kitt Peak Nat.Obs.), p.145

THE CHEMICAL COMPOSITION OF ASYMPTOTIC GIANT BRANCH STARS

David L. Lambert
Department of Astronomy, University of Texas, Austin, Texas

Abstract. Low resolution spectroscopic and photometric studies of Magellanic Cloud AGB stars have shown that the M → S → C sequence is the result of the third dredge-up of ^{12}C and s-process elements in AGB stars of low mass. In this paper, data on the chemical compositions of normal Galactic M → S → C stars are reviewed and shown to be broadly consistent with expectations for the third dredge-up on the AGB.

1. INTRODUCTION

The title assigned to my talk was 'The Chemical Composition of Real PRG Stars'. A definition of what constitutes a 'real PRG star' was not provided except for the laconic comment 'not Barium stars'. For this review, I consider 'real PRG star' to be synonymous with 'asymptotic giant branch star' (AGB); i.e., a star with a C-O degenerate core, a thin He shell, and a deep H-rich convective envelope. The appearance of the adjective 'peculiar' in the title of this colloquium is intended, I suppose, in the context of a talk on chemical compositions to refer to stars having a non-solar mix of the chemical elements, but just such a mix is the fate of all AGB stars; they dredge up nuclear processed material on the ascent of the giant branch and again on the AGB. Abundance anomalies created by mass transfer or exchange across a binary system are outside my terms of reference.

Even with my definition of a real PRG star, the assigned topic is so broad that I must limit my remarks to two principal topics.

Spectral Classification and Chemical Composition.

A cursory inspection of the complex spectra of the late-type or PRG stars may convince the novice that extraction of chemical compositions is a burdensome and lengthy task. Fortunately, there are important indicators of chemical composition (for example, the C/O ratio) that are obtainable by a mere inspection of low resolution spectra and, in some cases, of photometric indices. In other cases, examination of high resolution spectra may readily yield key information; the prime example is the presence or absence of technetium in AGB stars as judged by a search for the Tc I resonance lines in the blue. I comment on some of these possibilities for an "instant abundance analysis".

Chemical Compositions of AGB Stars in the Sequence $M \rightarrow S \rightarrow C$.

The label 'peculiar' seems inapposite when attached to these stars. The chemical compositions can now be determined with fair precision from high resolution optical and infrared spectra. Observed abundance differences with respect to the inferred initial (quasi-solar) compositions for these stars are attributable to mixing into the envelope of nucleosynthetic products from the He-burning shell. The principal products are carbon and those heavy elements synthesized by neutron capture through the s-process during a He-shell flash (thermal pulse). Although the observed and predicted changes of composition are not yet in full agreement, the measure of agreement surely suggests that these stars should no longer be marked out as 'peculiar'. I shall review recent explorations of the chemical compositions of stars in the sequence of increasing C/O ratio: $M \rightarrow S \rightarrow C$.

After the $M \rightarrow S \rightarrow C$ giants are culled from the lists of PRGs, there remains a mixed bag of peculiar stars whose evolutionary history is unclear. Examples include the remarkable star FG Sge, the hydrogen deficient carbon (HdC) stars, the R Coronae Borealis stars and especially the heavy element-enriched U Aqr, the warm carbon (R type) stars, the RV Tau and W Vir variables, and others. Some peculiar stars may be in a stage of rapid evolution experienced by the majority of AGB stars. Others may be following evolutionary paths avoided by the normal AGB stars. I shall make few references to peculiar giants.

2. SPECTRAL CLASSIFICATION AND CHEMICAL COMPOSITION

2.1 *The C/O Ratio*

Thanks to the preeminent role of the CO molecule in the dissociation equilibrium of carbon and oxygen, spectra of cool stars with an abundance ratio C/O < 1 (here, oxygen-rich) are readily distinguishable at even low resolution from spectra of cool stars with an abundance ratio C/O > 1 (here, carbon-rich). In the optical region, the spectra of oxygen-rich stars are dominated by metal oxide bands (e.g., TiO, ZrO, YO). In the same region, the spectra of carbon-rich stars are dominated by bands of C_2 and CN. Gross spectral differences between oxygen-rich and carbon-rich also exist in the infrared;

cool oxygen-rich stars show H_2O bands and carbon-rich stars show C_2 and CN bands and, in the coolest stars, bands of polyatomic molecules such as HCN and C_2H_2.

Such gross spectral differences allow observers to classify stars according to the abundance ratio C/O: C/O < 1 for oxygen-rich and C/O > 1 for carbon-rich stars. A finer classification for the oxygen-rich stars has been attempted through observations of band ratios of two oxides. Since the molecular bands of TiO and ZrO are visible even at low dispersion, the ratio ZrO/TiO is generally used. Although the principal reason for this ratio's dependence on O-C is well documented (see, for example, Wurm 1940 and Scalo 1974), misconceptions appear in the literature. Consider two metal oxides XO and ZO. If the molecular equilibrium for the metals involves only the competition between the metal and its oxide, it is readily shown that the partial pressure of the oxide is given by

$$P(XO) = \frac{\varepsilon_X P_T(H) P(O)}{(K_{XO} + P(O))}$$

where ε_X is the abundance of X relative to H, $P_T(H)$ is the "fictitious" total pressure of hydrogen ($P_T(H) \simeq P(H) + 2P(H_2)$) and K_{XO} is the equilibrium constant. Two limiting cases exist: (i) association of X into XO is nearly complete, then $P(XO) = \varepsilon_X P_T(H)$; (ii) the molecules XO are a minor constituent, then $P(XO) = \varepsilon_X P_T(H) P(O)/K_{XO}$. Consider now the ratio $R = P(ZO)/P(XO)$. If case (i) applies to both molecules, $R = \varepsilon_Z/\varepsilon_X$. If case (ii) is appropriate for both molecules, $R = (\varepsilon_Z/\varepsilon_X)(K_{XO}/K_{ZO})$. These expressions are *independent* of the abundance difference (O-C) but depend, as expected, on the abundance ratio Z/X. A dependence on (O-C) arises when the molecule with the higher dissociation energy (ZrO) approaches case (i) and case (ii) is more appropriate for the other molecule (TiO). Then, $R = (\varepsilon_Z/\varepsilon_X) K_{XO}/P(O)$ and $P(O) \approx (\varepsilon_O - \varepsilon_C) P_T(H)$, i.e., $R \propto (Z/X)/(O-C)$. This case arises when the dissociation energies $D_o(XO)$ and $D_o(ZO)$ differ significantly, e.g., $D_o(TiO) = 6.92 \pm 0.10$ eV and $D_o(ZrO) = 8.00 \pm 0.14$ eV (Pedley and Marshall 1983). Of course, this simple argument ignores many factors: competition from metal dioxides and other molecules, changes in the atmospheric structure as the densities of major molecules change with the O-C ratio, and variation of the molecular equilibrium with depth through the atmosphere. Nonetheless, it should be clear that R is controlled by Z/X and, perhaps, also (O-C). For the (O-C) dependence to be strong, a star's effective temperature (T_{eff}) and surface gravity (i.e., pressure) must be such that Z and X approach case (i) and case (ii) respectively in the photosphere.

In defining type S and introducing "an abundance class" defined by the relative intensities of ZrO and TiO bands, Keenan (1954) noted that "the questions of just which elements differ in abundance and of how large the differences are remain to be investigated on spectrograms of high dispersion". Ake (1979) introduced an abundance index defined by the relative intensities of TiO, ZrO, and YO bands and calibrated the index in terms of the C/O ratio by reference to synthetic spectrum calculations published by Scalo and Ross (1976). The YO/ZrO ratio, which should be less sensitive to the metal abundances because Y and Zr are s-process elements, derives some of its O-C sensitivity from the fact that dioxide formation is more important for Y than Zr. Keenan and Boeshaar (1980), who classified S and SC stars using primarily ZrO, TiO and C_2 bands, also provide a C/O index. These two abundance indices are assumed to be primarily controlled by the C/O ratio. This assumption must derive in large part from the emphatic statements made by Scalo and Ross. Piccirillo (1980), who calculated molecular column densities for a grid of M and S model atmospheres, argues that the ZrO/TiO strengthening seen in S stars "cannot be interpreted solely by C/O effects" and "the observed strength of ZrO in the spectra of S stars attests to Zr enhancement". These results from model atmosphere calculations confirm Boesgaard's (1970) demonstration that the abundance ratio Zr/Ti, as derived from a curve of growth analysis, is well correlated with a band index ZrO/TiO.

In short, the ZrO/TiO band ratio is sensitive to both the Zr/Ti and O-C abundances. However, if the dredge-up of C, the direct product of He-burning, and of Zr, a s-process product, occur simultaneously, the Zr/Ti at the surface will itself be a rough measure of the increased C abundance and, hence, the ZrO/TiO ratio will indirectly through Zr/Ti and directly through O-C reflect the O-C in the atmosphere. Today, we can calibrate the ZrO/TiO vs O-C dependence through abundance analyses of the brighter M and S stars (see Sec. 3.4). It should be possible using the C_2 bands to extend the O-C estimates to carbon stars. With a determined effort, it should be possible to calibrate Magellanic Cloud stars.

2.2 *Observation vs Theory: The Magellanic Clouds*

Searches of the Magellanic Clouds for carbon-enriched stars have yielded many new results on the evolution of AGB stars because reliable estimates of absolute luminosity are possible for Cloud members and the surface C/O ratio may be obtained from inspection of low dispersion spectra. These "instant abundance analyses" are generally limited to assigning a star to one of three C/O classes: M star with C/O $\leq \alpha$, S star with $\alpha \leq$ C/O < 1.00, and C star with C/O > 1.00 where $\alpha \simeq 0.95$ is usually assumed but our abundance analyses suggest a lower figure. Finer gradations are possible with the recognition of intermediate spectral classes and the calibration of an abundance index to MS and S stars. By searching the field and, in particular, the clusters in the Clouds for S and C stars, critical information is obtained on the third dredge-up phase when freshly synthesized carbon and s-process elements from the He-burning shell enter the outer deep convective envelope.

About a decade ago, the theory of stellar evolution and nucleosynthesis on the AGB (Iben 1975, 1976, 1977; Truran and Iben 1977; Iben and Truran 1978) predicted the following:

(1) stars within the range 3 to 8 $M_\odot$ (the intermediate mass stars) become C stars and are the dominant supplier of s-process elements in the Galaxy and a major supplier of ^{12}C. Significant dredge-up of ^{12}C and the s-process does not occur until luminosities of log $L/L_\odot$ ~ 4.3 are reached. The s-process elements are produced in approximately their solar system proportions and in amounts such that intermediate mass stars account for the solar system ratio of s-process to iron peak elements. The neutron source is $^{22}Ne(\alpha,n)^{25}Mg$ with the supply of ^{22}Ne controlled by the initial abundance of C, N, O. The ability of the ^{22}Ne source to generate a solar system mix of heavy elements has been questioned (Busso et al. 1988), but these doubts depend greatly on the (uncertain) estimate of the ^{22}Ne neutron capture cross section.

(2) stars less massive than about 2 $M_\odot$ do not become C stars and stars less massive than about 3 $M_\odot$ do not ignite a neutron source and synthesize heavy elements by the s-process.

The stars in the Galaxy that are luminous enough to be on the AGB and whose atmospheres are enriched in ^{12}C and the s-process elements are the S stars and the N-type carbon stars (Scalo 1976). Iben and Truran (1978) declined to recognize the abundance anomalies of these Galactic stars as being due to thermal pulses on the AGB. The two reasons cited were (i) the observed (uncertain) luminosities of S and N stars are below the predicted lower limit for dredge-up of ^{12}C by an intermediate-mass AGB star; (ii) the (qualitative) estimates of the Ba-peak to Sr-peak s-process abundances are too large to be consistent with production by the ^{22}Ne source or compatible with the solar system abundances. Iben and Truran speculated on an alternative origin for S and N stars: "Perhaps the simplest explanation of S stars and of many N-type carbon stars is that the surface composition characteristics originate during a helium [core] flash of an infrequently occurring nature". In such a core flash, the neutron source $^{13}C(\alpha,n)^{16}O$ may be ignited with a neutron supply exceeding that generated by the ^{22}Ne$(\alpha,n)^{25}$Mg reactions and, hence, higher Ba/Sr ratios may be produced.

The predictions that dredge-up of ^{12}C and s-process was the sole prerogative of intermediate-mass AGB stars were first shaken and then shattered by surveys of cool carbon stars in the Magellanic Clouds. The luminosity function of C stars in the clouds peaks at $M_{bol} \sim -4.8$ or $\log L/L_\odot \sim 3.8$ (Cohen et al. 1981; Richer 1981). Carbon stars are not found with $\log L/L_\odot > 4.3$ (Cohen et al. 1981; Wood, Bessell, and Fox 1983; Wood 1987); this luminosity was the predicted *lower* limit for carbon star production by intermediate mass stars. The low luminosity limit, ($\log L/L_\odot \sim 3.5$) for the cool carbon stars just exceeds the luminosity achieved at the He core flash at the tip of the red giant branch. The observed luminosity range is consistent with an identification of cool carbon stars as thermally pulsing low mass AGB stars. The near absence of lower luminosity carbon stars suggests that the He core flash is not primarily responsible for carbon enrichment. Less luminous carbon stars exist, but appear to be counterparts of the Galactic warm (R-type) carbon stars whose origin appears unrelated to thermal pulsing in a single PRG.

Clearly, the observed luminosity function of cool carbon stars in the Clouds is in stark conflict with the theoretical predictions current in 1981: (i) oxygen-rich (M) stars are converted to carbon-rich (C) stars at a much lower luminosity and, hence, at a lower mass than predicted, and (ii) the intermediate mass stars predicted to become C stars with $M_{bol} \lesssim -6$ are either not formed or, if formed, either evolve much more rapidly than predicted (e.g., shed their envelope in a "superwind"), or are reconverted to an O-rich star. Resolution of this conflict ("The Carbon Star Mystery: Why Do the Low Mass Ones Become Such and Where have the High Mass Ones Gone?" - Iben [1981]) is discussed by Iben (1988) and in several papers at this colloquium.

Between the observed upper limit ($M_{bol} \sim -6$) for cool carbon stars and the maximum luminosity for an AGB star ($M_{bol} \sim -7.1$), a limit set when the degenerate core reaches the Chandrasekhar limit, the AGB stars in the clouds are oxygen-rich. This sample defined first by Wood, Bessell, and Fox (1983) contains a few S stars. Wood et al. speculate that H-burning at the base of the convective envelope may through the cycling of C to N reconvert the carbon-star to an oxygen-star heavily enriched in nitrogen. Unfortunately, extraction of a N abundance for a cool oxygen-rich star calls for high resolution spectra or a thorough comparison of observed and synthetic low resolution spectra.

Spectral classification of stars in the Magellanic Cloud clusters is providing even more detailed results on the evolution of AGB stars; e.g., the minimum mass for which He-shell flashes can convert a star to a carbon star: Bessell, Wood, and Lloyd Evans (1983) estimate M > 0.9 $M_\odot$ for [Fe/H] ~ -1 and M > 1.3 $M_\odot$ for [Fe/H] ~ 0. The transition M → S → C occurs at a higher luminosity and a lower T_{eff} in more massive stars (Lloyd Evans 1983, 1984; Frogel and Blanco 1984). Thorough scrutiny of individual AGB stars and the frequency of occurrence of M, S, and C stars in the clusters promises to provide critical challenges of the theory of PRGs; for example, Lloyd Evans (1984) uses Keenan and Boeshaar's (1980) ZrO index ("calibrated" by them in terms of the C/O ratio) to make the tentative suggestions that "the build-up in the C/O value at the surface does not proceed at a uniform rate but accelerates abruptly after a slow start", "the S-star stage of evolution occupies a substantial period of time", "the observations are in better accord with an s-process build-up which precedes the main change in the C/O ratio". While "instant abundance analysis" of the Magellanic and other AGB stars will continue to provide novel results, more detailed abundance analyses of AGB stars in the Clouds and the Galaxy are surely necessary in order to complete the observational picture of the dredge-up of material from the He-burning shell.

In addition to insights into evolution on the AGB, low dispersion spectroscopy of Magellanic Cloud and Galactic stars yields cautionary hints; for example, Wood, Bessell, and Fox (1983) note that two SMC variables change from S to C stars as they vary. The atmospheric C/O ratio is presumed to be unchanging, but it may be so close to unity that alterations to the atmospheric structure exert a controlling influence on the partial pressure of carbon. These changes are reminiscent of longer-term trends from an oxygen-rich to a carbon-rich atmosphere reported for the Galactic stars TT Cen (Stephenson 1973) and BH Cru (Lloyd Evans 1985). Both are now SC stars but earlier showed ZrO bands. In these cases, the spectral classification appears to persist over several cycles.

2.3 *High Resolution Spectroscopy: Technetium*

Before I discuss some results obtained from rather detailed abundance analyses, I comment on extension of "instant abundance analyses" to high dispersion spectra. Undoubtedly the outstanding example is the detection of technetium through the Tc I resonance lines at 4238, 4262, and 4297 Å. Merrill's (1952) discovery of Tc in the atmospheres of S stars is rightly a landmark in studies of PRGs. Tc is the outstanding indicator of recent and, perhaps, current s-process nucleosynthesis in PRGs. It is not always appreciated that (i) Tc is not omnipresent in a sample of stars enriched to similar levels in the heavy elements that are products of the same neutron capture s-process that is responsible for the synthesis of technetium; (ii) Tc is present in some stars for which overabundances of s-process elements appear to be very small.

Scalo and Miller (1981) first pointed out that, of 30 stars of type MS, S, SC, and C in a survey and compilation by Little-Marenin and Little (1979), 9 stars did not contain Tc. These spectral types exhibit, in general, s-process and ^{12}C-enrichment. The absence of Tc in 30% of these stars led Scalo and Miller (1981) to suggest two possible explanations, both of which involved radical changes in the usual picture of AGB evolution: (i) the mixing of s-processed material to the surface does not occur at every shell flash - if many consecutive flashes occur without mixing into the envelope, the surface Tc decays steadily until it is replenished by a flash in which mixing occurs; and (ii) some stars above a certain core mass (say $M_{core} \gtrsim 0.8\ M_\odot$) never mix.

Iben and Renzini (1983) criticized Scalo and Miller's (1981) conclusions and suggested three alternative explanations for the Tc-poor stars. (1) Some stars may not be s-process enhanced; low-resolution classification spectra may be an inadequate basis from which to claim a definitive s-process enhancement. Indeed, we (Smith and Lambert 1986) found several stars classified as MS that did not exhibit measurable s-process enhancements. (2) The half-life of ^{99}Tc decreases from 2×10^5 yr at low temperatures ($T \lesssim 10^8$ K) to just 5 yr at $T \approx 3 \times 10^8$ K, the expected temperature for the ^{22}Ne neutron source (Cosner and Truran 1981; Schatz 1983). Then, Tc's absence in some AGB stars might be understood if the s-process operated at temperatures so high that Tc is destroyed very quickly. However, this conclusion is not supported by detailed studies of s-processing during thermal pulses; Mathews *et al.* (1986) show that Tc is produced at the higher temperatures because higher neutron fluxes offset the higher decay rates of ^{99}Tc and, hence, the absence of Tc in a s-process enriched star is not strong evidence for a hot s-process site. (3) Some stars may be cooler (or evolved) examples of the G and K giant barium stars which also exhibit s-process enrichments, yet contain no Tc. Such cooler barium stars have not yet begun to dredge-up fresh additions of s-processed material including Tc. The discovery by McClure, Fletcher, and Nemec (1980) and McClure (1983, 1985), that probably all barium stars are binaries and may have white dwarf companions has led to the hypothesis that the barium stars are the result of a transfer of mass from an AGB star (now the white dwarf) to a donor (now the barium star). This mass transfer could have easily happened so long ago that the Tc has decayed away.

When our recent survey (Smith and Lambert 1988a) for the TcI 4260 Å line in 40 MS/S stars is combined with published reports on an additional 19 stars, we confirm that MS/S stars enriched in the s-process but not containing Tc are common: about 40% of the sample do not show Tc. A comparison of the (uncertain) space densities suggests that the majority of these s-process enriched stars lacking Tc are probably evolved Barium GK giants that have not yet begun to dredge-up carbon and s-process elements on the AGB. This proposal is open to a simple observational test: if the Tc-poor S stars are descendants of Barium stars, they will show orbital radial velocity variations. Jorissen and Mayor (1988) observed 9 S stars with the CORAVEL spectrometer and report that at least 5 stars are binaries. An extension of this survey will test our proposal. If the test yields a negative result, it will be necessary to reconsider the Scalo and Miller (1981) ideas. Of course, as the M-type Barium stars evolve along the AGB, they will experience a dredge-up of additional carbon and s-processed material and presumably replenish their surface Tc abundance. However, the fraction of such stars among the total population of Tc-containing S stars should be small. Clearly predictions about the dredge-up in single stars should be tested against observations of single stars not of evolved binary/Barium stars. "Instant abundance analysis" through a search of high resolution spectra for the TcI resonance lines enables us to identify the thermally pulsing AGB stars.

Other examples of "instant abundance analysis" from high resolution spectra include the isolation of Li-rich cool carbon and S stars (e.g., WZ Cas), the classification of the ^{13}C-rich or J-type carbon stars, and the identification of the hydrogen deficient 'cool' carbon (HdC) and the warmer RCrB variables.

3. CHEMICAL COMPOSITION OF GALACTIC M, S, AND C STARS

3.1 *Spectroscopic Tests of An Evolutionary Scenario*

Observations of the cool luminous stars in the Magellanic Clouds suggest that the regular or normal M, S, and C stars are AGB stars of low mass ($M \sim 2\ M_\odot$, say) in which He-shell flashes (thermal pulses) occur and products of nucleosynthesis in the He-burning shell are dredged to the surface. Through a progressive enrichment of ^{12}C, the principal product of He-burning, the oxygen-rich M star is converted to an S star and then to a C star; of course, if the ^{12}C enrichment from a single pulse is sufficiently severe, the star may jump from M or MS to C without an intermediate step as a S star. In these low mass stars, predicted temperatures in the He-burning convective shell are too low for the ^{22}Ne(α,n) source to be the primary neutron source. The alternative neutron source for the lower temperatures is generally considered to be the ^{13}C(α,n)^{16}O reaction with an adequate supply of ^{13}C provided by a mixing or diffusing of protons into the ^{12}C-rich He shell; a small admixture of protons leads to ^{12}C(p,γ)^{13}N(β^+,ν)^{13}C and, if the protons are largely consumed in this chain, the ^{13}C avoids proton capture and is subsequently destroyed by ^{13}C(α,n)^{16}O.

As the discussion of Tc in MS and S stars indicated, evolved Barium stars may be mistaken for single ('real') MS, S, and C stars. Barium stars are most probably created by mass transfer across a binary when an s-process enriched AGB (MS, S, or C) star sheds its envelope onto its companion, the present peculiar giant. Thanks to the mass transfer, the compositions of 'real' PRGs and the Barium stars must be similar; the abundance anomalies of the latter are expected to be diluted forms of those found in the MS, S, or C stars. Observational confirmation of this expectation has been presented (Lambert 1985, 1988). Spectroscopic tests of the proposal that single ('real') MS, S, and C stars are thermally pulsing low mass stars should be applied to samples cleansed of evolved Barium stars. While an attempt to isolate such samples is now being made, some of the tests described below involve samples which are likely to contain evolved Barium stars.

The spectroscopic tests are most simply introduced as a series of questions:

(1) Are the elemental and isotopic abundances of C, N, and O in the M, S, and C stars interpretable as a sequence in which ^{12}C is added in an increasing amount? If the base of the convective envelope is cool, the ^{12}C from He-burning will not experience H-burning (i.e., conversion of ^{12}C to ^{14}N with a substantial amount of ^{13}C) as it is dredged-up. Also, the He-burning shell is not expected to synthesize a significant amount of ^{16}O. Then we expect (i) ^{12}C/^{16}O and ^{12}C/^{13}C to increase in a predictable way from M stars through S stars to C stars; (ii) the ^{14}N and ^{16}O abundances should be similar along the M $\rightarrow$ S $\rightarrow$ C sequence; (iii) the oxygen isotopes ^{17}O and ^{18}O should show a smooth progression in which a modest decrease of ^{18}O and, perhaps, ^{17}O is expected in the C stars.

(2) Are the enhancements of the s-process elements in the S and C stars consistent with predicted dredge-up from layers exposed to neutrons released by the ^{13}C(α,n)^{16}O reaction? Or is the ^{22}Ne(α,n)^{25}Mg reaction the dominant neutron supplier? Close examination of the s-process products may reveal details of the average conditions at the s-process site; for example, the total exposure and the mean neutron density: are these

quantities derived from observations consistent with predictions from model low-mass AGB stars? Are the AGB stars leading suppliers of s-process elements?

(3) Are there surprises in the observed chemical composition? Mild quantitative discrepancies between, for example, the predicted and observed C/O ratios for S stars of a given $^{12}C/^{13}C$ ratio are certainly a signal that revisions are necessary in the modeling of the thermal pulses or the abundance analyses or both. Such discrepancies are not surprises. They are an inevitable part of our current preliminary probing of these complex stars. A surprise is a major unexpected or rare signature in the chemical composition. A single example must here serve to illustrate my definition. Dominy and Wallerstein (1986) through empirical curves of growth derived the V, Zr, Nb, and Tc abundances for the long-period variables χ Cyg (known mild S star), o Cet, and R Lyr. For the traditional s-process, Tc, Nb, and Zr are predicted to be enhanced in abundance. The results for o Cet were a surprise: Zr has an approximately normal abundance, Tc is present (see also Little-Marenin and Little 1979), but Nb is depleted; i.e., the Nb/V ratio is substantially - a factor of nearly 5 - less than solar. Dominy and Wallerstein account for the surprise by supposing a very small and recent ($< 10^5$ yr) pulse of neutrons converted the original ^{93}Nb, the sole stable isotope, to ^{94}Mo, and that this supply has not been replenished because the progenitor ^{93}Zr has a long half-life ($\tau_{1/2} \sim 1.5 \times 10^6$ yr). An interesting task now is to examine additional stars for evidence of such mild pulses; a cobalt enrichment may be a more readily detected signature of such pulses (Smith and Lambert 1987). Long-period variables like o Cet with Tc but no reported overabundance of s-process elements are not rare (Little, Little-Marenin, and Bauer 1987). One may conjecture that such stars may be beginning that series of thermal pulses/shell flashes with which evolution on the AGB terminates. A reported abundance anomaly that fits my definition of a surprise is not always a true measure of the composition of a stellar atmosphere. Some surprises are the result of a defective analysis. I comment later on our earlier reports of anomalously low ^{17}O and ^{18}O abundances in S stars!

(4) Are all of the predicted phases of AGB evolution represented and then recognized in the sky? In posing this question, I am thinking of the proposal (Scalo, Despain, and Ulrich 1975; Renzini and Voli 1981) that the base of the convective envelope of very luminous AGB stars may be so hot that H-burning occurs such that carbon is converted to nitrogen, the C/O ratio is reduced, and the carbon star is reconverted to an oxygen-rich star. This luminous oxygen-rich star would remain enriched in the s-process elements and, hence, would probably be classified as a S star rather than a M star. Stars with a 'hot bottomed convective envelope' (HBCE) have been promoted as the explanation for several odd stars; e.g., very N-rich planetary nebulae, and the Magellanic Cloud AGB

Stars (oxygen-rich with a mixture of S stars) with luminosities below the AGB limiting luminosity and above that of the most luminous cool carbon stars (see Sec. 2.2). To check these and other identifications with HBCE stars, it seems desirable to establish the primary signature of a HBCE - a high N abundance and a low $^{12}C/^{13}C$ ratio. A high s-process abundance would establish the star's credentials as a former thermally pulsing AGB stars. Perhaps the more massive stars such as the long-period variable supergiants identified by Wood, Bessell, and Fox (1983) in the Magellanic Clouds that do not experience thermal pulses may also develop a HBCE. Then, an s-process enrichment would not accompany the signatures of H-burning.

(5) Are the derived compositions of the M, S, and C stars consistent with those of closely related objects? I shall comment on the symbiotic stars where the relative C, N, and O abundances can be determined from the emission line fluxes. Since the hot emitting gas was recently shed by the M giant, the C/N/O ratios from the emission lines should be identical to those ratios derived from molecular absorption lines in the M giant's spectrum. Such a comparison of C/N/O ratios offers a valuable check on the consistency of the two quite different analyses. Another and similar comparison of C, N, and O abundances (now with respect to H) is possible between the planetary nebulae and the M, S, and C stars.

3.2 Carbon Enrichment: $M \rightarrow S \rightarrow C$

Examination of the Magellanic Clouds and, in particular, of their clusters shows that the relative luminosities of stars in the $M \rightarrow S \rightarrow C$ sequence are consistent with the idea that the sequence arises through the dredge-up of ^{12}C from a He-burning shell. The simplest working hypothesis is that (i) the dredge-up adds ^{12}C but not ^{13}C and ^{16}O to the convective envelope; and (ii) the base of the envelope is cool so that hydrogen burning does not convert ^{12}C to ^{13}C and ^{14}N. On this hypothesis, the $^{12}C/^{16}O$ and $^{12}C/^{13}C$ ratios are predictable from the envelope's presumed composition prior to the dredge-up of ^{12}C. This presumed composition is taken to be that of the M giants whose ^{12}C, ^{13}C, and ^{14}N abundances have been rearranged during the first (and, perhaps, the second) dredge-up as the star ascended the giant branch prior to He-core ignition. Although dredge-up of ^{12}C is now the widely accepted scheme for production of carbon stars, it was not so long ago that a competing nucleosynthetic scheme could not be distinguished on the basis of observations or theoretical plausibility - hydrogen burning at a high temperature converts ^{12}C and ^{16}O to ^{14}N and achieves $^{12}C/^{16}O > 1$ in equilibrium. Spectroscopic determinations of C, N, and O serve not only to check the working hypothesis about ^{12}C dredge-up, but also to eliminate competing schemes, including H burning.

In the M, S, and C stars, the C, N, and O abundances are obtainable only from molecular lines. For a complete set of molecules, it is necessary to obtain infrared spectra. In particular, the vibration-rotation (V-R) lines of the CO molecule's ground electronic state are the sole source of the C abundance in a cool oxygen-rich atmosphere and of the O abundance in a cool carbon rich atmosphere. Often, a simultaneous analysis of two or more molecules may be required; e.g., the OH V-R lines in the spectrum of an oxygen-rich star effectively measure the abundance difference O-C so that OH and CO lines must be analysed simultaneously to yield the O and the C abundances. It is the development of high resolution infrared spectrometers that has made it possible to determine C, N, and O elemental and isotopic abundances for the AGB stars. The principal instrument has been the Fourier transform spectrometer at the KPNO 4m telescope. Other critical factors behind recent analyses include the development and distribution of model atmospheres for both O-rich and C-rich stars, and the increasing availability and accuracy of the necessary basic molecular data. My discussion concentrates exclusively on the derived abundances; the methods of abundance analysis and the accuracy of the results are not discussed for reasons of space. The reader is urged to consult the primary references for the omitted and vital discussions.

For the oxygen-rich stars (M, MS, S), we determined the C, N, O elemental abundances primarily from the CO 1.6 μm V-R, the OH V-R 1.6 μm, and the CN red system ($A^2\Pi$-$X^2\Sigma$, $\Delta v = -2$) 2 μm lines with NH V-R 3 μm lines providing a check on the N abundance. The ^{12}C/^{13}C, and ^{16}O/^{17}O/^{18}O ratios are determined from the CO lines. These model atmosphere high-resolution analyses of infrared spectra are described in detail by Smith and Lambert (1985, 1986, 1988b). For the cool carbon stars, I draw on our analysis of 30 stars (Lambert et al. 1986) in which the primary lines providing the elemental abundances were the CO 1.6 μm V-R, the C_2 Phillips ($A^1\Pi$-$X^1\Sigma^+$, $\Delta v = -2$), and the CN red system 2 μm lines - see our paper for a discussion of additional molecular transitions. The ^{12}C/^{13}C ratio was derived from the CO 1.6 and 2.3 μm and CN lines. The ^{16}O/^{17}O/^{18}O ratios were derived from the CO lines. To complete the sample of PRGs, I include the SC stars analysed by Dominy, Wallerstein, and Suntzeff (1986) and Dominy and Wallerstein (1987); note that the abundances of C and O were assumed to be equal for SC stars.

Tsuji (1986), who obtained ^{12}C abundances for 18 M giants from CO 2.3 μm lines, reports lower abundances: [C/H] = -0.65, but our recent sample of M, MS, and S stars gives [C/H] = -0.39 for M stars and -0.17 for MS and S stars. Our value for the normal M stars is in fair agreement with the C deficiency of K giants ([C/H] = -0.33, Lambert and Ries 1981). A contributing factor to the differences between our and Tsuji's C abundances may be our use of CO 1.6 μm lines and his use of CO 2.3 μm lines - see Tsuji (this conference) who reports on a systematic difference between the C abundances derived fron weak 1.6 and 2.3 μm lines. In our current sample, these abundances are quite similar except for a few stars for which the differences are up to 0.3 dex in the sense reported by Tsuji; i.e., the weak 2.3 μm lines give the lower abundance. These exceptions are stars exhibiting Tc lines and, hence, may be more luminous than the rest of the sample. Results given here are based on the CO 1.6 μm lines for the ^{12}C abundance and the ^{12}C/^{13}C ratio, and the CO 2.3 μm lines for the ^{16}O/^{17}O/^{18}O ratios.

The hypothesis that the progressive addition of ^{12}C controls the sequence M → S → C is checked in Figure 1. The M giants define a small area in the ^{12}C/^{16}O vs ^{12}C/^{13}C plane and represent the initial composition at the start of the thermal pulsing stage on the AGB. Note that the scatter of the M giants is dominated by the observational errors so that the intrinsic spread must be smaller than shown. The displacement of the M giants from their initial solar-like (^{12}C/^{16}O ~ 0.6, ^{12}C/^{13}C ~ 90) composition is in fair quantitative agreement with predictions for the first dredge-up occurring as a star ascends the giant branch prior to He-core ignition. Furthermore, the location of the M giants in Figure 1 is shared by the G and K giants (Lambert and Ries 1981; Kjaergaard et al. 1982) which have also experienced the first dredge-up but have yet to evolve to the AGB. This coincidence of G, K, and M giants is expected unless the M giants contain stars sufficiently massive to have experienced the second dredge-up (Becker and Iben 1979). The agreement is pleasing because different sets of lines and model atmospheres are employed for the two groups of stars.

With the addition of ^{12}C, the M giants would be displaced to the upper right in Figure 1. The solid lines describe the predicted change in composition as ^{12}C is added to the envelope of a M giant; the lines are labeled with the initial composition (^{12}C/^{16}O, ^{12}C/^{13}C) of the M giant. It is seen that the majority of the MS, S, SC, and C stars fall between these lines i.e., the ^{12}C/^{16}O and ^{12}C/^{13}C abundances are consistent with the hypothesis that these stars are evolved ^{12}C-enriched M giants. The ^{16}O abundances show a very similar mean value and scatter for the three principal groups: [O/H] = -0.18 ± 0.02 for the M,

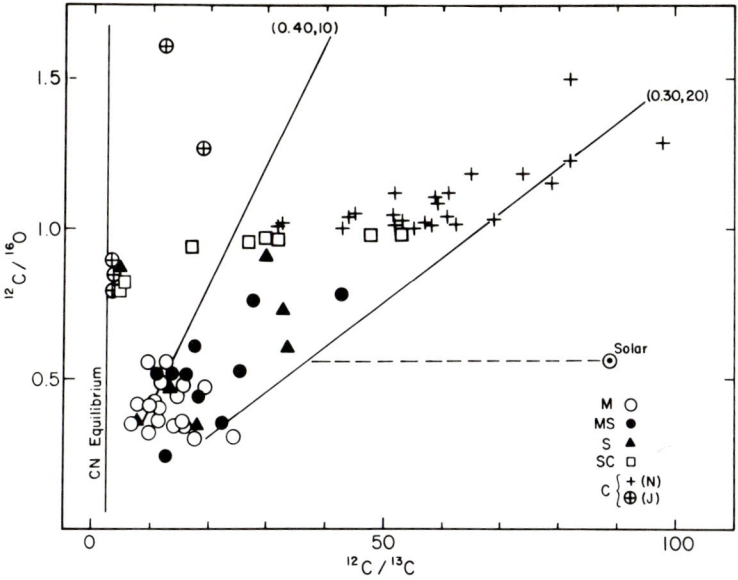

Figure 1. - $^{12}C/^{16}O$ versus $^{12}C/^{13}C$ for cool giants. Note that the majority of the MS, S, SC, and C stars fall between the solid lines that trace the change in composition as ^{12}C is added to the envelope of a M giant; the lines are labeled with the initial composition ($^{12}C/^{16}O$, $^{12}C/^{13}C$) of the M giant.

MS, and S stars and -0.22 ± 0.03 for the C stars. Hence, the differences in Figure 1 are due to ^{12}C, not to an odd combination of ^{12}C and ^{16}O changes. A few stars fall outside the loci; note especially a MS, two SC, and 3 C stars with a $^{12}C/^{13}C$ ratio close to the equilibrium value for the CN-cycle. Another obvious aspect (a surprise!) about Figure 1 is the absence of stars greatly enriched in ^{12}C; i.e., the area between the loci and above $^{12}C/^{16}O \simeq 1.2$ is unoccupied. Scalo and Miller (1985, private communication) give frequency distributions for the C/O ratio in simulated populations of AGB stars for various initial mass functions, birthrate histories, initial metallicity, and a simple representation of the dredge-up and mass loss. Although the observed distribution agrees quite well with the predicted one for the disk, this explanation would appear to be not the whole story because many C-rich PNs have C/O > 1.2. Other possible explanations include (i) the idea that carbon-rich envelopes may spawn copious amounts of graphite, which shroud the star, and hence, our sample may be biased toward low C/O ratios; and (ii) more evolved or more massive AGB stars may dredge-up some ^{16}O from the He shell and so flatten the predicted trend of $^{12}C/^{16}O$ with $^{12}C/^{13}C$.

If, as Figure 1 suggests, ^{12}C is added without significant exposure to the CN-cycle H-burning reactions, the ^{14}N abundances should be similar along the M → S → C sequence. When three very N-rich S stars are set aside, the N abundances for the M and the MS/S giants are quite similar; $^{14}N/^{16}O$ = 0.47 ± 0.06 from 18 M giants and 0.45 ± 0.06 from 12 MS or S stars. However, as noted by Lambert *et al.* (1986), the cool carbon stars appear to be N-deficient, not N-rich as expected: < [N/H] > = -0.27 ± 0.05 from a sample of 25 stars from which J-type stars are excluded. We suspect that systematic errors are responsible for this N deficiency. A possible source is the adopted CN dissociation energy and the f-values of the red system. However, since the same CN lines are used in the analysis of the M, MS, and S stars, likely adjustments owing to the cooler temperatures of the C stars appear to be small. Another candidate for the systematic error is the adopted T_{eff} scale for the C stars. If the T_{eff}'s were raised, the N deficiency could be erased. Currently, we are determining accurate excitation temperatures for CO and CN from high quality infrared spectra, and careful calibration of the T_{exc} - T_{eff} relation should provide a check on the T_{eff} scale.

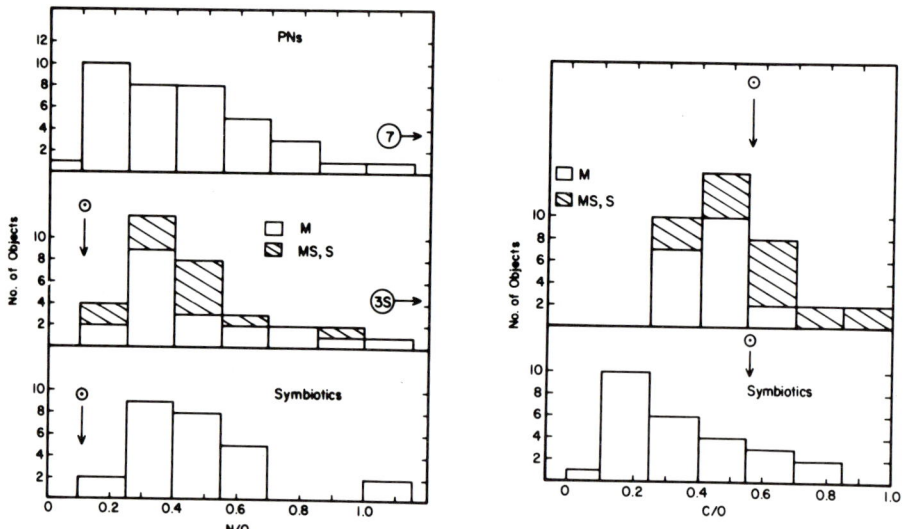

Figure 2 (left) - Distributions functions of N/O for symbiotic stars, AGB (M, MS, and S) stars, and planetary nebulae. Seven planetary nebulae and three S stars are indicated as having N/O > 1.3

Figure 3 (right). - Distribution functions of C/O for symbiotic stars and AGB (M, MS, and S) stars.

Indications that the N abundance in AGB stars is more closely represented by the M and S stars than the C stars are provided through comparison with related objects. First, the N enrichment of M and S stars is similar, as expected, to that found for the G and K giants. Second, analyses of emission lines in ultraviolet spectra of symbiotic stars (Nussbaumer *et al.* 1988) and in ultraviolet and optical spectra of planetary nebulae (PN) (Aller and

Czyzak 1983; Aller and Keyes 1987) confirm the N-enrichment - see Figure 2 for histograms compiled from the three samples. The sample of PN includes C-rich nebulae. For the three samples, the histograms are quite similar. A feature common to the MS/S and PN is the presence of very N-rich objects: N/O = 4.0 for the most N-rich S star and N/O = 3.3 for the extreme PN listed by Aller and Keyes (1987). The contrast with the C stars is especially striking when histograms are compared: the histogram (not illustrated!) for the C stars would be confined entirely to the two leftmost boxes in Figure 2. In Figure 3, I show histograms of C/O for the symbiotics and the M, MS, and S stars. Relative to the symbiotics, the stars seem to have systematically higher C/O ratios. If Tsuji's (1986) C abundances were substituted for the stars, the histograms would be quite similar. However, this substitution would result in higher stellar N abundances from the CN 2 μm lines such that the N/O histograms (Figure 2) of the stars and the symbiotics would be dissimilar. Of course, these checks on the different analytical techniques (emission lines of ions in hot gas, absorption lines of molecules in a cool photosphere) should be taken to their obvious conclusion: we shall analyse infrared spectra of the M stars in the symbiotics considered by Nussbaumer *et al.* (1988) who first drew attention to the similar C/N/O ratios!

Four S stars in the current sample are very N-rich. Three are among our most extreme S stars: [Y/Fe] ~ 0.7 to 1.3; the analysis is incomplete for the fourth star. The star with the lowest N/O ratio of the quartet shows no Tc and, therefore, we suggest that it has not yet begun its series of He-shell flashes. The other three stars have yet to be examined for Tc. Since the sum of C, N, and O exceeds the initial sum by about a factor of 2, it is likely that these stars have dredged up carbon and converted some of that carbon to nitrogen. However, the high $^{12}C/^{13}C$ (= 12, 30, and 33) ratios found in 3 stars appears at odds with this latter speculation. Perhaps we see these stars after the addition of fresh ^{12}C and before it is converted to ^{14}N in the long interval between He-shell flashes. The exception, TV Aur with $^{12}C/^{13}C = 5$, may possess the HBCE in which ^{12}C is converted to ^{14}N. These N-rich stars, which are, perhaps, galactic counterparts of the luminous S stars discovered in the Magellanic Clouds by Wood, Bessell, and Fox (1983) are likely progenitors of the N-rich PN.

3.3 *The $^{16}O/^{17}O/^{18}O$ Ratios*

Since the oxygen isotopes react with hot protons and helium nuclei, the isotopic abundances in regions of H and He burning differ from those found in the gas from the star formed (here assumed to be those of solar system material: $^{16}O/^{17}O = 2700$

and $^{16}O/^{18}O = 490$. The initial abundances endure at the surface until the star begins its first ascent of the red giant branch. Then, the convective envelope extends downward into regions which underwent hydrogen burning via the CNO cycles during the main-sequence phase. The abundance of ^{17}O is enhanced by the CNO cycle reactions, so that after mixing, its abundance at the surface will be enhanced. The abundance of ^{18}O in the interior is reduced slightly by $^{18}O(p,\alpha)^{15}N$. At its maximum extent, the convective envelope may reach layers slightly depleted in ^{16}O. The carbon isotopes behave similarly to ^{16}O and ^{17}O, with ^{12}C undergoing a rather larger depletion than ^{16}O, while ^{13}C is strongly enhanced by the CNO cycle like ^{17}O (and also ^{14}N). Characteristically, red giants in this phase of evolution exhibit $^{12}C/^{13}C$ ratios in the range 7-20 (Figure 1), $^{16}O/^{17}O$ ratios in the range 300-1000, and $^{16}O/^{18}O \approx 500$ (Harris and Lambert 1984a,b; Harris, Lambert, and Smith 1988), in approximate accordance with these expectations. Early models predicted a larger ^{18}O depletion than observed, but prediction and observation are now in good agreement because, as suggested (Harris and Lambert 1984a,b), a new measurement has provided a lower rate constant for the $^{18}O(p,\alpha)^{15}N$ reaction (Champagne and Pitt 1986). The predicted change in the $^{16}O/^{17}O$ ratio is dependent on mass. In low mass giants ($M \lesssim 1.5\ M_\odot$), the base of the convective envelope is not expected to reach the layers enhanced in ^{17}O. In higher mass stars ($M \gtrsim 3\ M_\odot$), layers enhanced in ^{17}O are mixed outwards and $^{16}O/^{17}O \sim 500$ is predicted (see Harris, Lambert, and Smith 1988). For stars of 1.5 to 3 $M_\odot$ mass, the predicted $^{16}O/^{17}O$ runs from close to the initial value (2700) to 500. The predicted changes are sensitive to the rate of the key reaction $^{17}O(p,\alpha)^{14}N$.

Following the end of core helium burning, a second dredge-up event is expected to occur in stars of above 4-5 $M_\odot$ (Iben and Renzini 1983). The convective envelope penetrates deeper down and into the hydrogen-exhausted region where the CNO cycle, burning in equilibrium, has transformed most of the original CNO nuclei in these layers into ^{14}N. Thus one effect of this second dredge-up is to enhance the ^{14}N abundance at the surface over and above the enhancement occurring after the first dredge-up. Of the oxygen isotopes, ^{16}O and ^{18}O are little affected; their surface abundances are reduced slightly by this dilution of their post-first dredge-up abundances. Since the ^{17}O abundance rises to a peak between the layer reached at the first dredge-up and the deeper layer reached at the second dredge-up, the $^{16}O/^{17}O$ ratio is expected to fall markedly after the second dredge-up. The predicted fall is sensitive to the rate adopted for $^{17}O(p,\alpha)^{15}N$. Harris, Lambert, and Smith (1985) attempted to estimate the rate by matching predicted and observed $^{16}O/^{17}O$ ratios for α Her, a star that may have undergone a second dredge-up. It was concluded that the true rate for $^{17}O(p,\alpha)$ is close to the "intermediate" rate of Fowler, Caughlan, and Zimmerman (1975).

Shortly after entry onto the AGB (which coincides with the second dredge-up in sufficiently massive stars), low- and intermediate-mass stars undergo a succession of dredge-ups resulting from thermal pulsing, which penetrate down to layers exposed to He-burning and which should be depleted of ^{17}O. It is immediately apparent that the first such dredge-up *must* bring up all the material in which the ^{17}O abundance peak lies (if the second dredge-up has not already done so). Harris, Lambert, and Smith (1985) performed a similar calculation to that above for α Her, in order to obtain a prediction for the $^{16}O/^{17}O$ following the onset of the third dredge-up process in AGB stars. ^{16}O and ^{17}O abundances were calculated for each zone of Becker and Iben's (1979) 3 $M_\odot$ ($Y = 0.28$, $Z = 0.01$) evolutionary model, estimating the third dredge-up depth from Iben and Renzini's (1983) heuristics. Using the "intermediate" rate for the reaction $^{17}O(p,\alpha)^{15}N$, as established above from α Her data, the ratio after dredge-up was estimated at $^{16}O/^{17}O \approx 280$, compared with $^{16}O/^{17}O \approx 440$ before the third dredge-up. The $^{16}O/^{18}O$ ratios around 500 left by the first dredge-up are expected to be reduced slightly by the second dredge-up. The effect of the third dredge-up depends on how far the ^{14}N left by the CNO cycle is processed to ^{22}Ne during helium burning. If the ^{14}N is only burnt as far as ^{18}O, then the ^{18}O abundance in the helium-burning region will be enhanced considerably. Stellar evolutionary model calculations suggest that in almost all cases the ^{14}N is burnt to ^{22}Ne or ^{25}Mg (Becker and Iben 1979; Boothroyd and Sackmann 1988). If this is the case, no ^{18}O will be brought up by the third dredge-up, and the $^{16}O/^{18}O$ ratio will fall from ~500 to ~600-700 as the ^{18}O in the envelope is diluted by the ^{18}O-free material brought up. If the reaction sequence upon ^{14}N proceeds no further than ^{18}O, a ratio $^{16}O/^{18}O \approx 100$ is possible.

Unfortunately, predictions of ^{17}O and ^{18}O for AGB models of stars of lower mass (say 2 $M_\odot$) are not available; however, the $^{16}O/^{17}O$ ratio is expected to be somewhere between the 3 $M_\odot$ prediction and the maximum pre-third dredge-up ratio, i.e., $280 \leq {}^{16}O/^{17}O \leq 650$. Since ^{17}O is not synthesized during the shell flashes, the material brought up by any subsequent dredge-up can only dilute the ^{17}O content of the envelope and increase the $^{16}O/^{17}O$ ratio. However, since the amount of material brought up is very small, this dilution effect should be very small. Unless destruction of ^{14}N is halted at ^{18}O, the $^{16}O/^{18}O$ ratio is not expected to depart significantly from the pre-third dredge-up value of ~600).

The ^{17}O and ^{18}O isotopic abundances are obtainable for AGB stars from the 2.3 μm CO lines. In the interval 4265-4295 cm^{-1}, several 2-0 R-branch $^{12}C^{17}O$ lines are readily

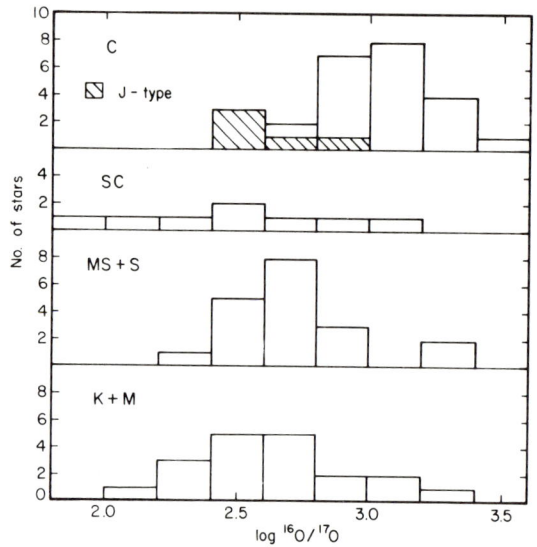

Figure 4. - Distribution functions of $^{16}O/^{17}O$ for samples of K and M, MS and S, SC, and C stars.

detectable. Most lines are blended with CN red system lines, and overlying telluric lines must be ratioed out. There is not a comparably suitable region of $^{12}C^{18}O$ lines. Our estimates of the $^{16}O/^{18}O$ ratio are based on syntheses of the region around the 2-0 $^{12}C^{18}O$ R-branch bandhead. A reliable determination of the $^{16}O/^{17}O$ or $^{16}O/^{18}O$ ratio requires that $^{12}C^{16}O$ lines be matched approximately in strength with the $^{12}C^{17}O$ and $^{12}C^{18}O$ lines. This requirement is met with varying degrees of success through use of high excitation 2-0 R-branch lines in the $^{12}C^{17}O$ window.

The present discussion of AGB stars draws on isotopic ratios from Dominy, Wallerstein, and Suntzeff (1986) and Dominy and Wallerstein (1987) for SC stars, Smith and Lambert (1988b) for M, MS, and S stars, and Harris et al. (1987) for C stars. A sample of K and M giants is drawn from Harris, Lambert, and Smith (1988). Oxygen isotopic ratios for these various samples are compared in Figures 4 and 5. Interpretation of the histograms of $^{16}O/^{17}O$ determinations (Figure 4) is not a simple matter. Selection effects affect all the histograms. In addition, the $^{16}O/^{17}O$ ratio for the K and M giants affected by the first (and possibly the second) dredge-up is expected to depend on stellar mass. The most massive stars in this sample presumably do not become S and C stars. The less massive stars, which are left with a high ($\simeq$ initial) $^{16}O/^{17}O$ ratio following the first dredge-up, are predicted to show a lower ratio as AGB stars. An additional complication is that the MS, S, and C samples likely contain evolved Barium stars in which the $^{16}O/^{17}O$ of the former AGB star (the mass donor in the binary) has been diluted with the ^{17}O-poor material of the companion (the mass gainer and the present giant star) and then the $^{16}O/^{17}O$ of the mixture may have been further changed as the first dredge-up occurred in the Barium star. Perhaps one should draw just a single conclusion from Figure 4: the C stars (excluding the J-type) do appear to be deficient in ^{17}O relative to the MS, S, and SC stars. I hesitate to insist that this is real difference - see below.

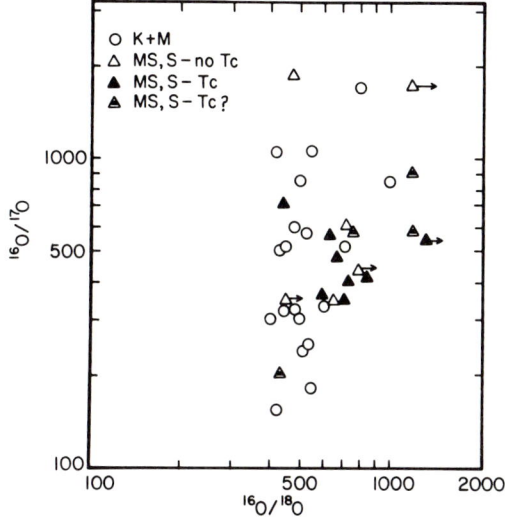

Figure 5. - $^{16}O/^{17}O$ vs $^{16}O/^{18}O$ for K, M, MS, and S stars.

In Figure 5, $^{16}O/^{17}O$ and $^{16}O/^{18}O$ ratios are compared for the samples of K, M, MS, and S stars. ^{18}O abundances are not available for the SC stars in Figure 4. The K and M stars span the predicted range in $^{16}O/^{17}O$ and, with few exceptions, show, as predicted, no significant depletion of ^{18}O below the presumed initial ($\simeq$ solar) ratio of $^{16}O/^{18}O = 490$. The exceptions, if real departures from the norm, may arise from mass loss or extensive mixing internal prior to the first or second dredge-up. On average, $^{16}O/^{18}O$ is reduced in the MS and S stars. In Figure 5, I distinguish stars with Tc from those without Tc; a search for Tc has yet to be attempted for four of the sample. The stars with Tc appear to be clustered near $^{16}O/^{17}O \sim 500$ and cover a range in $^{16}O/^{18}O$. Of the SC stars considered by Dominy, Wallerstein, and Suntzeff (1987) and Dominy and Wallerstein (1987), the two known to have Tc have $^{16}O/^{17}O \simeq 350$ and 200. The present range in the $^{16}O/^{17}O$ ratio for the Tc stars which we identify as stars undergoing the third dredge-up is consistent with the approximate predictions discussed above, i.e., $280 \leq {}^{16}O/{}^{17}O \leq 650$. However, this tentative conclusion must be checked using a larger sample of MS and S stars. The observed ${}^{16}O/{}^{18}O$ ratios appear, in some cases, to be outside the predicted range of ${}^{16}O/{}^{18}O \sim 500$ to 700. However, when the approximate nature of the present predictions and the difficulty of extracting the ${}^{18}O$ abundance are considered, this discrepancy should not be overemphasized. Results for the MS and S stars without Tc should be considered in light of the mass transfer hypothesis; higher than typical ${}^{16}O/{}^{17}O$ (and ${}^{16}O/{}^{18}O$) may be achievable by mixing of material from a S star with unprocessed material of a normal star. The two possibly Barium-like giants in our recent sample (Smith and Lambert 1988b) with a high ${}^{16}O/{}^{17}O$ ratio are HD 35155, the "neodymium" star (see below), and HD 96360. The former is also ${}^{18}O$ deficient: ${}^{16}O/{}^{18}O > 1200$. The sample of SC stars also includes one (BD+10° 3764) with a high ${}^{16}O/{}^{17}O$ (= 1200) ratio; its status with respect to Tc is undetermined. When the J types are excluded, the carbon stars (Harris *et al.* 1987) define a fairly narrow inclined strip running from (${}^{16}O/{}^{17}O$, ${}^{16}O/{}^{18}O$) $\simeq$ (600, 900) to (2500, 2000). The shrouded carbon star IRC+10216 has ${}^{16}O/{}^{17}O \simeq 650$ and ${}^{16}O/{}^{18}O \simeq 875$ (Keady, Hall, and Ridgway 1988) according to analysis of the CO absorption lines produced in the circumstellar shell. This (typical?) carbon star with ${}^{12}C/{}^{13}C \sim 35$ falls among the S stars in Figure 5 and at one end of the distribution of cool carbon stars.

The extreme ratios of ${}^{16}O/{}^{17}O \sim {}^{16}O/{}^{18}O \sim 1000$ to 2000 suggested for the cool carbon stars are difficult to explain in terms of evolution on the AGB - see Harris, Lambert, and Smith (1985, 1988) and Harris *et al.* (1988) for a discussion. Indeed, I would term the high ratios "a surprise" - see Sec. 3.1. Dominy and Wallerstein's (1987) reanalyses of two S stars gave much lower ratios than our previously obtained high ratios. Subsequently we have obtained new higher-quality spectra of several stars and analysed the ${}^{12}C{}^{17}O$ and ${}^{12}C{}^{18}O$ lines with respect to both ${}^{12}C{}^{16}O$ high excitation 2.3 μm lines and second overtone lines at 1.6 μm. Several stars have now been analysed independently. Table 1 shows that consistent results can be achieved. Our "new and improved" results as well as those of several additional MS and S stars show that the ${}^{16}O/{}^{17}O$ and ${}^{16}O/{}^{18}O$ ratios span a narrower range than we previously suggested; the previous extreme was ${}^{16}O/{}^{17}O = 3000$ (-1200, + 2500) for HR 363 (a star not yet reanalysed) and the current extreme is ${}^{16}O/{}^{17}O = 1900 \pm 600$ for HD 96360, an S star without Tc. This narrower range is probably within predicted limits. In light of this experience, we are presently reanalysing some of the C stars using higher quality spectra.

TABLE 1

Comparison of $^{16}O/^{17}O$ Ratios

Star	Type	$^{16}O/^{17}O$	Reference
α Her	M5II	190 ± 80	Tsuji (1985)
		200 ± 25^a	Harris and Lambert (1984b)
		$180^{-50^b}_{+70}$	
HR 1105	S3.5	500^{+300}_{-100}	Dominy and Wallerstein (1987)
		$350^{+150^c}_{-100}$	Smith and Lambert (1988b)
HR 8062	S4	930^{+400}_{-200}	Dominy and Wallerstein (1987)
		$500^{+200^d}_{-150}$	Smith and Lambert (1988b)
WZ Cas	SC7	465^{+150}_{-100}	Dominy and Wallerstein (1987)
		400^{+175}_{-100}	Harris et al. (1987)

[a] Based on the 2.3 μm CO lines.
[b] Based on the fundamental 5 μm CO lines.
[c] Previous results were $^{16}O/^{17}O$ = 2250 (+700, -550) by Harris, Lambert, and Smith (1985), and 600 (+200, -150) by Harris et al. (1987).
[d] Previous result was $^{16}O/^{17}O$ = 1850 (+550, -425) by Harris, Lambert, and Smith (1985).

3.4 The s-process

This discussion of the s-process enrichment of MS and S stars draws on our recent analyses (Smith and Lambert 1985, 1986, 1988b) in which intervals in the near-infrared (7400-7580 Å, 9980-10100 Å) were observed at a resolution of 0.2 Å. In all except the coolest stars, these intervals appear to be free from molecular line blanketing. Available atomic lines provide a useful selection of elemental abundances (e.g., Ti, Fe, Ni, Sr, Y, Zr, Ba, and Nd). The iron-peak abundances generally fall, as expected, in the range [Fe/H] = -0.2 to +0.2. Of the s-process elements, Y, Zr, and Nd are the most reliably determined. This trio provides a measure of the overall overabundance of s-process elements and the level of exposure to neutrons.

The overabundance as expressed by the mean of [Y/Fe] and [Nd/Fe] is correlated with the carbon enrichment ([^{12}C/Fe]) - see Figure 6. This correlation suggests that carbon and s-process synthesis are intimately related as is required by models of thermally pulsing AGB stars. An evolutionary relation between M, MS, S, and C stars is suggested by the fact that, in Figure 6, the MS and S stars connect the M to the N-type C stars. By contrast, the J-type (^{13}C-rich) C stars and the warm or R-type carbon stars are carbon rich but not significantly s-process enriched. A calibration of the "abundance" index appended to the spectral type of MS and S stars - see Sec. 2.1 - could now be provided from Figure 6.

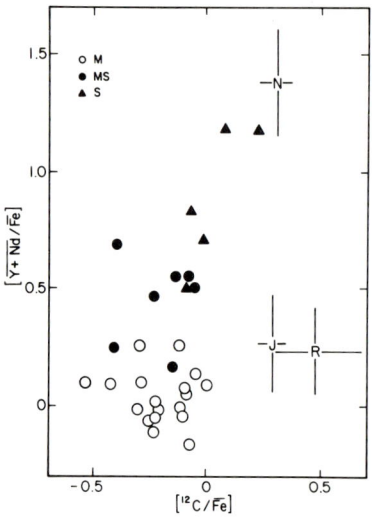

Figure 6. - The s-process overabundance [$\overline{Y+Nd}$/Fe] = <[Y/Fe] + [Nd/Fe}> versus the carbon overabundance [^{12}C/Fe]. Mean points for the N-type and J-type C stars are obtained by combining carbon abundances from Lambert et al. (1986) with s-process abundances from Utsumi (1985). The point for the R stars is obtained from Dominy (1984).

The C/O ratios for S stars span a wide range and are not clustered between C/O = 0.95 to 1.0 as some interpretations of the ZrO/TiO band intensities have suggested (see Sec. 2.1). The lower C/O ratios obtained by us apparently reflect the fact that (i) our sample does not include "pure" S (or SC) stars for which a C/O nearer unity is expected, and (ii) the existing calibrations of the "abundance" index defined by the ZrO/TiO bands place inappropriate stress on the role of the O-C abundance difference and ignore the role of the Zr/Ti abundance ratio. It should be added that our C/O ratios have not been determined from a self-consistent analysis; the derived C/O ratio differs from the ratio assumed in the construction of the model atmosphere. Elimination of this inconsistency is likely to increase our C/O ratios by a small amount.

Comparison of the heavy (e.g., Nd) and light (e.g., Zr) s-process elements provides an estimate of the total exposure to neutrons; high exposure leads to a high Nd/Zr ratio. As is common practice, I assume that the s-processed material mixed to the surface has been exposed to an exponential distribution of exposures $\rho(\tau) \sim \exp(-\tau/\tau_0)$ where the exposure $\tau = \int \Phi(t)dt$, $\Phi(t)$ is the neutron flux, and τ_0 is the average exposure. Malaney (1988) has estimated the elemental s-process abundances for a range of τ_0 (unit: mb^{-1} at kT = 30 keV). The predicted surface abundances depend on τ_0 and on the relative masses of the envelope (M_e) and the s-processed material (M_s). Predicted average enhancements of Y and Zr as a function of the heavy (Nd) to light (Y and Zr) s-process abundance ratio are shown in Figure 7 for a ratio M_s/M_e up to about 0.01 and a simple two-zone model. Comparison of the observations and predictions shows that $\tau_0 \sim 0.3$ accounts for the great majority of the MS and S stars; one star (HD 35155) is remarkably Nd-rich. If the origin for the predictions is displaced to the center of gravity of the M star distribution, the fit of

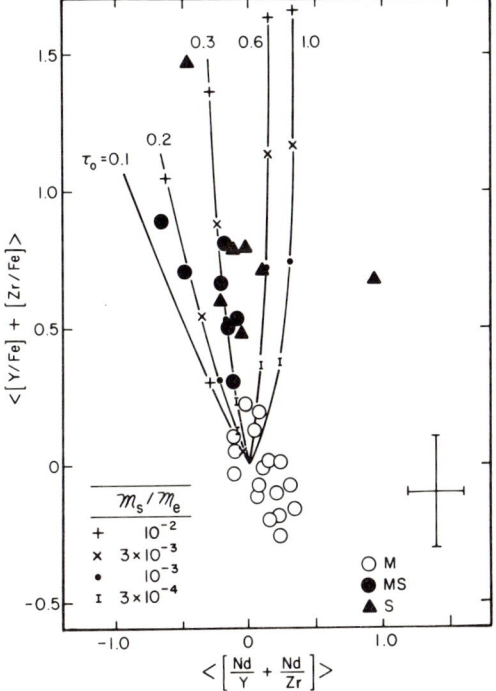

Figure 7. - The s-process overabundance < [Y/Fe] + [Zr/Fe] > versus < [Nd/Y] + [Nd/Zr] >. Predictions are given for the 2-zone mixing described in the text.

the $\tau_o = 0.3$ curve to the observations is improved. The derived mixing fraction is $M_s/M_e \lesssim 0.003$ with one star (TV Aur, a N-rich S star) requiring $M_s/M_e \simeq 0.01$. Cool carbon stars according to Utsumi's (1985) abundance analysis fall in Figure 7 at a mean location (-0.6, +1.7) which is about midway between the $\tau_o = 0.2$ and 0.3 predictions for $M_s/M_e \sim 0.02$. In light of the large uncertainty (± 0.4 dex) assigned to the abundances [s/Ti], this mean location derived using all four of the heavy s-elements (Ba, La, Nd, Sm) considered by Utsumi is approximately consistent with the suggestion that the $\tau_o = 0.3$ prediction fits the s-process enhancements of MS and S stars.

One very important conclusion is suggested by the discovery that the observed s-process enrichments correspond to synthesis at $\tau_o \simeq 0.3$ at $kT = 30$ keV. This estimate is identical to that derived from analysis of the solar system composition in the mass range $A = 90$ to 200: $\tau_o = 0.30 \pm 0.01$ at $kT = 30$ keV (Walter et al. 1986; Beer 1988). We suggest that the equal τ_o's, rather than being a chance coincidence, indicate that the solar s-process nuclides were synthesized in thermally pulsing AGB stars like the M $\rightarrow$ S $\rightarrow$ C stars discussed here. Several authors have shown that such or slightly more evolved AGB stars return sufficient mass to the interstellar medium to be major contributors to its enrichment in C and s-process elements (Jura - this book). Since the observed s-process enrichments along the sequence M $\rightarrow$ S $\rightarrow$ C appear to correspond to $\tau_o \sim 0.3$ over the entire sequence, it seems unlikely that further evolution prior to the onset of severe mass loss will change τ_o and so lead to the return to the interstellar medium of a non-solar mix of s-process elements.

By combining M_s/M_e with the observed ^{12}C enhancements, one may estimate the mass fraction of ^{12}C in the s-processed material: $Z(^{12}C) \sim 0.5$ is a *preliminary* estimate from the simple two-zone model. This is satisfactorily close to the predicted $Z(^{12}C) \sim 0.25$ for shell flashes in low-mass solar-metallicity stars (Boothroyd and Sackmann 1988). The He/H ratio at the surface is predicted to increase by no more than 1% of its initial value. Such a negligible increase is consistent with the observation that the vast majority of planetary nebulae are not enriched in He: $\log \varepsilon(He) = 11.06 \pm 0.02$ from 39 nebulae (Aller and Keyes 1987 - N-rich PBs excluded), and $\log \varepsilon(He) = 10.99 \pm 0.04$ for the Sun (Anders and Grevesse 1988). The question of a severe He-deficiency in cool carbon stars was addressed and dismissed by Lambert et al. (1986).

Analysis of the solar system abundances yields estimates of the neutron density and the temperature at the s-process site: $n(n) \simeq 1.3 \times 10^8$ cm^{-3} and $T \simeq 2.7 \; 10^8$ K (Beer 1988). A demonstration that similar conditions occur in AGB stars would strengthen our conclusion that these stars are major producers of s-process elements. The neutron density might be estimated from the ratio of the Rb abundance to that of neighboring elements Sr, Y, and Zr. This method, which exploits the branch in the s-process path at ^{85}Kr, has been applied to Barium K giants (Tomkin and Lambert 1983; Smith and Lambert 1984; Malaney and Lambert 1988): $2 \times 10^7 \lesssim n(n) \lesssim 5 \times 10^7$ cm^{-1} for 4 Barium giants. We are currently investigating whether the Rb I 7800 and 7947 Å lines are sufficiently free of blends to be useful in MS and S stars. A branch at the unstable isotope ^{95}Zr provides a second densitometer: if $n(n) \gtrsim 10^{10}$ cm^{-3}, stable ^{96}Zr is produced. The Zr isotopic abundances are obtainable from analysis of ZrO bands; a current attempt to derive the Zr isotopic ratios is summarized by Smith (1988). There is no evidence in our survey of about 20 S stars, mostly long-period variables, for an appreciable abundance of ^{96}Zr: $n(n) < 10^{10}$ cm^{-3}. As discussed by Malaney and Lambert (1988), the low neutron densities derived from the Barium giants appear to be considerably less than current predictions ($n(n) \sim 10^9$-10^{12} cm^{-3}) for the He-burning shell in AGB stars. One solution to this discrepancy is to discard the mass transfer hypothesis as the origin of the Barium giants. An alternative solution is to search for modifications to AGB models that reduce the neutron density in the He-burning shell. Gallino *et al.* (1988) have suggested that the s-process may operate in the He-burning shell of metal-poor low-mass stars at lower neutron densities. A successful extraction of Rb abundances in S stars could test this kind of prediction.

The elemental and isotopic abundances obtainable for AGB stars do not provide a thermometer for the s-process site. The temperature, $T \simeq 2.7 \; 10^8$ K, which is estimated from the solar system abundances seems inconsistent with that predicted for the active He-shell in low mass AGB stars (Boothroyd and Sackmann 1988). However, Gallino *et al.* (1988) remark that, according to their analysis of thermal pulses in a metal-poor low-mass star, the isotopes that serve as thermometers are synthesized during a short hot phase by a weak neutron pulse driven by the ^{22}Ne(α,n) reaction following the major neutron release from ^{13}C(α,n). At the rather higher temperature achieved in the shells of intermediate mass stars, the ^{22}Ne(α,n)^{25}Mg reaction replaces ^{13}C(α,n)^{16}O as the primary neutron supplier. A signature of s-processing driven by the ^{22}Ne source is an observable distortion in the Mg isotopic ratios (Truran and Iben 1977; Scalo 1978; Malaney 1987). In the Barium K, MS and S stars examined to date, the Mg isotopes have their normal ($\approx$ solar) abundance ratios - see Tomkin and Lambert (1979), McWilliam and Lambert (1988), Malaney and Lambert (1988), and Smith and Lambert (1986). In short, the signature of $^{22}Ne(\alpha,n)$ has yet to be recognized. By default, $^{13}C(\alpha,n)$ is presumed to be the neutron source.

Of several chronometers provided by the s-process, technetium is the most striking - see Little-Marenin's article in this book. ^{93}Zr with a half-life of 1.5×10^6yr is a longer-lived chronometer. In Sec. 3.1, I mentioned Dominy and Wallerstein's (1986) discovery that o Cet is Nb deficient yet contains Tc. This reduction of Nb is attributed to the recent ($< 10^5$ yr) exposure to a mild pulse of neutrons. Further investigations of abundances tied to such chronometers is likely to yield surprises.

The last few years have been exciting ones for students of PRGs. My intention in preparing this review was to illustrate the stimulating interplay between the application of "instant abundance analysis" to stars in the Magellanic Clouds and the interpretations of the comprehensive abundance analyses that are now possible for PRGs. As a result of this interplay, we may make the modest claim to understand the M $\rightarrow$ S $\rightarrow$ C AGB stars as victims of the third dredge-up in which the products of He-shell flashes are mixed into the envelope. But much observational and theoretical work remains to be done!

Many of the results discussed in this review are the fruits of collaborations with several colleagues; I thank Drs. K. Eriksson, B. Gustafsson, K. H. Hinkle, R. A. Malaney, and V. V. Smith for their major contributions. My research is supported in part by the U. S. National Science Foundation (grant AST86-14423) and the R. A. Welch Foundation of Houston, Texas.

REFERENCES

Ake, T. B. 1979, *Ap. J.*, **234**, 538.
Aller, L. H., and Czyzak, S. J. 1983, *Ap. J. Suppl.*, **51**, 211.
Aller, L. H., and Keyes, C. D. 1987, *Ap. J. Suppl.*, **65**, 405.
Anders, E., and Grevesse, N. 1988, *Geochim. Cosmochim. Acta*, in press.
Becker, S. A., and Iben, I., Jr., 1979, *Ap. J.*, **232**, 831.
Beer, H. 1988, in *Origin and Distribution of the Elements*, ed. G. J. Mathews, (Singapore: World Scientific), p. 505.
Bessell, M. S., Wood, R. P, and Lloyd Evans, T. 1983, *M.N.R.A.S.*, **202**, 59.
Boesgaard, A. M. 1970, *Ap. J.*, **161**, 163.
Boothroyd, A. I., and Sackmann, I.-J. 1988, *Ap. J.*, **328**, 653.
Busso, M., Picchio, G., Gallino, R., and Chieffi, A. 1988, *Ap. J.*, **326**, 196.
Champagne, A. E., and Pitt, M. L. 1986, *Nucl. Phys. A.*, **457**, 367.
Cohen, J. G., Frogel, J. A., Persson, S. E., and Elias, J. H. 1981, *Ap. J.*, **249**, 481.
Cosner, K., and Truran, J. W. 1981, *Ap. Space Sci.*, **78**, 85.

Dominy, J. F. 1984, *Ap. J. Suppl.*, **55**, 27.
Dominy, J. F., and Wallerstein, G. 1986, *Ap. J.*, **310**, 371.
_____. 1987, *Ap. J.*, **317**, 810.
Dominy, J. F., Wallerstein, G., and Suntzeff, N. B. 1986, *Ap. J.*, **330**, 325.
Fowler, W. A., Caughlan, G. R., and Zimmermann, B. A. 1975, *Ann. Rev. Astr. Ap.*, **13**, 69.
Frogel, J. A., and Blanco, V. M. 1984, in *Observational Tests of the Stellar Evolution Theory*, ed. A. Maeder and A. Renzini (Dordrecht: Reidel), p. 175.
Gallino, R., Busso, M., Picchio, G., Raiteri, C. M., and Renzini, A. 1988, preprint.
Harris, M. J., and Lambert, D. L. 1984a, *Ap. J.*, **281**, 739.
_____. 1984b, *Ap. J.*, **285**, 674.
Harris, M. J., Lambert, D. L., Hinkle, K. H., Gustafsson, B., and Eriksson, K. 1987, *Ap. J.*, **316**, 294.
Harris, M. J., Lambert, D. L., and Smith, V. V. 1985, *Ap. J.*, **299**, 375.
_____. 1988, *Ap. J.*, **325**, 768.
Iben, I., Jr., 1975, *Ap. J.*, **196**, 525.
_____. 1976, *Ap. J.*, **208**, 165.
_____. 1977, *Ap. J.*, **217**, 788.
_____. 1981, *Ap. J.*, **246**, 278.
_____. 1988, in *Astronomy in the Southern Hemisphere*, ed. V. M. Blanco and M. Phillips, in press.
Iben, I., Jr., and Renzini, A. 1983, *Ann. Rev. Astr. Ap.*, **21**, 271.
Iben, I., Jr., and Truran, J. W. 1978, *Ap. J.*, **220**, 980.
Jorissen, A., and Mayor, M. 1988, *Astr. Ap.*, **198**, 187.
Keady, J. J., Hall, D. N. B., and Ridgway, S. T. 1988, *Ap. J.*, in press.
Keenan, P. C. 1954, *Ap. J.*, **120**, 484.
Keenan, P. C., and Boeshaar, P. C. 1980, *Ap. J. Suppl.*, **43**, 379.
Kjaergaard, P., Gustafsson, B., Walker, G. A. H., and Hultqvist, L. 1982, *Astr. Ap.*, **115**, 145.
Lambert, D. L. 1985, in *Cool Stars with Excesss of Heavy Elements*, ed. M. Jaschek and P. C. Keenan (Dordrecht: Reidel).
_____. 1988, in *The Impact of Very High S/N Spectroscopy on Stellar Physics*, ed. G. Cayrel de Strobel and M. Spite (Dordrecht: Kluwer), p. 563.
Lambert, D. L., Eriksson, K., Gustafsson, B., and Hinkle, K. H. 1986, *Ap. J. Suppl.*, **62**, 373.
Lambert, D. L., and Ries, L. M. 1981, *Ap. J.*, **248**, 228.
Little, S. J., Little-Marenin, I. R., and Bauer, W. H. 1987, *A. J.*, **94**, 981.
Little-Marenin, I. R., and Little S. J. 1979, *A. J.*, **84**, 1374.
Lloyd Evans, T. 1983, *M.N.R.A.S.*, **204**, 985.
_____. 1984, *M.N.R.A.S.*, **208**, 447.
_____. 1985, in *Cool Stars with Excesses of Heavy elements*, ed. M. Jaschek and P. C. Keenan (Dordrecht: Reidel), p. 163.
Malaney, R. A. 1987, *Ap. J.*, **321**, 832.
_____. 1988, *Ap. Space Sci.*, **137**, 251.
Malaney, R. A., and Lambert, D. L. 1988, *M.N.R.A.S.*, in press.
Mathews, G. J., Takahashi, K., Ward, R. A., and Howard, W. M. 1986, *Ap. J.*, **302**, 410.
McClure, R. D. 1983, *Ap. J.*, **268**, 264.
_____. 1985, in *Cool Stars with Excesses of Heavy Elements*, ed. M. Jaschek and P. C. Keenan (Dordrecht: Reidel), p. 315.
McClure, R. D., Fletcher, J. M., and Nemec, J. M. 1980, *Ap. J. (Letters)*, **238**, L35.
McWilliam, A., and Lambert, D. L. 1988, *M.N.R.A.S.*, **230**, 573.
Merrill, P. W. 1952, *Ap. J.*, **116**, 21.
Nussbaumer, H., Schild, H., Schmid, H. M., and Vogel, M. 1988, *Astr. Ap.*, **198**, 179.

Pedley, J. B., and Marshall, R. M. 1983, *J. Phys. Chem. Ref. Data,* **12**, 967.
Piccirillo, J. 1980, *M.N.R.A.S.,* **190**, 441.
Renzini, A., and Voli, M. 1981, *Astr. Ap.,* **94**, 175.
Richer, H. B. 1981, *Ap. J.,* **243**, 744.
Schatz, G. 1983, *Astr. Ap.,* **122**, 327.
Scalo, J. M. 1974, *Ap. J.,* **194**, 361.
———. 1976, *Ap. J.,* **206**, 474.
———. 1978, *Ap. J.,* **221**, 627.
Scalo, J. M., Despain, K. H., and Ulrich, R. K. 1975, *Ap. J.,* **196**, 809.
Scalo, J. M., and Miller, G. E. 1981, *Ap. J.,* **246**, 251.
Scalo, J. M., and Ross, J. E. 1976, *Astr. Ap.* **48**, 219.
Smith, V. V. 1988, in *Origin and Distribution of the Elements,* ed. G. J. Mathews (Singapore: World Scientific), p. 535.
Smith, V. V., and Lambert, D. L. 1984, *Pub. A. S. P.,* **96**, 226.
———. 1985, *Ap. J.,* **294**, 326.
———. 1986, *Ap. J.,* **311**, 843.
———. 1987, *M.N.R.A.S.,* **226**, 563.
———, 1988a, *Ap. J.,* in press.
———. 1988b, in preparation.
Stephenson, C. B. 1973, *Pub. Warner and Swasey Obs.,* **1**, No. 4.
Tomkin, J., and Lambert, D. L. 1979, *Ap. J.,* **227**, 209.
———. 1983, *Ap. J.,* **273**, 722.
Truran, J. W., and Iben, I., Jr., 1977, *Ap. J.,* **216**, 797.
Tsuji, T. 1985, in *Cool Stars with Excesses of Heavy Elements,* ed. M. Jaschek and P. C. Keenan (Dordrecht: Reidel), p. 295.
———. 1986, *Astr. Ap.,* **158**, 8.
Utsumi, K. 1985, in *Cool Stars with Excesses of Heavy Elements,* ed. M. Jaschek and P. C. Keenan (Dordrecht: Reidel), p. 243.
Walter, G., Beer, H., Käppeler, F., Reffo, G., and Fabbri, F. 1986, *Astr. Ap.,* **167**, 186.
Wood, P. R. 1987, in *Late Stages of Stellar Evolution,* ed. S. Kwok and S. R. Pottasch (Dordrecht: Reidel), p. 197.
Wood, P. R., Bessell, M. S., and Fox, M. W. 1983, *Ap. J.,* **272**, 99.
Wurm, K. 1940, *Ap. J.,* **162**, 203.

THE ROLE OF TECHNETIUM IN THE EVOLUTION OF RED GIANTS

Irene R. LITTLE-MARENIN
Wellesley College, Wellesley, MA, USA 02181

ABSTRACT Tc is detected in many AGB stars providing unambiguous proof that recent nuclear s-processing and mixing (the third dredge-up) has taken place. During this evolutionary episode the atmospheres of AGB stars are progressively enhanced with helium burning products (primarily ^{12}C) and s-process elements as they evolve from M->MS->S->SC->C stars. The increase in s-process elements can be traced most easily by the presence and increasing strength of the Tc I lines accompanying this progression. We also find that the third dredge-up phase is accompanied by an increase in the amplitude of light variation since no non-variable or low amplitude variable M, MS, SC, or S (with one exception) have Tc lines. M star Mira variables show Tc if $P > 300^d$ (low mass Pop I stars). No Pop II star is known to have Tc. Nor do supergiants show Tc I lines. The significant fraction of MS, S and C stars that do not show Tc, are surmised to be cooler analogues to the Ba II stars, i.e. binaries. The source to provide the neutrons for the s-process is most likely the $^{13}C(\alpha,n)^{16}O$ reaction since most of the stars in which we observe Tc are thought to have masses less than 3 solar masses.

I. INTRODUCTION

The presence of Tc in the atmospheres of late type stars is an unambiguous tracer of recent s-process nuclear reactions in stellar interiors and subsequent outward mixing since all the isotopes of Tc are radioactive with half-lives much shorter than the lifetimes of stars. The longest lived isotope, ^{98}Tc, has a half-life of 4.2×10^6 years, however, the only isotope produced by the s-process, ^{99}Tc, has a half-life of 2.1×10^5 years. Both the s(slow) neutron capture and the r(rapid) neutron capture process are able to produce elements heavier than Fe by the addition of neutrons to the relatively abundant iron peak elements (Fe, Co, Ni). The r-process requires a large neutron flux (as for instance during a supernova explosion) so that successive neutron captures occur on the order of seconds to minutes; so fast that many radioactive daughter products do not have time to decay before the next neutron is captured. On the other hand, the s-process occurs under low neutron flux conditions. These low flux conditions can be found in the intershell region of thermally pulsing asymptotic giant branch stars, hereafter TP-AGB stars. The flux is so low that successive neutron captures occur on the order of days, allowing radioactive daughter products to decay before the next neutron capture takes place. The s-process proceeds along the neutron-rich edge of the valley of nuclear stability enhancing some isotopes relative to others. The process proceeds through the Tc region by a complex network of neutron captures and beta decays (Fig. 1) (Mathews et al. 1986) from ^{90}Zr through neutron captures and beta decays to Nb (where the only stable isotope ^{93}Nb is populated by the decay of ^{93}Zr) to Mo. Successive neutron captures on the stable isotopes of Mo, $^{95-98}$Mo, produce ^{99}Mo which decays with a half-life of 6^h to ^{99}Tc

$$^{98}Mo(n,\gamma)^{99}Mo(\beta^- \nu)^{99}Tc(n,\gamma)^{100}Tc(\beta^- \nu)^{99}Ru.$$

The other two long-lived isotopes, ^{98}Tc and ^{97}Tc, are by-passed by the

Path of s-process through Tc Region

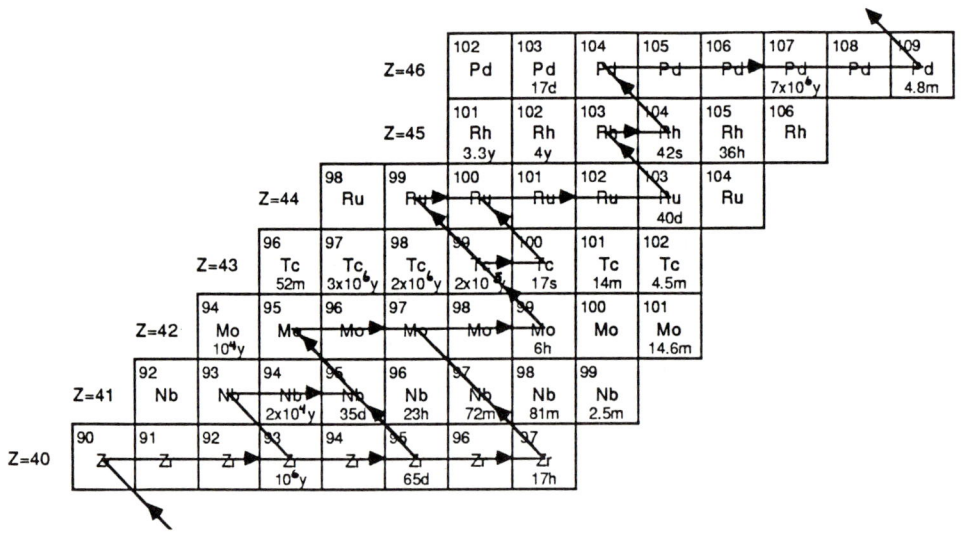

Figure 1 illustrates the s-process path in the Tc regions. The half-lives are listed for the radioactive isotopes of each element.

s-process. The isotope ^{98}Tc is shielded from beta and inverse beta decay and hence should only occur in very small quantities, and ^{97}Tc with a half-life of 2×10^6 years, can only be produced by the p-process (whose mechanism and site are not well understood) when ^{97}Ru decays by an inverse beta process. The isotope ^{99}Tc can be produced both by the s- and the r-process. Since stars with Tc give no indication of a recent violent event it is assumed that the isotope of Tc that is observed in TP-AGB stars is ^{99}Tc. With time, neutron capture and beta decay transforms ^{99}Tc into Ru. The enhancements of the various s-process elements depend on the details of the s-process path through this region and on the integrated neutron flux. Hence, observed abundance ratios can be used to determine the neutron exposures in a given star (Smith and Wallerstein 1983; Smith and Lambert 1985; Beer and Walter 1985; Dominy and Wallerstein 1986; Smith, Lambert and McWilliam 1987).

In order for ^{99}Tc to be produced and to be observable in the outer layers, one needs both a neutron source and a mixing mechanism. Two neutron sources have been postulated a) the ^{22}Ne$(\alpha,n)^{25}$Mg source and b) the ^{13}C$(\alpha,n)^{16}$O source. The conditions under which these two sources operate are very different. The ^{22}Ne source comes into play during helium burning when successive α-particle captures on ^{14}N lead to the production of ^{22}Ne and free neutrons

$$^{14}N(\alpha,\gamma)^{18}F(\beta^- \nu)^{18}O(\alpha,\gamma)^{22}Ne(\alpha,n)^{25}Mg.$$

In order to be efficient the process requires temperatures of 2-3 x 10^8 K, the presence of α-particles and abundant ^{14}N. The released neutrons are then available to power the s-process. These conditions are found in the intershell region between the hydrogen and helium burning shells of intermediate mass stars (> 3 solar masses). In these double shell source stars a degenerate carbon-oxygen core is surrounded by a helium burning shell, a small intershell region (which contains mostly He) and a hydrogen burning shell in which most of the ^{12}C and ^{16}O is burned into ^{14}N by the CNO cycle. Periodic runaway nuclear reactions, called helium shell flashes, occur in the helium burning shell which can lead to mixing and dredge-up of ^{12}C and s-process elements after several helium shell flashing episodes and the eventual birth of a carbon star (Iben 1987). Earlier theoretical evolutionary models by Iben and collaborators and many others (see Iben and Renzini 1983) predicted that the third dredge-up occurred in stars greater than about 5 solar masses but not in the lower mass stars. Since then it has been possible to produce dredge-up in lower mass models by modifying the input physics such as by taking semi-convection into account, by using improved carbon opacities, by considering mass loss and by changing the mixing length to scale height parameter (Iben 1988: this conference;Iben and Renzini 1982a,b; Wood 1980; Boothroyd and Sackmann 1988a,b,; Lattanzio 1986, 1987a,b, 1988:this conference). However, at the present time calculations of the third dredge-up have been confined to low metallicity models, i.e. those more appropriate to Pop II rather than Pop I stars.

The ^{13}C$(\alpha,n)^{16}$O source operates at much lower temperatures (around 10^8 K). In order to produce enough ^{13}C atoms, it is necessary first to mix hydrogen into a carbon-rich region in order to produce ^{13}C by proton capture [^{12}C$(p,\gamma)^{13}$N$(\beta^- \nu)^{13}$C], and second to mix the ^{13}C down into a helium burning region in order to activate the neutron source. These conditions should exist during and after helium flashing episodes in stars with masses between 0.7 and 2 solar masses. In a few models it has been shown that hydrogen from the convective envelope can be carried inward into a region in which He burning has previously taken place. ^{12}C will then be burned into ^{13}C during quiescent He shell burning. During the next He shell flash, the convective He burning shell grows outward in mass eventually engulfing the region of high ^{13}C and activating the ^{13}C source (Iben 1987; Hollowell 1987). The models by

Hollowell have not yet been able to dredge-up material from the intershell region after the ^{13}C source has been in operation.

The interpulse period for low mass stars is on the order of 10^5 years (it decreases to $<10^4$ years for intermediate mass stars) and the total duration of the thermal pulsing (TP) stage is only a few times 10^6 years for low mass stars (Iben 1983). From the observed abundances of Tc relative to other s-process elements in MS and S stars Smith and Lambert (1986) estimate that ^{99}Tc should be detectable for 6-7 half-lives or about $1-1.5 \times 10^6$ years after being mixed to the surface. Hence, it is expected that Tc should be observable in almost all TP-AGB stars.

At the temperatures of the intershell region in intermediate mass stars ($2-3 \times 10^5$ K), Cosner and Truran (1981) and Schatz (1983) have found that the half-life of ^{99}Tc decreases to < 40 years due to much shorter decay time of ^{99}Tc from an excited nuclear state. However, Mathews et al. (1986) have shown that by carefully considering the network of nuclear reactions that are responsible for the production and destruction of Tc that the thermally enhanced beta decay rate of ^{99}Tc at the higher temperatures is more than offset by the increased efficiency of the ^{22}Ne source at these temperatures leaving the Tc abundance almost unchanged. Hence, the ratios of Tc/Nb and Tc/Mo are a good indicator of the time a star has spend in the third dredge-up phase and these ratios are fairly independent of the temperature in the s-processing region. Mathews et al. (1986), Dominy and Wallerstein (1986), Winters and Macklin (1987) and Wallerstein and Dominy (1988) have used the observed abundance ratios in four M, MS and SC stars to estimate the time scale since processing as $10^5 - 10^6$ years.

Stars experiencing the third dredge-up, as expected, show an increasing amount of helium burning products, primarily ^{12}C, and s-process elements in their atmosphere with successive dredge-up episodes. During this time the stars evolve from being M stars into C stars. The location of stars on the AGB in intermediate age globular clusters in the Magellanic Clouds are observed to show a progression of spectral type from M->MS->S->(SC)->C with increasing luminosity (Wood 1985) providing direct support for this scenario. The SC stars are enclosed in parentheses in order to indicate that this stage may be by-passed since it requires a near-equality of C and O in the atmosphere and enough C may be dredged up during one helium shell flash episode to allow this phase to be bypassed.

Whereas stars in the Magellanic Clouds allow for the determination of good luminosities of AGB stars, the stars are so faint that it is not yet possible to obtain the high dispersion spectra needed to measure high quality abundances. Hence abundance analyses are confined to relatively few galactic AGB stars whose luminosities are not well known. With very few exceptions the stars analyzed for abundances have been non-Mira variable M, MS, S, and C stars since good atmospheric models are available for them. As expected the C/O ratio increases and the s-process elements are enhanced along the M->C sequence (Boesgaard 1970; Smith and Lambert 1985, 1986; Utsumi 1985; Dominy, Wallerstein and Suntzeff 1986; Lambert et al. 1986). M stars show marginal if any enhancements of s-process elements (Smith and Lambert 1985; Dominy and Wallerstein 1986), MS stars show enhancements from 2 to 5 (Smith and Lambert 1985, 1986) and S stars show enhancements by a factor of 3-8 (Smith and Wallerstein 1983; Smith and Lambert 1986). Therefore, the s-process enhancements correlate well with the increase of the ^{12}C abundance as predicted (Smith 1987). In SC stars Zr and Nb are enhanced by a factor of 5-10 with Mo and Ru being enhanced by about a factor of 50-100 over solar abundances (Smith and Wallerstein 1983). The C stars show s-process enhancements by factors of 10-100 (Utsumi 1985). The early R-type C stars apparently do not belong on this sequence being neither luminous enough to be AGB stars nor showing s-process enhancements (Dominy 1984). The carbon enrichment needed to make them C stars must have been produced by another mechanism than the third dredge-up, possibly an off-center He core flash.

II. OBSERVATIONS

Merrill's (1952) discovery of the resonance lines of the radioactive element technetium (Tc I) in several S, MS and Mira variable M stars at 4297 A, 4262 A and 4238 A, marked the beginning of our ability to provide observational tests of nucleosynthesis schemes. Merrill showed that the Tc I lines tended to be stronger in stars with more pronounced S-type characteristics. Since none of the isotopes are stable, Merrill was correct to suspect that the S stars (and MS stars) represent a transient stage of stellar evolution. In 1956 Merrill detected Tc in the carbon star TX (19) Psc, and this was followed by Peery's 1971 discovery of four more N-type C stars with Tc. The discovery of technetium stars has continued, and now the total number of stars known to have technetium stands at 86 with an additional 19 possible technetium stars out of a total of 301 stars searched (Little, Little-Marenin and Bauer 1987 and references therein, hereafter LLB; Smith and Lambert 1988). All types of AGB stars are represented in this sample of technetium stars: 34 M stars, 17 MS stars, 20 S stars, 3 SC stars and 12 C stars.

Stars with Tc have been identified primarily from the presence of the three resonance lines of Tc I in their spectra at 4297.06 A, 4262.27 A and 4238.19 A which have an intensity ratio of 5:4:3. The identification of the lines is by no means unambiguous since all three lines are blended at the cool temperatures found in AGB stars. The primary blending contributors to the 4297 A line are the lines of three s-process elements Zr II, Sm I and Ce II at about 4296.7 A; those blending the 4262 A line are Gd II, Cr I and AlH; and the 4238 A line lies in the wing of the Ca I 4227 A line (Little-Marenin and Little 1979). Many of the blending contributors are s-process elements and they strengthen as the Tc lines strengthen making identification even more difficult. Our analyses (done in conjunction with my husband and various other collaborators) measured precise wavelengths of the spectral lines on coudé spectra which usually had a resolution of about 0.2 A. Others, primarily Verne Smith and his collaborators, have calculated synthetic spectra in the 4262 A region in order to identify the lines and to determine abundances. We reached the same conclusions as Verne Smith about the presence of Tc for the stars in common to our analyses. Our search concentrated on variable stars, especially the Mira variables. These stars are estimated to be AGB stars and good atmospheric models are not yet available making spectrum synthesis calculations difficult.

The presence and absence of Tc in various types of stars can be summarized as follows. Individual categories are expanded in greater detail below.

(1) Nonvariable M stars do not show Tc.

(2) M supergiants do not show Tc.

(3) Irregular variable (Lb) M giants do not show Tc. In general these are low-amplitude variables (mean Δm ~0.2 mag).

(4) M star Mira variables tend to show the Tc I lines if their periods are longer than 300 days. The percent of Miras with Tc and P < 300 days is almost 0% (three Miras with possibly very weak Tc lines have periods of 229 days, 238 days and 264 days). It rises to 100% for Miras with 370 < P < 400 days (Figure 2). Too few Miras with P > 400 days have been observed in order to establish a firm pattern, but it is clear that some of the long period Miras do not show Tc.

(5) The semi-regular (SRb) M giants do not show Tc if P < 100 days or P > 150 days (except for TU CVn (50 days), RT Hya (290 days), and T Mic (347 days)). The SRb's in the 130-150 day range that show Tc may be stars which are the evolutionary equivalent to the Miras but pulsating in the first overtone rather than in the fundamental mode (Willson 1982). Only seven SRa's have been observed. In general the longer period SRa's show Tc and the shorter period ones do not. SRc variables are thought to be supergiants and do not show Tc.

(6) M stars with Tc have spectral types later than M2.

(7) The MS stars should be divided into two groups:

(7a) about 60 % of evolutionary MS stars show Tc. (Evolutionary MS stars are defined by LLB to be in an intermediate evolutionary phase between M and S stars. They show the spectroscopic peculiarities defined by Keenan (1954) as well an overabundance of s-process elements and usually show Tc).

(7b) spectroscopic MS stars do not show Tc. (Spectroscopic MS stars are defined by LLB as having the spectroscopic peculiarities defined by Keenan (usually weakly) but no overabundance of s-process elements can be determined (Smith and Lambert 1985, 1986, 1988))

(8) S stars should be divided into two categories:

(8a) Single S stars show Tc (usually very strongly).

(8b) binary S stars without Tc but other s-process enhancements (except for o^1 Ori (Johnson and Ake 1986)).

(9) SC stars show the resonance lines of Tc (but only three have been analyzed to date).

(10) Twelve out of 16 C stars (75%) show Tc. All 16 analyzed are N-type C stars. The four N-type C stars without Tc are SS Vir, UU Aur, X Cnc and Y CVn. In general J-type C stars do not show any s-process enhancements so that the absence of Tc in the J-type star Y CVn is not surprising.

(10a) All four irregular variable (Lb) C stars show Tc. The Lb C stars have larger amplitudes of light variation ($\Delta m \sim 1$) than the M star Lb's mentioned in (3).

(10b) Early R-type C stars were not observed but are known to have no s-process enhancements (Dominy 1984), hence no Tc is expected to be present.

(13) Ba II stars do not show either Tc I or Tc II lines.

(14) As expected two A stars (Vega and Sirius), one subgiant (HD 176021) and three K giants and supergiants do not show Tc I. These stars are neither AGB stars nor cool enough to have neutral Tc lines.

III. DISCUSSION

The pattern that emerges from the above data is the following: (A) the production and mixing of Tc with atmospheric material during the third dredge-up is associated with large amplitude light variation. Non-variables and low-amplitude variables are not observed to have Tc. The only exception known to date is the S star BS 8062. Possibly this star is temporarily in a quiescent phase. (B) Even when large light amplitudes are observed such as in Miras, the presence of Tc appears to occur only for stars with $P > 300$ days (Fig. 2). (C) The large percentage of MS, S, SC and C stars with Tc supports the scenario that M stars evolve into C stars. (D) Tc is a very sensitive indicator of s-processing. Its presence is detected in the atmospheres of stars even when it is not possible to measure the slight overabundances of other s-process elements either by spectrum synthesis calculations or by spectroscopic classification criteria i.e. the stars are classed as M stars rather than MS stars. Dominy and Wallerstein (1986) detected and determined abundances for Tc in Mira (o Ceti) but could not measure an enhancement of other s-process elements. (E) Not all TP-AGB stars show Tc. Roughly 40% of the MS and S stars (Smith and Lambert 1988) and about 25% of the N-type C stars (LLB) show no Tc.

I will discuss in greater detail some of the categories defined above starting with the M stars. The M supergiants searched for Tc do not appear to be experiencing the third dredge-up since they do not show the Tc I lines. They are likely to be in the core helium burning phase rather than the TP-AGB phase. No supergiant of any spectral class is known to have Tc. This may in part be due to that fact that it is very difficult to classify MS, S and C stars correctly as supergiants because reliable luminosity criteria have not yet been established for s-process enriched atmospheres. Non-variable M stars are not TP-AGB stars. Judging by the lack of Tc in low-amplitude variables and its presence in larger amplitude variables, I estimate that Miras are thermally pulsing AGB stars and that several He shell flashes are needed before a star becomes a long period Mira. However, only Miras with $P > 300$ days are able to dredge-up s-process elements as can be seen in Figure 2. The trend of Miras with Tc as a function of period can be seen easily in 2(a) where the percent of Miras with Tc is plotted in 15 day intervals. The number of stars analyzed in each 15 day interval can be seen in 2(b) and 2(c).

Miras form a well-defined group. Kinematic studies of Miras show that the shorter period Miras P ~ 150-200 days tend to be old Pop II, low mass objects with masses typically around 1 solar mass, whereas the longer period Miras P~300-400 days are somewhat younger (intermediate Pop I) stars with masses between 1.5 to 2.5 solar masses (Feast 1963; Clayton and Feast 1969). The analysis of Dean (1974) shows that C stars on the average have 1.2 solar mass progenitors and confirms the kinematic studies of Feast and Clayton and Feast since thermally pulsing M stars are expected to develop into C stars.

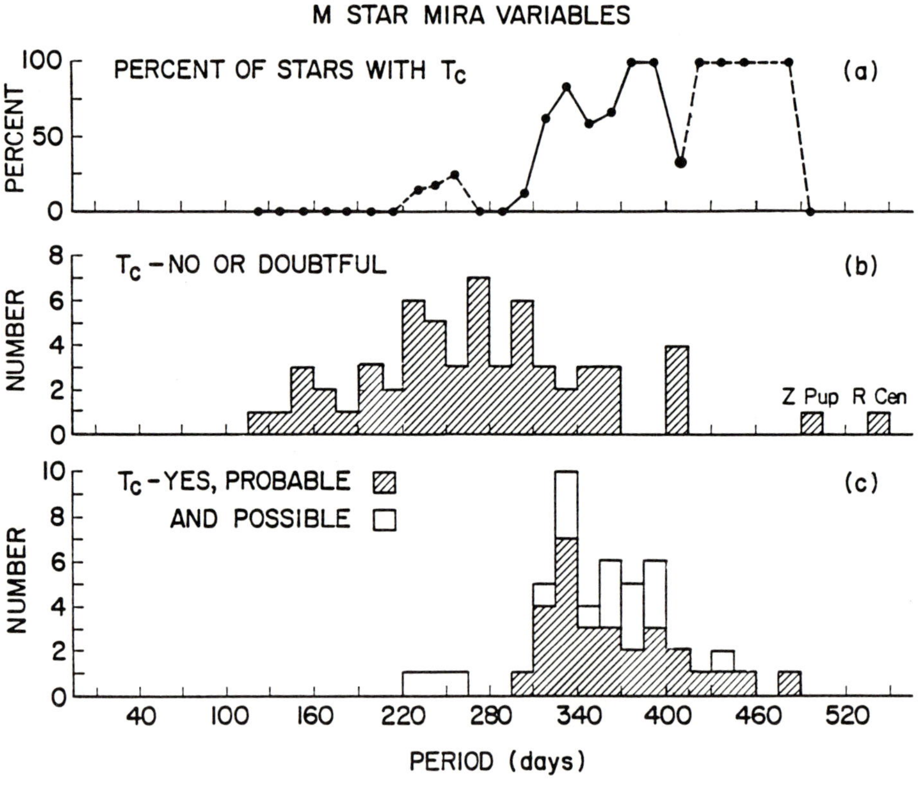

Figure 2. Histogram of M Mira variables with and without Tc. Reprinted from the Astron. J, 94, 981, (1987).

Miras define a period-luminosity relationship which is given by (Glass and Lloyd Evans 1981; Robertson and Feast 1981) as

$$M_{bol} = 0.76(+/- 0.11) - 2.09 \log P$$

for galactic Miras. This relationship indicates that M_{bol} lies between -4.4 to -4.9 for the Miras with Tc. Hence the data lead to the conclusion that stars showing Tc I lines are Pop I stars (Z ~ 0.01-0.02) with masses predominately in the 1.5 - 2.5 solar mass range and luminosities in the 5-6 x 10^3 L_o range. These stars have experienced s-processing and subsequent mixing. However, the third dredge-up can occur in higher mass stars, for example UV Aur has Tc and has an estimated mass of > 4 solar masses. The lack of Tc in the Mira variable V1 in the globular cluster 47 Tuc lends support to the conclusion that the third dredge-up operates mainly in low mass Pop I stars. No theoretical evolutionary model has yet been able to show a successful dredge-up in this mass, metallicity and luminosity range.

The stars without Tc are as intriguing as the stars that show Tc. Ba II stars are known to have s-process enhancements; however, no Tc I lines and more importantly (since Tc is estimated to be mainly ionized at the temperatures of the Ba II stars) no Tc II lines have been detected (Little-Marenin and Little 1987). Hence, the observed s-process enhancements must have occurred more than 10^6 years ago. The discovery by McClure, Fletcher and Nemec (1980) and McClure (1983;1988 this conference) that all Ba II stars appear to be binary led to the suggestion that the s-process enrichment is related to the binary nature of the system. Even though the exact mechanism is not yet well defined, the most likely suggestion is that a previous TP-AGB star, now a white dwarf, transferred part of its envelope more than 10^6 years ago onto the present day Ba II star transferring over both the s-process and ^{12}C enrichments (McClure 1984). The discovery of white dwarf companions for some Ba II stars by (Bohm-Vitense 1980; Bohm-Vitense, Nemec and Proffitt 1984; Bohm-Vitense and Johnson 1985) makes this scenario more plausible. Largely unsuccessful searches for white dwarf companions to technetium stars (Smith and Lambert 1987; Ake and Johnson 1988) except for o^1 Ori (Johnson and Ake 1986) diminish the possibility that Tc as well as the s-process enhancements observed in MS and S stars resulted from mass transfer from a more highly evolved TP-AGB star. But if does demonstrate that if any such mass transfer occurred, it happened very long ago (so that the companion is no longer detectable) at a time when the present day red giant was still on the main sequence. Therefore the MS and S stars with Tc do represent examples of thermally pulsing AGB stars in which the third dredge-up is operating.

The data listed in LLB and the careful abundance analyses of Smith and Lambert (1986, 1988) clearly show that not all stars estimated to be TP-AGB stars show Tc. Specifically about 40% of the evolutionary MS and S stars and about 25% of the C stars do not have Tc I lines but show other s-process enhancements and ^{12}C enrichments. This confirms the suggestions of Scalo and Miller (1981) that among the AGB stars about 30% of the stars are enriched by s-process elements but are without Tc. From the fact that the estimated space densities of M giants to MS/S stars without Tc is very similar to the ratio of G/K giants to (G/K) Ba II stars, Smith and Lambert (1988) conclude that the s-process-enriched MS/S stars without Tc are evolved Ba II stars, i.e. are members of binary systems in which the s-process enrichment occurred more than 10^6 years ago allowing for the decay of Tc. A search for radial velocity variations among Ba II and S stars (largely those without Tc) by Jorissen and Mayor (1988) finds that at least 5 out of 9 S stars (over 50%) show radial velocity variations indicative of orbital motion and hence are likely to be members of a binary system. Hence, the work of

Jorissen and Mayor lends support to the conclusion that only single S and MS stars show Tc in their spectrum. I have not included in the discussion the MS stars defined by LLB as spectroscopic MS or as very mild MS stars by Smith and Lambert (1988) which show neither Tc nor have measurable s-process enhancements. Their mild MS characteristics seen on spectrograms must be produced by another mechanism than enhanced abundances of Zr and Ba.

Four N-type C stars do not have any observable Tc in their spectrum. This is surprising since the progenitors of N-type C stars are TP-AGB M stars. Utsumi (1985) has found that C stars rich in ^{13}C (type J) show no s-process enhancements. Hence, the origin of the carbon enrichment in the J-type carbon star Y CVn, which shows no Tc, and its low ^{12}C/^{13}C ratio of 3.5 implies a compositional change during an earlier evolutionary phase and not the third dredge-up. Unfortunately, no other J-type C star has been searched for Tc. The carbon isotope ratio increases during the third dredge-up as can be seen from Figure 2 of LLB. Stars with stronger s-process enhancements show progressively larger ^{12}C/^{13}C ratios. UU Aur and X Cnc do show s-process enhancements but no Tc. If these are cooler analogues to the Ba II stars and are more evolved than the MS and S stars without Tc is not yet known. We were surprised to find the carbon Mira variable SS Vir without Tc lines in its spectrum since it is a prime candidate for the third dredge-up. No information about s-process enhancements in this star is known. None of the three stars have not been searched for radial velocity variations.

IV. CONCLUSIONS

Stars with Tc trace the onset and continuation of thermal pulses in asymptotic giant branch stars. These stars experience the third dredge-up and are predicted to show increasing amounts of carbon and s-process elements in their atmospheres with successive helium shell flashes as a star evolves from being an M->MS->S->(SC)->C star. Observations of stars in globular clusters in the Magellanic Clouds lend support to this evolutionary spectral sequence and detailed abundance analyses of M, MS, SC and C stars confirm the increasing trend of s-process enhancements and carbon abundances along this sequence. Tc is a very sensitive tracer of the mixing of recent s-processed material with the atmosphere of a star since it can be detected before increases in the abundances of other s-process elements can be measured. The mere detection of Tc is enough to establish that s-processing and mixing have occurred and is more easily accomplished than the determination of abundances by spectral analyses. The third dredge-up is accompanied by an increase in the amplitude of light variation since neither non- nor low-amplitude variables show the Tc I lines. In the Mira variables Tc is observed almost exclusively in stars with P > 300 days, indicative of 1.5-2.5 solar mass Pop I stars. Tc is not observed in Pop II stars.

A surprisingly large percent (25%-40%) of the C, S and MS do not show Tc even when other s-process elements are enhanced. Based on the space density of evolved stars, Smith and Lambert (1988) argue that

MS and S AGB stars without Tc are the cooler, more evolved analog of the Ba II stars. The observed s-process enrichments are estimated to have been produced by mass transfer from a companion so long ago that Tc has had time to decay away.

V. RECOMMENDATIONS

Our understanding of the late stages of evolution have progressed greatly during the last decade. High quality, high dispersion spectra have become available making it possible to establish the presence of Tc in a large cross-section of stars; better model atmospheres have allowed the determination of abundances of Tc and other s-process elements; more detailed stellar evolution models are able to match many of the observed characteristics; precise measurements of radial velocity variations has discovered many new binary systems and established that probably all Ba II stars and many MS and S stars are members of binary systems. Spacecraft observations have detected binary white dwarf companions to some of these systems. With all these accomplishments what problems are left unsolved?

1. More carbon stars need to be searched for technetium in order to define a statistical basis for the number of C stars with and without Tc.
2. Similarly more Miras with P > 400 days need to be analyzed in order to define the percentage of stars that show Tc in this category.
3. No Pop II stars have yet been found to have Tc even though evolutionary models indicate that dredge-up is more easily accomplished in these lower metallicity stars. Known TP-AGB stars of Pop II should be searched for Tc.
4. More s-process abundance determinations of SC and C stars are needed in order to allow detailed comparisons between predictions of evolutionary models and the observations.
5. Model atmospheres for the coolest stars and Mira variables are not yet available making it difficult to perform spectral synthesis calculations on these important objects.
6. The stellar evolutionary models are not yet able to produce the observed abundances of Tc and s-process elements in Pop I stars in which they are observed.

The usefulness of Tc is not confined to the late stages of stellar evolution. A new proposal has been made to use Tc occurring naturally in ores on earth as a way of detecting the ever elusive solar neutrinos!

REFERENCES

Ake, T.B and Johnson, H.R. 1988, Ap.J., 327, 224.
Beer, H., and Walter, G. 1985, in Cool Stars with Excesses in Heavy

Elements, eds M. Jaschek and P.C. Keenan, (Reidel: Dordrecht), p. 373.
Boesgaard, A.M. 1970, Ap.J., 161, 163.
Bohm-Vitense, E. 1980, Ap.J. (Letters), 239, L79.
Bohm-Vitense, E. and Johnson, H.R. 1985, Ap.J., 293, 288.
Bohm-Vitense, E., Nemec, J., and Proffitt, C. 1984, Ap.J., 278, 726.
Boothroyd, A.I., and Sackmann, I.-J., 1988a, Ap.J., 328, 653.
------------------------ 1988b, Ap.J., 328, 671.
Clayton, M.L, and Feast, M.W. 1969, Mon.Not.R.Astron.Soc., 146, 411.
Cosner, K., and Truran, J.W. 1981, Ap. Space Sci., 78, 85.
Dean, Ch. A. 1974, Astron.J., 81, 364.
Dominy, J.F. 1984, Ap. J. Suppl., 55, 27.
Dominy, J.F., and Wallerstein, G. 1986, Ap.J., 310, 371.
Dominy, J.F., Wallerstein, G., and Suntzeff, N.B. 1986, Ap.J., 300, 325.
Feast, M.W. 1963, Mon.Not.R.Astron.Soc., 125, 367.
Glass, L.S., and Lloyd Evans, T. 1981, Nature, 291, 303.
Hollowell, D.E. 1987, in Late Stages of Stellar Evolution, eds S. Kwok and S.R. Pottasch, (Reidel: Dordrecht), p.239.
Iben, I., Jr. 1983, Ap.J. (Lett), 275, L6.
Iben, I., Jr. 1987, in Late Stages of Stellar Evolution, eds S. Kwok and S.R. Pottasch, (Reidel: Dordrecht), p.175.
Iben, I. Jr., and Renzini, A. 1982a. Ap.J. (Letters), 259, L79.
------------------- 1982b, Ap.J. (Letters), 263, L23.
------------------ 1983, Annu. Rev. Astron. Astrophys., 21, 271.
Johnson, H.R. and Ake, T.B. 1986, New Insights in Astrophysics -- 8 Years of UV Astronomy with IUE, ed E. Rolfe, ESA-SP-263, p. 395.
Jorissen, A. and Mayor, M. 1988, Astron. Ap., (in press).
Keenan, P.C., 1954, Ap.J., 120, 484.
Lambert, D.L., Gustafsson, B., Eriksson K., and Hinkle, K.H. 1986, Ap.J. Suppl., 62, 373.
Lattanzio, J.C. 1986, Ap.J., 311, 708.
----------------1987a, Ap.J. (Letters), 313, L15.
----------------1987b, in Late Stages of Stellar Evolution, eds S. Kwok and S.R. Pottasch (Dordrecht: Reidel), p.235.
Little-Marenin, I.R., and Little, S.J. 1979, Astron.J., 84, 1374.
------------------1987, Astron.J., 93, 1539.
Little, S.J., Little-Marenin, I.R., and Bauer, W.H. 1987, Astron. J., 94, 981.
Mathews, G.J., Takahashi, K., Ward, R.A., and Howard, W.H. 1986, Ap.J., 302, 410.
McClure, R.D. 1983, Ap.J., 268, 264.
------------------- 1984, Publ. Astron. Soc. Pac., 96, 117.
McClure, R.D., Fletcher, J.M., and Nemec, J.M. 1980, Ap.J. (Letters), 238, L35.
Merrill, P. 1952, Ap.J., 116, 21.
Peery, B.F. 1971, Ap.J., 163, L1.
Robertson, B.S.C., and Feast, M.W. 1981, Mon.Not.R.Astron.Soc., 196, 111.
Scalo, J.M., and Miller, G.E. 1981, Ap.J., 246, 251.

Schatz, G. (1983), *Astron. Astrophys.*, *122*, 327.
Smith, V.V. 1987, in *Late Stages of Stellar Evolution*, eds S. Kwok and S.R. Pottasch (Dordrecht: Reidel), p. 241.
Smith, V.V. and Lambert, D.L. 1985, *Ap.J.*, *294*, 326.
---------------------- 1986, *Ap.J.*, *311*, 843.
---------------------- 1987, *Astron.J.*, *94*, 977.
---------------------- 1988, *Ap.J.*, (in press).
Smith, V.V., and Wallerstein, G. 1983, *Ap.J*, *27*, 742.
Smith, V.V. Lambert, D.L., and McWilliam, A. 1987, *Ap.J.*, *320*, 862.
Utsumi, K. 1985, in *Cool Stars with Excesses of Heavy Elements, Ap. Space Sc. Library, Vol 114.*, (eds M.Jaschek and P.C. Keenan) (Reidel:Dordrecht), p. 243.
Wallerstein, G. and Dominy, J.F. 1988, *Ap.J.*, *330*, 937.
Willson, L.A. 1982, in *Pulsations in Classical and Cataclysmic Variable Stars*, ed. J.P. Cox and Carl J. Hansen (Boulder, CO:JILA), p. 269.
Winters, P.R., and Macklin R.L. 1987, *Ap.J.*, *313*, 808.
Wood, P.R., 1980, in *Physical Processes in Red Giants*, eds I. Iben, Jr and Renzini, A. (Dordrecht: Reidel), p.135.
---------------------- 1985, in *Cool Stars with Excesses of Heavy Elements, Ap. Space Sc. Library, Vol 114.*, (eds M.Jaschek and P.C. Keenan) (Reidel:Dordrecht), p.357.

Long-Period Radial-Velocity Variations of Arcturus

Alan W. Irwin, Bruce Campbell   C.L. Morbey
University of Victoria   Dominion Astrophysical Observatory

G.A.H. Walker   S. Yang
University of British Columbia

We have measured the relative radial velocity of Arcturus using the HF absorption cell technique on 43 occasions from 1981 through 1985. The range of our velocities is 500 m s^{-1}, which is much larger than our estimated internal errors (typically 10 m s^{-1}). This confirms the radial velocity variability of Arcturus that has been previously reported by our group and others based on shorter observational time spans.

Using a non-linear least squares technique, we have determined a number of multi-periodic models which give a good representation of our data as well as those of Smith, McMillan, and Merline. The small-amplitude short-period components of these models presumably result from non-radial oscillations of Arcturus. More observational data are required to sort out aliasing problems of these components. All least squares models consistently require the largest amplitude component to have a period of 650 days or longer regardless of which aliases are chosen to represent the short-period oscillations.

There are two general possibilities for explaining the long-term component. The component could be the result of motions of the Arcturus surface or could be the result of an orbital companion to Arcturus with m sin i in the range from 1 to 8 Jupiter masses.

S-Process Dispersion in G and K Field Giants

Andrew McWilliam
Cerro Tololo Inter-American Observatory

A model atmosphere abundance analysis has been performed for a large number of G and K giants in a limited set of lines. The study was aimed at measuring the frequency of mild barium stars among GK giants in order to provide a test for the evolutionary status of the mild barium stars.

Published effective temperatures (Teff), determined via the infrared flux method, and broad band photometry were used to calibrate several color-Teff relations. Using these and standard methods the model atmosphere parameters for over 600 stars were determined, and are presented here.

It is shown that accurate abundance ratios, which are insensitive to errors in the model atmosphere parameters, can be measured if the lines used in the abundance analysis are carefully selected.

The final s-process abundances indicate that a large fraction of previously identified mild barium stars are probably normal. A group of stars in this survey are provisionally identified as being s-process enriched. This group has a high incidence of duplicity and radial velocity variation. The frequency of this group, if they are truly s-process enriched, constrains the mass transfer scenario explanation for the origin of mild barium stars.

Heavy Element Abundances In FG Satittae

T. Kipper and M. Kipper
Tartu Astrophysical Observatory
202444 Toravere, Estonian SSR, USSR

ABSTRACT. The spectrogram of FG Sge obtained by the 6-m telescope in 1986 is analyzed. The iron peak elements are underabundant relative to the sun, and the heavy elements (Z>39) have abundances up to 3 dex (1,000 times) higher than in the Sun.

1. Introduction

The spectrum of the variable star FG Sge has been intensively observed in recent years as its spectral type has been rapidly changing, and since 1955 growing progressively later, while its atmosphere has become very strongly enriched with heavy elements. The authors have tried to follow the changes in FG Sge in every possible case (Kipper and Kipper 1977; Kipper 1978, 1981, 1984). In this note we report line identifications in the spectrum of FG Sge in the yellow-red spectral region and our estimates of elemental abundances in its atmosphere in 1986.

2. Observations

In 1986 (August 21) at the 6-m telescope a 14 Å/mm spectrogram of FG Sge was obtained with the main stellar spectrograph. The spectral region registered was $\lambda\lambda$ 5200-6700. As the star was quite faint ($m \sim 9.3$) for that spectrograph-telescope combination and due to rather bad weather conditions, the signal-to-noise ratio at the continuum level was not much higher than 10. But as the possible changes in the spectrum of FG Sge are of great interest, this spectrogram was nevertheless analyzed.

3. Analysis of the Spectrum

The spectrum was reduced using the semiautomatic spectrophotometric reduction system developed at the Tartu Astrophysical Observatory (Kipper 1986).

There are no great changes in the spectrum compared with that in 1980. At that time we estimated the spectral type of FG Sge as K0 Ib - K2 Ib. The recent infrared photometric observations (1.25 - 3.5 μm) by Taranova (1986, 1987) show that the much earlier spectral type G2 I and even F5 I should be assigned to FG Sge for the moment of maximum light. Differences in spectral types determined using the visual and infrared spectral regions are obviously caused by the extremely heavy blanketing at shorter wavelengths due to the unusual chemical composition.
Taranova also estimated, at minimum light, the effective temperature,

Tmin = 5500 K and the bolometric absolute magnitude Mbol = -5.12, if the distance of FG Sge was 4.1 kpc. Using these rather reliable data and the estimates by Fadeyev and Tutukov (1981), on the basis of the stellar evolution theory, of r = 4.2 kpc and Mbol = -5.73, we adopted for the following analysis the model atmosphere with Teff = 5500 K and log g = 1.0.

Table 1

Z	Element	Solar abundance	[ε] 1980	[ε] 1986	Number of lines	References for log gf
11	Na	-5.7		-0.6	4 Na I	3
12	Mg	-4.4		-1.1	1 Mg I	20
20	Ca	-5.7	-0.7	-0.7	5 Ca I	20
21	Sc	-8.9	0	-0.5	6 Sc II	8
22	Ti	-7.0	-1.5	-0.2	9 Ti I	3
23	V	-7.9	-1.5	0.4	8 V I	15
24	Cr	-6.3	-1.5	-0.6	2 Cr I	3
25	Mn	-6.6	-1.0	-0.5	2 Mn I	15
26	Fe	-4.4	-1.6	-2.1	8 Fe I	15
				-0.7	9 Fe II	15
27	Co	-7.0	-1.0	-0.4	5 Co I	5
28	Ni	-5.7	-1.0	-1.2	3 Ni I	3
38	Sr	-9.1		1.8	2 Sr I	3
39	Y	-9.9	2.2	2.0	3 Y I	9
				1.8	10 Y II	9,5
40	Zr	-9.2	2.1	1.2	4 Zr I	7
42	Mo	-9.8	2.1	2.0	6 Mo I	2
44	Ru	-10.1	2.0	3.4	4 Ru I	3
45	Rh	-10.5		3.1	2 Rh I	3
50	Sn	-10.0		3.3	1 Sn I	3
56	Ba	-9.9	2.4	0.5	2 Ba II	15
57	La	-10.9	2.2	2.7	30 La II	3
58	Ce	-10.4	2.1	1.8	26 Ce II	3
59	Pr	-11.2	2.2	1.4	12 Pr II	1
60	Nd	-10.8	2.1	1.7	59 Nd II	5
62	Sm	-11.3	2.4	2.3	12 Sm II	3
63	Eu	-11.3	2.6	1.6	4 Eu II	3
64	Gd	-10.9	2.7	1.4	16 Gd II	3
70	Yb	-11.8		3.1	6 Yb II	3
71	Lu	-11.2	3.0	1.6	4 Lu II	3
72	Hf	-11.1	2.0	2.7	1 Hf II	3
90	Th	-11.8		1.8	1 Th II	4
92	U	-11.4		2.1	1 U II	5

As the spectrogram obtained in 1986 has twice the dispersion of any of our previously analyzed spectra, we had to re-identify the lines. This was done by using the list of lines in the solar spectrum (Moore et al., 1966) and the tables by Meggers et al. (1975). A line was considered to be identified if it followed reasonably well the curve of growth for the given element. As most of the found lines have not been classified, and we do not know their transition probabilities, the number of reliably identified lines is small. Even a cursory examination of the results of identifications reveals that most of the lines belong to heavy elements, and there is a very small number of iron-peak-element lines. The largest number of identified lines belong to Nd II.

The abundances of elements were estimated from the curves of growth computed for the model (5500/1.0). Using the lines of Nd II, we estimated the microturbulent velocity to be ξ_t = 10 km/s. This value coincides with the result found for FG Sge in 1980 (Kipper 1981). The radial velocity of FG Sge estimated from the tracings was -41 ± 5 km/s.

4. Results

The results of the abundance determination are given in the table. The errors in the abundances obtained are quite large, reaching 1 dex as can be seen by comparing the results for the first and second spectra of Fe and Y. Such a low precision is due to the errors in equivalent widths derived from the low signal-to-noise spectrum and especially to the errors in identifications and the oscillator strengths used. If these errors are considered, it appears that the abundances in FG Sge have not changed since 1980. Thus one can see the underabundances of metals up to Ni and the large overabundances for the heavy elements beginning with Sr.

The Na I lines D1,2 have strong circumstellar components which are shifted bluewards by approximately 46 km/s. If one assumes that the ionization in the circumstellar shell corresponds to that at the top of the atmosphere and the radius of FG Sge is around 100 $R_\odot$, one can estimate the lower limit to the mass loss of M $\sim 5.10^{-7}$ $M_\odot$/year. Fadeyev and Tutukov (1981) estimated from the modeling of the FG Sge pulsations an upper limit to the mass loss rate of 10^{-5} $M_\odot$/year.

References

1. Biemont, E., Grevesse, N., Hauge, O. 1979, Solar Phys. **61**, 17.
2. Biemont, E., Grevesse, N., Hannaford, P., Lowe, R.M., Whaling, W. 1983, Astrophys. J. **275**, 889.
3. Corliss, C.H., Bozman, W.R. 1982, NBS Monograph No. 53.
4. Corliss, C.H. 1979, Monthly Not. Roy. Astron. Soc. **189**, 607.
5. Cowley, C.R., Corliss, C.H. 1983, Monthly Not. Roy. Astron. Soc. **203**, 651.
6. Fadeyev, Yu.A., Tutukov, A.V. 1981, Monthly Not. Roy. Astron. Soc. **195**, 811.
7. Gurtovenko, E.A., Kostik, R.I. 1980, Ukrainian Acad. Sci., preprint ITF-79-138P.
8. Gurtovenko, E.A., Kostik, R.I., Orlova, T.F. 1985, Kinematika Fiz. Nebesn Tel. **1**, 75.
9. Hannaford, P., Lowe, R.M., Grevesse, N., Biemont, E., Whaling, W. 1982, Astrophys. J. **261**, 736.
10. Kipper, T.A., Kipper, M.A. 1977, Sov. Astron. Lett. **3**, 410.
11. Kipper, T.A. 1978, Sov. Astron. Lett. **4**, 280.
12. Kipper, T.A. 1981, Sov. Astron. Lett. **7**, 428.
13. Kipper, T.A. 1984, Sov. Astron. Lett. **10**, 219.

14. Kipper, T.A. 1986, <u>Estonian</u> <u>Acad.</u> <u>Sci.</u>, Preprint A-2.
15. Kostik, R.I. 1987, private communication.
16. Meggers, W.F., Corliss, C.H., Scribner, B.F. 1975, <u>NBS</u> <u>Monograph</u> No. 145.
17. Moore, Ch.E., Minnaert, M.G.J., Houtgast, J. 1966, <u>NBS</u> <u>Monograph</u> No. 61.
18. Taranova, O.G. 1986, <u>Astrofizika</u> **25**, 453.
19. Taranova, O.G. 1987, <u>Sov.</u> <u>Astron.</u> <u>Lett.</u> **13**, 891.
20. Wiese, W.L., Smith, M.W., Miles, B.M. 1969, <u>Atomic</u> <u>Transition</u> <u>Probabilities</u> <u>NSRDS-NBS</u>, 2.

HD 39853: A High Velocity K5III Star With an Exceptionally Large Li Content

Raffaele G. Gratton   Franca D'Antona
Osservatorio Astronomico di Roma, ITALY

High dispersion spectra of the high velocity star HD 39853 show that it is a slightly metal-poor ([Fe/H]=−0.5±0.1) giant of the old disk population, with exceptionally strong lithium lines. The abundance of Li, derived by synthetic spectra of the 6103.6 and 8126.4 Å lines, is log N(Li)=2.8±0.2. Abundances for the other observed elements are typical for mildly metal-poor giants: oxygen is slightly overabundant ([O/Fe]=+0.25±0.15); the C/N ratio is large (10±2); the $^{12}C/^{13}C$ ratio is small (6.6±1.2); and light elements (Na, Mg, Al, Si, Ca and Ti) are enhanced with respect to Fe by about a factor of 2.5. Observed s-elements (Zr and La) are not overabundant. Finally, no variation in the radial velocity of the star were detected at a level of 1 Km s^{-1}.

We explored several possible scenarios which may have produced the abnormally large surface Li abundance in HD 39853. Preservation of the primordial Li content is very unlikely, since the low $^{12}C/^{13}C$ ratio indicates that mixing of surface material with regions of uncomplete H-burning has occurred. Pollution of the outer convective envelope by material expelled by the nova explosion of a possible white dwarf companion is also improbable, due to the absence of variation in the radial velocity. Engulfing of a brown dwarf having a mass M ≤0.065 M$_o$ may explain the observed Li abundance in the

atmosphere of HD 39853 only if the original Li abundance was log N(Li)≥3.3, in contrast with the old age indicated by the large space velocity and low metal content. It is also difficult to invoke mixing of material after or in connection with a He-shell flash, in view of the low luminosity and of the small mass inferred from the space velocity.

We conclude that the present knowledge of stellar evolution seems to preclude any of the proposed scenarios. Further theoretical work is required to provide a plausible interpretation of this star.

Hydrogen Deficiency in Peculiar Red Giants

Robert F. Wing   Pedro Saizar
Astronomy Department, The Ohio State University

    Hydrogen abundances (or H/He ratios) are hard to determine in stars cooler than the Sun because the Balmer lines, when visible at all, are formed largely in the chromosphere, while the bands of CH and NH are often strongly saturated and badly blended with atomic lines. A few stars (the hydrogen-deficient carbon or HdC stars) are known to be extremely hydrogen-deficient, as their G bands of CH are absent despite an overabundance of carbon. A means of detecting less extreme cases of hydrogen deficiency would improve our understanding of red giant evolution. Minor variations in hydrogen content may be expected as the result of the mixing of processed material to the surface, and more radical changes might result from a star's shedding its entire hydrogen-rich envelope, say in the course of binary-star evolution.

    The hydrogen negative ion H^- is the primary source of continuous opacity in the optical and infrared regions of most late-type stars, and it has a pronounced effect on the shapes of their infrared energy distributions. Calibrated scans of infrared energy distributions therefore offer the possibility of detecting differences in hydrogen content among late-type stars.

    The cooled grating spectrometer at Kitt Peak National Observatory has been used to measure fluxes in 13 narrow bands between 1 and 4 μm chosen to avoid atomic and molecular lines as well as telluric absorption. Data have been obtained for several categories of carbon stars, S stars, barium stars, metal-poor stars, Cepheids, and RV Tauri variables, as well as for sequences of normal dwarfs, giants, and supergiants. The effect of H^- appears as a departure from a blackbody curve and is clearly seen from spectral type F to at least M7, reaching a maximum effect of about 0.4 mag near type M0. The effect is greatly reduced in the HdC star HD 182040. The RV Tau stars R Sct and U Mon also have very little H^-, but this may be a result of their relatively warm photospheric temperatures and low pressures. Barium stars have roughly normal H^- content for their temperatures and gravities.

    We intend to analyze the data by comparison with energy distributions calculated from a grid of model atmospheres computed at Indiana University for varying degrees of hydrogen deficiency. These comparisons should enable us to calibrate our measured H^- indices in terms of actual H/He ratios.

Relative CNO Abundances in Upper AGB Stars of the Magellanic Clouds.
A Search for Envelope Burning

J.M. Brett
Astronomy Centre, School of
Mathematical & Physical Sciences
University of Sussex

M.S. Bessell
Mt. Stromlo & Siding Spring Obs.
Inst. of Advanced Studies
Australian National University

We have investigated the atmospheric abundances of upper AGB stars of the SMC, searching in particular for evidence of the hypothetical envelope burning process. To this end we have computed synthetic spectra with varying C, N, and O abundances selected by considering the effect of the processes of the 3rd dredge up and envelope burning of a degree sufficient to prevent C star formation. The synthetic spectra (covering 0.5 μm to 2.5 μm) were analysed for observable effects of these two atmospheric enrichment processes. By analysis of band strengths of TiO, CO and CN we have found that substantial envelope burning is detectable for stars with $T_{eff} \geq 3000$ K but not below, due to the temperature dependence of CN bands. The synthetic spectra were compared to near-infrared and infrared observations of a small sample of SMC upper AGB stars thought to be prime candidates for the occurrence of envelope burning. This comparison indicated that envelope burning, of the extent considered here, is not occurring in these stars but rather the spectra are consistent with mild C enhancements produced by the 3rd dredge up alone.

However, the use of CN bands limits us to $T_{eff} \geq 3000$ K and our model atmospheres restrict us to C/O ratios below 1.0. Until these restrictions are lifted (in future work) it is not possible to discount in general the occurrence of envelope burning on the upper AGB.

We present a condensed version of this work, the full account of which is currently being prepared for journal submission.

Fluxes in M Giants With Improved Water Vapor Opacity

D.R. Alexander   G.C. Augason
Wichita State University   NASA/Ames Research Center
J.A. Brown   H.R. Johnson
          Indiana University

In the atmospheres of cool oxygen-rich stars, water vapor is the most important molecular absorber. Yet no laboratory or theoretical spectrum exists for water vapor of sufficient detail to be useful for radiative transfer calculations in a stellar atmosphere. Because of the complexities of its bent triatomic structure, this deficiency is not likely to the be resolved soon. We have, therefore, undertaken the preparation of a synthetic spectrum for water vapor.

For this synthesis, we produce hypothetical lines of water vapor which accurately reproduce narrow-band laboratory observations. The calculations are based upon the laboratory observations of Ludwig, C.B., et al. (NASA SP-3080, 1973), for the determination of straight mean opacities and mean line spacings, and Rothman, L.S., et al. (AFCRL, private communication, 1987), for the determination of the distribution of energy levels in the hot water vapor spectrum. Lines are placed randomly in intervals determined by the mean line spacing. Their strengths are taken randomly according to an exponential distribution to reproduce the observed straight mean for that interval. The lower energy level for each line is determined randomly from the distribution of lower energy levels for lines of similar strength and wavelength.

The resulting spectrum contains over four and a half million lines. The average opacity computed from this synthetic spectrum accurately reproduces the laboratory data. Altogether, our line archive for the computation of model stellar atmospheres now contains nearly 21 million lines. Compared to our previous models, which included only straight mean opacities for water vapor, these models have generally lower opacities, higher temperatures, and higher pressures over most of the atmospheres, with a reversal of that trend in the very shallowest layers.

Emergent flux curves for models calculated using the new opacities lack the large-scale discontinuities caused by the stop-wise nature of the straight-mean opacities seen in previous flux curves. The H^- flux peak at 1.65 μm is greatly reduced due to a correction in the calculation of the H^- bound-free opacity. The strength of the water vapor bands is also diminished. While the models fit observations better than previous ones, they show marginally too strong TiO features, insufficient opacity around 1 μm, and too much visual flux.

S-Process Deficiencies in Low-Mass Supergiant Variables

Howard E. Bond
Space Telescope Science Institute

R. Earle Luck
Case Western Reserve University

We have carried out abundance analyses of four low-mass supergiant variable stars (the RV Tauri or RV Tau-like variables AI Cmi, RU Cen, and U Mon, and the Type II Cepheid Kappa Pav) and two Population I Cepheids (CO Aur and V378 Cen). We used model atmospheres in which hydrostatic equilibrium, plane-parallel geometry, and local thermodynamic equilibrium (LTE) were assumed. Discussion of the results, and of published analyses of additional low-mass variables, leads to the following conclusions. (1) The Population I Cepheids show normal, solar elemental abundance ratios (except for the CNO elements, which have been altered by hydrogen burning), lending some support to the validity of the above assumptions for analyses of luminous variable stars. (2) The low-mass variables show metallicities ranging from solar down to [Fe/H] values typical of thick-disk and, in a few cases, of halo stars. (3) Most low-mass variables show a systematic underabundance of the heavy s- and r-process elements. In a few cases this may indicate that the stars were initially of extremely low metal content, and are now hydrogen deficient. However, most of the variables do not appear to belong to the halo population, nor do they show other abundance patterns seen in halo stars. The origin of these underabundances, and their apparent confinement to luminous variables, are difficult to understand in the context of nuclear processing. (4) The heavy-element underabundances correlate with second ionization potential in a manner suggesting that they are non-LTE phenomenan arising from overionization by Lyman-continuum photons. Why a similar effect is not seen in Population I Cepheids is unclear, but may be related to their generally weaker hydrogen emission. (5) Several low-mass variables, including RU Cen and V553 Cen, show carbon enhancements and solar s-process abundances. Relative to the majority of the Type II variables, these stars are s-process enhanced, and we argue that they are related to the Ba II and CH stars.

The Spectrum of TX Psc

Uffe G. Jorgensen
Niels Bohr Institute, Blegdamsvej 17, DK-2100 Copenhagen, Denmark

TX Psc is one of the very few carbon stars for which spectroscopic data from the blue toward the near infrared can be found in the literature. We present calculations of the first synthetic spectra that are in agreement with all the major spectral features in the observed spectrum, including the H^- peak and the 3 µm band. The spectra are calculated on the basis of available observational estimates of the basic stellar parameters and model atmospheres that include absorption from HCN, C_2H_2 and C_3 in the opacity calculation.

Three papers have dealt with the comparison of synthetic spectra with the observed spectrum of TX Psc (Johnson et al., Ap.J., 270, L63, 1983 and Ap.J., 292, 228, 1985; Lambert et al., Ap. J. Suppl., 62, 373, 1986). These point out the difficulties in fitting the H^- peak at 1.65 µm by model atmospheres with diatomic molecules alone, the discrepancies between observed and calculated H_2 quadrupole lines, and show that inclusion of HCN and C_2H_2 opacity in the model atmosphere overcomes much of these problems but show a far too strong 3 µm band of HCN and C_2H_2.

The strength of the 3 µm feature cannot be reduced by assuming that the calculated absorption coefficient should be systematically overestimated, since the absorption coefficient used for the fundamental (3µm band) is in agreement with laboratory values. Reducing the absorption coefficient from the part of the bands that are not experimentally verified will only strengthen the 3 µm band. This happens also if only diatomic molecules are considered in the opacity calculation. Reducing the hydrogen abundance also does not reduce the strength of the 3 µm band.

Because bands from polyatomic molecules are generally formed higher in the atmosphere than the diatomic bands and the metallic lines, these bands are much more sensitive to basic stellar parameters. We present results of calculations of the 3 µm band for different choices of parameters, and conclude that the parameter set Teff = 3100 K (from lunar occultation measurements), log(g) = -0.5 and Log(C/O) = 0.01 (which are all in agreement with the parameter set adopted by Lambert et al. within the errors quoted by them) gives good agreement between calculated and observed 3 µm band. This parameter set also predicts a spectrum that matches well with the rest of the gross features of the observed spectrum, including the H^- peak and the H_2 lines.

BD-21.3873: An Heavy Element-Rich Symbiotic?*

A. Jorissen
Institut d'Astronomie,
d'Astrophysique et de Geophysique
Universite Libre de Bruxelles

The recent discovery (Jorissen and Mayor, 1988) that not only Ba II but also S stars appear to belong predominantly to binary systems with periods of several hundred days raises the question of the similarity between these peculiar red giants and symbiotic stars. That question has been addressed by looking whether symbiotics have enhanced s-element lines in their spectra. In most cases, definite conclusions are hampered by the composite nature of such spectra. Nevertheless, in the case of the "yellow symbiotic" BD -21.3873, the spectral classification criteria provided by Keenan and Wilson (1977) allow to assign without ambiguity the type GIII-II to BD -21..3873, the symbiotic nature of that star being apparent only through weak Fe II emission lines. Some heavy element lines (such as Sm II $\lambda 4220.7$, $\lambda 4221.1$ and $\lambda 4566.3$) are clearly enhanced. Comparison with a spectrum of a G3Ib supergiant ensures that luminosity is not responsible for that effect. Moreover, some of these spectral features are also enhanced in Barium stars, as seen on a comparison spectrum of a Barium Star.

If abundance effects were really involved - a fact that needs to be confirmed by more detailed spectroscopic studies, - this should constitute a further indication that Barium, S and (some) symbiotic stars represent the same kind of binary objects, seen at different evolutionary stages.

In that respect, it is interesting to note that the other yellow symbiotic AG Dra also seems to show some evidence for heavy element peculiarities (Lutz et al., 1987), and, on the other hand, that the S star HD 35155 and ER Del display some "symbiotic-like" behavior in their UV spectrum (Johnson and Ake, 1983, 1988).

References

Johnson, H.R., Ake, T.B. 1983, in Proceedings of the Third Cambridge Conf. on Cool stars, Stellar Systems and the Sun, eds. S.L. Baliunas, L. Hartmann, Heidelberg: Springer, p. 362.
Johnson, H.R., Ake, T.B. 1988, this volume.
Jorissen, A., Mayor, M. 1988, Astron. Astrophys. 198, 187.
Keenan, P.C., Wilson, O.C. 1977, Astrophys. J. 24, 399.
Lutz, J.H., Lutz, T.E., Dull, J.D., Kolb, D.D. 1987, Astron. J. 94, 463.

*Based on observations carried on at the European Southern Observatory (ESO, La Silla, Chile).

Comparison of Blanketing in a 3000 K, Oxygen Rich, Spherically Symmetric
Model with the Blanketing in Non-Mira, M Giant Stars

G.C. Augason	J.A. Brown	D.R. Alexander
NASA/Ames	Indiana University	Wichita State

A flux curve has been computed using a preliminary, spherically symmetric model for a 3000 K, oxygen rich, giant star. The model was computed using the opacity sampling method with an improved frequency set, improved molecular equilibrium data and an improved set of opacities. In addition, a continuum flux curve is computed using the same model and only continuum opacity sources. The relative and to some extent the absolute blanketing used to compute both the model and the flux curve derived from the model may be illustrated by dividing the normal flux curve by the continuum flux curve. This same procedure is used to illustrate the blanketing in an observed star by dividing the observed flux curve by the continuum flux curve. When this is done, the blanketing in an observed flux curve may be compared with the blanketing in a model. When this comparison is made, it is obvious that the treatment of blanketing in the "new" flux curve, computed using the spherically symmetric model and using new parameters, is superior to the flux curves based on earlier models. This is especially true in the regions of the fundamental, the first overtone and the second overtone of CO. Also, the new water vapor opacity is much improved. The new water vapor opacity is based on actual measurements of high temperature water vapor. Correct representation of water vapor opacity is extremely important for oxygen rich stars because it forms a psuedo continuum because of its many lines. The TiO opacity does not fit the observations well. When the spherically symmetrical model flux curve is compared directly to an observed flux curve, the new flux curve gives a better fit than do flux curves computed from previous models. There is still (at least for the non-Mira stars) a serious flux excess in the model flux curves at 1.6 microns in the region of the H minus b-f and f-f crossover. However, this excess is not as great for the spherically symmetric model as it is for earlier plane parallel models. It is not determined if this improvement is due to spherical symmetry or due to the new model parameters.

3. Evolution of Peculiar Red Giant Stars

EVOLUTION AND MIXING ON THE AGB

John C. Lattanzio
Institute of Geophysics and Planetary Physics
Lawrence Livermore National Laboratory
L-413, P.O. Box 808, Livermore, Ca., 94550

Abstract It is now well known that Nature can make Carbon stars at lower luminosities than can (human) theorists. A number of workers, stimulated by this challenge, have been attracted to the problem. In this paper I review recent evolutionary models of relatively low mass AGB stars, with emphasis placed on the mixing of carbon to the stellar surface. In particular I discuss some recent improvements in the physics used to construct stellar models. These topics include: breathing pulses of the convective core found during core helium exhaustion; the effects of carbon recombination; the occurrence of semiconvection in the region between the two nuclear burning shells; and the importance of mass loss. Recent calculations have successfully produced models of low luminosity Carbon stars. The strengths and weaknesses of these models will be contrasted.

INTRODUCTION

An asymptotic giant branch (AGB) star is one which has recently exhausted its core helium supply. Outside the carbon-oxygen core (the most recent $^{12}C(\alpha,\gamma)^{16}O$ rate predicts about 80% ^{16}O and 20% ^{12}C in this core) is a helium burning shell. On top of this is material which has been processed by the hydrogen burning shell, and thus contains primarily helium, with enhancements of ^{14}N from CN(O) cycling. Surrounding this is the hydrogen burning shell itself, which is eating its way into the envelope, whose composition is that of the ZAMS star, with abundance changes due to the first and, possibly, second dredge-up (for details see Becker & Iben 1979, 1980, Iben & Renzini 1983, hereafter IR83).

Pioneering studies by Iben (1975a, 1975b, 1976) initiated a systematic investigation of AGB evolution. It was known that the helium burning shell suffered periodic thermal instabilities, called "shell flashes" or "thermal pulses". During these the luminosity from helium burning, L_{He}, reaches $\sim 10^7 L_\odot$. The deposition of such large amounts of energy causes a convective zone to form at the base of the helium burning shell, and this intershell convection extends almost to the hydrogen burning shell. This convective pocket contains about 20% carbon by

Work performed under the auspices of the U.S.Department of Energy by the Lawrence Livermore National Laboratory under contract No. W-7405-ENG-48.

mass. During later pulses the temperature at the base of the convective region exceeds 300×10^6K, and the ^{14}N$(\alpha,\gamma)^{18}$F$(\beta^+\nu)^{18}$O$(\alpha,\gamma)^{22}$Ne$(\alpha,n)^{25}$Mg reactions take place, forming free neutrons which can be captured by ^{56}Fe, and result in the formation of s-process elements (*e.g.* Iben 1975a,b). Following the pulse, rapid expansion extinguishes hydrogen burning. As the envelope cools, convection reaches inward, often beyond the extinct hydrogen shell, with the result that fresh helium is "dredged" to the surface. This deep convective envelope can also reach into the ^{12}C-rich region formed by the erstwhile intershell convection, and carbon can be mixed to the surface. This is called the third dredge-up. These models appear, qualititatively, to provide a natural explanation for carbon stars (whose atmospheres show n(^{12}C)/n(^{16}O) > 1.0), together with MS and S stars which show enhancements of s-process elements. Indeed, it was shown by Iben (1975a, 1975b) that AGB stars with hydrogen depleted core masses $M_H \gtrsim 0.95$ could form s-process elements in the same relative distribution as seen in the solar system.

The problems came when a quantitative comparison was made between theoretical AGB star distributions (Iben 1981; Renzini & Voli 1981) and observations of Magellanic Cloud stars (Blanco *et al.* 1978, 1980). Theory predicted no carbon stars less luminous than $M_{bol} \simeq -5$ or brighter than -7.5. This contradicts the observed range in M_{bol} of -3.5 to -6. It was clear that carbon stars must be produced at lower luminosities than predicted. Since there is a (linear) relation between M_H and the total quiescent luminosity before a pulse, this is equivalent to saying that the third dredge-up must operate at lower M_H. Although the luminosity varies slightly during a pulse cycle, the "extended post-flash dip" is a little under one magnitude in size, and cannot account for all of the discrepancy, especially since this dip lasts for only about 20% of the pulse cycle.

Later observations by Wood *et al.* (1983) showed that very bright AGB M stars ($M_{bol} < -6.5$) do exist. Possibly hydrogen burning at the base of the convective envelope is turning carbon stars back into M stars at these luminosities (*e.g.* Renzini & Voli 1981). But the number of these stars is small, and it is believed that most AGB stars reduce their envelope mass $M_e (= M - M_H)$ via stellar winds as they ascend the AGB. As M_e approaches zero the star leaves the AGB to become a white dwarf. This may remove from the AGB the only stars shown capable of producing a solar system distribution of s-process elements, *i.e.* those with $M_H \gtrsim 0.95$. The burden then falls on AGB stars of low M_H (and other sources ?) to produce s-process elements *and* low luminosity carbon stars. Recent calculations addressing these problems will be the topic of this review.

INPUT PHYSICS
Core Helium Burning

Before discussing AGB evolution it is prudent to backstep to the core helium burning phase, which immediately precedes the AGB. Calculations by Lattanzio (1986, hereafter JL1) showed that a semiconvective region formed at the edge of the convective core in the (1-3 $M_\odot$) models of that study. This is analogous to the situation known to exist for $M \lesssim M_\odot$, but often ignored in more massive stars (*e.g.* Becker & Iben 1979). Briefly, a semiconvective zone is one which is marginally unstable to convection, but where the act of mixing creates a structure which is radiatively stable (according to the Schwarzschild criterion; for details see Castellani *et al.* 1971a, 1971b). The result is a region where the abundances are adjusted to give precise convective neutrality. The consequences of semiconvection are twofold. Firstly more helium is mixed into the convective core, and upon core helium exhaustion the helium depleted core is larger. Secondly, as a consequence of the increased helium supply, the time spent burning helium is significantly larger ($\sim$ 50% or more), so that the hydrogen burning shell has time to eat further into the envelope. Thus the hydrogen exhausted core is also larger, upon helium depletion, than in models which ignore semiconvection. Obviously, semiconvection will greatly alter the structure of the star at the start of the AGB.

Core Breathing Pulses

An instablility has been found to occur during the final stages of core helium burning (*e.g.* Sweigart & Demarque 1972, 1973; Gingold 1976; JL1). It has been seen that the convective zone grows rapidly, mixing large amounts of helium into the burning region. Sweigart & Demarque (1973) provided a linear stability analysis of the phenomenon, known as "core breathing pulses", and found that the instability is due to the strong dependence of the triple alpha reactions on the helium abundance. A small growth in the size of the convective core causes an increase δY_c in the central helium content Y_c. When $Y_c \lesssim 0.12$ even a small δY_c can result in a large increase in the energy generation, due to the extreme sensitivity of helium burning to the helium content. This extra energy results in the growth of the convective core, which continues the runaway. The instability is quenched when either Y_c is sufficiently large that a small δY_c has a negligible effect, or when the helium stratification in the surrounding layers is no longer capable of providing sufficiently large values of δY_c in response to growth of the convective core. Thus we see that the instability has a physical basis. One should note that it appears in different formulations of semiconvection (*e.g.* Gingold 1976; JL1; Castellani *et al.* 1985a). Note also that it occurs both when convective (and semiconvective) boundaries are obtained implicitly, as in the Robertson & Faulkner (1972) method used by Gingold and others, or explicitly (as in JL1 and Castellani *et al.* 1985a).

A detailed study by Castellani *et al.* (1985a) shows that a model typically experiences three core breathing pulses before finally exhausting its core helium supply. This agrees with JL1 (see Lattanzio 1984 for details) and Mazzitelli & D'Antona (1986). The effects of these convective pulses are analogous to those of semiconvection: an increase in the amount of helium burnt, and consequently an increase in both the hydrogen and helium exhausted cores at the start of AGB evolution. A further consequence of each breathing pulse is a rapid blueward loop in the HR diagram. Typically $\Delta log(L/L_\odot) \simeq \Delta log T_e \simeq 0.1$, with the loop taking some 10^5 years. Such small variation is not open to observational detection, unfortunately.

Convective Overshooting

Almost all stellar structure codes use a mixing length formulation of convection. This "local" theory makes no allowance for the kinematics of the convective motions (velocity, momentum *etc.*), and consequently ignores the possibility of overshoot beyond the formally convective region. Recently Bressan et al. (1981) addressed this problem, and included overshooting in their models (see Chiosi et al. 1987 for a summary). They find that their models develop neither semiconvection nor core breathing pulses, which they claim are due to using "local" theories of convection. Interestingly, their models complete core helium burning with hydrogen and helium depleted core masses which are very similar in size to models which include both semiconvection and convective pulses. The models seem to be demanding a certain structure. In any event, these models have not yet been evolved into the thermally pulsing regime, and will not be discussed further.

Observations

One can appeal to observations in an attempt to discriminate between various mixing scenarios. A simple test is to determine the ratios of the lifetimes on the AGB and the horizontal branch. This ratio should be equivalent to ratio of the number of stars in these two phases. As summarized by Renzini and Fusi Pecci (1988), the observations seem to favor semiconvection without core breathing pulses, or possibly the overshooting models of Chiosi and co-workers (see Chiosi et al. 1987). Why breathing pulses seem to be required, yet do not match the observations, is unknown. Perhaps a better test would be to construct luminosity functions for models with the various forms of convection, and compare these with the observations. This work is in progress (Bertelli et al. 1988). Note that it has been recently suggested that core breathing pulses are caused by the assumption of instantaneous mixing (Chieffi and Renzini, private communication).

Carbon Recombination

Thermal pulse calculations by Sackmann (1980) found that the expansion engendered by the thermal pulse could push the carbon-rich region out to very low temperatures ($\sim 10^4$K in that study). The older opacity tables of Cox & Stewart (1970) were then widely in use, yet these only provided opacity of carbon-rich regions for $T > 10^6$ K. Carbon begins to recombine a little below this temperature (*e.g.* Sackmann & Boothroyd 1985), and this would greatly increase the opacity beyond estimates based on the Cox & Stewart tables.

Iben & Renzini (1982a) investigated the effect of this recombination on the electron pressure P_e and the adiabatic gradient. They found P_e decreased by 9% at $T = 2 \times 10^5$K, but by only 1% at $T = 4 \times 10^5$K, with variation being insignificant for higher temperatures. Thus it appears that the effect of recombination in the equation of state may only be important for $T \lesssim 4 \times 10^5$K. The changes in ∇_{ad} were even smaller than the changes in P_e, and can be safely ignored.

Iben & Renzini included, in an approximate way, recent opacity calculations by Art Cox for carbon-rich mixtures of low temperatures. They found that when the carbon-rich pocket (formed by flash-driven convection during the previous pulse) experienced temperatures below $\sim 5 \times 10^6$K, a semiconvective zone appeared at the outer edge of the carbon-rich zone. Note that for this temperature (and density $\sim$few$\times 10^{-3}$g cm^{-3}) we are still in a region of the (T,ρ) plane which is covered by Cox & Stewart tables. Presumably an important effect operating here is the increase in recent opacity calculations of $\sim 50\%$ compared to the Cox & Stewart values (Magee et al. 1975). Also, Cox & Stewart give tables for only three mixtures relevant to the carbon-rich region: $(Y,^{12}C)=(0.0,1.0)$, $(0.5,0.5)$, $(1.0,0.0)$. Interploation in these sparse values of the carbon content might also be responsible for previous investigators missing this phenomenon. It is clear that although carbon recombination may be important at later times (the carbon-rich material does cool to few $\times 10^5$K), the initial appearance of semiconvection *may not* be due to the recombination of carbon. Iben and Renzini (1982a) showed that 99.97% of the carbon is still fully ionized at the temperatures and densities where semiconvection first appears. Of course, although the amount of carbon recombined is small, it may still be the primary opacity source.

In any event, the semiconvection mixes carbon outward and hydrogen inward, the latter being the more extensive. Some carbon is mixed outward by $\sim 2 \times 10^{-5} M_\odot$, sufficiently far to make contact with the (inward moving) convective envelope, with the result that carbon is mixed to the surface by the usual dredge-up (Iben & Renzini 1982a,b; hereafter IR82a, and IR82b, respectively). The entry of hydrogen into a region rich in carbon causes the $^{12}C(p,\gamma)^{13}N(\beta^+\nu)^{13}C(p,\gamma)^{14}N$ reactions to consume all the hydrogen when this region is heated later in the pulse cycle. As a consequence, a few $\times 10^{-4} M_\odot$ now shows the abundance of ^{13}C exceeding that of ^{14}N. During the next thermal pulse temperatures rise enough to ignite $^{13}C(\alpha,n)^{16}O$, which releases free neutrons for capture by ^{56}Fe with the potential for forming s-process elements (*e.g.* IR82b).

Unfortunately, Iben & Renzini were forced to terminate their calculations because of convergence difficulties caused by instantaneous mixing of regions of vastly different composition. Hollowell (1987, 1988, hereafter DH1 and DH2, respectively) has repeated these calculations with detailed opacity and an algorithm designed to overcome the mixing problems. This work will be discussed below.

While discussing opacity, it should be noted that dredge-up of carbon significantly alters the composition of the envelope. One can find the "metallicity" $Z = 1 - X - Y$ increasing by a factor of 2 (*e.g.* Boothroyd & Sackmann 1988d). In calculating the opacity of these mixtures the significant abundance of carbon should be included. One should also check on the effects of carbon recombination on the equation of state. We have seen that this can be ignored for mixtures containing as much as $\sim 20\%$ carbon provided that temperatures remain above a few $\times 10^5$K (IR82a). But the envelope will reach down to a few thousand degrees. Of course, although carbon may be the main contributor to Z, it is still a small amount of the total mass of the envelope ($Z \approx 10^{-3}$).

EVOLUTIONARY RESULTS

A brief description of the structure of an AGB star and the evolution through one pulse was given in the Introduction. Because this evolution is well understood, the reader is referred to other reviews for details (*e.g.* IR83; Iben & Renzini 1984; Iben 1984, 1987). We will concentrate on the results of the recent calculations of Lattanzio (1986, 1987a, 1987b, 1988, hereafter JL1–4, respectively), Boothroyd & Sackmann (1988a–d, hereafter BS1–4, respectively), and Hollowell (DH1 and DH2). Differences between the physics used by these investigators will be discussed below, as will the differences in the carbon stars that resulted.

One very important parameter is the mass of the hydrogen exhausted core at the first thermal pulse, M_H^{TP}. The models of JL, BS and Mazzitelli & D'Antona (1986) are all in agreement, showing that M_H^{TP} is approximately independent of mass for solar metallicity and masses in the range 1–3$M_\odot$, as stressed by JL1. A second important result is the strong dependence on total mass for the $Z = 0.001$ models (JL1, BS3). Note also that attempts by IR83 to estimate M_H^{TP} for stellar masses in the range 1–3$M_\odot$ based on the results for more massive and less massive stars fail to give accurate results (JL3, JL4, BS3). Since this is an important parameter in synthetic AGB star distributions (Iben 1981; Renzini & Voli 1981) these calculations will be significantly altered because of the new results.

Another critical number is the maximum (quiescent) luminosity L_{TP} reached by the models just prior to the first thermal pulse (see figures in JL2, JL3 and BS3). The new determinations are 1 to 2 magnitudes below the IR83 estimate, based on the only models available at that time, which were outside the required mass range. This is encouraging, as it allows some time for the models to experience dredge-up before reaching the luminosity of carbon stars. Again, synthetic distributions using these results should be better able to explain the observations.

An important effect included in the BS models is mass loss. During the latter stages of AGB evolution mass loss is more important in reducing the envelope mass than is the advance of the hydrogen shell (Schonberner 1979), even before the "superwind" phase (*e.g.* IR83). The reduction of envelope mass complicates the prediction of evolutionary behaviour. It has been shown (*e.g.* Wood 1981) that models with smaller envelope masses are less likely to experience dredge-up than are stars with a larger M_e (given the same M_H). But when/if carbon dredge-up begins, the reduced envelope mass means a smaller dilution of the added carbon. Hence less carbon is needed to form a carbon star, and thus fewer pulses are necessary than when mass loss is ignored.

A second significant effect is that a star can only remain on the AGB provided $M_e > 0$! Consequently mass loss will completely terminate the AGB phase, for a given mass, much earlier than when mass loss is ignored. BS3 note that their $3M_\odot$ model with $Z = 0.001$ shows M_H^{TP} larger than the expected final mass for such a star, based on the observed initial–final mass relation. Obviously, if this is correct, such a star should not experience the third dredge-up at all. BS3 believe, however, that they have overestimated the rate of mass loss, probably by as much as a factor of 2. BS used the Reimers formula for mass loss, and took 0.4 as the value of the parameter η, which enters this formula (the larger value of 1.4 was used for the $3M_\odot$ models). This was determined by previous calculations as the value needed to match the mass loss occurring during the ascent of the giant branch. But this calibration was made for an α, the ratio of the mixing length to the pressure scale height, of 1.5, whereas BS used $\alpha = 1.0$. Perhaps more importantly, BS included the effects of some molecules in the envelope opacities which they used. Both of these effects will directly alter the stellar radius, which enters the Reimers formula. It seems probable that, under these circumstances, a smaller value of η would be needed.

In summary, it is clear that mass loss is important for (at least) three reasons: 1) the reduction in M_e makes dredge-up less likely; 2) with a smaller envelope mass fewer pulses are needed to produce a carbon star; 3) mass loss terminates the AGB evolution at some stage, thus limiting the number of thermal pulses which a star can experience. We should also note the apparent dependence of some characteristics of the evolution on the total mass and past history of the star (BS3, DH2). This makes it dangerous to neglect mass loss and artificially alter the envelope mass (for a given core mass) in an attempt to explore parameter space.

DREDGE-UP AND CARBON STARS

Before discussing the dredge-up found in recent models we should recall some basics already well understood. For example, dredge-up is more likely to occur in models with higher core masses. For a given core mass, dredge-up is favored by lower Z, larger α, or increased envelope mass (*e.g.* Wood 1981, IR83).

The Models of Lattanzio

These included semiconvection and core breathing pulses during the core helium burning phase. The opacities are from Huebner et al. (1977). No allowance is made for the effect of any extra carbon which may be dredged into the envelope. No opacities are calculated for carbon-rich mixtures below 10^6K, but the models of JL never entered this regime. The opacity of the carbon-rich matter is calculated with the Huebner et al. code, but at the same (T, ρ) and abundances as the older Cox & Stewart values (see above). Care was taken to accurately obtain the core masses at the core helium flash (see JL1 for details).

It was found that many variables previously thought to depend only on core mass (*e.g.* luminosity, interpulse period) also showed a dependence on abundance. For example, M_H^{TP} increases by $\sim 0.05 M_\odot$ when Y increases from 0.2 to 0.3. Including these effects in calculations of AGB star luminosity functions will be necessary.

The encouraging results from these models include the reduction of L_{TP} and the new M_H^{TP}, discussed in detail above. Calculations of the thermally pulsing evolution of $1.5M_\odot$ models, without mass loss, showed dredge-up at core masses as low as $0.62M_\odot$ (JL3). Three of the four models of that study ($Z = 0.003$ and 0.006, each with $Y = 0.2$ and 0.3, $M = 1.5, \alpha = 1.5$) were found to dredge carbon to their surfaces. Calculations were stopped when one of the models became a carbon star, with M_{bol} dropping to -4.4 after the pulse, and $M_H = 0.65$ at this time. This luminosity agrees very well with estimates of the transition luminosity between M and C stars (see JL3 for details). (There is no reason to believe that the fourth model, whose quiescent luminosity had reached $M_{bol} = -4.8$ when calculations were stopped, would not also have experienced dredge-up in the future.)

These models may be criticized on a number of counts. Firstly the effective temperature of JL3's carbon star is $\log T_e \simeq 3.56$, but the observations show $\log T_e \lesssim 3.5$ (*e.g.* Richer 1981). Ignoring uncertainties in the temperature calibration, agreement could be forced if $\alpha \simeq 1.25$. Note that JL1 found no dredge-up for $\alpha = 1.0$, while the carbon star of JL3 used $\alpha = 1.5$. It is unknown if dredge-up would occur at this intermediate value of α. Of course, the inclusion of molecular opacities and other neglected envelope effects, including the enhanced carbon abundance, may aid this situation (JL4), so perhaps the disagreement is not serious.

Secondly, these models ignored mass loss. Consequently the large envelope mass will be acting in favour of dredge-up at lower luminosities. But we have shown earlier how the larger M_e also hinders carbon star formation because of the much larger dilution factor. It is not clear which of these effects dominates.

Thirdly, temperatures in the carbon rich region were always $< 200 \times 10^6$K, and hence too low to ignite the ^{22}Ne source. No semiconvection has been found, and thus the Iben/Renzini/Hollowell mechanism is not operating. Consequently the ^{13}C neutron source is not operating in these models either. Thus these calculations would produce carbon stars without s-process elements, contrary to the (available) observations (Smith & Lambert 1986). The resolution of this discrepancy (if, indeed, there are no C-rich stars without s-process element enhancements) is unclear. We have already discussed the possibility that inaccuracies in interpolation within opacity tables may be responsible. Also, it may be that stronger pulses are needed to push the carbon rich mixtures out to temperatures low enough for semiconvection to occur. That Wood and Lattanzio, using essentially the same code, obtain less violent pulses than do Iben and Hollowell, using essentially the same code, indicates that a detailed comparison between codes may be required. Thermal pulses are a demanding phase of stellar evolution, and small differences in codes may cause large differences in calculated behaviour. Note also that it appears that semiconvection does not occur over a wide range of mass and composition (Iben 1983), and seems to require total masses < 1.0 and $Z \simeq 0.001$ (see also DH2).

It is clear that a larger survey of parameter space is required to make definite conclusions. Note also that JL1 found no dredge-up for the $1.5M_\odot$ models over a wide range of Z, with $\alpha = 1.0$. It seems likely that these models would experience the third dredge-up for $\alpha = 1.5$, at least for the lower metallicity models.

The Models of Boothroyd & Sackmann

BS1–4 investigate AGB evolution of models with $Z = 0.001$ and 0.02, and initial masses of 1–3 $M_\odot$. These calculations included semiconvection during the core helium-burning stage, but experienced convergence difficulties during the core breathing pulses. Consequently these were suppressed in most calculations. They used the latest Los Alamos opacities, with effects of some molecules included at the lowest temperatures (see Sackmann & Boothroyd 1985). The opacity of any carbon dredged to the surface was included approximately by using an opacity table for a metallicity $Z = 1 - X - Y$, even though much of the "Z" is pure carbon. Nevertheless, this is a better approximation than ignoring the effect, as everyone else has done. Opacities for carbon-rich mixtures were included for a wide range of temperatures, thus allowing for carbon recombination during the expansion following a thermal pulse. (Again, the opacity tables used here were available for a much denser sampling in carbon abundance than used by Lattanzio and Wood.) Likewise, the effect of carbon recombination on the equation of state has been included in the carbon-rich region (although it is believed to be small, IR82a). A Reimers mass-loss formula was included, although the value of η used is believed to be too large, possibly by as much as a factor of two. As in the calculations of JL2 (and some of JL1), the models are evolved from the ZAMS. Note that BS evolved their models through the core helium flash, whereas JL "jumped over" it, after obtaining the core mass and envelope abundance changes. This difference is negligible. Incidentally, it is worth noting that BS3 find that the helium flash is ignited in the center of their models, contrary to all other calculations including neutrino losses (except for Mazzitelli & D'Antona 1986). Although in subsequent calculations one model ignited carbon off-center (Boothroyd, private communication), the reason for this difference with virtually all other codes is unknown. (Unpublished calculations by Lattanzio also showed off-center ignition.) BS3 state that a possible reason is the use of different screening corrections.

The calculations of BS show the importance of mass loss, as mentioned earlier. Even though they have overestimated its effectiveness, the reduction of the envelope mass severely limits the total number of pulses experienced on the AGB. They showed that only models with initial masses $\lesssim 2 M_\odot$ can become low luminosity carbon stars and still satisfy the observed initial-final mass relation of Weidemann & Koester (1983). They also show that the flash strength, as measured by the maximum luminosity due to helium burning, L_{He}^{max}, is not simply a function of M_H, but depends on the total mass and the composition. This, unfortunately, means one cannot adjust the envelope mass of a given model in the hope that the result will accurately reflect the effects of mass loss. This is especially true of the dredge-up phase, which depends very sensitively on the flash strength.

BS3 showed that the composition of the carbon rich region is virtually independent of both stellar mass and composition, being approximately 20% carbon and only

~ 2% ^{16}O, even with the increased rate for the ^{12}C$(\alpha,\gamma)^{16}$O reaction. For models with $Z = 0.001$ a semiconvective region formed at the top of the carbon pocket, as found by IR82a,b and DH1,2. However the models of BS lacked the fine resolution (in mass co-ordinate) obtained by Hollowell, and this may be why they find an order of magnitude less ^{13}C than DH2.

For models with $\alpha = 1.0$ BS4 find no dredge-up of carbon. For their model with $(M, Y, Z) = (2.0, 0.24, 0.001)$ on the ZAMS it was necessary to increase α to 1.5 to obtain dredge-up. This change was made between the 9th and 10th pulses, and resulted in a carbon star of 1.72$M_\odot$ with a minimum luminosity of $M_{bol} = -4.68$ during the extended post-flash dip. Interestingly, and possibly quite importantly, no subsequent pulses produced dredge-up. BS claim that this is due to the increased metallicity (carbon) in the envelope, and also the reduction of envelope mass due to mass loss. Yet neither of these effects were included in the models of Iben (1983), who observed similar behaviour.

No dredge-up had been found during the evolution of the model with ZAMS values $(M, Y, Z) = (1.2, 0.24, 0.001)$. The evolution was repeated with α=1.5, 2.0 and 3.0, the change to new values of α being made between the fifth and sixth pulses. Dredge-up was obtained only for $\alpha = 3.0$, with the model becoming a carbon star on the next pulse and with M_{bol} dropping to -3.59. Again, further dredge-up did not occur. It is worth noting that in both of the sequences which became carbon stars, no semiconvection was seen when carbon was actually dredged to the surface, although it was seen in both subsequent and previous pulses.

Although BS succeeded in making low luminosity carbon stars, the requirement of α as large as 3.0 is disturbing. It seems that $\alpha \simeq 1.5$ is capable of matching models to observations over a surprisingly large range of evolutionary stages (e.g. Wood 1981, JL1). Nevertheless, BS4 remind us that various constants of order unity enter the mixing length theory of convection, and different implementations may not match exactly. BS4 state that $\alpha_{Iben} = 1.5$ is equivalent to $\alpha = 2.5$–3.0 in "all other codes that we are aware of". Yet Hollowell, using an Iben code, does not find that $\alpha_{Iben} = 1.5$ is *sufficient* to induce dredge-up in a model similar to the BS4 model which required $\alpha = 3$ (actually, all one can say is that the critical value of α is between 2 and 3). In summary, there does seem to be some variation in models constructed with different codes but "identical" α's. A source of calibration could be the construction of a standard solar model. Lattanzio needs $\alpha = 1.42$ (Lattanzio 1984), BS3 needs $\alpha \simeq 2$, while it is unknown what value the current Iben code would require. VandenBerg (1983) needs α =1.4–1.5. Perhaps a calibration with the same opacities and abundances would be a worthwhile exercise.

Encouraging results of the BS study include confirmation of the composition dependence found by Lattanzio. Also note that the models of JL1 which were claimed to have reached full flash amplitude probably have not, as noted by BS2 and BS3. BS find L_{TP} values in good agreement with JL1 and JL2, and the importance of this has been discussed above. On the negative side is the fact that BS may need large values of α to obtain dredge-up in lower mass stars (but see below). Finally, the importance of mass loss has been stressed by BS, although this can work both for and against carbon star formation.

The Models of Hollowell

Hollowell (see DH1 and DH2) has carefully investigated the effect of carbon recombination on opacities and the semiconvection first obtained by IR82a,b. These models use the latest Los Alamos opacities, including (fits to) tables for carbon-rich mixtures (which are provided for many carbon abundances, probably allowing for more accurate determination of the carbon dependence than in Lattanzio's models). No allowance is made for the effect of carbon recombination on the equation of state (believed to be small, IR82a), nor for any carbon added to the envelope as a result of dredge-up. Mass loss is not included. The model studied by Hollowell was previously studied by Iben (1982, 1983), where it is stated that the model had been evolved from the ZAMS by Despain, although we are not told if semiconvection (or core breathing pulses) were found. Note that Despain (1981) did include semiconvection in calculations of a 0.6 $M_\odot$ model, and it would be expected to occur in a 0.7$M_\odot$ star also.

DH2 shows the opacity for carbon rich mixtures for $Z = 0.001$ and $Z = 0.02$. The opacity bump due to recombination of carbon is seen near 10^6K in the $Z = 0.001$ mixture but is not obvious in the more metal rich mixture. This explains why semiconvection has so far only been seen in models with metallicity $Z = 0.001$.

In an attempt to minimize convergence difficulties found by IR82a, due to instantaneous mixing of large regions of very different abundance, Hollowell has developed a "random walk" model for time-dependent convection. In this formulation convection mixes abundances only over a finite region L_{mix}, determined by the time-step and the convective velocity (see DH2 for details), rather than instantaneously throughout the entire convective zone. While one may criticize this (or any) particular formulation, it is physically motivated and not unreasonable.

Basically, Hollowell confirms the picture painted by Iben & Renzini. After a pulse the top of the carbon rich pocket becomes semiconvective (at $T \simeq 5 \times 10^6$K), mixing carbon outward by a small distance (in mass). Hollowell does not find that carbon is mixed sufficiently far for the inner edge of the convective envelope to penetrate the carbon enhanced zones, as necessary for the third dredge-up. Note that this is despite using $\alpha_{Iben} = 1.5$. The semiconvection does, however, mix hydrogen inward quite a distance (see DH1 and DH2) This results in the ignition of the ^{13}C neutron source, and the formation of s-process elements during subsequent AGB evolution. (Note that DH2 and Hollowell & Iben 1988 provide a detailed analysis of the nucleosynthesis occurring in these models, including the formation of s-process elements. This will not be discussed in this paper.)

Motivated by the fact that low luminosity carbon stars do exist, DH2 then repeated the calculations but with convective motions overshooting by a distance L_{mix}, which was never allowed to exceed one pressure scale height. In this case Hollowell obtained dredge-up of both carbon and s-process elements. His model became a carbon star with a post-flash luminosity reaching down to $M_{bol} = -4.3$ and $M_H = 0.639$. While it seems unfortunate that some convective overshooting is required to obtain dredge-up, there can be little doubt that this is a real phenomenon. Only the precise extent and details are unknown. Nevertheless, it is encouraging that stars of such low mass $(0.7 M_\odot)$ can become carbon stars at the luminosities required by the observations, and can indeed be a source of s-process elements.

SUMMARY

One thing is clear from the calculations discussed above. Theory is now in a much better position to confront the observations. The parametrized input used in the synthetic AGB star distributions of Iben (1981) and Renzini & Voli (1981) has been shown to be inaccurate for (initial) masses in the range 1–3 $M_\odot$, which probably form the bulk of the Magellanic Cloud carbon stars.

Each of the recent sets of calculations (Lattanzio; Boothroyd & Sackmann; Hollowell) succeeds in producing carbon stars of quite low luminosity. The models of Lattanzio have larger envelope masses than appropriate, however. This favors dredge-up but delays carbon star formation. The Boothroyd & Sackmann models may require large values of α to obtain dredge-up at low luminosities. But recall that BS included molecular opacity sources in their envelopes. This cools the envelope substantially. To return the envelope to the original temperature would require a larger α. That BS need $\alpha \approx 2$ to make a solar model while other codes need 1.4–1.5 is consistent with this picture. Perhaps we should not be too hasty in criticizing larger α in the BS code. Hollowell also forms carbon stars (and s-process elements), but requires some form of convective overshooting. That overshooting is real is not denied, but the extent is unknown and a good understanding is lacking.

Since different evolutionary codes differ in some respects (such as L_{He}^{max}), it may be worth making a detailed comparison between codes. One identical model should be distributed to each investigator, and the subsequent evolution compared. The insights gained would make this a valuable exercise, and aid future comparisons.

Let us now discuss the physics which should be included in the ideal AGB star stellar structure code. Firstly, the models should be evolved from the ZAMS, to accurately find the core mass at ignition of the core helium supply. Mass loss should be included, as should semiconvection. We must determine if core breathing pulses occur, and if so include them. (If not, then we must understand why our present understanding predicts them.) Of course, the most accurate opacities and nuclear reaction rates must be included. One should include opacities for carbon rich regions, and also allow for the opacity of any carbon which is mixed into the envelope. The effect of carbon recombination on the equation of state will be important for the carbon pocket at temperatures below few$\times 10^5$K, and must be included if these temperatures are reached. Although the carbon added to the envelope constitutes only a small fraction by mass, it will be experiencing temperatures down to a few thousand degrees. At these temperatures the recombination of carbon may effect the equation of state. The formation of various carbon molecules in the envelope will effect the opacity, and must be included. (With these opacity sources it will probably be necessary to recalibrate the values of α and η needed to match the observations.) From a numerical point of view, it should be noted that some authors allow a model to converge *before* finding the convective boundaries (*e.g.* BS3; Becker & Iben 1979). Mixing of the abundances is then performed between those boundaries. The resultant model is somewhat inconsistent, because the abundances feed into the structure equations, which have been solved for a different composition than is finally indicated after mixing. Although the errors will probably be small for small time-steps, a better procedure (somewhat more expensive in computer time, but not excessively) is to calculate the convective boundaries after each iteration, and perform any mixing at that time (JL1). Thus the model is internally consistent.

To include all these modifications in any stellar structure code would be a laborious exercise, but it is also bound to be fruitful. Note that some of these effects have been included in recent calculations, but no-one has considered all of them. In this sense, the works of Lattanzio, Boothroyd & Sackmann, and Hollowell are complimentary, each addressing some different aspect of AGB star evolution.

Finally, in the hope that we have now solved the low luminosity carbon star problem (or that the "ideal" code described above will solve it), what can we say about the absence of high luminosity carbon stars ? Mass loss has long been suspected to be at least partly to blame, the implication being that the rate of mass loss must be higher than used in the synthetic distributions of Iben (1981) and Renzini & Voli (1981). A recent, and intriguing, development has been the suggestion that previous estimates of $M_{up}(=M_5$ in JL4), the maximum (initial) stellar mass which will develop a degenerate carbon-oxygen core, have been too large (Renzini et al. 1985). Because stars more massive than this will ignite their core carbon supply, M_5 is the maximum mass star which can appear on the AGB. Certainly core breathing pulses and convective overshooting both act to reduce M_5 (Castellani et al. 1985b, Chiosi et al. 1987). Once a star reaches the AGB, of course, it will be mass loss which determines the maximum luminosity which it will attain. But with the most massive stars denied passage to this phase we may now understand why there are so few very bright ($M_{bol} < -6$) AGB stars.

I would like to thank Arnold Boothroyd, Dave Hollowell, and Peter Wood for their assistance in assembling (and helping me to digest) the information in this paper.

References

Becker, S.A., and Iben, I., Jr., 1979, *Astrophys. J.*, **232**, 831.
Becker, S.A., and Iben, I., Jr., 1980, *Astrophys. J.*, **237**, 111.

Bertelli, G., Chiosi, C., and Lattanzio, J.C., 1988, in preparation.

Blanco, B.M., Blanco, V.M., and McCarthy, M.F., 1978, *Nature*, **271**, 638.

Blanco, V.M., McCarthy, M.F., and Blanco, B.M., 1980, *Astrophys. J.*, **242**, 948.

Boothroyd, A.I., and Sackmann, I.-J., 1988a, *Astrophys. J.*, **328**, 632 (BS1).

Boothroyd, A.I., and Sackmann, I.-J., 1988b, *Astrophys. J.*, **328**, 641 (BS2).

Boothroyd, A.I., and Sackmann, I.-J., 1988c, *Astrophys. J.*, **328**, 653 (BS3).

Boothroyd, A.I., and Sackmann, I.-J., 1988d, *Astrophys. J.*, **328**, 671 (BS4).

Bressan, A., Bertelli, G., Chiosi, C., 1981, *Astron. Astrophys.*, **102**, 25.

Castellani, V., Chieffi, A., Pulone, L., and Tornambe, A., 1985a, *Astrophys. J. Lett.*, **296**, 204.

Castellani, V., Chieffi, A., Pulone, L., and Tornambe, A., 1985b, *Astrophys. J.*, **294**, L31.

Castellani, V., Giannone, P., and Renzini, A., 1971a, *Astrophys. Space. Sci*, **10**, 340.

Castellani, V., Giannone, P., and Renzini, A., 1971b, *Astrophys. Space. Sci*, **10**, 355.

Chiosi, C., Bertelli, G., and Bressan, A., 1987, in *Late Stages of Stellar Evolution*, Ed. S. Kwok & S.R. Pottasch, (Dordrecht:Reidel), 213.

Cox, A.N., and Stewart, J.N., 1970, *Astrophys. J. Suppl. Ser.*, **19**, 261.

Despain, K.H., 1981, *Astrophys. J.*, **251**, 639.

Gingold, R.A., 1976, *Astrophys. J.*, **204**, 116.

Hollowell, D.E., 1987, in *Late Stages of Stellar Evolution*, Ed. S. Kwok & S.R. Pottasch, (Dordrecht:Reidel), 239 (DH1).

Hollowell, D.E., 1988, Ph.D. thesis, University of Illinois (DH2).

Hollowell, D.E., and Iben, I., Jr., 1988, preprint.

Huebner, W.F., Merts, A.L., Magee, N.H., Jr., and Argo, M.F., 1977, *Los Alamos Scientific Laboratory Report LA-6760-M*.

Iben. I., Jr., 1975a, *Astrophys. J.*, **196**, 525.

Iben. I., Jr., 1975b, *Astrophys. J.*, **196**, 549.

Iben. I., Jr., 1976, *Astrophys. J.*, **208**, 165.

Iben. I., Jr., 1981, *Astrophys. J.*, **246**, 278.

Iben, I., Jr., 1982, *Astrophys. J.*, **260**, 821.

Iben, I., Jr., 1983, *Astrophys. J. Lett.*, **275**, L65.

Iben, I., Jr., 1984, in *Observational Tests of the Stellar Evolution Theory*, Ed. A. Maeder & A. Renzini, (Dordrecht:Reidel), 3.

Iben, I., Jr., 1987, in *Late Stages of Stellar Evolution*, Ed. S. Kwok & S.R. Pottasch, (Dordrecht:Reidel), 175.

Iben, I., Jr., and Renzini, A., 1982a, *Astrophys. J. Lett.*, **259**, L79 (IR82a).

Iben, I., Jr., and Renzini, A., 1982b, *Astrophys. J. Lett.*, **263**, L65 (IR82b).

Iben, I., Jr., and Renzini, A., 1983, *Ann. Rev. Astr. Astrophys.*, **21**, 271 (IR83).

Iben, I., Jr., and Renzini, A., 1984, *Phys. Rept.*, **105**, 329.

Lattanzio, J.C., 1984, Ph.D. thesis, Monash University, Australia (JL1).

Lattanzio, J.C., 1986, *Astrophys. J.*, **311**, 708 (JL2).

Lattanzio, J.C., 1987a, in *Late Stages of Stellar Evolution*, Ed. S. Kwok & S.R. Pottasch, (Dordrecht:Reidel), 235 (JL3).

Lattanzio, J.C., 1987b, *Astrophys. J. Lett.*, **313**, L15 (JL4).

Lattanzio, J.C., 1988, in *Origin and Distribution of the Elements*, Ed. G.J. Mathews, (Singapore:World Scientific), 398

Magee, N.H., Jr., Merts, A.L., and Huebner, W.F., 1975, *Astrophys. J.*, **196**, 617.

Mazzitelli, I., and D'Antona, F., 1986, *Astrophys. J.*, **311**, 762.

Renzini, A., Bernazanni, M., Buonanno, R., and Corsi, C.E., 1985, *Astrophys. J. Lett.*, **294**, L7.

Renzini, A., and Fusi Pecci, F., 1988, preprint.

Renzini, A., and Voli, M., 1981, *Astron. Astrophys.*, **94**, 175.

Richer, H.B., 1981, *Astrophys. J.*, **243**, 744.

Robertson, J.W., and Faulkner, D.J., 1972, *Astrophys. J.*, **171**, 309.

Sackmann, I.-J., 1980, *Astrophys. J. Lett.*, **241**, L37.

Sackmann, I.-J., and Boothroyd, A.I., 1985, *Astrophys. J.*, **293**, 154.

Schonberner, D., 1979, *Astron. Astrophys.*, **79**, 108.

Smith, V.V., and Lambert, D.L., 1986, *Astrophys. J.*, **311**, 843.

Sweigart, A.V., and Demarque, P., 1972, *Astron. Astrophys.*, **20**, 445.

Sweigart, A.V., and Demarque, P., 1973, in *Variable Stars in Globular Clusters and Related Systems*, Proc. IAU Coll. No. 21, Ed. J.D. Fernie, (Dordrecht:Reidel), 221.

VandenBerg, D.A., 1983, *Astrophys. J. Suppl. Ser.*, **51**, 29.

Weidemann, V., and Koester, D., 1983, *Astron. Astrophys.*, **121**, 77.

Wood, P.R., 1981, in *Physical Processes in Red Giants*, Ed. I. Iben, Jr., & A. Renzini, (Dordrecht:Reidel), 135

Wood, P.R., Bessell, M.S., and Fox, M.W., 1983, *Astrophys. J.*, **272**, 99.

s-PROCESS ENRICHMENT IN LOW-MASS AGB STARS

R.Gallino

Istituto di Fisica Generale dell'Università, Torino, Italy

Abstract. After a brief description of the developments of the theory of s-process nucleosynthesis, the difficulties recently encountered in envisaging reliable astrophysical conditions for obtaining a solar-system distribution of s-isotopes are discussed. In particular, while the reaction $^{22}Ne(\alpha,n)^{25}Mg$ may account for the nucleosynthesis of the weak s-component in massive stars, it fails to reproduce the main s-component in intermediate mass stars. The efficiency of the alternative reaction $^{13}C(\alpha,n)^{16}O$ occurring in low mass stars during recurring thermal instabilities of the He shell is then analyzed. It is shown that, contrary to previous expectations, the ^{13}C source well reproduces the main component, provided that realistic physical conditions are assumed for the temporal behaviour of the pulse and the effect of the light n-absorbers (especially ^{12}C) is properly taken into account. The results satisfactorily compare with the constraints of the classical s-analysis. Key observational evidences also appear to be in agreement with this scenario.

1. INTRODUCTION

The analysis of Clayton et al. (1961) and Seeger et al. (1965) of the products $\sigma_i N_{s,i}$ of s-only isotopes, where σ_i is the n-capture cross section of a given isotope and $N_{s,i}$ its solar-system abundance, demonstrated that the s-isotopes cannot be reproduced by a unique neutron irradiation, but more likely by an exponential distribution of neutron exposures

$$\rho(\tau)d\tau = Ge^{(-\tau/\tau_0)}d\tau$$

where $\rho(\tau)d\tau$ represents the number of iron seeds irradiated with a neutron exposure between τ and $\tau + d\tau$. The neutron exposure $\tau = \int N_n v_T \, dt$ is the time integrated product of neutron density and thermal velocity for a given irradiation. The fitting constant τ_0 is called the *neutron exposure parameter*, and $G = f N_{56}/\tau_0$, being f the fraction of iron seed nuclei N_{56} that suffered n-captures. Actually, as first suggested by Clayton and Rassbach (1967), and extensively discussed by Ward and Newman (1978), Käppeler et al. (1982), Beer (1986a,b), and Walter et al. (1986), at least two different distributions of neutron exposures are required: a *weak* component, which is responsible for the synthesis of the s-nuclei with $A \leq 90$, and a *main* component to account for the heavier isotopes; a third *strong* component is not excluded, in order to reproduce the heaviest Pb isotopes. For the weak, as well as for the strong component, a single irradiation appears more suitable to reproduce the data than an exponential distribution of neutron exposures (Beer, 1986b; Beer and Macklin, 1988b). Fig. 1 shows the best fit of the

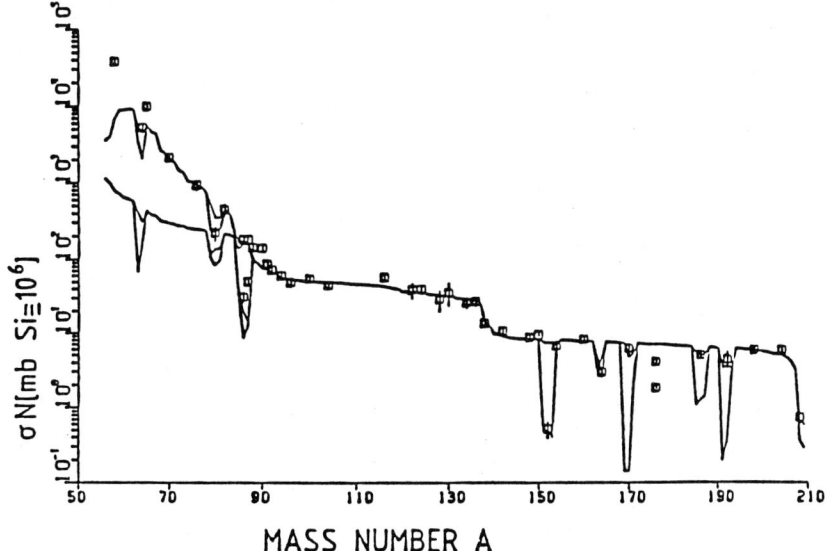

Fig. 1. Solar-system σN distribution of s-only isotopes fitted by a *main* pulsed s-component. Below $A = 90$ the *weak* s-component with a single irradiation is superimposed and the main component is shown separately. In the region of the heaviest isotopes the *strong* s-component is effective. [From Beer and Macklin, 1988b].

Table 1
Parameters of the three s-process components.

	Fraction of iron seeds(%)		Exposure τ_0 (mb^{-1})		n_c
	SF	EE	SF	EE	EE
Main	—	0.048 ± 0.003	—	0.30 ± 0.01	11.2 ± 0.7
Weak	0.32	0.26	0.23	0.06 ± 0.01	1.4 ± 0.4
Strong	$0.9\,10^{-4}$	$1.2 \pm 0.7\,10^{-4}$	≥ 2.5	≥ 6	≥ 150

SF: single flux s-process

EE: s-process with exponential exposure distribution

Data are from:

Beer (1986a,b), Beer et al. (1984), Walter et al. (1986), Beer and Macklin (1988b).

σN_s distribution, and Table 1 shows the characteristic parameters for the three components.

The classical analysis of s-processing can be undertaken independently of the true astrophysical sites and neutron sources. Concerning n-sources, for a long time the most promising ones are known to be the two reactions ^{22}Ne$(\alpha,n)^{25}$Mg and ^{13}C$(\alpha,n)^{16}$O (Cameron, 1955; Burbidge et al., 1957; Reeves, 1966). Both reactions are typical of He-burning environments. Large amounts of ^{22}Ne are expected from the chain of reactions ^{14}N$(\alpha,\gamma)^{18}$F$(\beta^+\nu)^{18}$O$(\alpha,\gamma)^{22}$Ne that occurs at the very beginning of He-burning. Nevertheless, the ^{22}Ne n-source requires rather high temperatures ($T \gtrsim 3 \ 10^8$ K) to be activated. On the contrary, the ^{13}C source can easily take place at lower temperatures ($T \simeq 1.5 \ 10^8$ K); more difficult is to envisage how sufficient amounts of ^{13}C should be present there. This requires some mixing process to dredge down a small amount of protons from the envelope into the ^{12}C-rich region, in order to make the formation of ^{13}C possible through the chain ^{12}C$(p,\gamma)^{13}$N$(\beta^+\nu)^{13}$C, without further proceeding to the reaction ^{13}C$(p,\gamma)^{14}$N. Actually, the lifetime of ^{13}C against p-captures being about 4 times less than that of ^{12}C, the production of a certain fraction of ^{14}N is unavoidable; this works against the synthesis of s-isotopes, since ^{14}N itself is a strong n-poison, via the reaction ^{14}N$(n,p)^{14}$C.

As for the astrophysical sites for s-processing, a long series of researches have been addressed to the analysis of the He-burning phases in stars of different masses. In *massive stars*, ($M \gtrsim 15 \ M_\odot$), the ^{22}Ne source can only be activated near core He-exhaustion or in the subsequent He-shell phase. The nucleosynthesis of s-nuclei has been followed within the framework of different stellar evolutionary scenarios and nuclear reaction networks, with substantial agreement in the results. Since one is faced with a single neutron irradiation, with a fairly low number of neutrons captured per iron seed, significant overproductions are obtained only for s-isotopes with $A < 80$ (Peters, 1968; Couch et al., 1974; Lamb et al., 1977). Recent revisions in the nuclear reaction rates, particularly the increased rate of the ^{12}C$(\alpha,\gamma)^{16}$O reaction, and the increased importance of the major light n-poisons ^{22}Ne and ^{25}Mg, reduce the efficiency of s-processing in He-burning cores (Busso and Gallino, 1985; Thielemann and Arnett, 1985; Gallino and Busso, 1985; Prantzos et al., 1987). Nevertheless, it is still commonly believed that the weak s-component has to be ascribed to massive stars so that they can account for the observed s-enhancements in halo Pop. II stars (Spite and Spite, 1978; Barbuy et al., 1985; Krishnaswamy-Gilroy et al., 1988).

In stars less massive than about 8 $M_\odot$ the He-shell burning phases are potentially more attractive for the synthesis of the bulk of s-elements. These stars, while ascending the asymptotic red giant branch (AGB), suffer a double-shell burning phase, which acts through recurring thermal instabilities. During a thermal pulse, a convective shell grows in the region between the carbon-oxygen core and the hydrogen-helium discontinuity. This behaviour was discovered by Schwarzschild and Härm (1965) for a 1 $M_\odot$ star and confirmed for stars in a large mass range up to about 8 $M_\odot$ (Weigert, 1966; Iben, 1975a; Sugimoto and Nomoto, 1975; Becker, 1981 and references therein). Schwarzschild and Härm (1967) found that a small amount of proton-rich material can be dredged down from the external envelope into the convective He shell, thus activating the ^{13}C source. The corresponding s-processing was investigated by Sanders (1967). In a similar context, Ulrich (1973) demonstrated that, independently of the neutron source, an exponential distribution of neutron exposures is naturally achieved by a pulsed mechanism in which the s-processed material undergoes repeated n-irradiations, each followed by a mixture of fresh Fe-seeds and fresh n-producing nuclei. Actually, Iben (1976) pointed out that an entropy barrier driven by radiation pressure prevents the penetration of the convective He-shell into the hydrogen-rich envelope, and consequently the ^{13}C source would never operate. Nevertheless, following Iben (1975a,b), in *intermediate mass stars* (IMS: $3 \ M_\odot \lesssim M \lesssim 8 \ M_\odot$), during AGB phases the temperature at the bottom of the convective shell increases from one pulse to another up to a rather high asymptotic value: $T_b \gtrsim 3.5 \ 10^8$ K. In this scenario, the repeated process of n-exposures envisaged by Ulrich would indeed occur in IMS, with the ^{22}Ne source replacing the ^{13}C one. On the basis of this hypothesis, Truran and Iben (1977) showed that in IMS an asymptotic distribution of s-process abundances is obtained that fits well the solar abundance pattern for $80 \lesssim A \lesssim 200$.

Since then, many authors (see Mathews and Ward, 1985 and references therein) have examined the details of the s-process chain, emphasizing the importance of taking into account several branches. Besides, more detailed calculations on s-processing were made possible by the strong improvements both in the available experimental set of neutron-capture cross sections at stellar energies and in the evaluation of the solar-system abundances. The combined information coming from network calculations in stellar models and from analyses of the s-process soon showed that the ^{22}Ne source suffers serious problems. Indeed, according to Iben (1977), the maximum temperature reached at the bottom of the convective shell increases from $T_b = 264$ to 415 million degrees for C−O core masses ranging from 0.8 to 1.36 $M_\odot$; consequently, the maximum neutron density varies from 10^8 to 10^{11} cm^{-3}. In these conditions, only for core masses near the minimum limit for the activation of the ^{22}Ne source could the neutron density be reconciled with the s-process conditions, but in that case the neutron exposure parameter τ_0 is too low. For higher core masses, the high n-densities imply a neutron irradiation intermediate between the s- and the r-process (Despain, 1980), and that would lead to a non-solar distribution of s-isotopes. The possibility was envisaged that the final decline of the n-density could help in overcoming this difficulty, since the resulting abundances are strongly influenced by the n-density tail (Cosner et al., 1980). However, as pointed out by Howard et al. (1986), the final drop in neutron density is too rapid for stars with high core masses. Another problem arises from the increased effect of the light n-poisons, particularly ^{22}Ne. Indeed, when adopting the set of cross-sections recommended by Bao and Käppeler (1987: hereafter BK87), the ^{22}Ne source in IMS becomes inefficient in producing the heavy s-isotopes (Busso et al., 1988a). The very recent reduction in the cross sections of Ne isotopes suggested by Winters and Macklin (1988) could lead to a possible revision of this scenario, but the achievement of solar isotopic ratios would still be precluded by the above considerations. Besides these intrinsic nuclear problems, the activation of the ^{22}Ne source in thermally pulsing IMS is also faced with increasing observational difficulties. Indeed, the number of AGB stars with $M_{bol} < -6$, at which the ^{22}Ne source is supposed to be effective, have not been found in the expected frequency in LMC surveys (Blanco et al., 1980; Cohen et al., 1981; Mould and Reid, 1987; see also Habing, 1987 for a similar problem in the Galaxy). Moreover, together with s-isotopes, the ^{22}Ne source produces high overabundances of ^{25,26}Mg, but Smith and Lambert (1986, 1988) did not observe in S stars any significant overabundance of these n-rich Mg isotopes relative to ^{24}Mg. These facts work against the operation of ^{22}Ne source in IMS. Finally, the observation in low mass AGB stars of the long-lived unstable isotopes ^{99}Tc (Smith and Wallerstein, 1983; Dominy and Wallerstein, 1986; Little et al., 1987, Wallerstein and Dominy, 1988), and ^{93}Zr (Peery and Beebe, 1970; Zook, 1978, 1985), clearly indicates that s-processing is operating *in situ* during these phases by a source which has to be different from the ^{22}Ne one. A thorough discussion of the above observational constraints and of their implications for stellar models is given in Hollowell and Iben (1988a).

2. THE ^{13}C SOURCE

The problems encountered by the ^{22}Ne source in the synthesis of the main s-component in IMS brought many authors to reanalyze the conditions for the activation of the alternative source ^{13}C$(\alpha,n)^{16}$O in *low mass stars* (LMS). As previously mentioned, the main difficulty concerning this reaction is that it requires some mixing between the p-rich and the C-rich zones. Iben and Renzini (1982a; 1982b, hereafter IR82) suggested that a suitable mechanism for this process to occur can indeed operate in LMS of low metallicity. After the occurrence of each thermal instability, in the interpulse phase the C-rich material is pushed to low temperatures and the local opacity strongly increases owing to partial recombination of carbon (Sackmann, 1980). This fact allows the formation of a semiconvective zone where a small amount of hydrogen is dredged down into carbon enriched zones. The entropy barrier problem discussed above is thus overcome. About $2\ 10^4$ yr later the region heats up and the reaction chain ^{12}C$(p,\gamma)^{13}$N$(\beta^+\nu)^{13}$C$(p,\gamma)^{14}$N consumes all the hydrogen, producing a ^{13}C- (and ^{14}N)-pocket of a few $10^{-4}\ M_\odot$. During the next pulse, this pocket is engulfed in the convective He-shell, where ^{13}C nuclei easily suffer α-captures and release neutrons. The effectiveness of this mechanism critically depends on the complex physics of convective boundaries and on the treatment of ^{12}C opacities in the recombination phase. Actually, Boothroyd and Sackmann (1988a) did not find this process to be important in their LMS models; nevertheless, Hollowell and Iben (1988b,c) confirmed the possibility of formation of a consistent ^{13}C-pocket. Again, first estimates of neutron densities were in the range of 10^{11} to 10^{12} cm^{-3} (IR82; Malaney, 1986a,b, 1987; Picchio et al., 1988a), thus implying that the ^{13}C source would produce an s-isotope distribution far from the solar one. In particular, Malaney (1986c, 1987) pointed out that a too high overabundance of ^{96}Zr would result with respect to ^{90}Zr. This fact is at odds with the observations of S stars, showing little or no ^{96}Zr (Peery and Beebe, 1970; Zook, 1978). However, from other points of view the reaction ^{13}C$(\alpha,n)^{16}$O

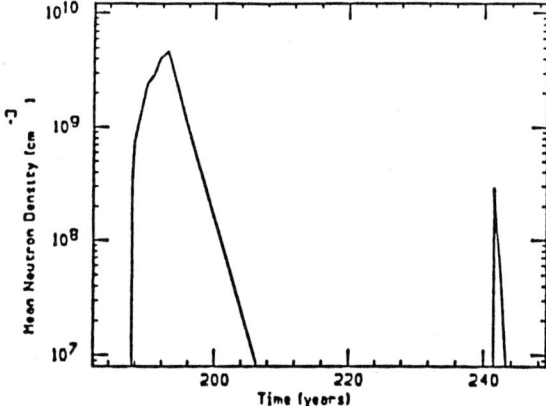

Fig. 2 - Temporal evolution of the neutron density in a low mass AGB star. Time is calculated from the beginning of the thermal pulse. The first neutron burst is caused by the reaction $^{13}C(\alpha,n)^{16}O$ operating at $T_b \simeq 1.5\ 18^8$ K, the second neutron burst is caused by the reaction $^{22}Ne(\alpha,n)^{25}Mg$ operating at $T_b \simeq 3.0\ 10^8$ K.

Table 2
Relative importance of various n-absorbers

a) Cross sections from BK87

	^{12}C	^{14}N	^{20}Ne	^{22}Ne	^{25}Mg	^{56}Fe
$\sigma_i(30keV)$	0.2	1.7	1.1	0.9	6.5	13.1
$10^4 \sigma_i X_i/A_i$	50	0.9	0.05	0.4	0.09	0.04

b) New cross sections

	^{12}C	^{14}N	^{20}Ne	^{22}Ne	^{25}Mg	^{56}Fe
$\sigma_i(30keV)$	0.2	0.81	0.12	0.045 – 0.24	6.5	13.1
$10^4 \sigma_i X_i/A_i$	50	0.4	0.04	0.02 – 0.10	0.09	0.04

would offer some important advantages. Indeed: (i) it would not imply a too large production of n-rich Mg isotopes; (ii) it would be consistent with the observation of ^{99}Tc (and ^{93}Zr) in LMS of M_{bol} as low as -4.0 (corresponding to a core mass $M_c \simeq 0.6$ $M_\odot$).

In this framework, following the IR82 suggestion for the formation of the ^{13}C-pocket, the pulse model by Iben (1982) for a 0.7 $M_\odot$, $Z = 0.001$ star has been reanalyzed by Gallino et al. (1988a), incorporating a large network of reactions that includes charged-particle processes from He to Si and n-captures from C to Po (see Picchio et al., 1988b for details). The network takes into account more than 100 branching points along the s-chain as well as the reaction channels of the most important isomeric states and adopts the recommended neutron cross sections by BK87, further improved by Ratynski and Käppeler (1988) and Beer and Macklin (1988b). The set of β-decay lifetimes at stellar conditions is from Takahashi and Yokoi (1987), Fuller et al. (1983) and Klay and Käppeler (1988). The adopted solar-system abundances are from Anders and Ebihara (1982). The process of engulfment of the ^{13}C-pocket formed during the interpulse period has been followed in detail, assuming the 11th pulse by Iben (1982) as representative of the asymptotic thermal instabilities. The duration of the entire ingestion phase is of about 5.5 yr; the rate of ingestion (7 10^{-5} $M_\odot/yr$) corresponds to the growing rate of the external profile of the convection zone and was assumed to be constant. As long as ^{13}C is ingested, it is mixed over the convective region and suffers α-captures. Next, the surviving ^{13}C is brought to exhaustion as He-burning proceeds. The ^{13}C(α,n)^{16}O reaction takes place at a mild temperature: $T_b \simeq 1.5\ 10^8$ K. The entire neutron irradiation lasts for about 30 years, reaching an exposure $\Delta\tau(^{13}C) \simeq 0.11$ mb^{-1}.

The resulting mean neutron density $\bar{N}_n$ in the shell is far from being constant (contrary to what is currently assumed in the classical analysis of the s-process), reaching a peak of about 3 10^9 cm^{-3} at the end of the ingestion phase and then decreasing smoothly (Fig. 2). The strong reduction of the neutron density with respect to the previous estimates is obtained thanks to the damping effect of three moderating phenomena: (i) the temperature and density stratification of the convective zone, (ii) the realistic (low) ingestion rate of ^{13}C, (iii) the neutron recycling introduced by n-captures on ^{12}C. Concerning this last effect, one has to notice that ^{12}C nuclei are very abundant in the He-shell, so that a very small n-capture cross section would be sufficient to let them become the major neutron poisons. The neutrons captured by ^{12}C are not lost, but recycled through the sequence ^{12}C(n,γ)^{13}C(α,n)^{16}O. The importance of this n-recycling by ^{12}C for reproducing a solar-system distribution of s-isotopes has to be underlined. This chain acts as a reservoir for neutrons, allowing them to be released at delayed times, the peak $\bar{N}_n$ to be reduced and the $\bar{N}_n$ tail to decline smoothly. Furthermore, the n-recycling on ^{12}C has the effect of reducing the consequences of the somewhat uncertain ingestion rate. Actually, the n-capture cross section of ^{12}C at astrophysical energies is poorly known: the theoretical evaluation by Fowler et al. (1967) at 30 keV is as low as 0.003 mb, while the experimental value by Allen et al. (1971) is higher, $\sigma_{12} = (0.2 \pm 0.4)$ mb, though affected by a large uncertainty. A recent theoretical estimate (Reffo, 1988) gives $\sigma_{12} = (0.02 \pm 0.01)$ mb. The relative importance of n-captures on ^{12}C with respect to the other major light poisons can be evaluated by comparing the various products $N_i \langle \sigma v \rangle_i$ or, apart from a common factor, the products $(X_i/A_i)\,\sigma_i$, which are shown in Table 2. Typical abundances by mass X_i at the $\bar{N}_n$ peak have been used. From Table 2a it appears that ^{12}C dominates over all other n-poisons whenever $\sigma(^{12}\text{C}) \geq 0.006$ mb. The effect is even greater when adopting the most recent estimates for the n-capture cross sections on ^{14}N (Brehm et al., 1988) and on Ne isotopes (Winters and Macklin, 1988), as shown in Table 2b. In this case ^{12}C dominates whenever $\sigma(^{12}\text{C}) \geq 0.002$ mb. Also n-captures on ^{14}N could give rise to a n-recycling effect (Jorissen and Arnould, 1986; Brehm et al., 1988), through the reaction $^{14}\text{N}(n,p)^{14}\text{C}$, followed by $^{12}\text{C}(p,\gamma)^{13}\text{N}(\beta^+\nu)^{13}\text{C}(\alpha,n)^{16}\text{O}$. Nevertheless, during the first ingestion phase, that lasts for about 3.4 years and accounts for about half of the ^{13}C dredged-down, the abundance of ^{14}N is low; in the second phase of ingestion, where ^{14}N is an important n-poison, its recycling accounts for less than 15% of the neutrons released, owing to the concurrent n-poisoning by ^{22}Ne and to the fact that more than about 50% of protons get lost via other p-channels, mainly $^{18}\text{O}(p,\alpha)^{15}\text{N}$.

Near the end of the pulse, the convective shell grows rapidly to its maximum extension, and the s-processed zone is remixed with 40% of He-rich matter containing fresh ^{14}N and fresh iron seeds. There, the temperature at the bottom of the shell sharply increases up to $T_b = 3\ 10^8$ K and then decreases exponentially. All the ^{14}N nuclei are transformed into ^{22}Ne via the chain $^{14}\text{N}(\alpha,\gamma)^{18}\text{F}(\beta^+\nu)^{18}\text{O}(\alpha,\gamma)^{22}\text{Ne}$, about 1% of ^{22}Ne further suffering α-captures. Consequently, a second burst of neutrons at high temperature is released by the reaction $^{22}\text{Ne}(\alpha,n)^{25}\text{Mg}$, which lasts for about 2 years, $\bar{N}_n$ reaching a maximum value of about $4\ 10^8$ cm^{-3}. The total neutron exposure of the pulse is not significantly affected, being $\Delta\tau(^{22}\text{Ne}) \simeq 1\ 10^{-3}$ mb^{-1}. Subsequently, in the interpulse period, which in Iben's 0.7 $M_\odot$ model lasts for about $2\ 10^5$ yr, the long-lived nuclei produced during the two irradiations are allowed to decay according to their lifetimes. The process of nucleosynthesis was followed for 20 subsequent pulses. Typically, an asymptotic distribution of s-isotopes is reached after about 10 pulses.

3. ANALYSIS OF s-PROCESS CONTRIBUTIONS

The behaviour of the n-density sketched in Fig. 2, and in particular its long tail, is exactly what one needs for producing a solar s-isotope distribution, as stressed by Howard et al. (1986). Indeed, the resulting overabundances of s-only and prevailing-s nuclei for $A > 80$ are remarkably flat (Fig. 3), with a strong overproduction factor of about 2000. The various isotopes shown in Fig. 3 have been indicated with different symbols according to the analysis of Käppeler, Beer and Wisshak (1988) of their relative s-contributions. Overall, the agreement is very stringent; the only exceptions are the so-called r-only isotopes and the two isotopes ^{86}Kr, ^{87}Rb. Concerning the former ones, the choice of "a priori" r-only isotopes might be uncorrect, since these nuclei are not *shielded* by the s-process chain. In Fig. 3 no correction has been applied for the very long-lived nuclei with $\tau > 10^{10}$ yr and their by-products. When this correction is introduced,

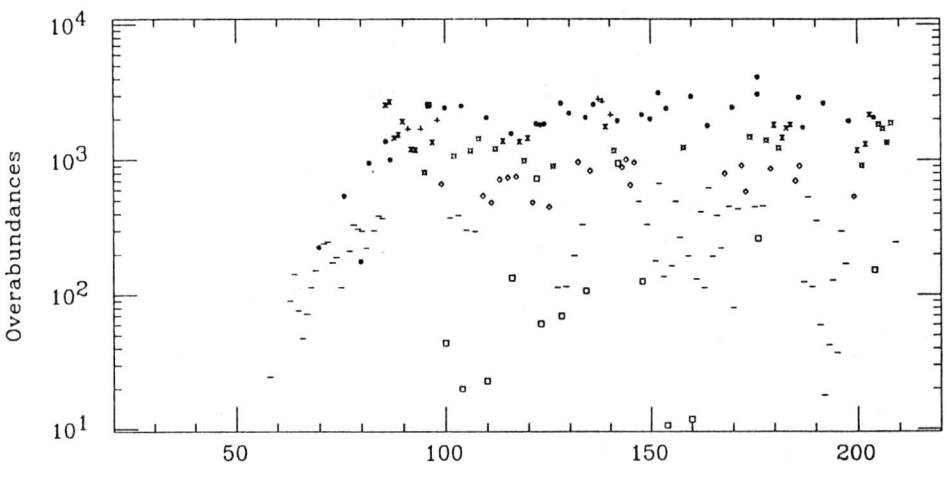

Fig. 3. Overabundances of heavy isotopes in a low mass star following the IR82 model. The n-capture cross sections at 30 keV for the major light n-poisons are indicated at the top (in mb). The s-only isotopes are denoted by a $*$. The prevailing-s isotopes are denoted by a $+$ ($80 \leq s\text{-MAIN} < 100$), or by a $\times$ ($60 \leq s\text{-MAIN} < 80$), where s-MAIN is the relative contribution for the main s-component, according to the analysis of Kappeler, Beer and Wisshak (1988). Isotopes with minor s-contributions are denoted by a *starburst* ($40 \leq s\text{-MAIN} < 60$), by a $\diamond$ ($20 \leq s\text{-MAIN} < 40$), or by a $-$ ($0 \leq s\text{-MAIN} < 20$). The "$r$-only" isotopes are denoted by an *open square*. Initial abundances are from Anders and Ebihara (1982). A quite similar distribution is obtained by reducing the n-cross section on ^{12}C down to $\sigma \simeq 0.003$ mb.

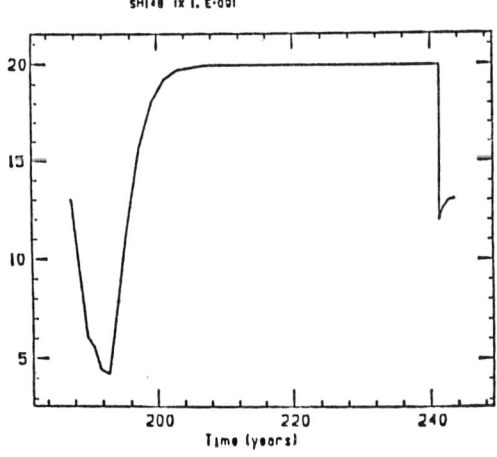

Fig. 4 - Temporal evolution of the mass abundance of ^{148}Sm, a typical branching-dependent s-only isotope. The initial decrease is an effect of the high peak $\bar{N}_n$. The mass abundance is frozen out near ^{13}C-exhaustion when $\bar{N}_n \simeq 2\ 10^8$ cm^{-3}. At $t = 241$ y the convective shell rapidly expands up to its maximum extent, dredging down 40% of the He-rich matter containing original seeds. At the right the small effect of the high temperature phase is shown.

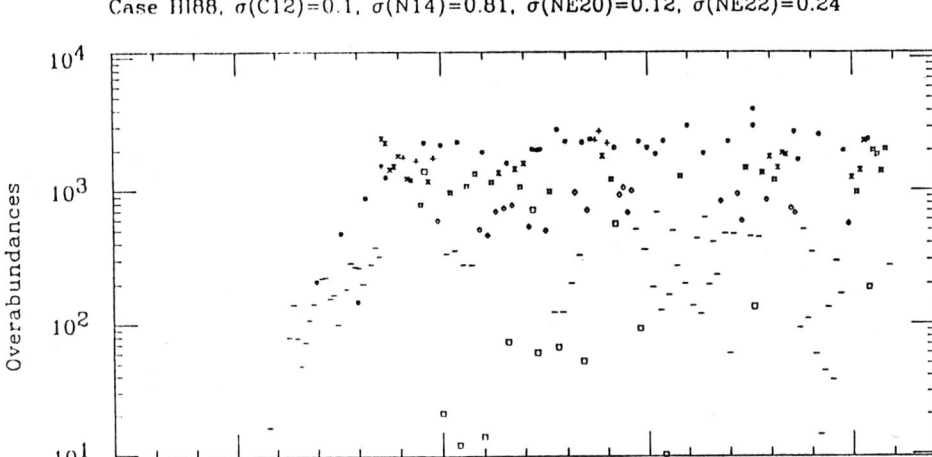

Fig. 7. Distribution of the overabundances of heavy isotopes in the model of Hollowell and Iben (1988), adopting the reduced n-cross sections of light n-poisons shown at the top (in mb). The same distribution is obtained when reducing the n-capture cross section on ^{12}C down to $\sigma_{12} \simeq 0.003$ mb. Initial abundances are from Anders and Ebihara (1982). Notations for the various isotopes are the same as in Fig. 3.

TABLE 3

CHARACTERISTICS OF THE NEUTRON IRRADIATION

Parameter	HR82	HI88
Δn_c	7.89	11.30
$(\bar{N}_n)_M$	4.7(09)	3.1(09)
$\Delta \tau$	0.110	0.154
n_c	11.83	11.30
τ_0	0.215	0.222

lating the s-nucleosynthesis with these new prescriptions and adopting the most recent data for the n-capture cross sections of light n-poisons is shown in Fig. 7. Despite the differences about the formation and engulfment of the ^{13}C-pocket and the reduced effect of light n-poisons, the distribution of s-isotopes is hardly distinguishable from that of Fig. 3. However, an improved reduction in the abundances of r-only isotopes, such as ^{96}Zr, is now obtained.

6. CONCLUSION

So far, the possibility of formation of the ^{13}C-pocket has been envisaged only for low mass stars of low metallicity. This is presently the main theoretical problem concerning the ^{13}C source. A low metallicity favours the occurrence of semiconvection in the intershell region and makes the thermal instability stronger (Iben, 1983; Boothroyd and Sackmann, 1988b). Consequently, both the third dredge-up and the formation of the ^{13}C-pocket are favoured. Stars of 1.5 $M_\odot$ with metallicity 0.003 and 0.006 have been followed by Lattanzio (1987) up to AGB phases, both showing carbon dredge-up. The MS, S, CS, C, CH and Ba stars showing s-enhancements are typically disc population stars, with metallicities of the order of the solar one or slightly lower (particularly the CH and Ba stars, see Cowley and Downs, 1980; Luck and Bond, 1982; Lambert, 1985; Tomkin and Lambert, 1986). In all these peculiar stars the observed ^{12}C is enhanced, thus indicating that the third dredge-up did occur. Then, between the intershell and the envelope a hydrogen/carbon discontinuity forms, which may favour the penetration of a small amount of protons into the carbon-rich zone (Iben and Renzini, 1983). More detailed stellar evolutionary calculations of LMS with different metallicities in AGB phases with improved carbon opacities and treatment of convection are highly desirable. Anyway, to have a first idea of the expectations of the s-abundance distribution in stars of different metallicity, one has to outline the importance of the initial abundance of light n-poisons. Indeed, while the ^{13}C and ^{14}N nuclei produced in the semiconvective zone derive from p-captures on the newly formed ^{12}C and are independent of the original metal content, the Fe seeds and the light n-poisons increase with metallicity. As a first guess, one can assume that the general characteristics of the thermal pulses and of the ^{13}C-pocket remain unchanged while varying the metallicity. The results so far obtained (Raiteri et al., 1988) show that, going towards high Z values, the efficiency of the s-processing continuously decreases due to the increased light n-poisons. The distribution of overabundances remains flat up to about $Z \simeq Z_\odot/3$, whereas for higher metallicities a sharp discontinuity occurs near the magic-neutron nucleus ^{86}Kr. For very metal poor stars the situation is different. Indeed, on one side the decrease of the initial metallicity reduces the amount of Fe seeds, and this tends to favour the process of s-nucleosynthesis; on the other hand, one has to take into account the effect of the primary ^{14}N-nuclei produced together with ^{13}C, that do not vary with the metal content. Consequently, again keeping the efficiency of the ^{13}C-pocket unchanged, an asymptotic condition is reached for $Z \lesssim 10^{-4}$, with $\Delta \tau \simeq 0.15$ and overproduction ratios increasing from 3500 up to 7000 for the heaviest s-only isotopes.

A few noticeable effects concerning isotope ratios are obtained with increasing metallicity. First of all, the ratio ^{96}Zr/^{90}Zr decreases by a factor of 3 in disc population stars. This allows for a full explanation of the ^{96}Zr controversy. More interesting, the relative ratios ^{86}Kr/^{82}Kr and ^{80}Kr/^{82}Kr decrease with increasing metallicity, whereas ^{83}Kr/^{82}Kr and ^{84}Kr/^{82}Kr ratios remain unchanged. This fact may provide a full explanation (Gallino et al., 1988b) of the strange behaviour shown by the exotic s-Kr inclusions in the Murchison meteorite (Ott et al., 1988; Clayton, 1988; Jorgensen, 1988).

Reproduction of the solar-system distribution of the s-isotopes is the first goal to be achieved. Even if the efficiency of s-processing in LMS with increasing metallicity may be questionable, in the framework of current theories of chemical galactic enrichment the main contribution of LMS to the solar s-enrichment has to come from stars of lower metallicity, formed near the end of the galactic halo phase (Busso and Gallino, 1983). For them ($Z \simeq 5 \; 10^{-3}$) a consistent and flat distribution of the heavy s-isotopes in the He-shell can still be obtained.

The second main task is an attempt to reproduce the s-enhancements shown by AGB stars. Smith and Lambert (1985, 1986), and Smith et al. (1987) showed that the s-process abundance distributions in MS and S stars are described by a large range of neutron exposures: $\tau_0 \simeq 0.1 - 0.6 \text{ mb}^{-1}$, with an average value well consistent with the classical s-analysis and with the results discussed above. A possible spread of neutron exposures is expected in LMS, when several aspects like the variation of the temporal profile of the convective zone in the first pulses, the temporal variation of the overlapping factor (see Iben, 1982), the amount of the ^{13}C ingested and the metallicity are more carefully considered. For example, the higher s-enhancements shown by Ba stars could be related to their relatively lower metallicity. Furthermore, the MS and S stars could have suffered only a few thermal pulses and the asymptotic conditions for s-processing could have not been achieved.

ACKNOWLEDGMENTS

I deeply acknowledge Maurizio Busso, Franz Käppeler, Guido Picchio, Claudia M. Raiteri, Gianni Reffo and Alvio Renzini, for continuous discussions and collaboration in obtaining the new results presented here. In particular I thank G. Picchio and M. Busso for help in the discussion and redaction of the present paper. I thank Hermann Beer for the permission of reproducing Fig. 1. I am very indebted to Icko Iben, Jr. and Dave Hollowell for providing me their new results on the formation of the ^{13}C-pocket and for a full and encouraging discussion during this Conference.

REFERENCES

Allen, B. J., Gibson, J. H., and Macklin, R. L. 1971, *Adv. Nucl. Phys.*, **4**, 205.
Anders, E., and Ebihara, M. 1982, *Geochim. Cosmochim. Acta*, **46**, 2263.
Bao, Z. Y., and Käppeler, F. 1987, *Atomic Data and Nucl. Data Tables*, **36**, 411.

Barbuy, B., Spite, F., and Spite, M. 1985, *Astr. Ap.*, **144**, 343.

Becker, S. A. 1981, in *Physical Processes in Red Giants*, ed. I. Jr. Iben and A. Renzini (Dordrecht: Reidel), p.141.

Beer, H. 1986a, in *Nucleosynthesis and its Implications on Nuclear and Particle Physics*, ed. by J. Audouze and N. Mathieu (Dordrecht: Reidel), p. 263.

Beer, H. 1986b, in *Advances in Nuclear Astrophysics*, ed. E. Vangioni-Flam, J. Audouze, M. Cassé, J. P. Chieze and J. Tran Thanh Van (Paris: Editions Frontières), p. 375.

Beer, H., and Macklin, R. L. 1988a, *Ap. J.*, **331**, 1047.

Beer, H., and Macklin, R. L. 1988b, Measurements of the ^{86,87}Rb Capture Cross Section for s-Process Studies, *Ap. J.*, in press.

Beer, H., Walter, G., and Macklin, R. L. 1985, in *Capture Gamma-Rays Spectroscopy and Related Topics*, ed. by S. Raman (New York: American Institute of Physics), p. 778.

Beer, H., Walter, G., Macklin, R. L., and Patchett, P. J. 1984, *Phys. Rev.* **C30**, 464.

Blanco, V. M., McCarthy, M. F., and Blanco, B. M. 1980, *Ap. J.*, **242**, 938.

Boothroyd, A., and Sackmann, I.-J. 1988a, *Ap. J.*, **328**, 653.

Boothroyd, A., and Sackmann, I.-J. 1988b, *Ap. J.*, **328**, 671.

Brehm, K., Becker, H. W., Rolfs, C., Trautvetter, H. P., Käppeler, F., and Ratynski, W. 1988, *Z. Physik* **A330**, 167.

Burbidge, G. R., Burbidge, E. M., Fowler, W. A., and Hoyle, F. 1957, *Rev. Mod. Phys.*, **29**, 54.

Busso M., and Gallino, R. 1983, *Ap. and Space Sci.*, **94**, 273.

Busso M., and Gallino, R. 1985, *Astron. Astrophys.*, **151**, 205.

Busso, M., Picchio, G., Gallino, R., and Chieffi, A. 1988a, *Ap. J.*, **326**, 196.

Busso, M., Gallino, R., Käppeler, F., Picchio, G., and Raiteri, C. M., 1988b, Comparison of s-Processing in Low Mass AGB Stars with the Classical s-Analysis, in preparation.

Cameron, A. G. W. 1955, *Ap. J.*, **121**, 1446.

Clayton, D. D. 1988, *Nature*, **332**, 683.

Clayton, D. D., Fowler, W. A., Hull, T. E., and Zimmerman, B. A. 1961, *Ann. Phys.*, **12**, 331.

Clayton, D. D., and Rassbach, M. E. 1967, *Ap. J.*, **148**, 69.

Cohen, J. G., Frogel, J. A., Persson, S. A., and Elias, J. H. 1981, *Ap. J.*, **249**, 481.

Cosner, K., Iben, I., Jr., and Truran, J. W. 1980, *Ap. J. Letters*, **238**, L91.

Couch, R. G., Schmiedekamp, A. R., and Arnett, W. D., 1974, *Ap. J.*, **190**, 95.

Cowley, C. R., and Downs, P. L. 1980, *Ap. J.*, **236**, 648.

Despain, K. H. 1980, *Ap. J. Letters*, **236**, L165.

Dominy, J. F., and Wallerstein, G. 1986, *Ap. J.*, **310**, 371.

Fowler, W. A., Caughlan, G. R., and Zimmerman, B. A. 1967, *Ann. Rev. Astron. Astrophys.*, **5**, 525.

Fuller, G. M., Fowler, W. A., and Newman, M. J. 1983, *Ap. J. Supp.*, **329**, 943.

Gallino, R., and Busso, M. 1985, in *From Nuclei to Stars*, ed. A. Molinari and R. A. Ricci (Amsterdam: North-Holland), p.309.

Gallino, R., Busso, M., Picchio, G., Raiteri, C. M., and Renzini, A. 1988a, *Ap. J. Lett.*, **334**, L45.
Gallino, R., Busso, M., Picchio, G., Raiteri, C. M., 1988b, An Interpretation of Ne, Kr and Xe Isotopic Anomalies in the Murchison and Murray Meteoritic Inclusions, 1988, in preparation.
Habing, H. J. 1987, in *The Galaxy*, ed. G. Gilmore and B. Carswell (Cambridge: Cambridge Univ. Press), p.173.
Hollowell D. E., and Iben, I., Jr. 1988a, in *Atmospheric Diagnostics of Stellar Evolution*, Proc. IAU Coll. no. 108 (Berlin: Springer-Verlag), in press.
Hollowell, D. E., and Iben, I., Jr. 1988b, *Ap. J. Letters*, **333**, L25.
Hollowell, D. E., and Iben, I., Jr. 1988c, Neutron Production and Neutron-Capture Nucleosynthesis in a Low-Mass, Low-Metallicity AGB Star, preprint.
Howard, W. M., Mathews, G. J., Takahashi, K., and Ward, R. A. 1986, *Ap. J.*, **309**, 633.
Iben, I., Jr. 1975a, *Ap. J.*, **196**, 525.
Iben, I., Jr. 1975b, *Ap. J.*, **196**, 549.
Iben, I., Jr. 1976, *Ap. J.*, **208**, 165.
Iben, I., Jr. 1977, *Ap. J.*, **217**, 788.
Iben, I., Jr. 1982, *Ap. J.*, **260**, 821.
Iben, I., Jr. 1983, *Ap. J. Lett.*, **275**, L65.
Iben, I., Jr., and Renzini, A. 1982a, *Ap. J. Lett.*, **259**, L79.
Iben, I., Jr., and Renzini, A. 1982b, *Ap. J. Lett.*, **263**, L23.
Iben, I., Jr., and Renzini, A. 1983, *Ann. Rev. Astr. Ap.*, **21**, 271.
Jorgensen, U. G. 1988, *Nature*, **332**, 702.
Jorissen, A., and Arnould, M. 1986, in *Nucleosynthesis and its Implications on Nuclear and Particle Physics*, ed. by J. Audouze and N. Mathieu (Dordrecht: Reidel), p. 303.
Käppeler, F. 1986, in *Advances in Nuclear Astrophysics*, ed. E. Vangioni-Flam, J. Audouze, M. Cassé, J. P. Chieze and J. Tran Thanh Van (Paris: Editions Frontières), p. 355.
Käppeler, F., Beer, H., Wisshak, K., 1988, private communication.
Käppeler, F., Beer, H., Wisshak, K., Clayton, D. D., Macklin, R. L., and Ward, R. A. 1982, *Ap. J.*, **257**, 821.
Klay, N., and Käppeler, F. 1988, *Phys. Rev. C***38**, 295.
Krishnaswamy-Gilroy, K., Sneden, C., Pilachowski, C.A., and Cowan, J.J. 1988, *Ap. J.*, **327**, 298.
Lamb, S.A., Howard, W.M., Truran, J.W., and Iben, I., Jr. 1977, *Ap. J.*, **217**, 213.
Lambert, D. L. 1985, in *Cool Stars with Excesses of Heavy Elements*, eds. M. Jaschek and P. C. Keenan, (Dordrecht: Reidel), p.191.
Lattanzio, J. C. 1987, *Ap. J. Lett.*, **313**, L15.
Little, S. J., Little-Marenin, I. R., and Bauer, W. H. 1987, *A. J.*, **93**, 1539.
Luck, R. E., and Bond, H. E. 1982, *Ap. J.*, **259**, 792.
Malaney, R. A. 1986a, *M.N.R.A.S.*, **223**, 683.
Malaney, R. A. 1986b, *M.N.R.A.S.*, **223**, 709.
Malaney, R. A. 1986c, *Advances in Nuclear Astrophysics*, ed. E. Vangioni-Flam,

J. Audouze, M. Cassé, J. P. Chieze and J. Tran Thanh Van (Paris: Editions Frontières), p. 407.

Malaney, R. A. 1987, *Ap. J.*, **321**, 832.

Malaney, R. A., and Lambert, D. L. 1988, *M.N.R.A.S.*, in press.

Mathews, G. J., and Ward, R. A. 1985, *Rep. Progr. Phys.*, **48**, 1371.

Mould, J., and Reid, N. 1987, in *Late Stages of Stellar Evolution*, ed. S. Kwok and S.R. Pottasch, (Dordrecht: Reidel), p.209.

Ott, U., Begemann, F., Yang, J., and Epstein, S. 1988, *Nature* **332**, 700.

Peery, B. F., and Beebe, R. F. 1970, *Ap. J.*, **160**, 619.

Peters, J. G. 1968, *Ap. J.*, **154**, 224.

Picchio, G., Busso, M., Gallino, R., and Raiteri, C. M. 1988a, in *Mass Outflows from Stars and Galactic Nuclei*, ed. L. Bianchi and R. Gilmozzi (Dordrecht: Kluwer Academic Publ.), p. 279.

Picchio, G., Busso, M., Gallino, R., and Raiteri, C. M. 1988b, The Neutron Source $^{13}C(\alpha,n)^{16}O$ in Thermally Pulsing Stars and the s-Processing, preprint.

Prantzos, N., Arnould, M., and Arcoragi J.-P. 1987, *Ap. J.*, **315**, 209.

Raiteri, C. M., Busso, M., Gallino, R., Picchio, G., and Renzini, A. 1988, The Effect of Light n-Poisons on the s-Nucleosynthesis of AGB Low Mass Stars of Different Metallicity, in preparation.

Ratynski, W., and Käppeler, F. 1988, *Phys. Rev.* **C37**, 595.

Reeves, H. 1966, *Ap. J.*, **146**, 447.

Reffo, G. 1988, private communication.

Sackmann, I.-J., 1980, *Ap. J. Lett.*, **241**, L37.

Sanders, R. H. 1967, *Ap. J.*, **150**, 971.

Schwarzschild. M., and Härm, R. 1965, *Ap. J.*, **142**, 855.

Schwarzschild. M., and Härm, R. 1967, *Ap. J.*, **150**, 961.

Seeger, P. A., Fowler, W. A., and Clayton, D. D. 1965, *Ap. J. Suppl.*, **11**, 121.

Smith, V. V., and Lambert, D. L. 1985, *Ap. J.*, **294**, 326.

Smith, V. V., and Lambert, D. L. 1986, *Ap. J.*, **311**, 843.

Smith, V. V., and Lambert, D. L. 1988, s-Process Enriched Cool Stars with and without Technetium: Clues to AGB and Binary Star Evolution, *Ap. J.*, in press.

Smith, V. V., Lambert, D. L., and McWilliam, A. 1987, *Ap. J.*, **320**, 865.

Smith, V. V., and Wallerstein, G. 1983, *Ap. J.*, **273**, 742.

Spite, M., and Spite, F. 1978, *Astr. Ap.*, **67**, 23.

Sugimoto, D., and Nomoto, K. 1975, *Publ. Astr. Soc. Japan*, **27**, 197.

Takahashi, K., and Yokoi, K. 1987, *Atomic Data and Nucl. Data Tables*, **36**, 375.

Thielemann, F.-K., and Arnett, W. D. 1985, *Ap. J.*, **295**, 589.

Tomkin, J., and Lambert, D. L. 1986, *Ap. J.*, **311**, 819.

Truran, J. W., and Iben, I., Jr. 1977, *Ap. J.*, **216**, 797.

Ulrich, R. K. 1973, in *Explosive Nucleosynthesis*, ed. D. N. Schramm and W. D. Arnett (Univ. of Texas: Austin), p. 139.

Wallerstein, G., and Dominy, J. F. 1988, *Ap. J.*, **330**, 937.

Walter, G., Beer, H., Käppeler, F., Reffo, G., and Fabbri, F. 1986, *Astron. Astrophys.*, **167**, 186.

Ward, R. A., and Newman, M. J. 1978, *Ap. J.*, **219**, 195.
Weigert, A. 1966, *Z. Physik*, **64**, 395.
Winters, R. R., and Macklin, R. L. 1988, *Ap. J.*, **329**, 943.
Zook, A. C. 1978, *Ap. J. Lett.*, **221**, L113.
Zook, A. C. 1985, *Ap. J.*, **289**, 356.

THE ROLE OF BINARITY IN THE EVOLUTION OF PECULIAR RED GIANTS

R.D. McClure
Dominion Astrophysical Observatory, National Research Council, Canada

INTRODUCTION

There are several types of Peculiar Red Giants (PRG's) which have enhanced s-process elements and/or carbon, but which cannot be explained by mixing during helium shell-flashing in the late stages of Asymptotic Giant Branch (AGB) evolution. These are the BaII, CH, sgCH, and the hotter R–type carbon stars. All these PRG's have absolute magnitudes which range down to zero and fainter. The BaII stars (Bidelman & Keenan 1951) are G–K giants whose spectra have strong CN and CH bands, and in extreme cases bands of C_2 become noticeable. In addition they have strong lines of s-process elements such as BaII and SrII. The CH stars (Keenan 1942) are Population II equivalents of the BaII stars, and show similar spectral features, but with weaker metal lines, and usually stronger carbon bands. Bond (1974) introduced a new class of late F and G stars named the CH subgiants, which he suggests are fainter than the classical CH and BaII stars. They appear to be of mixed population (Luck & Bond 1982), and they are probably subgiants which will eventually evolve up the giant branch to become classical CH and BaII stars. The R–type carbon stars, at least the hotter (R0–R4) ones, differ from many of the other peculiar red giants in that they do not show enhanced s-process element abundances. The R stars have strong carbon bands, but relative to the N-type carbon stars they have less blue and ultraviolet absorption in their spectra. They range in absolute magnitude from near zero, similar to the BaII stars, up to several magnitudes brighter than this.

In 1976, McClure & Norris (1977) discovered a CH star in the globular cluster M 22, and noted that ω Cen (which also contains CH stars) and M 22 possess the common feature that they are very loose clusters for their mass. McClure (1979) speculated that these two clusters along with M55, another low concentration cluster in which a CH star had been discovered, perhaps contain the CH stars because these are binary systems. Therefore, an observational program was begun to monitor the radial velocities of faint PRG's to look at the binary frequency among them. The radial-velocity spectrometer on the 1.2 m telescope coudé at Dominion Astrophysical Observatory (McClure et al. 1985) was used for these observations, giving an accuracy for a velocity of better than 1 km s^{-1}. In 1979, observations were begun of a sample of BaII stars since they tended to be brighter, and a few years later samples of CH, R0–R4, and sgCH stars were added to the program. Preliminary results of these programs have been reported by McClure et al. (1980), McClure (1983, 1984a,b), and McClure et al. (1985). The conclusions reached in these papers were that probably all BaII and CH stars are binaries, but that the R stars have a normal binary frequency for giants. The BaII, CH, and sgCH stars are too faint to contaminate their own atmospheres by mixing, in the same manner as the stars in the late stages of AGB evolution. It has been suggested widely in the literature, therefore, that they may have been contaminated by mass exchange from a companion which was formerly an AGB star. The latter component of the binary system should now be a white dwarf. However, numerous stars have been observed with IUE (e.g. Böhm-Vitense et al. 1984; Bond 1984) to look for UV light from white dwarf companions, with only a very few found.

BINARY FREQUENCY

Observations have now been obtained over a period of 10 years for the BaII stars, and somewhat less than this for the other types. Relatively good statistics can now be reported on the binary frequency among these stars as follows:

BaII	CH	sgCH	R
17-18/20	11/12	7/10	4/15

where the number of binaries is given relative to the total number in each sample. The conclusions of the preliminary studies mentioned above still stand. The BaII stars and CH stars are almost certainly all binaries. One or two stars in each group not showing velocity variations can be explained by the probability of observing a star pole on. In addition, one of the BaII stars, HD 130255, which has a constant radial velocity, has been shown by Keenan (this conference) not to exhibit the enhanced CN bands characteristic of BaII stars. Keenan has suggested that this may be a dwarf star.

New observations by Jorissen & Mayor (1988) have now confirmed the binary nature of the BaII stars. In a sample of 27 stars, they have found that 24 show significant velocity variations in their observations which span over three seasons.

Although observations of sgCH stars commenced several years after the other groups, there are now enough observations to say tentatively that they too are all binaries. One or two of the nonvariables may yet turn out to be variable with further observations. Finally, the result still stands that the R0–R4 stars appear to have a binary frequency normal for giant stars. Some different mechanism must be found to explain the high carbon abundances for the R stars. It is important to note that these stars do not show s-process enhancement (Dominy 1984), and therefore it should not be surprising if they are not the result of mass transfer from an AGB star like the other faint PRG's.

THE CASE OF THE S STARS

Two recent studies have indicated that the S stars may not all be AGB stars which are in the process of helium shell-flashing in the M–MS–S–SC–C sequence as has been suggested widely in the literature. Smith & Lambert (1988) have found that 38% (22 out of 58) of these s-process enriched stars do not show Tc. Similar results are reported by Little-Marenin (this conference). Since Tc has a half-life of 2×10^5 yr, and this is an s-process element produced in AGB evolution, it would be expected that it should be present in all S stars. They suggest that those stars not showing Tc in their spectra may be binaries which have undergone mass transfer in the past like the BaII stars, which also lack Tc. Indeed, they find that the space density of these Tc-poor S stars relative to M giants is comparable to the space density, relative to normal G and K giants, of the BaII stars.

Secondly, Jorissen & Mayor (1988) have reported that eight out of a sample of nine S stars have shown velocity variations in observations over three observing seasons. These observations may suggest that a very high percentage of S stars are binaries, contrary to expectations that most are single AGB stars. The evidence for S stars being in the M–S–C sequence on the AGB is strong, however (see Bessell *et al.* 1983; Lloyd Evans 1984). The Jorissen & Mayor result can perhaps be explained by the selection criterion for their stars. Since they picked only photometrically constant S stars for their sample, it is likely that they missed those stars that are in the late stages of AGB evolution. Most nonvariable S stars are hotter, and perhaps these are the ones that are evolved BaII stars with no Tc. Obviously, this problem is ripe for solution within the next year or two.

ORBITAL ELEMENTS OF BINARY PRG'S

Webbink (1986) has made a comparison between the orbital eccentricities and mass functions of BaII stars and normal G–K giant stars using the data on seven BaII star orbits available at that time (McClure 1983). He found that the BaII stars have smaller eccentricities on average than a sample of normal G–K giant stars. He concluded that there has been *some* dissipation of the orbit indicating that one of the binary components must, in the past, have filled or nearly filled its Roche Lobe. Because the eccentricities are not zero, however, there must not have been common envelope evolution. Webbink also looked at the distribution of mass functions for the seven BaII stars and concluded that they all have very similar mass-ratios, and for the expected masses for BaII star primaries, the secondary masses should lie between 0.45 AND 0.86 $M_\odot$, a reasonable range for white dwarfs.

The periods of many of the binary PRG's are very long, but orbits have now been computed for 16 of the BaII stars and 8 CH stars. In order to compare the binary PRG's and normal G–K giant stars, a sample of orbital elements for 43 normal binaries was collected from the work of Griffin (1983, 1984 a,b, 1985 a,b,c, 1986 a,b, 1988 a,b). This sample was limited to those stars which had MK spectral classification indicating giant luminosity and G or K spectral type, and orbital periods in the same range (greater than 80 days) as the BaII and CH stars. Griffin's velocity accuracy is similar to that for the PRG binaries discussed here, and this high accuracy as well as his systematic search and long baseline has minimized the severe biases usually found in binary orbit compilations (Griffin 1985c).

In order to compare the eccentricities of the PRG's and normal giants, their cumulative distributions are plotted in Figure 1. As evident in this figure, the BaII stars have significantly lower eccentricies than the normal giant stars, and the CH stars have even lower eccentricies. As pointed out by Webbink (1986), significantly non-zero eccentricities are very rare among semi-detached binaries. Thus mass transfer or common envelope evolution seems to damp quite strongly any orbital eccentricity.

Fig. 1. The fraction of BaII stars, CH stars, and normal giant binaries with orbital eccentricities less than a given eccentricity.

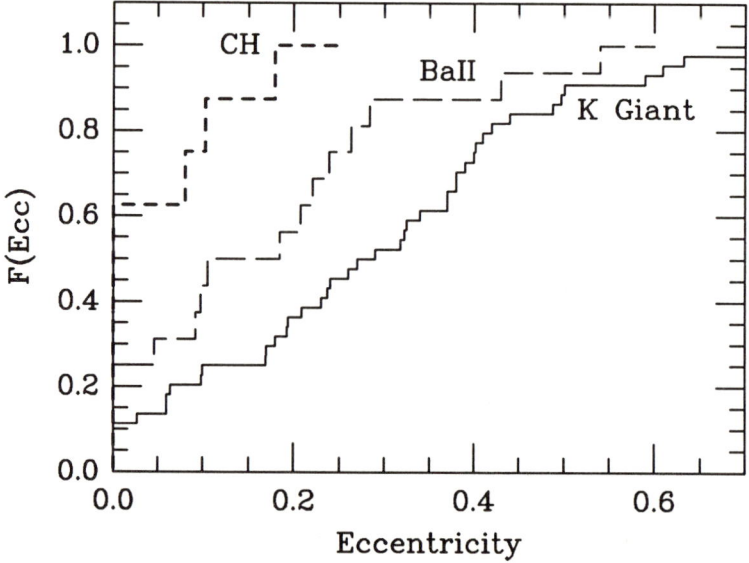

Theoretical studies of tidal dissipation (Zahn 1977), or of the dynamics of mass transfer streams (Piotrowski 1964; Kruszewski 1966) also indicate this. See also the review by Shu & Lubow (1981). For the case of the CH stars, one would be hard pressed to say that they are not a sample of binaries with circularized orbits, which suggests common-envelope evolution could have taken place. The BaII stars, however, exhibit decidedly non-zero eccentricities for the most part, which is surprising if they are post-mass-transfer binaries. On the other hand, the fact that they have significantly lower eccentricities than normal stars is clear evidence for dissipation of the original orbital energy in these systems by some mass-transfer mechanism. Two hypotheses have been put forward for this mechanism. The first is tidal dissipation (Webbink 1986), where the companion of the BaII star has, in the past, filled or nearly filled its Roche Lobe. The second mechanism, suggested by Boffin & Jorissen (1988) to circumvent the problem of rapid circularization of orbits in the case of Roche Lobe overflow, is transfer of mass by a stellar wind and planetary nebula. In this case the system always remains detached, and their model allows eccentric orbits to survive the mass transfer process. On the other hand, the exact effect on orbital eccentricity is uncertain.

Further evidence for mass transfer in BaII and CH stars comes from an examination of the distribution of mass functions of these systems, as defined by $f(m) = sin^3 i \cdot m_2^3/(m_1 + m_2)^2$. The cumulative distribution of mass functions for the sample of normal giants, BaII stars and CH stars are shown in Figure 2. As in the case of the eccentricities, the distributions of mass functions for these three samples is very different. Notice that the BaII stars have considerably lower mass functions than the CH stars, and that the distribution of mass functions for the normal giants has a very different shape. Whereas the BaII and CH star distributions rise quickly to a relatively low maximum value, the distribution for the normal giants has an excess of large mass functions.

Fig. 2. The fraction of BaII stars, CH stars, and normal giant binaries with mass functions less than a given value.

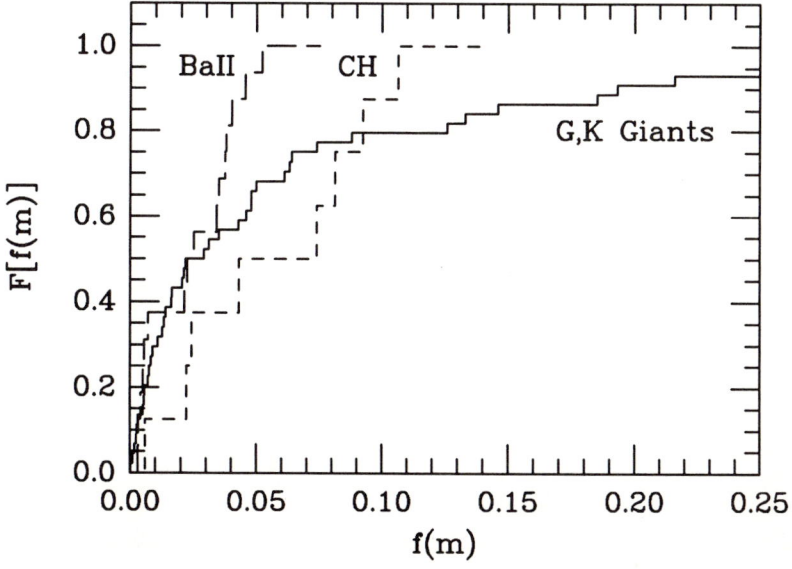

To understand the nature of these differences, let us first examine the cumulative distribution of mass functions expected for a sample of binaries that are homogeneous with respect to mass ratio of the components. In Figure 3, the cumulative distribution is plotted (solid curve) for a sample of binaries which have a constant value of the mass ratio, or true mass function $m_2^3/(m_1 + m_2)^2 = 0.04$ $M_\odot$. This, and all other comparisons described below were calculated by modeling the distribution of the mass function assuming a random distribution of orbital inclinations on the sky. In any *real* sample of stars, even a homogeneous sample, it would be expected that there would be *some* dispersion in mass ratio of the components. The long-dashed curve in Figure 3 displays the distribution of mass functions modeled for a sample of stars with true mass-functions near $m_2^3/(m_1 + m_2)^2 = 0.04$, but with a small dispersion of 0.01 about this value, as shown by the inset in the bottom right side of the figure. The distribution of mass functions for a similar model with twice the dispersion is shown by the short-dashed curve. Notice that, unlike the distribution of mass functions for normal giants (Figure 2), these models do not exhibit the large values of the mass function. The mass functions for normal giants, as expected, are not distributed like those of a homogeneous sample.

Figure 4 illustrates a model (inset figure) that does fit the cumulative distribution of mass functions for normal giants. The dashed curve in Figure 4 represents the modeled distribution, while the solid lines represent the observed distribution for Griffin's sample. As expected, there is a broad distribution of true mass-functions, representing a sample of binaries with a large spread in masses and mass ratios. In order to reproduce the curved cumulative-distribution for normal giants a majority of the systems have to have small values of the true mass-function (relatively small masses and large mass ratios), but with a tail toward large values of the true mass-function (large and near equal masses). Small variations from this will produce a noticeably poorer fit to the observed distribution.

Fig. 3. The fraction of stars with mass functions less than a given value for models of the true mass-function as shown in the inset figure and for a sample with random distribution of orbital inclinations.

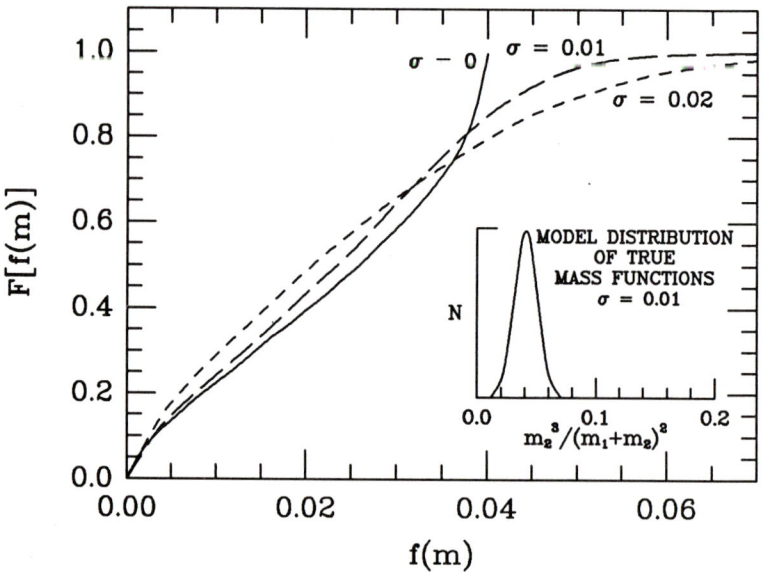

Fig. 4. The fraction of normal giant binaries with mass functions less than a given value (solid lines), and for a model (dashed curve) assuming random orbital inclinations and a distribution of the true mass-function as shown in the inset figure.

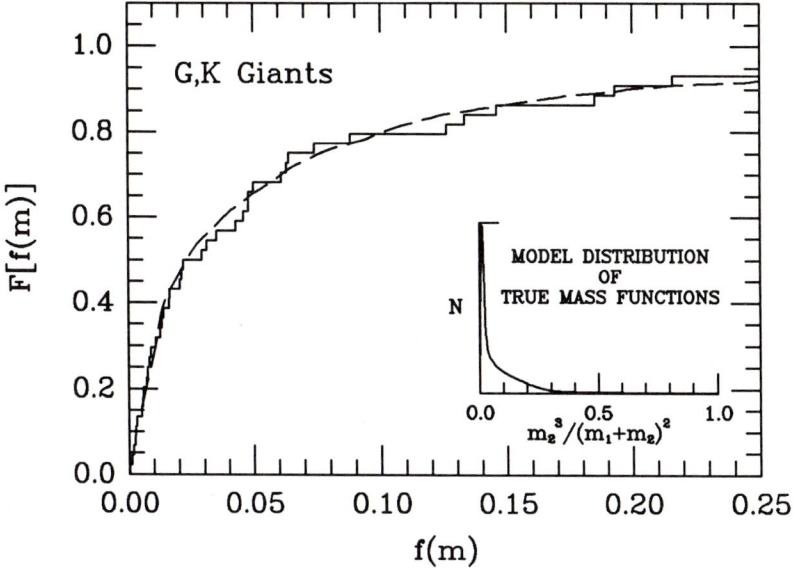

Fig. 5. The fraction of BaII binaries with mass functions less than a given value (solid lines), and for a model (dashed curve) assuming random orbital inclinations and a distribution of the true mass-function as shown in the inset figure.

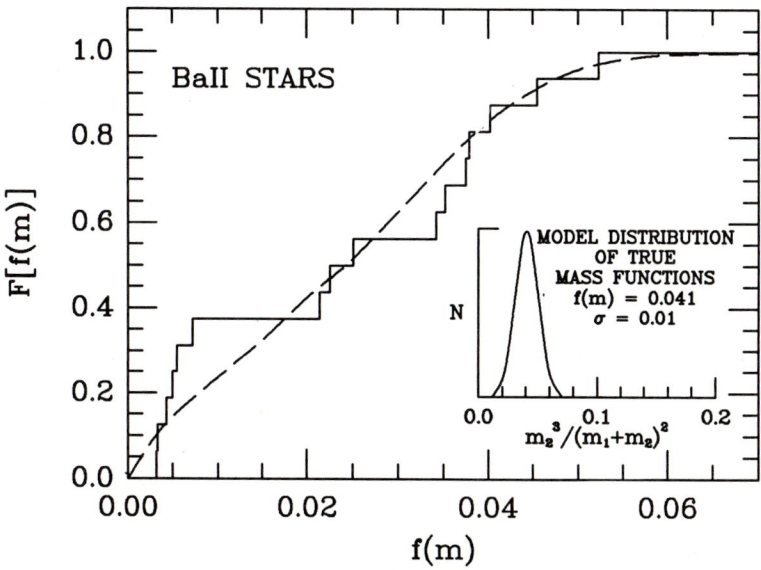

The situation for the BaII and CH stars is quite different; the distribution of their mass functions is amazingly like those of homogeneous samples. Figure 5 shows the cumulative distribution of mass functions for the BaII stars, and a model for a sample of stars of true mass-function of 0.041 $M_\odot$ and dispersion $\sigma = 0.01$ about this value (illustrated in the inset). Within the errors of the rather small sample, the fit of the model to the observed distribution is excellent. The conclusion drawn by Webbink (1986) still stands, therefore, that the masses of the companions to the barium stars must be highly correlated with those of the barium stars themselves, and that it is possible that both are virtually single-valued. A similarly good fit is obtained for the CH stars (Figure 6) by a model with true mass-function of 0.095 $M_\odot$ and dispersion of 0.015.

BaII stars have been thought to be members of the old disk population based on their kinematics (Eggen 1972), with the confirming evidence of the membership of two BaII stars in an old disk cluster NGC 2420 (McClure et al. 1974; Suntzeff & Smith 1988). Williams (1975) and Hakkila (this conference) suggest, however, that the BaII stars are younger than this, and Catchpole et al. (1977) suggest that there may be an age spread in the sample. The masses of the BaII stars probably lie in the range of about 1.0 $M_\odot$ (from the discussion of kinematics by Eggen 1972) to <3 $M_\odot$ (as suggested from the highest masses found from visual binary membership by Culver & Ianna (1976, 1980) and Culver et al. (1977), and by the discussion of kinematics by Hakkila). An estimate for the *average* mass of a BaII star of about 1.5 $M_\odot$, therefore, cannot be far wrong, this relatively low value being suggested because of the secure evidence of membership in the old disk cluster. Given this mass for the primaries of the BaII stars and the value found from orbits for the true mass function of 0.041 $M_\odot$, the mass calculated for the companions is $m_2 = 0.56$ $M_\odot$, a quite reasonable value for the mass of a white dwarf.

Fig. 6. The fraction of CH star binaries with mass functions less than a given value (solid lines), and for a model (dashed curve) assuming random orbital inclinations and a distribution of the true mass-function as shown in the inset figure.

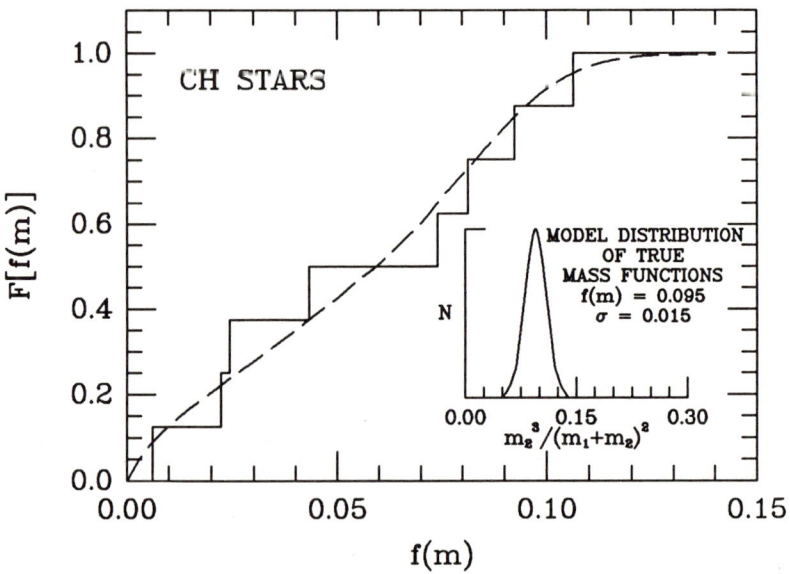

CH stars, being members of Population II, likely have masses of 0.8 ± 0.1 M$_\odot$. The corresponding masses of the companions, given the value of the true mass-function of 0.095 M$_\odot$ found above, is again 0.56 M$_\odot$. The fact that the secondary masses for both BaII stars and CH stars comes out the same is a coincidence, given the uncertainty in the masses of the primaries. The similarity of the resultant masses for the secondaries to the value we expect for white dwarfs, however, is quite surprising, especially considering the different stellar populations, and different mass ratios of the binaries in these two samples. This gives us confidence that the mass-transfer hypothesis for contamination of the atmospheres of the low-luminosity PRG's is very plausible.

QUESTIONS FOR DISCUSSION

Although binarity among BaII and CH stars helps to explain how the atmospheres of some of the PRG's may have been contaminated, there are numerous questions raised by this discovery. Some of these that come to mind are the following.

1) Where do the R-type carbon stars fit into the picture of the evolution of PRG's? These stars differ from other PRG's in that they do not appear to have enhanced s-process elements. The hotter members of this class are not bright enough to be on the asymptotic giant branch where helium shell-flashing is occurring, and the contamination of the stellar atmosphere with carbon cannot be blamed on mass transfer in a binary system because they appear to have a normal binary frequency. The peculiar abundances in the R stars must be explained by a different evolutionary process than those used to explain any other stars discussed at this conference. It is likely that the helium core flash is responsible (see discussion by Dominy 1984), but more theoretical work must be done to show how this works.

2) What is the status of the S stars? Smith and Lambert (1988) and Little-Marenin (this conference) suggest that S stars which do not exhibit Tc lines (about 38% of the sample) may be mass-transfer binaries, with the BaII stars being the precursors. Jorissen and Mayor (1988), however, have found that almost all of the S stars they have observed have variable radial velocities. Are there two sequences of S stars, and why do Jorissen and Mayor seem to find almost all S stars are binaries? They picked photometrically non-variable stars for their sample, and as they mention, this may bias the sample towards evolved BaII stars as opposed to stars in the M-S-C sequence.

3) If all BaII and CH stars are binaries with white dwarf companions, as the orbital parameters discussed here strongly suggest, why is there so little evidence for white dwarfs found in the IUE observations of their ultraviolet spectra? Can the lack of white dwarfs be explained by their having cooled to luminosities below the detection limit, and is this telling us something about time scales for the evolution and binary mass-transfer process?

4) If binarity is necessary for the BaII star phenomenon, why are the only two *cluster* BaII stars known found in the same cluster (NGC 2420)? Also, why is one of these two in such a peculiar position in the color-magnitude diagram, off the giant branch towards the blue (see McClure 1984a)?

5) Why are the CH stars that have been found in globular clusters located mainly in those clusters with low central concentration as well as in the very loose dwarf spheroidal galaxies? If this correlation is not just a coincidence, are more binaries formed to begin with in the low concentration clusters, or is it a result of *disruption* of binaries in concentrated clusters?

I wish to thank Dr. Andy Woodsworth, who has contributed significantly to the program of measuring velocities of PRG's during the last year.

REFERENCES

Bessell, M.S., Wood, P.R., & Lloyd Evans, T. (1983). Mon. Not. Roy. Astron. Soc., 202, 59.
Bidelman, W.P., & Keenan, P.C. (1951). Astrophys. J., 114, 473.
Boffin, H.M.J., & Jorissen, A. (1988). preprint.
Böhm-Vitense, E., Nemec, J.M., & Proffitt, C. (1984). Astrophys. J., 278, 726.
Bond, H.E. (1974). Astrophys. J., 194, 95.
Bond, H.E. (1984). Future of Ultraviolet Astronomy based on Six Years of I.U.E. Research, N.A.S.A. Conf. Publ. No. 2349, p. 289.
Catchple, R.M., Robertson, B.S.C., & Warren, P.R. (1977). Mon. Not. Roy. Astron. Soc., 181, 391.
Culver, R.B., & Ianna, P.A. (1976). Publ. Astron. Soc. Pacific, 88, 41.
Culver, R.B., & Ianna, P.A. (1980). Publ. Astron. Soc. Pacific, 92, 829.
Culver, R.B., Ianna, P.A., & Franz, O.G. (1977). Publ. Astron. Soc. Pacific, 89, 397.
Dominy, J.F. (1984). Astrophys. J. Suppl., 55, 27.
Eggen, O.J. (1972). Mon. Not. Roy. Aston. Soc., 159, 403.
Griffin, R.F. (1983). Observatory, 103, 273.
Griffin, R.F. (1984a). Observatory, 104, 6.
Griffin, R.F. (1984b). Observatory, 104, 268.
Griffin, R.F. (1985a). Observatory, 105, 9.
Griffin, R.F. (1985b). Observatory, 105, 128.
Griffin, R.F. (1985c). In Interacting Binaries, ed. P.P. Eggleton & J.E. Pringle, p. 1. Dordrecht: Reidel.
Griffin, R.F. (1986a). Observatory, 106, 35.
Griffin, R.F. (1986b). Observatory, 106, 108.
Griffin, R.F. (1988a). Observatory, 108, 17.
Griffin, R.F. (1988b). Observatory, 108, 90.
Jorissen, A., & Mayor, M. (1988). Astron. Astrophys., 198, 187.
Keenan, P.C. (1942). Astrophys. J., 96, 101.
Kruszewski, A. (1966). Adv. Astron. Astrophys., 4, 233.
Lloyd Evans, T. (1984), Mon. Not. Roy. Astron. Soc., 208, 447.
Luck, R.E., & Bond, H.E. (1982). Astrophys. J., 259, 792.
McClure, R.D. (1979). Mem. Soc. Astron. Italiano, 50, 15.
McClure, R.D. (1983). Astrophys. J., 268, 264.
McClure, R.D. (1984a). Publ. Astron. Soc. Pacific, 96, 117.
McClure, R.D. (1984a). Astrophys. J. (Letters), 280, L31.
McClure, R.D., Fletcher, J.M., Grundmann, W.A., & Richardson, E.H. (1985). In I.A.U. Colloq. No. 88, Stellar Radial Velocities, ed. A.G.D. Philip, & D.W. Latham, p. 49, Schenectady: L. Davis Press.
McClure, R.D., Fletcher, J.M., & Nemec, J.M. (1980). Astrophys. J. (Letters)., 238, L35.
McClure, R.D., Forrester, W.T., & Gibson, J. (1974). Astrophys. J., 189, 409.
McClure, R.D., & Norris, J. (1977). Astrophys. J. (Letters), 217, L101.
Piotrowski, S.L. (1964). Acta Astron., 14, 251.
Shu, F.H., & Lubow, S.H. (1981). Ann. Rev. Astron. Astrophys., 19, 277.
Smith, V.V., & Lambert, D.L. (1988). Astrophys. J. in press.
Suntzeff, N.B., & Smith, V.V. (1988). Astron. J., 93, 359.
Webbink, R.F. (1986). In Critical Observations versus Physical Models for Close Binary Systems (Proc. Beijing Colloq. on Close Binary Systems), ed. K.-C. Leung & D.S. Zhai, in press, New York: Gordon & Breach.
Williams, P.M. (1975). Mon. Not. Roy. Astron. Soc., 170, 343.
Zahn, J.P. (1977). Astron. Astrophys., 57, 383.

PECULIAR RED GIANTS — WHAT KIND OF WHITE DWARFS DO THEY BECOME?

Icko Iben, Jr.
University of Illinois

<u>Abstract</u>. After a brief commentary on the place of "peculiar red giants" in the overall scheme of stellar evolution, an outline is given of the various possibilities for post asymptotic giant branch (AGB) evolution. The behavior of a post-AGB model star is crucially dependent on where in a thermal pulse cycle the mass of the hydrogen-rich envelope is reduced to such an extent that departure from the AGB must follow on a thermal time scale. If departure from the AGB occurs while the model is still burning hydrogen, post-AGB behavior depends on the mass of the helium buffer zone (= zone containing predominantly helium which has been processed through the hydrogen-burning shell following the last thermal pulse on the AGB). If departure occurs at an arbitrary time during the hydrogen-burning phase, then: (1) in ~ 25% of all cases, the post-AGB model will experience a final helium shell flash, and, in consequence of additional mass loss, may become a non-DA white dwarf; (2) in ~ 60% of all cases, the model will cease burning hydrogen when the mass in its hydrogen-rich envelope is reduced to ~ $10^{-4} M_\odot$ and will evolve into a DA white dwarf; and (3) in ~ 15% of all cases, the model will experience a final hydrogen shell flash, but the outcome with regard to spectroscopic type is unclear. If departure from the AGB occurs while the model is burning helium, the result is either the same as in option (3) just described, or mass loss during the post-AGB helium-burning phase may turn the star into a non-DA white dwarf.

1 PRELIMINARY REMARKS

I will begin with a few philosophical comments about "peculiar" red giants (PRGs) and then narrow my remarks to a discussion of their fate "after death". In particular, I will ask the question: what happens to PRGs <u>as</u> they depart from the giant branch, or what happens to their descendants <u>after</u> they have departed from the giant branch, or both, that determines which descendants become DA white dwarfs and which become non-DA white dwarfs. But, first, a few remarks.

That we should call red giants with non-solar distributions of the
heavy elements at their surfaces "peculiar" is ironic: most stars
which are massive enough to evolve off the main sequence in a Hubble
time and to retain their hydrogen-rich envelope until after they have
developed an electron-degenerate core composed of carbon and oxygen and
have entered the helium shell flashing stage develop surface abundance
"peculiarities" in consequence of dredging up material processed
through hydrogen burning and partial helium burning. Thus, all stars
of initial mass between about $1.5 M_\odot$ and $8 M_\odot$ pass through the "peculiar"
phase. The term "peculiar" is a consequence of time scales: the
duration of the shell flashing stage is of the order of only 10^6 yr or
less, and it is this which makes PRGs appear to be such a rare phenomenon. The helium shell flashing, or thermally pulsing stage, as it is
often called, does not begin until the star has developed a carbon-
oxygen core of mass larger than $\sim 0.5 M_\odot$ and dredge up does not begin
until the star is brighter than at the tip of the "first red giant
branch" through which stars of initial mass less than $\sim 2 M_\odot$ pass before
igniting helium in an electron degenerate core composed of helium. It
was not until studies of the brightest red stars in globular clusters
in the Magellanic Clouds had progressed far enough in this past decade
that a true understanding of the nature of the "asymptotic giant
branch" as distinct from "the first giant branch" became clear in the
context of PRGs. The PRGs that have developed peculiaries of their own
making are on the asymptotic giant branch (AGB). Those which appear to
be on the first red giant branch probably have a white dwarf companion
which was once an AGB star, as Robert McClure and his collaborators
have so elegantly demonstrated (see McClure, this conference).

As has been detailed by David Lambert and others at this conference,
the chemical "peculiarities" are of two main types: C/O ratios greater
than solar (carbon stars show this feature most prominently); and
overabundances of s-process isotopes (stars with technicium show this
most dramatically). Although not as straightforward a theoretical
consequence as we would like, particularly in the case of low-mass AGB
stars (see Lattanzio, this conference), it is generally agreed that the
overabundances of C relative to O are due to dredge up of material
which has experienced partial helium burning (X_{12} = abundance by mass
of $^{12}C \sim 0.25$, X_4 = abundance by mass of $^4He \sim 0.75$). It is further
agreed, both on observational and theoretical grounds, that the source
of neutrons required to produce overabundances of s-process isotopes is
the $^{13}C(\alpha,n)^{16}O$ reaction. However, the manner in which ^{13}C is produced
is subject to controversy. It is still possible that, during a flash,
hydrogen enters the convective shell which is sustained by helium
burning at its base by some sort of "extra-mixing" process (in all
extant models in which radiation pressure is included, formal contact
between the outer edge of the convective shell and hydrogen-containing
material is prevented by an entropy barrier). Should hydrogen find its
way into the convective shell, it would react with the highly abundant
^{12}C there to form ^{13}N, which promptly beta decays into ^{13}C; the ^{13}C
would then be convected down to the center of the convective shell
where it would release neutrons on reacting with α particles.

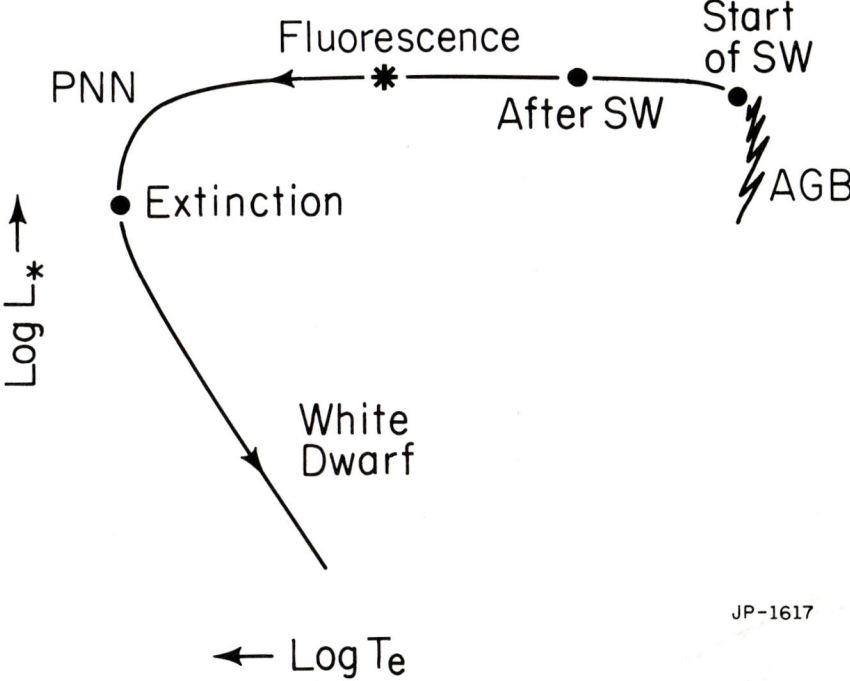

Figure 1. Schematic of evolution in the HR diagram from the asymptotic giant branch (AGB), through the planetary nucleus (PNN) phase, to the white dwarf (WD) phase.

On the other hand, it has been explicitly demonstrated that, following the peak of a thermal pulse in low mass models of low metallicity, semiconvection followed by hydrogen burning produces a layer of ^{13}C near the outer edge of the ^{12}C-rich region left behind by the retreating helium convective shell, and this ^{13}C is swept up by the convective shell which is formed during the next thermal pulse. The rate at which ^{13}C enters the convective shell is determined by the rate at which this shell grows and this, in turn, determines the rate at which neutrons are released near the base of the shell by the $^{13}C(\alpha,n)O^{16}$ reaction. A poster paper at this meeting by David Hollowell and myself (see also Hollowell and Iben 1988) shows how this works. That this process can lead to the production of s-process isotopes in solar system proportions is demonstrated beautifully by Roberto Gallino in his talk at this conference.

2 THE QUESTION OF WHEN A STAR LEAVES THE AGB
2.1 Overview

The standard picture of the evolution of a star from the AGB to the white dwarf stage is illustrated in Figure 1. Because of wind mass loss, the mass M_e of hydrogen-rich material above the burning zones in the AGB star continues to decline, until, when M_e decreases below a critical value (which depends on the mass of the CO core, on which nuclear fuel is burning, and possibly on other things as well), nuclear burning can be sustained only if the outer layers of the star can contract enough to maintain matter in the burning zone at high temperature. The star evolves rapidly toward the blue through the Hertzsprung gap and then evolves more slowly to the blue once the gap has been traversed. Continued nuclear burning in a shell and possibly continued mass loss act to reduce M_e still further, forcing the star to contract steadily and move to the blue.

When the surface temperature of the star becomes high enough (> 30,000K), photons of energy larger than the ionizing potential of hydrogen are emitted at a sufficiently high rate that surrounding material, which was blown off by the precursor both while it was a bonafide AGB star and while it was in the process of leaving the AGB, fluoresces as a planetary nebula. The central star continues to burn nuclear fuel and therefore to contract. Eventually, the central star approaches white dwarf dimensions and, ultimately, the weight of unburned fuel above the burning shell can no longer maintain large enough temperatures for nuclear burning to continue to supply the loss of energy from the surface.

The central star then evolves as a white dwarf to ever smaller luminosities and temperatures, with energy losses being supplied by the thermal energy of the ions in its interior. The emission measure of the expanding nebula drops and the nebula ultimately becomes invisible. As it continues to expand outward, the nebular material becomes incorporated into interstellar clouds, thus enriching the interstellar medium in carbon and s-process elements which were once in the envelope of the precursor PRG.

Note that I have not specified as yet which nuclear fuel is supplying the energy during the planetary nebula stage. Ever since Shklovski first proposed this general picture over three decades ago (Shklovski 1956), the general consensus has been that the energy source of most planetary nebulae is hydrogen. This point of view became solidified when, almost two decades ago, Paczynski (1970, 1971) constructed explicit models based on this scenario by stripping hydrogen from the surface of model AGB stars during the quiescent hydrogen-burning phase until the models were forced to evolve off the AGB while still burning hydrogen. About one decade ago, Härm and Schwarzschild (1975) repeated the Paczynski excercise, but stripped mass from the stellar surface during the high surface luminosity phase of a thermal pulse cycle when the AGB model was burning only helium, and showed that one obtains essentially the same evolutionary track and roughly the same lifetime for the helium burning central star as for a central star of the same mass which burns only hydrogen. Schonberner (1979) carried the exercise one step further by stripping matter from the AGB model star at a rate which is more or less independent of where in a thermal pulse cycle the AGB star is, showing that, depending upon precisely where in this cycle the envelope mass M_e decreases below the critical one (for the fuel which happens to be burning at the time), a very complicated post-AGB evolutionary behavior is possible. A frequently occurring sequence consists of a post-AGB central star which at first burns hydrogen at high surface temperature, then ignites a final helium shell flash, returns to the AGB, and finally reverses direction in the HR diagram once again to reach high surface temperatures during the remainder of the quiescent helium-burning phase.

Building upon these numerical results, Renzini (1979, 1983) developed a comprehensive scenario for planetary nebula formation and evolution, and broadened the scope of the inquiry to postulate a causal connection between the nature of mass loss on the AGB and the spectral characteristics of white dwarfs. Iben (1984), Iben and Tutukov (1984), Iben and MacDonald (1986), Wood and Faulkner (1986), and Schonberner (1986, and references therein) constructed models useful in extending Renzini's ideas.

In recent years, there has been a tremendous amplification in our quantitative understanding of the variation of white dwarf spectral characteristics with respect to surface temperature and magnitude, wrought by technological advances and by the energy and dedication of a large group of observational and theoretical white dwarf "buffs" including Fontaine, Greenstein, Heber, Holberg, Koester, Kudritzki, Liebert, Mendes, Wesemael, Schonberner, Shipman, Sion, Wegner, Weidemann, and Winget, and their minions. One interpretation of these observations suggests that the picture I am about to paint in the remainder of this section and in the next cannot possibly be true. In the final section, I will outline the salient features of the relevant observations and the currently popular interpretation of the meaning of these observations, leaving it to the reader to make his own judgement.

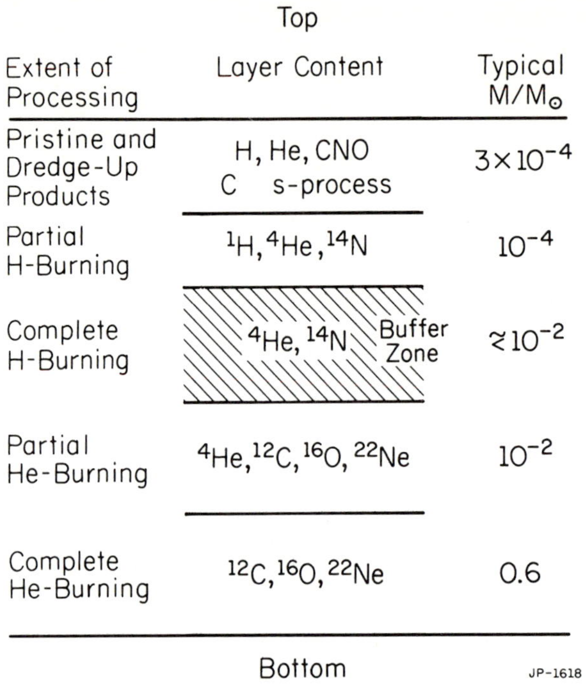

Top
Extent of Processing
Layer Content
Typical M/M⊙
Pristine and Dredge-Up Products
H, He, CNO
C s-process
3×10⁻⁴
Partial H-Burning
¹H, ⁴He, ¹⁴N
10⁻⁴
Complete H-Burning
⁴He, ¹⁴N
Buffer Zone
≳10⁻²
Partial He-Burning
⁴He, ¹²C, ¹⁶O, ²²Ne
10⁻²
Complete He-Burning
¹²C, ¹⁶O, ²²Ne
0.6
Bottom
JP-1618

Figure 2. The compositional structure of an AGB star. The strata of different compositions are shown schematically, not to scale.

Light Curve
Log L*
Buffer
Zone Mass
ΔM Buffer
0
Time
JP-1619

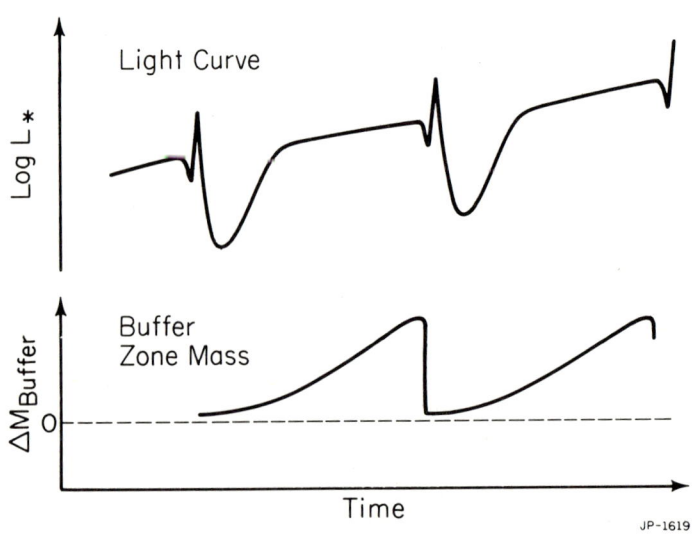

Figure 3. The variation with time of the surface luminosity and of the mass of the helium buffer zone. Schematic only.

2.2 Mass Loss and The Composition of a Post-AGB Star

The distribution of the most abundant elements in an AGB model star of core mass $\sim 0.6 M_\odot$ is shown schematically in Figure 2. As time progresses during a quiescent hydrogen-burning phase, the hydrogen-burning shell, of thickness $\sim 10^{-4} M_\odot$ adds ^{4}He and ^{14}N to an underlying "buffer" zone. When the buffer zone grows in mass to about $10^{-2} M_\odot$, temperatures at its base become large enough to ignite helium and a helium shell flash takes place. The mass M_{buf} of the buffer layer is correlated with the surface luminosity of the AGB model in the fashion illustrated schematically in Figure 3. The behavior of a model post-AGB star depends strongly on the mass of the buffer zone when departure from the AGB is assumed to occur. What this mass should be is not known from first principles. At this point one can only guess. In time, with enough observational facts to explain, the answer may be forced upon us, if, as discussed in the section IV, this forcing has not already occurred.

The dependence on time of critical masses is shown schematically in Figure 4. The location of the hydrogen-burning shell is given by the curve labled M_H, the boundary of the convective shell formed during a helium shell flash is given by the outline of the shaded region, and two of many possibilities for the location of the surface are sketched as the curves labled M_*. In placing these latter two curves, it has been assumed that the mass of the hydrogen-rich surface layer, given by $M_e = M_* - M_H$, decreases below the critical value of $M_{e,crit}$ during the quiescent hydrogen-burning phase. $M_{e,crit}$ is about one-tenth of the mass ΔM_H through which the hydrogen-burning shell moves during the quiescent hydrogen-burning phase between pulses. For a core mass of $0.6 M_\odot$, $\Delta M_H \sim 0.01 M_\odot$, so that $M_{e,crit} \sim 0.001 M_\odot$. The maximum buffer mass is, of course, equal to ΔM_H.

For the typical core mass of $0.6 M_\odot$, the time between thermal pulses is $\sim 200,000$ yr. If, between pulses, the mass loss rate is of the order of 10^{-5}-$10^{-4} M_\odot \text{yr}^{-1}$ (a so-called fast or "superwind" rate), it seems likely that the envelope mass will be reduced below the critical value at some arbitrary point during the interpulse phase, and so it makes sense to consider the post-AGB evolution of model stars with initial $M_e < M_{e,crit}$ and $M_{buf} \sim (0-1)\Delta M_H$. Since the duration of the quiescent helium-burning phase is roughly 10% of the duration of the interpulse phase, approximately 10% of all central stars of planetary nebulae should be powered from the start by helium burning. Further, the critical mass $M_{e,crit,a}$ necessary for a helium-powered AGB star to depart from the AGB is smaller than $M_{e,crit}$ by a factor of 3-10, depending on how far into the helium-burning phase it has progressed. The reason for this is that the base of the hydrogen-rich envelope does not need to be compressed by the weight of overlying layers in order to supply enough energy to maintain the fluxes necessary to bloat these layers; the fluxes are maintained by helium burning and the base of the hydrogen-rich envelope is spatially much further from the electron-degenerate core than during the quiescent hydrogen-burning phase.

If wind mass loss during the interpulse phase occurs on a time scale long compared with the time between pulses (an "ordinary", "slow", or

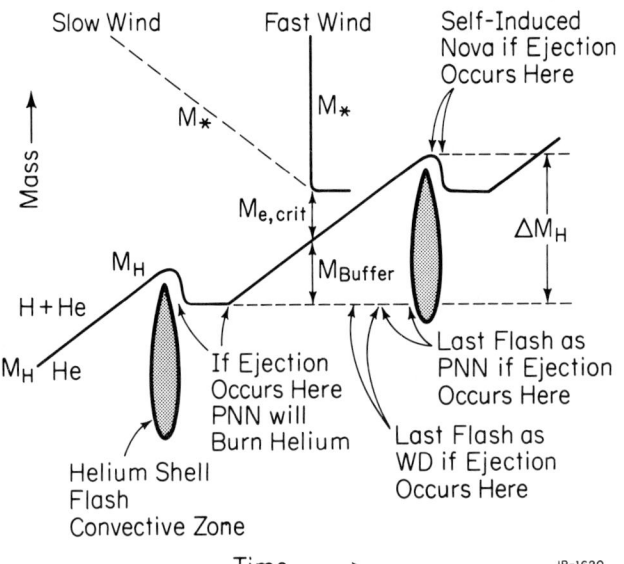

Figure 4. Schematic showing time variation of the location in mass of: (1) the boundaries of the helium convective shell formed during a thermal pulse [bordering shaded regions]; (2) the center of the hydrogen-burning shell (during active hydrogen burning), or the hydrogen-helium discontinuity (after dredge up and during the quiescent helium-burning phase) [M_H]; (3) the surface of the star [M_*].

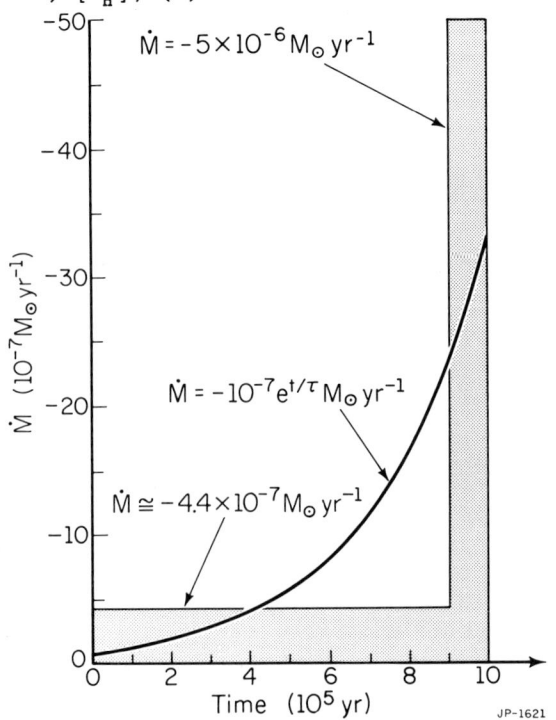

Figure 5. Hypothetical mass-loss rates as a function of time. The area under the exponential ($\tau = 2.86 \times 10^5$ yr) is the same as under the two-component curve consisting of a "slow" wind followed by a "superwind".

"Reimer's" wind), one might guess that the sharp jolt that the star experiences during a thermal pulse (see the jump in surface luminosity in Fig. 3) might trigger mass loss at a much higher rate than average, and that departure from the AGB would be likely to occur most often at the time of a pulse. In this case, most central stars of planetary nebulae should be powered by helium burning.

Very strong feelings are sometimes aroused when those of us in the model-making business use the terms "superwind" and "slow wind" or their variants. This useage is, of course, an oversimplification, but unless and until the observers can provide us with the rate at which an AGB star of a given total mass and core mass loses mass, or can tell us where in a thermal-pulse cycle final departure from the AGB occurs, I cannot appreciate the objections raised to an admitted oversimplification of a complicated physical process that neither observers nor theoreticians can yet quantitatively describe. To extend this thought, consider an AGB model star of initial mass $1.5 M_\odot$ and initial core mass $0.5 M_\odot$ which loses mass at the hypothetical rate of $10^{-7} M_\odot \mathrm{yr}^{-1} \times \exp(t/2.86 \times 10^5 \mathrm{yr})$. At the end of 10^6 yr it will have lost $0.9 M_\odot$ and its core mass will have grown to $0.6 M_\odot$, so that it must of necessity leave the AGB. The hypothetical mass-loss rate is shown by the solid curve in Figure 5. Shown also in Figure 5 is another hypothetical mass-loss rate consisting of a "slow" wind of magnitude $4.44 \times 10^{-7} M_\odot \mathrm{yr}^{-1}$ which operates for 0.9×10^6 yr and a "superwind" of magnitude $0.5 \times 10^{-5} M_\odot \mathrm{yr}^{-1}$ which operates for 10^5 yr. The area under the two-mode function is identical with that under the exponential curve, and the outcome, insofar as we are interested in when departure from the AGB occurs, is the same in the two cases. I am not convinced that the two-mode function is less likely to eventually fit the facts than the exponential; as far as I understand it, there are, in fact, insufficient facts to be able to draw observationally based curves in the plane of Figure 5.

3 POST-AGB HELIUM SHELL FLASHES AND THE BORN AGAIN AGB PHASE

There are six main types of evolution for model post-AGB stars, four for those which depart from the AGB while burning hydrogen, and two for those which depart while burning helium. In describing the hydrogen burners, it is convenient to define an angle $\phi = M_{buf}/\Delta M_H$. Although all illustrations here will be made for models of $\sim 0.6 M_\odot$ and metallicity $Z = 0.001$, the behavior of models of larger or smaller mass is qualitatively the same, and the classification in terms of ϕ is expected to be essentially mass invariant. It is, however, composition dependent.

Model stars which depart from the AGB with $\phi < 0.75$ evolve for $\sim 10^4$ yr as luminous, hot central stars of planetary nebulae until the mass M_e in their hydrogen-rich envelope drops below about $10^{-4} M_\odot$ (see Figure 6 for evolution in the HR diagram). At this point, hydrogen-burning by CN cycle reactions is extinguished, and the model dims by about a factor of ten in luminosity in a matter of only 10^3 yr. Although mass loss at rates observed for PN central stars ($\sim 10^{-9}$-$10^{-7} M_\odot \mathrm{yr}^{-1}$) abstracts mass from the surface and thus accelerates the rate of

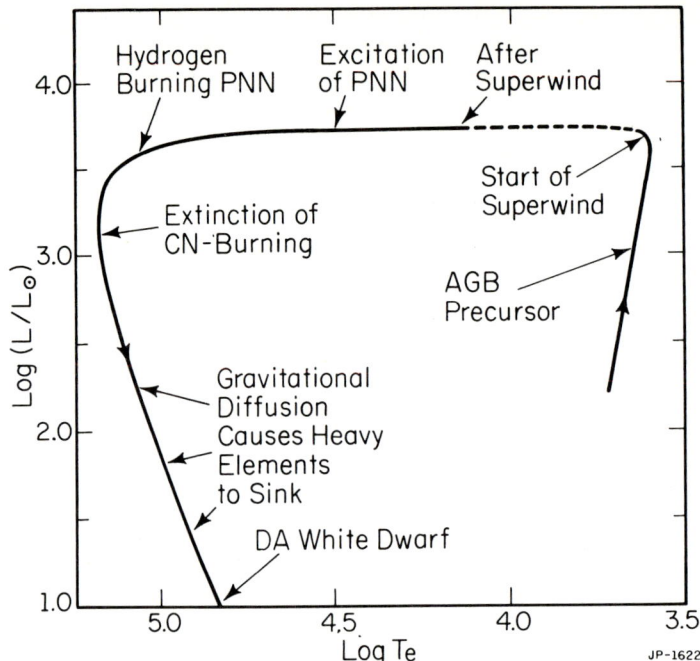

Figure 6. Evolution in the HR diagram for models with buffer masses between ~ 0.15 and 0.75 times the maximum possible.

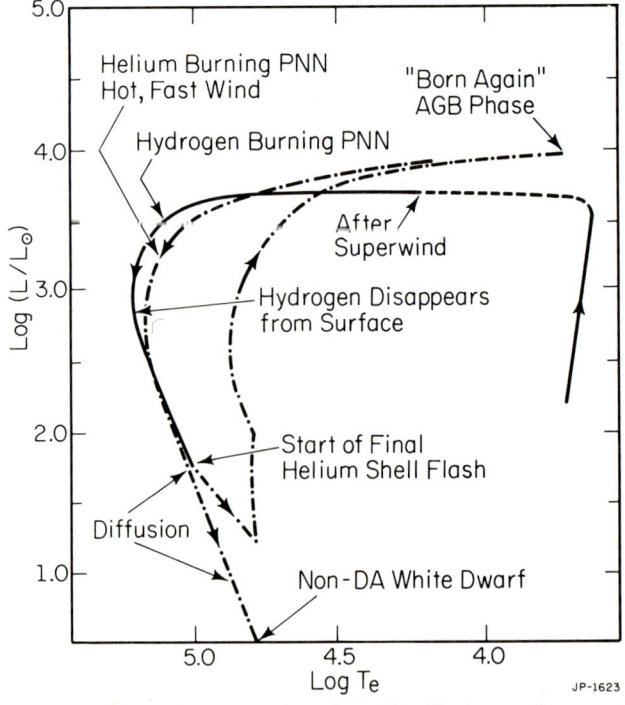

Figure 7. Evolution in the HR diagram for models with buffer masses between 0.75 and 0.85 times the maximum possible.

evolution in the HR diagram, the model star will in any case arrive at the turning point of highest surface temperature with $M_e \sim$ few x 10^{-4} $M_\odot$, a value which has come to be known as a "thick" hydrogen envelope. After the phase of rapid dimming, the model then proceeds to cool as a white dwarf, with its luminosity being supplied primarily by the thermal energy of ions in the interior. Gravitational settling forces helium and the heavy elements to sink below the surface. However, diffusion also insures that there will be a tail of hydrogen that extends deeply into the star. As long as ϕ is larger than a critical value (estimated to be ~ 0.15), no further episode of hydrogen burning will occur, and, even if mass loss from the surface continues, the inner tail of hydrogen will be shielded from loss. To reduce the total amount of hydrogen in the star to less than $\sim 10^{-13} M_\odot$, as is argued in some quarters (see section IV), would require a most unusual wind indeed. If such a wind does not occur, then the real analogues of the models with $0.15 < \phi < 0.75$ should become DA white dwarfs.

Up to the point that hydrogen burning by CN-cycle reactions is extinguished, the evolution of models with $0.75 < \phi < 0.85$ (shown in Figure 7) is essentially identical with that of models with $0.15 < \phi < 0.75$. Following the extinction of nuclear burning in both cases, as the flux of energy through it is reduced, the hydrogen-rich layers contract rapidly, increasing the weight on the helium-rich buffer zone below, thus compressing and heating matter in this zone. When $\phi < 0.75$, the compression and heating is not sufficient to ignite helium-burning reactions, and the story of nuclear burning is by and large over, except for a very mild burning of hydrogen via the pp chains (provided mass loss does not abstract too much more of the hydrogen surface layer). For $\phi > 0.75$, howevever, the compression and heating is sufficient; the star experiences a final helium shell flash, as first predicted by Masayuki Fujimoto over a decade ago (Fujimoto 1977). Since hydrogen is not burning, there is no entropy barrier to overcome, and the convective shell which is formed in the helium buffer zone extends into hydrogen-rich layers, further decreasing M_e.

What happens after this is not known precisely. The star may return to the AGB (see Figure 7) as an R CrB-like star burning helium quiescently in a shell of mass $\sim 0.01 M_\odot$ and hydrogen at the base of an envelope of mass $\sim 0.001 M_\odot$ in which the abundance by mass of hydrogen is very small (say, $< 10^{-4} M_\odot$, giving a total mass of hydrogen in the star $< 10^{-7} M_\odot$). Or, helium burning could continue to be the only source of surface luminosity, and the same wind which caused departure of the progenitor from the AGB could abstract more hydrogen-rich matter from the star.

In any case, the "born again" AGB phase must be of very short duration, and the model, now with a total mass of hydrogen much less than $10^{-4} M_\odot$, quickly moves to high temperatures, nearly retracing the path it followed earlier as a hydrogen-burning PN central star. Since helium burning keeps the model at high luminosities and surface temperatures for $\sim 3 \times 10^4$ yr, wind mass loss at rates typical of real PN central stars may be expected to abstract all of the remaining hydrogen from the real analogue of the model before diffusion inward has had a chance to hide it. The real analogue becomes a luminous DO star and then,

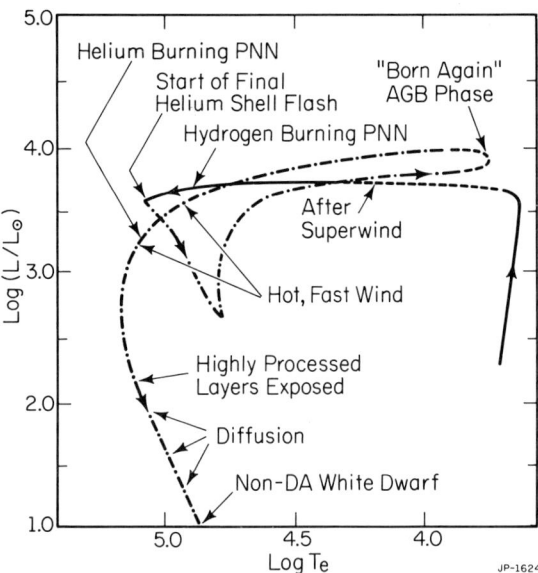

Figure 8. Evolution in the HR diagram for models with buffer masses between 0.85 and 1.00 times the maximum possible.

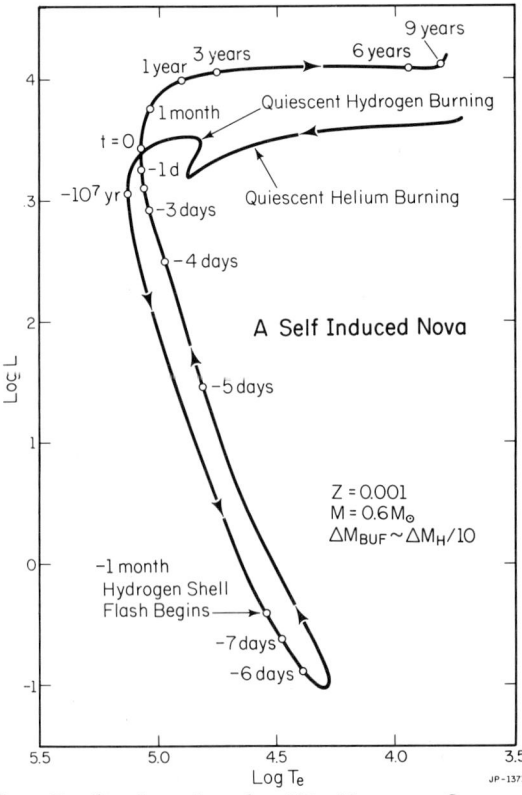

Figure 9. Evolution in the HR diagram for models which depart the AGB during a helium flash or during the quiescent helium-burning phase with a thick layer of hydrogen at their surfaces. Models which depart from the AGB while burning hydrogen, but with $\phi < \phi_{min} \sim 0.15$ (?) will also experience the "self induced" nova phenomenon.

after helium burning is extinguished, cools to become a non-DA white dwarf.

Models with $0.85 < \phi < 1.0$ have a large enough buffer mass as they depart from the AGB that, during their tenure as hydrogen-burning PN central stars, the buffer mass grows to be as large as the critical value (ΔM_H) which leads to a shell helium flash on the AGB. But, since hydrogen is still burning, an entropy barrier prevents contact between the convective shell formed in the buffer layer and hydrogen-rich regions. Therefore, model evolution can be followed in a straight-forward fashion, with the result shown in Figure 8. The model returns to the AGB, formally remaining there for a time proportional to M_e. Since the surface characteristics of the model are similar to those of its AGB precursor, a superwind should operate once again, and since hydrogen-rich layers now occupy a region in the density-temperature plane which was occupied by the hydrogen-rich matter lost in the first superwind phase, this superwind should reduce M_e considerably. But, reduction of M_e again forces the model to return to the blue, where it continues to burn helium on a long time scale and lose mass from its surface by a hot, fast wind. Eventually, a real analogue of this model star should evolve into a non-DA white dwarf.

Up to the point that hydrogen-burning by CN-cycle reactions is first extinguished (at the bluest point along an evolutionary track) and diffusion becomes important, the evolution of models with $\phi \sim 0.15$ or less is the same as that of models with $0.15 < \phi < 0.75$. Thereafter, howevever, carbon diffusing through the small buffer layer from below and hydrogen diffusing inward through the buffer layer from above (see Figure 2) meet within the buffer zone at such high temperatures and at such high abundances that CN-cycle burning is activated again and the model experiences a hydrogen shell flash which carries it back to the giant branch in the fashion of a slow nova, as shown by the dash-dot portion of the evolutionary track in Figure 9 (from Iben and MacDonald 1986).

The calculation whose results are described in Figure 9 has actually been done for a model which departs from the AGB while it is burning helium quiescently. The mass in the envelope at departure is $M_e \sim 10^{-3} M_\odot$, and this illustrates the statement made earlier that the value of M_e required for departure from the AGB during helium burning depends on how far into the quiescent helium-burning phase the model has progressed. As a consequence of helium burning, the mass of the buffer zone is steadily reduced, but the hydrogen-rich envelope of the model continues to contract and the temperature at the base of this envelope continues to rise until hydrogen-burning is again ignited, and this adds mass to the buffer zone. Once both hydrogen- and helium-burning are extinguished, the mass of the buffer zone is $\sim 10^{-3} M_\odot$, giving $\phi \sim 0.1$; the mass of the hydrogen-rich layer remaining is the canonical few $\times 10^{-4} M_\odot$. The configuration of the model at this point could also have been achieved by waiting until the end of the quiescent helium-burning phase before removing all but $\sim 10^{-3} M_\odot$ of hydrogen-rich matter from the surface.

In any case, the "self induced" nova phenomenon occurs: (1) when the model star departs from the AGB shortly after the end of the quiescent helium-burning phase with $(M_{buf} + M_e - \text{few} \times 10^{-4} M_\odot) < M_{buf,crit} \sim 0.15$; and (2) when departure from the AGB occurs either during the helium-flashing stage or during the quiescent helium-burning stage, <u>if</u> enough hydrogen, say $M_e > M_{e,dif}$, is retained at the surface. If M_e is too small, then there will not be enough fuel to produce a thermal runaway along the white dwarf cooling sequence. Obviously, calculations need to be carried out to determine both $M_{e,dif}$ and $M_{buf,crit}$.

A major consequence of the final hydrogen shell flash is, of course, the reduction in the total amount of hydrogen in the star. Nuclear burning will destroy the hydrogen tail that has been built up by diffusion prior to the flash. As a giant, more hydrogen-rich material will be abstracted from the surface by a stellar wind. The net result is that, when the flash has run its course and the star again becomes a white dwarf, the mass of the hydrogen envelope will be smaller (perhaps much smaller) than when the star first became a white dwarf.

The final type of theoretical behavior is obtained by abstracting mass from a helium shell flashing or quiescent helium-burning model until so little hydrogen is left in surface layers that no final hydrogen-burning phase can occur. Obviously, if all of the hydrogen were abstracted, the existence of DA stars would be difficult to explain, except perhaps by some convoluted scenario involving accretion from the interstellar medium. However, something close to the almost complete abstraction of hydrogen is receiving increasingly serious attention, as will be described in the next section.

4 THE INCREDIBLY THIN HYDROGEN ENVELOPE SCENARIO

There are several lines of evidence which may be used to argue against the scenario I have just described. These arguments have been summarized recently by Fontaine and Wesemael (1987), Koester (1987), and Shipman (1988). Most of the arguments call into question the possibility that DA white dwarfs have "thick" hydrogen envelopes, as follows from evolutionary theory when it is assumed that most central stars of planetary nebulae are burning hydrogen and that stellar winds are not very effective after the extinction of hydrogen-burning by CN-cycle reactions.

For example, it is often stated that a DA white dwarf model cannot pulsate as a ZZ Ceti star if the mass of hydrogen in its envelope exceeds $\sim 10^{-7} M_\odot$, a value which is three orders of magnitude smaller than suggested by the evolutionary calculations, given the assumptions just made. A more accurate statement is that model white dwarfs of the appropriate luminosity do not pulsate at surface temperatures consistent with the observations (even though they may indeed pulsate) unless, when a "standard" treatment of convection is assumed, the hydrogen envelope is chosen to be "thin" (namely less massive than $\sim 10^{-7} M_\odot$). It is not out of the question, however, that the standard theory of convection is the culprit, and not the mass of the hydrogen envelope. For that matter, some other aspect of the input physics, such as the opacity or the ionization equation of state, could be at fault.

On the other hand, there may be no reason why a radiatively driven wind might not continue to operate after hydrogen-burning by CN-cycle reactions ceases and reduce the mass of hydrogen remaining in surface layers to the amount suggested by current calculations with "standard" physics. Indeed, by adopting a hypothetical wind which does not violate energy and momentum conservation laws (e.g., Mv = L/c, where M is the mass loss rate, L is the stellar luminosity, c is the velocity of light, and v is some factor times the escape velocity from the star), it is possible to abstract most of the mass of hydrogen remaining after the cessation of hydrogen burning by the time the ZZ Ceti region is reached (Iben and Tutukov 1986).

Thus, this particular argument for a thin hydrogen envelope does not destroy the overall scenario described in section III, particularly when it is remembered that winds already play a major role in this scenario. In fact, by invoking a wind which continues during the cooling phase, one acquires an additional advantage: the potential for explaining why the frequency of non-DA dwarfs increases with decreasing luminosity in spite of the fact that all white dwarfs in the solar vicinity should experience episodes of accretion as they pass through interstellar clouds.

A much more formidable argument against the thick shell scenario (even as modified by invoking winds during the cooling phase) is based on the fact that there are no known DA white dwarfs with surface temperatures larger than 75000K, although DO white dwarfs with surface temperatures well in excess of 100000K are numerous, and on the fact that there is a distinct absence of non-DA white dwarfs in the surface temperature range 30000-45000K. The argument assumes that all white dwarfs have a common origin, and therefore must begin their cooling phase as hot DO white dwarfs, turn into DA white dwarfs when their surface temperature drops below some critical value between 75000K and 45000K (presumably depending on differences in initial mass and composition, and so forth), then reemerge as non-DA white dwarfs as they cool to surface temperatures below 30000K (Liebert, Fontaine, and Wesemael 1987; Fontaine and Wesemael 1987).

The detailed argument requires that the total mass of hydrogen in surface layers of a DO white dwarf be less than ~ $10^{-13} M_\odot$. Presumably, the white dwarf is born with this incredibly small mass of hydrogen, most of which is apparently hidden in a long tail below the surface. As the star cools, gravitational diffusion brings hydrogen to the surface, and drives heavier elements into the interior, so that the star becomes a DA white dwarf (but, why is the transition not sharp in the DA distribution at 45000K as well?). Then, as it cools still further to a surface temperature of about 30000K, a convective zone that appears in the region of helium ionization below the thin hydrogen layer is clever enough to reach up to grab hydrogen and devour it, making some stars (not all!) turn once again into non-DA white dwarfs.

The specifics of this scenario clearly have some loose ends to tidy up. However, the "extremely thin" (as opposed to thin) hydrogen envelopes demanded by the scenario have some support from current interpretations of the X-ray spectra of DO white dwarfs (Holberg 1988). If the hydrogen envelopes really are this thin, then it is very difficult to escape the conclusion that AGB evolution is invariably terminated either during a helium shell flash or during the ensuing quiescent helium burning phase, and that all central stars of bright planetary nebulae must be sustaining themselves by burning helium rather than hydrogen. A linear pulsation analysis by Kawaler (1988) suggests that hydrogen-burning central stars must pulsate at some point (although a linear analysis cannot predict amplitudes) and a search for pulsations in real central stars (Robinson 1988) reveals that none are pulsating (above some noise limit); the interpretation favored by the authors is that hydrogen-burning central stars are ruled out, an interpretation which to me seems a bit premature when based on their results alone. However, when coupled with the current interpretation of X-ray spectra, with the absence of DA white dwarfs with surface temperatures greater than 75000K, and with the gap in the non-DA distributions at surface temperatures in the range 30000-45000K, the evidence for "extremely thin" hydrogen envelopes becomes very impressive. The evidence is not yet compelling, because there are other arguments, based on other observational data, which can be used to defend the thick envelope scenario. One of these has been constructed by Schonberner (1986), who presents evidence in the number-luminosity distribution for DA white dwarfs for the rapid envelope contraction phase that, in models with thick envelopes, follows immediately upon the extinction of hydrogen burning by CN-cycle reactions. This rapid phase of envelope contraction does not occur in helium-burning models.

REFERENCES

The press of time has prevented a compilation of an extensive bibliography. For this, the reader may consult a review by the author in LATE STAGES OF STELLAR EVOLUTION, ed. S. Kwok and S. R. Pottasch (Dordrecht: Reidel), p. 175, 1987, and relevant papers by other authors in the just-cited work, in THE SECOND CONFERENCE ON FAINT BLUE STARS, ed J. liebert and A. G. Davis-Phillips (Schenectady: L. Davis Press), 1988 and in WHITE DWARFS, Proceedings of I. A. U. Colloquium No. 114, ed. G. Wegner (Springer-Verlag), in press, 1988.

Fontaine, G., and Wesemael, F. 1987. In The Second Conference on Faint Blue Stars, ed. A. G. Davis Philip, D. S. Hayes, and J. W. Liebert (L. Davis Press: Schenectady), p. 319.
Fujimoto, M. Y. 1977. Publ. Astr. Soc. Japan, 29, 331.
Gallino, R. 1988. This Conference.
Härm, R., and Schwarzschild, M. 1975. Ap. J., 200, 324.
Holberg, J. B. 1988. In White Dwarfs, ed. G. Wegner, in press.
Hollowell, D., and Iben, I. Jr. 1988. Ap. J. Lett., 333, 125.
Iben, I. Jr. 1984. Ap. J. 1984. Ap. J., 277, 333.
Iben, I. Jr., and MacDonald, J. 1986. Ap. J., 301, 164.
Iben, I. Jr., and Tutukov, A. V. 1984. Ap. J., 282, 615.

---------- 1986. Ap. J., **311**, 742.
Kawaler, S. 1988. Ap. J., **334**, 000.
Koester, D. 1987. In The Second Conference on Faint Blue Stars, ed. A. G. Davis Philip, D. S. Hayes, and J. W. Liebert (Schenectady: L. Davis Press), p. 329.
Lambert, D. L. 1988. This conference.
Lattanzio, J. 1988. This conference.
Liebert, J., Fontaine, G., and Wesemael, F. 1987. Mem. Soc. Astr. Ital., **58**, 17.
McClure, R. D. 1988. This conference.
Paczynski, B. 1970. Acta Astron., **20**, 47.
------------- 1971. Acta Astron., **21**, 417.
Renzini, A. 1979. In Stars and Star systems, ed B. E. Westerlund (Dordrecht: Reidel), p. 155.
-------------- 1983. In Planetary Nebulae, ed. D. R. Flower (Dordrecht: Reidel), p. 267.
Robinson, E. L. 1988. In White dwarfs, ed G. Wegner, in press.
Schonberner, D. 1979. Astron. Ap., **79**, 108.
--------------- 1986. In Late Stages of Stellar Evolution, ed. S. Kwok and S. R. Pottash (Dordrecht: Reidel), p.337.
Shipman, H. 1988. In White Dwarfs, ed. G. Wegner (Berlin: Springer-Verlag), in press.
Shklovski, I. S. 1956. Astron. Zh., **33**, 315.
Wood, P. R., and Faulkner, D. 1986. Ap. J., **307**, 659.

A Statistical Study of Spectroscopic Binaries Containing A Late-Type Giant Star

H.M.J. Boffin[1]
Institut d'Astrophysique
Universite Libre de Bruxelles

[1] Boursier I.R.S.I.A.

Binarity seems to be a feature shared by various classes of Peculiar Red Giants (PRG). This observational fact has led to the general agreement that those stars result from a mass transfer originating from an asymptotic giant branch companion star.

To discover which of the two mass exchange scenarios (i.e., a stellar wind or Roche lobe overflow) actually operates necessitates a careful study of the PRG's orbits. That study should compare the orbital parameters of PRG's with those of normal red giants. Understanding the behaviour of so-called normal stars is thus a pre-requisite to the study of the distribution of the PRG's orbital elements distribution. To this end, have we constructed, from the available literature, a catalogue of 195 spectroscopic binaries containing at least one late-type giant. From a statistical study of this catalogue can we extract some data of high interest in view of the comparison with a sample of PRGs orbits.

We deduce the following mean values from our catalogue:
$\langle P \rangle = 1098 \pm 232$ days, $\langle e \rangle = 0.205 \pm 0.015$, $\langle e^2 \rangle = 0.09 \pm 0.01$,
$\langle f(M) \rangle = 0.124 \pm 0.015 \; M_\odot$.

The well-known orbital period-eccentricity correlation (physically (un)meaningful?) is also present in our sample and can be written $\langle e \rangle \approx 0.16 \log P - 0.16$.

Other statistical properties are given, namely the distributions of eccentricity, period and mass function. We also attempt to determine the distribution of the mass ratio in such systems.

Theoretical and Observational Tests for the Mass Transfer Scenario of
Ba II stars*

H.M.J. Boffin[1]
Institut d'Astrophysique
Université Libre de Bruxelles

* Based on observations carried out at the Observatoire de
  Haute-Provence (France)
[1] Boursier I.R.S.I.A.

Ba II stars are red giants showing an enhancement of carbon and s-process elements. The elucidation of their nature seems to require a mass transfer, either by wind or Roche lobe overflow, during their past evolution. Were it really the case, all Ba II stars would be binaries with a white dwarf as companion. To better understand the exact role of their binarity, more orbits are definitely needed. They can be obtained by monitoring the radial velocity variations of those stars. However, a quicker way to find new Ba II stars with orbital elements would be to search for their existence among known spectroscopic binaries. This would also crucially test whether mass transfer is a necessary and sufficient condition to explain Ba II stars. If it is indeed the case, then all spectroscopic binaries, made of a giant and a white dwarf, in a reasonable range of periods, would exhibit the Ba II pecularity. However, the discovery of a peculiar giant+main sequence binary system would imply a revision of our ideas about Ba II stars. To this end have we begun a systematic spectral survey of spectroscopic binaries with orbital periods in the range characteristic of known Ba II stars and containing a red giant. The realization that some stars of the catalogue we compiled were already identified as semibariium stars encourages us to pursue our investigation. Coude spectra were taken with the 152 cm telescope, at a dispersion of 12 Å mm^{-1}. Until now, 2 stars out of a sample of 31 present a slight enhancement of s-process elements (their anomaly being in the range Ba 0.3 to 0.5), and 2 more appear to be good candidates. The study of a larger sample is currently in progress. A discussion of the nature of the companion to the 2 newly discovered semibarium stars is presented on grounds of their mass function and photometric indices.

The Evolution of Stars on the AGB: The Mass Loss Intensity and the Formation of Carbon Stars

Yu. L. Frantsman

Radioastrophysical Observatory, Latvian Academy of Sciences, Riga

Simulated populations of white dwarfs and N type carbon stars were generated for a Salpeter initial mass function and constant stellar birth rate history. The effect of very strong mass loss on the mass distribution of white dwarfs and the luminosity distribution of carbon stars is discussed and the results are compared with observations. A significant mass loss by stars on the TP-AGB occurs besides regular stellar wind and planetary nebulae ejection. Thus it is possible to explain the luminosity functions of carbon and M stars in the Magellanic Clouds (with very few stars brighter than $M_{bol} = -6.0$), the very narrow mass distribution of white dwarfs, and the very small number of white dwarfs with $M > 1.0\ M_\odot$. The luminosity of some AGB stars in the SMC is so high that they may be supernova of type 1 1/2 precursors. There are no such stars in the LMC. Comparison of the theoretical and observed luminosity distributions of high-luminosity AGB stars in the Magellanic Clouds shows that the mass-loss rate of these stars in the LMC is about an order of magnitude larger than in the SMC. In the Galaxy carbon stars may form only from stars with initial mass less than $1.5\ M_\odot$ due to the relatively small initial heavy element abundance in these stars; this is perhaps the main reason for the absence of carbon stars in open clusters in the Galaxy.

Evolution of a Star of 7 $M_\odot$ With Mass Loss

Huang Runqian   Jiang Suyun
Yunnan Observatory, Kunming, China

The evolution of a Population I star of 7 $M_\odot$ with mass loss has been followed. The mass-loss rates of the star are calculated by using the empirical relation of Waldron (1984) and the formula introduced in this paper. In the red supergiant stage there occur thermal pulses of the He-burning shell source either for the evolution with constant mass or for the evolution with the mass loss rate calculated by Waldron's relation. The rate of mass loss will become very large before the occurence of thermal pulses of He-burning shell source and the envelope of the star can be lost in a short period of time, if the introduced formula of the mass loss rate is used. The effects of mass loss on the internal structure and the evolution of the star, especially on the He-burning shell source are considered in detail.

Red Giant Mass Loss and Planetary Nebula Formation

P.J. Huggins A.P. Healy
New York University

It is generally believed that red giant mass loss plays some role in the formation of planetary nebulae (PNe), but the connection is not clearly understood. To investigate this issue we have undertaken an extensive search for molecular gas in PNe, which is likely to be remnant material of the red giant wind not yet ionized by the central star. The search has been carried out with the NRAO 12 m telescope in the 1.3 mm line of CO which is widely observable in the molecular winds of red giants.

About 100 PNe have been observed, and CO has been detected in 19, a few being tentative. New detections (with LSR radial velocities in km/s and line strengths in K km/s) include: IC 5117 (-10, 4.6), NGC 2440 (+44, 3.6), NGC 2474 93" NW (-70, 3.2), NGC 6072 (+15, 29), NGC 6445 (+20, 11), NGC 6563 (-27, 19), IRAS 21282 + 5050 (+18, 40), M1-7 (-11, 17), M1-16 (+50, 26) and M4-9 (-16, 13). Earlier results on NGC 2346, NGC 6720, and NGC 7293 have been given elsewhere (Healy and Huggins 1988, A.J., 95, 866, and references therein). The number observed is now large enough that some general conclusions can be drawn. First, molecular envelopes are a fairly common property of PNe. Second, nebulae with massive molecular envelopes are almost exclusively young population objects as evidenced by their morphological types, nitrogen abundances, and positions of the central stars on the H-R diagram. Third, the PNe detected cover a wide range in molecular mass (thousandths to tenths of solar mass) and nebular size (from compact objects to very extended nebulae); their mass ratio of molecular to ionized gas decreases systematically over four orders of magnitude with increasing radius, indicating that the mass of the optical nebula grows as the molecular gas becomes ionized. For these objects the molecular envelope plays a key role in the evolution of the nebulae.

Neutron Nucleosynthesis in a Low-Mass, Low-Metallicity AGB Star

David Hollowell and Icko Iben, Jr.

Astronomy Dept., University of Illinois, Urbana, Illinois, U.S.A.

Stellar evolution calculations confirm that semiconvection will occur below the convective envelope of a low-mass, low-metallicity AGB star, after a thermal pulse. These calculations show how semiconvection leads to the creation of a "^{13}C layer" in the star, which can provide a potent source of neutrons (via the $^{13}C[\alpha,n]^{16}O$ reaction) in a convective shell during later evolution. The rate at which neutrons are released is largely determined by the rate at which the ^{13}C layer is introduced into the convective shell. The ^{13}C neutron source maintains neutron densities of 10^9–10^{10} n/cm^3 for ~ 10 years. This provides a neutron exposure $\tau \approx 0.15$ mb^{-1} during most of the pulses calculated. Because of the strong filtering effect by light elements, only 10–20% of the neutrons produced will be captured by iron-seed nuclei, each such nucleus capturing 4–5 neutrons per pulse. Approximately one half of the irradiated material in a convective shell survives to be irradiated during a subsequent pulse, and this material is characterized by a neutron exposure parameter $0.2 \le \tau_0(\text{mb}^{-1}) \le 0.3$, which is remarkably similar to the exposure parameter characterizing solar system material.

Dredge up of interior processed material does not occur unless some form of convective overshoot is adopted. Dredge up will occur when the model star is dimmer than $M_{bol} = -4.5$, creating either MS-, S-, or C-star surface characteristics. Besides inducing dredge up, overshoot leads to the production of more ^{13}C, as well as to the production of more neutron filters (such as ^{14}N and ^{22}Ne) near the ^{13}C layer. The neutron irradiation in overshoot models also occurs at neutron densities of 10^9–10^{10} n/cm^3, but leads to a neutron exposure that is approximately 1.5 times that of models without overshoot.

It is found that ^{99}Tc, a very temperature sensitive radioactive isotope, can survive the temperatures in excess of $300 \cdot 10^6$ K that are reached at the base of the convective shell. This is because ^{99}Tc spends most of its time in much lower temperature regions of the convective shell.

The Effects of Main-Sequence Mass Loss on Surface C/N Abundance Ratios During the Ascent of the First Giant Branch

J. A. Guzik and T. E. Beach (Iowa State University)

The surface C/N abundance ratios of many cluster and field G and K giants following the 1st dredge-up phase are much lower than predicted from standard stellar evolution modeling. The occurrence of substantial mass loss, either during or immediately after the main-sequence phase would both reduce the mass fraction of the unprocessed envelope necessary to contaminate with CN-cycle products, as well as allow CN-processing of a greater amount of core material during the earlier high-mass phase. Willson, Bowen and Struck-Marcell (1987) have proposed that a combination of pulsation and rapid rotation could drive substantial mass loss in main-sequence stars of initial mass 1-3 $M_\odot$. We evolved a grid of 16 mass-losing models from the zero-age main sequence through 1st dredge-up. The models have initial masses of 1.25, 1.5, 1.75 and 2.0 $M_\odot$, and exponentially decreasing mass-loss rates with e-folding times 0.2, 0.4, 1.0 and 2.0 Gyr; all models evolve toward a final mass of 1.0 $M_\odot$. Since the mass-loss epoch is short-lived, most of the models reach 1.0 $M_\odot$ rapidly, and follow the evolutionary track of a standard 1 $M_\odot$ model redward away from the main sequence and up the 1st giant branch. The convective envelope deepens during 1st dredge-up to homogenize the outer 3/4 of the star's final mass.

The initial ratio of C to N mass fractions for all models is 3.07. The final C/N ratios depend primarily upon initial mass, but also decrease somewhat with increasing mass-loss timescale. Compared to a final C/N ratio of 2.12 for a standard 1 $M_\odot$ model without mass loss, the final C/N ratios, averaged for models with different mass-loss timescales, decrease to 2.0, 1.3, 0.65, and 0.32 for initial mass 1.25, 1.5, 1.75, and 2.0 $M_\odot$, respectively. These lower values can easily accomodate the observed abundance ratios for low-mass giants. The surface C/N ratios also begin to change at lower luminosities with increasing initial mass, with only a slight dependence upon mass-loss timescale. The average log $L/L_\odot$ at which the change commences is 0.48 for the standard 1 $M_\odot$ model without mass loss, compared to 0.45, 0.32, 0.26, and 0.18 for initial mass 1.25, 1.5, 1.75, and 2.0 $M_\odot$, respectively. Because the mass-loss phase is short-lived, stars with differing initial mass and nearly equal final mass may have nearly equal main-sequence lifetimes, enabling giants with a spread of post dredge-up C/N abundance ratios to be present in the same cluster. Since in the main-sequence mass loss scenario, cluster ages cannot necessarily be determined from main-sequence turnoff mass, the magnitude and onset location of C/N abundance ratio changes in cluster giants ascending the first giant branch may be a useful indicator of their initial mass and mass-loss timescale, and hence the cluster age.

The Production of Low Mass Carbon Stars:
Carbon-rich Dredge Up or Oxygen-rich Mass Loss?

R.E. Stencel and J.E. Pesce (CASA, Univ. Colorado) and
K.M. MacGregor (High Altitude Observatory/NCAR)

ABSTRACT:

Conventional theory explains the origin of carbon stars as due to dredge up of carbon enriched material from the stellar core during helium flash events late in the life of solar mass AGB stars (e.g. Boothroyd and Sackmann 1988). This relatively efficient process however, seems to produce a larger C/O ratio than observed (Lambert et al. 1987). A secondary effect which could contribute to the appearance of carbon stars, is the selective removal of oxygen from the atmosphere by radiative force expulsion of oxygen rich dust grains (e.g. silicates like $[Mg, Fe_2SiO_4]$). We present calculations for this scenario which evaluate the degree of momentum coupling between the grains and gas under the thermodynamical conditions of AGB star atmospheres.

In their pioneering analysis of the composition of the atmospheres of carbon stars, Lambert et al. (1986) state that there is a surprisingly thin margin separating carbon stars from their oxygen rich progenitors. Moreover, they also determined that ^{13}C, the preferred product of the dredge–up process, is lower than theory predicts. This problem appears to be especially serious for lower mass carbon stars, where large values of mixing length have to be invoked to produce carbon dredge up (Boothroyd and Sackmann 1988). We suggest an additional process: that oxygen in the star's atmosphere could be *selectively removed* by the formation of oxygen–rich dust grains, which are driven off by radiative forces. The importance of this process with respect to dredge–up will depend sensitively on stellar mass.

The advent of the IRAS infrared sky survey and low resolution spectroscopy, enabled Little–Marenin (1986), Willems and deJong (1986) and Nakada, Deguchi and Forster (1988) to discover and analyze carbon stars *with silicate features*, suggestive of oxygen–rich circumstellar envelopes. This juxtaposition of features suggests an evolutionary connection in which the oxygen–rich characteristic of the carbon star progenitor persists in the outflow of material, which left the star on an outflow timescale. Ultimately, it is surmised that such objects will replenish their circumstellar envelopes with carbon–rich material. Hence, it seems established that evidence of the composition history can survive in the dusty stellar wind.

In their analysis of time–dependent models for the expanding atmospheres of carbon rich stars, where the driving force was radiation on grains, Woodrow and Auman (1982) noted that the drift velocity of the carbon–rich grains, with respect to the gas, was large, implying that while the grains shared momentum with the gas, the two species were not positionally coupled. Thus, the carbon was leaving the atmosphere *faster than* the other stellar material and the surface layers were being *differentially depleted of carbon*. They further note that the drift velocities were not so large as to destroy the grains by sputtering. Hence, they noted that the dust–forming process can selectively alter the composition of the stellar atmosphere.

Composition anomalies have been noticed by several authors. Eaton and Johnson (1988) studied oxygen–rich M stars using ultraviolet spectroscopy, and came to the unsettling conclusion that silicon was somehow depleted in these chromospheres. Judge (1986) did not find silicon underabundance in similar stars, however. Luck and Lambert (1985) performed CNO abundance analysis for Cepheids and non-variable F supergiants and conclude that a significant oxygen underabundance exists which cannot be attributed exclusively to dredge–up of ON cycled material. An analysis of the infrared characteristics of similar stars by Stencel, Pesce and Bauer (1988, 1989) indicates that some of these stars probably have evolved blueward from red supergiant phases where extensive dust production dominated their mass loss history. Snijders et al. (1984) argued that the ejecta in Nova Aql 1982 underwent grain formation which may have led to gas–phase element depletions. Similar arguments have been advanced by Snow et al. (1987) who examined cimcumstellar reddening toward the red supergiant α Sco, using techniques adapted from interstellar work, and concluded that the grains are large and silicon–rich, on the basis of extinction and elemental depletion. Hence, in several instances we find indication that for stars within which dust formation is occuring or has occurred, that composition anomalies occur as well.

The degree of coupling between dust grains and gas atoms in a stellar atmosphere can be evaluated in terms of the ratio of drag forces to radiative acceleration. The drag force per grain is given by (Gilman 1972):

$$f_{drag} = \rho_{gas} \sigma_{gr} v_D^2 \qquad (1)$$

where ρ_{gas} is the gas density, σ_{gr} is the grain cross section [cm^2] and v_D is the drift velocity:

$$v_D = |v - v_{gr}| \qquad (2)$$

v is the wind velocity, derived from the formalism originally discussed by Parker (1958) where mass and momentum conservation in an isothermal flow require that the wind solution pass through a critical point. For red giants with dust in their outer atmospheres, we add a term for the radiation force. With appropriate values, we were able to reproduce the Parker solutions for the solar wind, and to compute representative wind solutions for red giant winds with non–zero opacity.

The grain terminal velocity can be evaluated in the limit of rarified gas dynamics and constant flow velocity. Gilman (1972) estimated this to be approximately 40 km sec^{-1} for red giants. The radiation force is given by:

$$\dot{f}_{rad} = \rho \sigma L/(4\pi r^2 c) \qquad (3)$$

where σ is the atmospheric opacity in $cm^2\ gm^{-1}$.

For three isothermal models, with radii and luminosities comparable to α Ori, and with surface temperatures of 4000K, 5000K and 6000K, $\sigma = 400\ cm^2\ gm^{-1}$, and photospheric densities of $10^8\ cm^{-3}$, we compute that the ratio of drag to radiation forces is much smaller than 0.5, even at the photosphere, where it is largest. Hence, the gas and grains are not coupled positionally and differential depletion of oxygen–rich solids may occur. We assume that a "molecular catastrophe", where the rapid cooling effect of simple molecules like CO and SiO in the upper photosphere, can give rise to grain formation at low altitudes (Muchmore, Nuth and Stencel 1987; Tsuji 1988), and hence, directly affect derived photospheric abundances over evolutionary timescales. For higher photospheric densities, the degree of coupling is larger. Hence the de-oxygenation of stellar photospheres is sensitive to effective gravity and therefore evolutionary state. Low mass stars near the end of their AGB may be the most affected by this process, in addition to dredge–up (if any).

While this low coupling argues for differential depletion, it also raises the question of the role of the grains in the overall dynamical state of the atmosphere. This issue requires further quantitative study with two component wind solutions. We are grateful for support of this research from NASA grants JPL 957632 and NAG5-816 to the University of Colorado.

REFERENCES

Boothroyd, A. and Sackmann, I.J. 1988 *Ap.J.* **328**, 671.
Eaton, J. and Johnson, H. 1988 *Ap.J.* **325**, 355.
Gilman, R. 1972 *Ap.J.* **178**, 423.
Judge, P. 1986 *M.N.R.A.S.* **223**, 239.
Kwok, S. 1975 *Ap.J.* **198**, 583.
Lambert, D., Gustafsson, B., Eriksson, K. and Hinkle, K. 1986 *Ap.J.Suppl.* **62**, 373.
Luck, R. and Lambert, D. 1985 *Ap.J.* **298**, 782.
Muchmore, D., Nuth, J. and Stencel, R. 1987 *Ap.J.* **315**, L141.
Parker, E. 1958 *Ap.J.* **128**, 664.
Snijders, M., Batt, T., Seaton, M., Blades, J. and Morton, D. 1984 *M.N.R.A.S.* **211**, 7P.
Snow, T., Buss, R., Gilra, D. and Swings, J. 1987 *Ap.J.* **321**, 921.
Stencel, R. Pesce, J. and Bauer, W. 1988 *A.J.* **95**, 141.
Stencel, R. Pesce, J. and Bauer, W. 1989 *A.J.* submitted.
Tsuji, T. 1988 *A.& A.* in press.
Woodrow, J. and Auman, J. 1982 *Ap.J.* **257**, 247.

Early Shaping of Asymmetric Planetary Nebulae

Noam Soker
Department of Astronomy, University of Virginia
P.O. Box 3818 University Station, Charlottesville, VA 22903

We suggest that the shape of a young asymmetric planetary nebulae may be influenced by a close binary star located at its center. This binary is a relic of the common envelope phase, presumably through which the asymmetric planetary nebula evolved. We assume that for a short period of time, shortly after the cession of the slow wind and long before the fast wind becomes effective, the binary ejects a small amount of mass, mainly in the equatorial plane. In this work we do not discuss the exact mechanism for the ejection of this pulse of mass. In the case in which the cooling is very efficient, (i.e., high-Mach-number isothermal flow), we can solve the problem analytically by using a few simplifying assumptions. In this case the high density region is shaped like a ring. We use two-dimensional hydrodynamics for the more general case. We find that at late times the high density region has a "horseshoe" shape, as viewed in the symmetry plane. There is an instability in the maximum density region. Finally we compare our results with the shape of the planetary nebula M2-9.

IRAS 21282+5050: A Transitional Planetary Nebula

L. Likkel   M. Morris   A. Omont   T. Forveille
     UCLA   Grenoble

IRAS 21282+5050 is very compact planetary nebula[1] with a substantial molecular shell[2]. It is carbon-rich and exhibits infrared features attributed to PAH molecules[3]. It has a far infrared color temperature higher than almost all other planetary nebulae, with IRAS flux ratios similar to low color temperature evolved stars. We present VLA images of IRAS 21282 at wavelengths of 2 and 6 cm (with FWHM beamsizes of 0.5" and 0.9", respectively).

The images show rather clumpy optically thin emission, with a ring-like appearance. The ionized region is slightly elliptical, with a full extent of about 4.5" by 3.4" at both wavelengths. Assuming a distance of 2 kpc, the ionized region extends to a radius of about 6×10^{16} cm, and CO (J = 1-0) is observed out to about 15×10^{16} cm. The data give an emission measure of about 3×10^5; this implies an electron density of about 3000 cm^{-3}. This is larger than that deduced from optical data[1]; since the optical size is smaller than the radio size, this suggests that the mass loss rate has declined. The measured electron density implies a mass loss rate of about 10^{-5} $M_\odot$/year, which is somewhat smaller than the mass loss rate deduced from CO observations. This suggests that the mass loss rate has either declined since the formation of the H II region or the mass loss rate deduced from CO observations is overestimated, perhaps because of the neglect of heating of the molecular gas by shocks associated with the H II region. The ionized mass (assuming $Te = 10^4$ and D = 2 kpc) is 0.003 $M_\odot$, which is very low for a planetary nebula. We interpret this as an indication of its extreme youth.

References:

[1] Cohen, M. and Jones, B. 1987, Ap. J. Lett. 321, L151.
[2] Likkel, L., Forveille, T., Omont, A., and Morris, M. 1988, Astron. Ap., in press.
[3] de Muizon, m., Geballe, T.R., D'Hendecourt, L.B., and VBaas, F. 1986, Astrophys. J. Letters 306, L105.

5 Ceti: A Long-Period Binary Evolving Through Mass Exchange*

Joel A. Eaton
Indiana University

Binaries with very wide spearations are thought to evolve to small separations through a catastrophic form of mass exchange/loss known as common-envelope evolution. The theory of this process is fairly well developed, but proper tests remain elusive. Simply put, the theory argues that the rapidly shrinking Roche lobe of the mass losing giant will strip away the giant's main-sequence companion. Loss of mass from the system during the process carries away orbital angular momentum, thereby strengthening the effect.

In 5 Cet we have a binary, containing a K giant, that seems to be caught in this stage of mass loss from a more massive giant to a less massive companion. Mass-loss rates do not seem to be extreme enough for common envelope evolution, however.

Critical properties are inferred from a variety of evidence. Ellipsoidal light variation requires the K giant to be near its Roche lobe. The K giant's companion, however, is too bright in the ultraviolet to be in the main-sequence. This and its apparently random light variations argue it is an accretion disk around a 1.1 $M_\odot$ star. The luminosity requires a mass transfer rate of no more than 5×10^{-7} $M_\odot$/yr. The more extreme mass loss from the system expected under common-envelope evolution is not detected in the ultraviolet, although it could still exist in the plane of the orbit outside the line of sight. Although a 260 km/s wind from the system is detected in Mg II h + k profiles, its greater strength when the K giant is in front of the hot companion argues that this is a wind of the K giant itself, analogous to the high-speed winds in ζ Aur systems. In this case, it carries away only 10^{-9} $M_\odot$/yr, about the same mass-loss rate per unit surface area as in a ζ Aur star.

*This is a summary of a paper by J.A. Eaton and S.C. Barden accepted for publication by <u>Acta Astronomica</u>.

The N/O–Core Mass Relation in Planetary Nebulae

J.B. Kaler
University of Illinois

R.A. Shaw
Lick Observatory

K.B. Kwitter
Williams College

We define the relationship between nitrogen enrichment in planetary nebulae and the mass of the nucleus. N/O remains flat at about 0.3 (double solar) from a core mass of 0.55 M(sun) to 0.8 M(sun), whereupon it rises quickly to values that approach and may exceed 2. The rate of increase of N/O with core mass exceeds that predicted for giant stars by standard dredge-up and mass-loss theories.

Models of AGB Stars Envelopes and Atmospheres

M. Forestini[1]
[1]Aspirant of the Fonds National pour la Recherche Scientifique
Institut d'Astronomie, d'Astrophysique et de Geophysique
Universite Libre de Bruxelles, C.P. 165, av. F.-D. Roosevelt, 50
B-1050 Bruxelles, Belgium

The possible existence of a hot CN-cycle at the bottom of the AGB convective envelopes ("hot-bottom burning" or HBB) encounters some revealing difficulties with observations (e.g. Reid and mould, 1985). On the other hand, existing models disagree with each other about it.

In a first step, we have studied in detail the convective properties of AGB stars by developing a new envelope code (the classical mixing-length theory being assumed). Molecules and grains contributions are included in the radiative opacity calculation. We have introduced an atmospheric temperature profile incorporating blanketing effects (the optical depth being corrected for the sphericity in thick atmospheres). Such a code is required in order to obtain accurate effective temperatures, which determine the structure of the ionization zones, and, in this way, the convection extension. The H and He ionic abundances are calculated in a self-consistent way, resolving the Saha equations together with the charge neutrality constraint by a minimization method. Previous iterative methods do not give accurate electronic number densities, affecting the adiabatic gradient, and thus, the stellar convective structure. As firstly emphasized by Sackmann and Boothroyd (1985), electronic number density variations could have similar effects in the region where CNO elements are produced, below the bottom of the convective envelope. We have consequently calculated all the ionization stages of these elements.

We obtain convective bottom temperatures between 5×10^5 and 2.5×10^6 K (the corresponding densities being about 10^{-6} to 10^{-3} g/cm^3) in models having total luminosity corresponding to the beginning of the thermal pulses period (following Iben and Truran, 1978). Such conditions are inhospitable for the HBB process. This conclusion is roughly insensitive to a change of the core mass-luninosity relation used. More luminous models have progressively less deep convective envelopes. This conclusion is hardly strengthened by introducing a realistic mass loss rate for AGB stars (at constant luminosity). Such a stellar wind, probably activated before the thermal pulses epoch, could substantially modify this phase (as suggested by the observations of Mould and Reid, 1987).

References

Iben, I. and Truran, J.W. 1978, Astrophys. J. 220, 980.
Mould, J. and Reid, N. 1987, Astrophys. J. 321, 156.
Reid, N. and Mould, J. 1985, Astrophys. J. 299, 236.
Sackmann, J. and Boothroyd, A.I. 1985, Astrophys. J. 293, 154.

Comparison of s-Processing Occurring in a Low Mass AGB Star of Low Metallicity and the Results of s-Classical Analysis.

M. Busso, G. Picchio
Osservatorio Astronomico

R. Gallino
Instituto Di Fisica
C.M. Raiteri
International School for Advanced Studies

F. Kappeler
Kernforschungszentrum

We examine the results of s-Processing occurring in a low mass star of low metallicity during the pulsed He-instability in AGB phases by comparing them with the s-Classical analysis. Neutron exposures are provided by the C13(Alpha,N)O16 reaction, according to the mechanism suggested by Iben and Renzini (1983) for the formation in the interpulse phase of a small zone rich of C13 and its subsequent ingestion in the next pulse.

As far as the s-Classical process is concerned, the analysis of several branchings of the s-Flow provides stringent constraints on neutron density, temperature, electron density and duration of the pulsed process. In particular, the neutron density has to be low enough [Nn = (1.0-0.5)E8 N/CM**3] and the temperature has to be sufficiently high [T = (2-3)E8 K].

These conditions apparently seem to exclude the C13-source as efficiently producing the main component of heavy s-Isotopes in a solar system composition. Indeed, the bulk neutron densities during C13 exhaustion are much higher, of the order of 5E9 N/CM**3, and the temperature at the bottom of the He-burning shell is too low: typically 1.5E8 K.

Nevertheless, the realistic neutron flow is not constant, as assumed in the s-Classical analysis, reaching a maximum of 5E9 N/CM**3 and then decreasing slowly until C13 is exhausted. At the beginning of neutron exposure, the n-Processing is far from an s-Processing, but the n-rich isotopes are frozen out when the average neutron density is decreasing. This density practically coincides with the s-Classical constraints. The main bulk temperature during neutron exposure is fairly low; nevertheless, a short phase of high temperature (of the order of 3E8 K) is met near the end of the pulse, when convection extends over its maximum extension. In these conditions, the abundant isotope NE22 undergoes small alpha-captures through the NE22(Alpha,N)MG25 reaction, thus releasing a small flux of neutrons. The high temperature branching points in the s-Flow are now open, and a consistent production of just the few "thermometers-isotope" is obtained.

We conclude that the C13-source in low mass AGB stars may fulfill all the constraints provided by the s-Classical analysis.

Evolution of a $10 M_\odot$ Star With LMC Metallicities

N. Rathna Sree   A. Ray
Tata Institute of Fundamental Research

Why stars become red giants has been a subject of investigation in many contexts, most recently in the discovery of the progenitor of the Supernova 1987A. Sanduleak $69°202$ was found to be a blue supergiant star although it was generally presumed that type II SNe arise from red supergiants. Immediately after SN1987A, it was suggested that the blue spectral nature was due to lower metallicity in the LMC ($Z = Z_\odot/3 - Z_\odot/4$) although the existence of many red supergiants in the 30 Doradus region where SN1987A took place and in particular the observation of low-velocity nitrogen-rich gas presumed to be a circumstellar shell indicates that mass loss may also have played a significant role in bringing SK $69°202$ from red to blue. We report here work in progress on the evolution of massive stars in the LMC with and without mass loss which can ultimately produce type II SNe. The number of red supergiants in a homogeneous group of stars at a given time depends on t_{RSG}/t_{Total} and the mass function $N(M)$. With a view to seeing how the evolutionary behaviour on the colour-magnitude diagram changes with and without mass-loss, composition, and convection criteria, we evolved a $10 M_\odot$ star with initial composition of $X = 0.715$, $Y = 0.28$ and $Z = 0.005$, with updated nuclear reaction rates implemented in a code originally due to Icko Iben. We report here the structure, chemical composition, and other thermodynamic variables throughout the constant $10 M_\odot$ star at different stages of evolution. At the time of central carbon exhaustion, the model has $\log(L/L_0) = 4.47$ and $\log T_c = 4.07$. The value of t_{RSG}/t_{Tot} is 0.15 as compared with the value 0.08 obtained by Maeder (1987) for a $20 M_\odot$ star evolved with substantial mass loss.

4. The Variability–Evolution Connection

OBSERVATIONS OF LONG PERIOD VARIABLE STARS

T. Lloyd Evans
South African Astronomical Observatory
P.O. Box 9   Observatory 7935
South Africa

1. INTRODUCTION.

The study of long period variable stars has been transformed in recent years by two observational developments. Large samples of stars have been observed at infrared wavelengths, providing knowledge of the intrinsic properties of the star as well as of circumstellar dust shells, and these observations have been extended to the variables in well defined stellar systems to allow their properties to be studied in relation to the stellar population to which they belong. Spectroscopic determinations of chemical composition have also provided several crucial insights.

2. TYPES OF OPTICALLY SELECTED RED VARIABLES

The fourth edition of the General Catalogue of Variable Stars (Kholopov et al. 1985) lists over 11,000 red variables. Several hundred, most of them M giants, have published spectral types or infrared photometry. The stars are classified on the basis of light curves in the photographic or visual regions as irregular (L), semiregular (SR) or Mira Ceti type (M). The distinction between SR and M is primarily one of amplitude: the frequency of amplitudes of red variables has a minimum at 2.5 mag in blue or yellow light and the Miras have larger amplitudes (Payne-Gaposchkin 1951). The Miras of spectral class M or S usually have much larger amplitudes, typically 5-8 magnitudes, because TiO and other bands which strengthen towards minimum light absorb preferentially in B and V (e.g. Smak 1966). The amplitudes in J (1.2 μ), K (2.2 μ) and in m_{bol} are only in the ranges 0.3-1.4 mag, 0.2-1.0 mag and 0.3-1.2 mag, respectively (Feast et al. 1982). The carbon variables usually have quite small amplitudes in V, with more rounded light curves (Alksne et al. 1981; Wing 1985). The blue amplitude of a variable carbon star might be enhanced relative to the amplitude in V by variation of the violet opacity. The classification of the carbon stars is correspondingly less certain than for M stars.

The SR variables are divided into SRa stars which have persistent periodicity although with variable amplitude and light curve shapes and SRb stars which have less marked periodicity which may be interrupted by intervals of irregularity or even constancy. An important subgroup of the SRb stars comprises those which show two periods simultaneously, so that the mean magnitude may change in a much longer period than the normal variation. Houk (1963) lists over 100 such stars, including 66 of spectral class M and 20 of class N. The ratio of the two periods,

P_2/P_1, is approximately 9.4 for M stars and 12.2 for N stars; there are some very discrepant values (Payne-Gaposchkin 1954; Houk 1963).

The Mira variables show H emission during part of the light cycle. Those semiregular variables with weak emission lines belong to a population of higher velocity and presumably greater age than the stars without emission (Feast et al. 1972). Hydrogen emission in stars of relatively small amplitude is also found among the SRd variables of Population II (Rosino 1951, Feast 1965) and in some variables in globular clusters (Joy 1949, Lloyd Evans 1983a).

The classes described here refer specifically to giant stars. M supergiants, which are often found in associations or young clusters and are believed to have $m \gtrsim 9m_0$, are classed as SRc or Lc; there are no Miras.

3 POPULATIONS OF RED VARIABLES

3.1 Galactic Globular Clusters

The richest clusters contain sufficient stars to show how the variability of a star develops as it ascends the giant branch. Optical studies especially of 47 Tuc (Arp et al. 1963; Eggen 1972; Lloyd Evans & Menzies 1973; Lloyd Evans 1974, 1983b; Fox 1982), show that a star varies with increasing amplitude and lengthening period until it becomes a Mira variable (Feast 1973). There are also variables of small or moderate amplitude which lie above, or perhaps more accurately to the blue of, the giant branch. These are still quite close to the tip of the giant branch and are not to be confused with the Type II Cepheids and RV Tauri stars which lie in the Cepheid instability strip. They have been referred to as Pec (Frogel 1983) or supra-giant branch (SGB) (Lloyd Evans 1983a). Infrared JHK photometry of 47 Tuc (Frogel et al. 1981) confirmed the picture obtained from VI photometry and in addition showed that the luminosity of the three Mira variables as well as the semiregular V4 placed them above the tip of the first giant branch and so presumably on the asymptotic giant branch (AGB). Infrared studies of several clusters (Frogel 1983; Frogel & Elias 1988) show that this is generally true of Mira variables (which they denote LPV), while the fainter irregular and semiregular variables could equally well be on their first ascent of the giant branch. The Miras have infrared excesses at 3.4μm and 10μm as well as enhanced H_2O absorption; this may indicate mass loss. These properties are also exhibited, less markedly, by the stars which fall to the blue of the giant branch. The latter generally have luminosities close to that of the top of the first giant branch.

Galactic globular clusters cover a wide range of metal content while all are of a similar great age (e.g. Buonanno 1986, Burstein 1985) so that intercomparing the variables in these clusters shows the effect of varying metal content in stars of a similar low mass. The most metal deficient clusters contain red variables with small amplitudes and short periods; they are not confined to the immediate tip of the giant branch (Eggen 1972). Variability of small amplitude sets in at redder V-I colour and hence later spectral type the higher the metal content in the more metal rich clusters (Lloyd Evans & Menzies 1973, 1977; Lloyd Evans 1983b). The periods of the Mira variables are longer the more metal-rich the cluster (Lloyd Evans & Menzies 1973; Lloyd Evans 1983a). There is no case known of Mira variables of significantly different period belonging to the same cluster. Frogel & Elias (1988) find that the more metal-poor clusters with $[Fe/H] < -1.0$ have no stars with the infrared excesses, H_2O absorption and high luminosity of Miras, although the variables displaced to the blue of the giant branch are found over the whole range of $[Fe/H]$, -2.0 to -0.3. The two cluster Miras of shortest period, V42 (P = 149 d) in ω Cen which has a spread of metal abundance, and V16 (P = 135 d) in NGC 362 which has $[Fe/H] = -1.3$, were not observed because of crowding, however.

Frogel et al. (1981, 1983) found from observations of a large sample of globular cluster giants that the large amplitude variables (Miras) are disployed from the locus defined by the other stars in the J-H, H-K diagram. This displacement is closely correlated with the strength of the H_2O band at 1.9μm.

Menzies & Whitelock (1985) obtained JHKL photometry for twelve Miras in galactic globular clusters and derived a period-lumiunosity relation, subject to some uncertainty arising from the determination of the relative distances of the clusters.

The globular cluster variables are generally of type K or M, depending on temperature and metal content (Feast 1973). Marginal S stars have been reported in NGC 6723 but were not found in 47 Tuc (Lloyd Evans 1984). Clusters which are much more metal deficient than NGC 6723 contain no stars with molecular bands such as TiO in any case. ω Cen contains several S stars of rather low luminosity which probably owe their enhanced ZrO to the unusual primordial chemical composition of the cluster rather than to the appearance of the products of reactions within these stars (Lloyd Evans 1983c). These would be "spectroscopic S stars" as defined by Little et al. (1987). Observations of Tc and binarity studies would be valuable. Three of the stars are variable; V6 belongs to the class of semiregular variables which resemble Miras in having strong H emission (Dickens et al. 1972) and lies at or just above the luminosity of the tip of the first giant branch (Frogel 1983). Although no S type Mira is known in a globular cluster, the example of the field Se variable NT Tel (P = 252 days) which has a very high radial velocity (Catchpole & Feast 1971; Andrews 1975) suggests that such stars might be looked for in clusters.

3.2 The Solar Neighbourhood

The solar neighbourhood contains a large number of Mira variables bright enough for detailed study, although individual distances and relative luminosities are usually not obtainable.

The classical division of the red variables into irregular, semiregular and Mira types may be tested against infrared photometry.

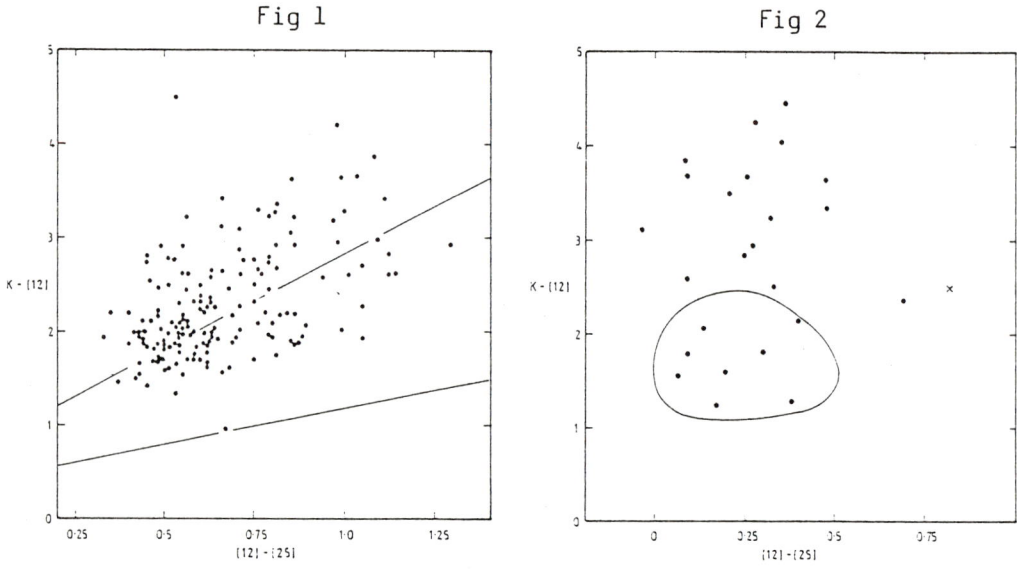

Fig 1. M-type Miras. The L and SR stars fall between the two straight lines.

Fig 2. Carbon Miras. The L and SR stars lie in the region indicated, with a few exceptions such as R Scl (cross).

Figs 1 and 2 show the K-[12] against [12]-[25] for M and N stars, respectively, combining 2μm photometry by Neugebauer & Leighton (1969) and Catchpole et al. (1979) with 12μm and 25μm photometry from the IRAS Point Source Catalogue (Beichman et al. 1985). The Miras are plotted individually while almost all other variables, except for a few SRa stars most of which have H emission, fall in the regions indicated. There is no such distinction between the other types of red variable. This is consistent with the conclusions of Frogel & Elias (1988) for globular cluster stars.

The Mira variables of type M are well separated from the small amplitude variables in the J-H, H-K diagram (Feast et al. 1982), reflecting the occurrence of strong H_2O absorption in the H band in the Mira variables but not in the constant M stars or those with small amplitudes of variation (Hyland 1974, Frogel et al. 1981). Miras which are known to have OH or H_2O masers are displaced by more than the average amount in this diagram, and are also displaced towards larger J-K and K-L in the J-K, K-L diagram. The Se variables do not show such displacements as a group, nor do they exhibit H_2O absorption bands or OH or H_2O maser emission (Hyland 1974). However the Se stars are displaced from the S stars as a group, more nearly parallel than perpendicular to the black body line. The carbon Miras are less well separated from variables of small amplitude, except for a few very red stars. Catchpole & Feast (1985) show that the latter include V Hya, a spectacular example of the doubly periodic stars (Houk 1963; Mayall 1965).

Kinematic studies of the Mira variables (Feast 1963) showed that the longer period stars have lower velocities relative to the Sun, suggesting that period depends on age and mass. M-type Mira variables with $P \lesssim 200$ days have a large systematic motion with respect to the local standard of rest which associates them with Population II, consistent with the presence of Miras with $P \sim 200$ days in metal-rich globular clusters. Stars with $P > 400$ days have a small systematic motion and they were identified with a Hyades population, corresponding to an age of $\sim 5 \times 10^8$ years and a mass of 2.5 $m_\odot$ on the main sequence.

Robertson & Feast (1981) analysed the infrared data of Catchpole et al. (1979) to find $M_{bol} \sim -3.9$ for the whole period range 150-600 days from statistical parallaxes, whereas the stars in clusters and binaries showed a trend from -3.8 at 200 days to -4.9 at 350 days. The Type II OH/IR sources fall on an extension of the period-luminosity relation, drawn primarily from Miras in the LMC (Feast 1984), to periods in the range 600-1800 days (Engels et al. 1983; Feast 1985). The masses must be quite small (Feast 1986).

The S and carbon Miras are few in number and do not lend themselves to statistical analysis. Their periods are much longer than those of the M stars in the mean; Merrill's (1960) study shows that the S stars are common only for $P > 280$ days while most N stars have $P > 350$ days. Analogy with the Magellanic Cloud data (Section 3.4) suggests correspondingly higher bolometric luminosities and perhaps masses than for the shorter period Miras of spectral type M.

Lloyd Evans (1985a) noted that the proportion of Miras among the red stars is at a maximum for the extreme S, SC and weak-banded carbon types. This was attributed to these stars being formed by the envelope-burning process at high luminosity but the situation may be more apparent than real. Factors contributing to this are: (1) the bias towards cool stars of the main search technique for pure S stars (Keenan 1954) results in a bias towards the discovery of Mira variables (Lloyd Evans & Catchpole 1988), (2) hot SC stars are difficult to find, because of the weakness of the molecular bands, and are not obviously distinguishable from the Ba stars which are probably not AGB objects (Yorka & Keenan 1985), (3) Tsuji's (1981a, b) temperature calibration for small amplitude variables among the N stars shows that the C_2 bands increase in strength as the effective temperature decreases from 3200 to 2400 K; the Mira variables must be cooler than this and are probably beyond the peak in the C_2-T relation (Morris & Wyller 1967) so that the weakness of the C_2 bands may be a temperature effect.

3.3 The Galactic Centre

The red variables in the central bulge of the Galaxy have been observed via the Baade windows in the foreground dust clouds. Visible light studies revealed 113 Mira variables (Lloyd Evans 1976); a few additional stars, with periods in excess of 400 days, were discovered on the original plates at the position of IRAS sources but the search appears to have been complete for shorter period stars (Feast 1986; Glass 1986). The very long period stars are only observable in visible light for a short period around maximum light and are not identifiable as Miras on the basis of amplitude.

Some of the Miras with P $\gtrsim$ 400 d have exceptionally red colours (Glass & Feast 1982; Wood & Bessell 1983; Glass 1986). Whitelock et al. (1986) have interpreted these as stars with very thick circumstellar shells, similar to WX Ser, IK Tau and WX Psc in the solar neighbourhood. These extreme Miras, which have very long periods, are considered as intermediate between typical optically-selected Miras and the Type II OH/IR variables which can only be studied at longer wavelengths.

Lloyd Evans (1976) considered two extreme possibilities for the galactic centre Miras: the period-age relationship for field stars (Feast 1963) suggested that the long period Miras might be of age similar to the Hyades, while extrapolation of the period - [Fe/H] relationship for those in globular clusters (Lloyd Evans & Menzies 1973) implied that they are metal rich but possibly very old. The mounting evidence for a very metal-rich component in the central bulge of the Galaxy (e.g. Whitford & Rich 1983) and the lack of evidence for the many main sequence A stars implied by the first possibility suggested that the second case is more nearly correct. Wood & Bessell (1983) argued for much higher masses and a correspondingly young age for the bulge Miras, but their arguments have been disputed by Feast (1986) and Frogel & Whitford (1987). The latter present evidence that these Miras are not different from those found elsewhere and that they have masses and ages similar to globular cluster stars.

There appear to be no carbon stars among the Miras, nor are any N stars known in the galactic centre fields (Lloyd Evans 1976; Blanco et al. 1984). This is an important difference between the variable star population in the central bulge of the Galaxy and that in the vicinity of the Sun.

The period-luminosity relation for Mira variables in the galactic bulge has been studied on the basis of JHK photometry (Glass & Feast 1982, Wood & Bessell 1983, Feast 1986). The slope is not significantly different from that in other samples and the zero point is the same as in galactic globular clusters and the Magellanic Clouds if the same assumptions are made about the luminosity of the RR Lyrae stars in each system (Feast & Whitelock 1987).

3.4 The Magellanic Clouds

The Magellanic Clouds are of particular importance because they contain a wide range of stars of different age and composition at the same distance, so that relative luminosities can be obtained and evolutionary conclusions drawn. There are two types of large amplitude variables. The Shapley-Nail variables are bright visually and were discovered by the Harvard observers (Shapley & Nail 1951, Payne-Gaposchkin & Gaposchkin 1966, Payne-Gaposchkin 1971). They have a period-visual luminosity relation in the opposite sense to that of galactic Miras (Buscombe et al. 1954), visual amplitudes $\Delta V \gtrsim 5$ mag and strong H emission lines (Lloyd Evans 1971a). They are believed to be massive AGB stars, $m \lesssim 7\ m_o$, on the basis of surface distribution (Lloyd Evans 1971a, 1985a) as well as comparison with lines of constant pulsation mass in the M_{bol}, P diagram (Wood et al. 1983; Wood et al. 1985). The frequent occurrence of ZrO indicating s-process enhancement and probably a raised C/O ratio in the envelope also suggests they are AGB stars (Wood et al. 1983).

Stars comparable to galactic Mira variables have been found by using V and I plates taken with larger telescopes (Lloyd Evans 1971b; Glass & Lloyd Evans 1981; Wood et al. 1985; Glass & Reid 1985; Glass et al. 1987; Lloyd Evans et al. 1988; Reid et al. 1988). These stars fall on a P - M_{bol} locus which is similar to and perhaps identical with that seen in the solar neighbourhood (Robertson & Feast 1981) and globular clusters (Menzies & Whitelock 1985; Feast 1986). The high luminosity stars fall on a steeper locus which may intersect that of the ordinary Miras near P = 250 days; however most of the stars are clumped between M_{bol} = -5.8 and -6.8 with only a few to define the lower part of the locus (Wood et al. 1983; Lloyd Evans 1985a; Lloyd Evans et al. 1988; Reid et al. 1988). The two CS stars (Lloyd Evans 1980, 1985a) may both belong to the ordinary Mira population (Lloyd Evans et al. 1988). The stars of high luminosity are estimated to be $\sim 10^8$ years old and their separation from the much older ordinary Miras may result from episodic star formation in the Magellanic Clouds which has resulted in partial filling of a broad instability region in the period-luminosity plane (Wood et al. 1983; Wood et al. 1985; Reid et al. 1988).

The M, S and C stars follow an almost identical locus in the P, K diagram but the carbon stars are fainter at given period in the P, M_{bol} diagram. The difference is much smaller than would be expected from current understanding of the T_{eff}, J-K relation for cool stars (Wood et al. 1985) but the bolometric luminosity may not be accounted for satisfactorily by observations extending only to 2.2 μ (Glass et al. 1987).

The inventory of the ordinary Miras may be incomplete for the visually fainter long period stars of M and S type especially. The periods of some of the stars are poorly known and there may be a few SR variables in the current lists. Glass et al. (1987) find for the LMC a total of 32 probable M stars, 3 S stars and 37 carbon stars. The more fragmentary data for the SMC comprises one M star, plus two of short period which may not be Miras (Lloyd Evans 1985a), five C stars and the two CS stars referred to above (Lloyd Evans et al. 1988). The SMC lacks the 200-day Miras which are common in the other systems, perhaps because the heavy element content was too low to produce this 47 Tuc component some 14×10^9 years ago. The mean periods of the carbon Miras in the various systems are: SMC, 310 days (Lloyd Evans et al. 1988; LMC, 320 days (Wood et al. 1985; Reid et al. 1988); solar neighbourhood, 400 days (Kholopov 1985); galactic central bulge, none found (Lloyd Evans 1976). This matches the relative metal contents of these stellar systems.

The intrinsic colours of the high luminosity stars show the same shift from the intrinsic line towards the blackbody line in the J-H, H-K diagram as galactic Miras, with some detailed differences attributable to peculiarities in composition (Elias et al. 1985; Lloyd Evans 1985a). The absence of galactic counterparts of these stars, which are especially common in the SMC, has been attributed to the development of obscuring dust shells around the more metal rich galactic stars (Lloyd Evans 1971a; Elias et al. 1985). The relatively low luminosity of the OH/IR stars, which lie on a linear extension of the P-L relation for the Mira variables (Engels et al. 1983; Feast 1985) and are therefore nearly a magnitude fainter than the bright Magellanic Cloud stars at P ~600 days, may rule them out of consideration as possible counterparts. Their masses, ~ 1.3 m_0 (Feast 1986), are much less than those considered likely for the bright Magellanic Cloud stars (Lloyd Evans 1971a; Wood et al. 1983; Lloyd Evans 1985a). The indirect (kinematic) nature of the luminosity estimates for OH/IR stars may leave room for an unrecognised population of more luminous OH/IR stars in the Galaxy. The Magellanic Cloud stars are mostly of spectral type MS (Wood et al. 1983), with C/O ~1. Thus most of the O may be taken up in the CO molecule leading to a weakness of OH emission in dusty galactic counterparts which would otherwise appear as OH/IR stars.

4 CIRCUMSTELLAR DUST SHELLS

The early ground-based infrared studies showed that Mira variables, as well as M supergiants which are generally variables of small amplitude, have large infrared excesses and 10μm emission bands which are different in carbon and oxygen stars (Woolf & Ney 1969; Gehrz & Woolf 1971; Dyck et al. 1971).

The development of circumstellar shells from the optically thin shells of typical Miras to the increasingly thick dust shells of OH/IRC, OH/AFGL and OH/IR stars was traced by ground based work (e.g. Engels et al. 1983) and has been studied in more detail using IRAS photometry (Olnon et al. 1984; Bedijn 1987; van der Veen & Habing 1988). The reddest and coolest shells of all belong to non-variable OH/IR stars and are considered the last stage in the evolution of an AGB star before the development of a planetary nebula. The variable objects lie on a smooth track in the [12]-[25], [25]-[60] diagram, which can be interpreted in terms of a single dust shell of increasing optical thickness whose inner radius lies at the point where dust condensation occurs in the case of the variable objects. Mass loss occurs at an increasing rate along the track (Rowan-Robinson et al. 1986; Bedijn 1987), matching the increasing amplitude of the driving pulsation of the central star, from $\Delta K \sim 1$ mag for a typical Mira (Feast et al. 1982) to 4 mag for some of the OH/IR stars (Engels et al. 1983). The [25]-[60] colours of the irregular, semiregular and optical Mira variables lie in the range -0.4 to +0.2 and as yet show no increase with increasing [12]-[25].

Van der Veen & Habing (1988) give the [12]-[25], [25]-[60] diagram for all reliable Point Source Catalogue data and find a gap in the distribution of [12]-[25] values, from 0.15 to 0.45 (applying the zero point corrections, which they omit, to match our Fig 1). These correspond to the Rayleigh-Jeans point for a cool photosphere and the blue edge of the Mira zone, respectively. The gap is populated by some of the small amplitude variables whose colours range from the Rayleigh-Jeans point to [12]-[25] ~ 1.1 which is also the limit for optically selected Miras. The globular cluster data show that the small amplitude variables are more numerous than the Miras which they precede in evolution (Feast 1973) and the prominence of the gap must arise from a combination of the contribution of more common stars of a range of temperature to the Rayleigh-Jeans sources coupled with the much greater luminosity at $60 \mu m$, the selection flux, of the Mira variables compared to the irregular variables. The density of points along the track does not correspond to the rate of evolution, as the evolution is probably not monotonic towards redder [12]-[25] in the early stages and flux selection plays a vital role.

De Gioia-Eastwood et al. (1981) have shown that the rate of mass loss increases monotonically with the period of an optical Mira, while Whitelock et al. (1987) show that it is also related to the amplitude of pulsation at a given period. The outflow velocity measured in the 1612 MHz line of OH emission in circumstellar envelopes increases with the period (Dickinson et al. 1978). Vardya et al. (1986) find that the infrared spectra (IRAS Science Team 1986) and hence the dust properties depend on the shape of the light curve which in turn may be an indicator of the strength of the atmospheric shock wave. An asymmetric light curve tends to be associated with stronger $9.7 \mu m$ silicate emission.

Lloyd Evans (1987) found that the [12]-[25] and to a lesser extent K-[12] colours were redder for the doubly periodic stars than for singly periodic SRb stars of similar amplitude. They have the strongest $10 \mu m$ silicate emission of any of the red giant variables. This parallels the enhanced infrared excess in RV Tauri stars which show a long wave as well as the usual short period variation (Lloyd Evans 1985b). The RV Tauri stars lie far to the blue of the giant branch, in and to the red of the Cepheid instability strip. This tempts the speculation, in the absence of direct evidence, that there is a connection between the doubly periodic variables and those globular cluster variables which fall to the blue of the giant branch and have enhanced infrared emission and H_2O absorption relative to ordinary irregular and semiregular variables (Frogel & Elias 1988). The physical reason for the slow secondary variation is unknown (Wood 1975).

Similar properties are found for the S, SC and N stars, although with some differences in detail. The Miras have enhanced K-[12], considered as a group (Fig 2); the relatively small values of some of the carbon stars may indicate difficulties in classification arising from the smaller amplitudes and the red B-V of carbon variables. Jura (1986) and Claussen et al. (1987) used the IRAS data to show that mass loss increases with period and amplitude among C stars and Jura (1988) obtained the same result for S stars. These studies do not distinguish between Miras and SR variables, except by way of the 2μm variation (Neugebauer & Leighton 1969), so a large part of the period dependence results from comparing Miras with semiregular variables which generally have shorter periods.

Feast et al. (1984) reported that R For (a carbon Mira of P = 388 d) became unusually faint in both visible and infrared radiation during 1983; this was attributed to a change in circumstellar obscuration. Le Bertre (1988) was able to explain this with a model in which the dust grains in the inner part of the circumstellar shell are alternatively formed and destroyed as the central star fades and brightens. The 1983 event resulted from an unusually faint minimum of the central star.

The distribution of the S stars in [12]-[25] is similar to that of the M stars but carbon stars are largely confined to the range 0.0 to 0.5 (Fig 2). Willems (1986) finds that the carbon SR and Lb variables are more scattered than the Miras in the [25]-[60], [12]-[25] diagram: he attributes this to episodic mass ejection in the variables of small amplitude while the Miras undergo steady mass loss leading to more uniform shells. R Scl (SRa, P = 370 d) is a case in point: Rowan-Robinson et al. (1986) deduce from the IRAS photometry that ejection of a shell ended about 100 years ago, leaving a cool distant shell today, or even that we see a primordial dust shell; even at 12μm the dust emission is unusually strong for a SR variable (Fig 2).

Secondary periods are at least as common among N stars as among the M stars (Houk 1963). V Hya, with periods of 530 and 6500 days, is one of the most spectacular stars of this type and has even been misclassified as a Mira because of the large total amplitude, $\Delta V \sim 6$ mag. Mayall (1965) gives a light curve. Forrest et al. (1975) discovered the large infrared excess. Spatial variations in the 2.6 mm CO emission reveal a bipolar outflow in the outer envelope which extends to at least 20 arcsec from the star (Tsuji et al. 1988; Kahane et al. 1988) and this may be related to the unusually rapid rotation of this star.

Knapp & Morris (1985) have deduced from CO observations of the envelopes of a large sample of long period variables that they lose mass at rates of 3×10^{-8} to 2×10^{-4} $M_\odot$ yr^{-1}. The higher rates are for stars with optically thick envelopes, several of which are optically bright C or S stars including R Scl (SRa, P = 370 d) and the doubly periodic V Hya as well as Mira variables.

5 ABUNDANCE INDICATORS OF EVOLUTION

The occurrence of S and N stars among the longer period Miras (Section 3.2, 3.4) indicates that the third dredge-up (Iben & Renzini 1983) occurs in them. Little et al. (1987) find that the radioactive ^{99}Tc (half-life = 2×10^5 years) occurs in most M type Miras with P > 300 days, indicating that the third dredge-up has begun even though C/O < 1. ^{99}Tc is also seen in almost all S and carbon Miras, but not in M stars which are constant or variable with a small amplitude. $^{12}C/^{13}C$ increases along the sequence M with no Tc, M with Tc, MS, S, indicating the dredge-up of increasing quantities of triple-alpha helium. Some MS and S stars which are small-amplitude variables show no Tc and may be cool Ba stars (Little et al. 1987). Dominy & Wallerstein (1986) have shown how quantitative abundances of several elements may be used to deduce details of the s-process events in Mira variables.

6 ATMOSPHERIC STRUCTURE

Mira variables show a spectacular range of emission features, both of lines and bands, as well as doubling of absorption lines (Merrill 1960). Observational work in recent years has concentrated on studying the kinematic structure of the deep atmosphere of M and S stars. This is best done at wavelengths of 2μm because of the lower Rayleigh scattering opacity (Willson et al. 1982; Hinkle et al. 1984). Additional absorption by TiO bands obscures other features in the near infrared so that line doubling can be seen in this region in S stars (Merrill & Greenstein 1958) but only in a few M stars which have weak TiO near maximum light (Willson et al. 1982.

The picture revealed by studies of absorption lines in the visible (Willson et al. 1982; Wallerstein 1985) and infrared regions (Hinkle, et al. 1982; Hinkle et al. 1984) and of emission lines (Fox et al. 1984; Gillet 1988) is of several distinct levels. Circumstellar gas in a slowly expanding shell at a temperature of not more than 300 K is revealed by resonance lines. The SiO maser emission may arise in a hotter layer below this and above the photosphere, so-called although it has a low velocity of infall, which is indicated by the absorption lines of excited levels in the blue spectral region; it shows little change of velocity with phase. Stellar oscillation drives a shock which passes through the stellar atmosphere near maximum light. The infrared absorption lines are double near maximum light, from which phase a continuous velocity curve can be traced, from about -15 km s^{-1} to +15 km s^{-1} (in χ Cyg, see Hinkle et al. 1982) at the next light maximum when the curve reverses shortly before that component disappears. Ionization occurs immediately below the shock and Balmer emission lines of H are produced by recombination below this; their profiles are distorted by atomic and molecular absorption in the cooler overlying layers. Some of the structure in the emission lines may result from the visibility of both advancing and receding parts of the shock zone near the limb (Gillet 1988). Emission lines of Mg I and Si I arise from thermal excitation (Fox et al. 1984), whereas the selective excitation of certain lines of Fe I is attributed to optical pumping by strong Mg II emission (Thackeray 1937).

Emission bands of AlO have been seen at a faint maximum of o Ceti (Merrill 1940) while Herbig (1956) identified many emission lines seen at minimum light in several stars as arising from AlH. The normal band structure is lost in this case because the AlH molecules are formed directly in the A'Π state by inverse predissociation and only short sequences of rotational levels are populated.

The less-studied carbon stars show the same characteristic emission lines of metals and hydrogen (Merrill 1960), although the relative intensities are different because of the different overlying absorption. A new feature is the appearance of C_2 emission at 4737Å and 4715Å (Lloyd Evans, to be published). This has been seen on the rising branch of the light curve in R Lep (Mira) but also appears in R Scl (SRa) and V Hya (doubly periodic SR) and other stars. The principal common factors are a very red colour or very strong SiC_2 absorption bands. The absence of the emission bands at minimum light in R Lep argues against one possible explanation, that they are circumstellar or chromospheric features seen against a weak continuum.

REFERENCES

Alksne, Z.K., Ikaunieks, Y.Y. & Baumert, J.H. (1981). Carbon Stars. Tucson, Ariz.: Pachart.

Andrews, P.J. (1975). NT Tel, a halo population S-type Mira variable. Mon. Not. R. astr. Soc., 173, no. 3, 701-7.

Arp, H., Brueckel, F. & Lourens, J. v. B. (1963). Long-period and red variables in 47 Tucanae. Astrophys. J., 137, no. 1, 228-48.

Bedijn, P.J. (1987). Dust shells around Miras and OH/IR stars. Astr. Astrophys., 186, no. 1, 136-52.

Beichman, C.A., Neugebauer, G., Habing, H.J., Clegg, P.E. & Chester, T.J. (1985). IRAS Point Source Catalogue. Pasadena, Calif.: Jet Propulsion Lab.

Blanco, V.M., McCarthy, M.F. & Blanco, B.M. (1984). Giant M stars in Baade's window. Astr. J., 89, no. 5, 636-47.

Buonanno, R. (1986). Turn offs and ages of globular clusters. Mem. Soc. astr. Ital., 57, no. 3, 333-43.

Burstein, D. (1985). Observational constraints on the ages and abundances of old stellar populations. Publs. astr. Soc. Pacific, 97, no. 588, 89-103.

Buscombe, W., Gascoigne, S.C.B. & de Vaucouleurs, G. (1954). The Magellanic Clouds. Aust. J. Sci. Suppl., 17, no. 3.

Catchpole, R.M. & Feast, M.W. (1971). An Se variable of the halo population. Observatory, 91, no. 980, 29-30.

Catchpole, R.M. & Feast, M.W. (1985). The distributions and motions of peculiar red giants. In Cool Stars with Excesses of Heavy Elements, ed M. Jaschek & P.C. Keenan, pp 113-32. Dordrecht: Reidel.

Catchpole, R.M., Robertson, B.S.C., Lloyd Evans, T.H.H., Feast, M.W., Glass, I.S. & Carter, B.S. (1979). JHKL infrared photometry of Mira variables. SAAO Circ., 1, no. 4, 61-97.

Claussen, M.J., Kleinmann, S.G., Joyce, R.R. & Jura, M. (1987). A flux-limited sample of galactic carbon stars. Astrophys. J. Suppl., 65, no. 3, 385-404.

De Gioia-Eastwood, K., Hackwell, J.A., Grasdalen, G.L. & Gehrz, R.D. (1981). A correlation between infrared excess and period for Mira variables. Astrophys. J., 245, no. 2, L75-8.
Dickens, R.J., Feast, M.W., & Lloyd Evans, T. (1972). Photometry and spectroscopy of red variables in Omega Centauri. Mon. Not. R. astr. Soc., 159, no. 3, 337-48.
Dickinson, D.F., Reid, M.J., Morris, M. & Redman, R. (1978). Long period variables: stellar and expansion velocities. Astrophys. J., 220, no. 3, L113-16.
Dominy, J.F. & Wallerstein, G. (1986). Quantitative Technetium abundances in two long-period variables. Astrophys. J., 310, no. 1, 371-77.
Dyck, H.M., Forrest, W.J., Gillett, F.C., Stein, W.A., Gehrz, R.D., Woolf, N.J. & Shawl, S.J. (1971). Visual intrinsic polarization and infrared excess of cool stars. Astrophys. J., 165, no. 1, 57-66.
Eggen, O.J. (1972). Narrow and broad-band photometry of red stars. Astrophys. J., 172, no. 3, 639-77.
Elias, J.H., Frogel, J.A. & Humphreys, R.M. (1985). M supergiants in the Milky Way and the Magellanic Clouds. Astrophys. J. Suppl., 57, no. 1, 91-131.
Engels, D., Kreysa, E., Schultz, G.V. & Sherwood, W.A. (1983). The nature of OH/IR stars. Astr. Astrophys., 124, no. 1, 123-138.
Feast, M.W. (1963). The long period variables. Mon. Not. R. astr. Soc., 125, no. 5, 367-415.
Feast, M.W. (1965). Long period variables in globular clusters. Observatory, 85, no. 944, 16-20.
Feast, M.W. (1973). Observational aspects of slow variables in globular clusters. In Variable Stars in Globular Clusters and in Related Systems, ed. J.D. Fernie, pp 131-144. Dordrecht: Reidel.
Feast, M.W. (1984). The period-luminosity relation for Mira variables and the distance of the Large Magellanic Cloud. Mon. Not. R. astr. Soc., 211, no. 3, 51-5 P.
Feast, M.W. (1985). The bolometric luminosities of Type II OH/IR sources. Observatory, 105, no. 1066, 85-9.
Feast, M.W. (1986). Variables, the galactic bulge and IRAS. In Light on Dark Matter, ed F.P. Israel, pp 339-48. Dordrecht: Reidel.
Feast, M.W., Robertson, B.S.C., Catchpole, R.M., Lloyd Evans, T., Glass, I.S. & Carter, B.S. (1982). The infrared properties of Mira-type variables and other cool stars. Mon. Not. R. astr. Soc., 201, no. 2, 439-50.
Feast, M.W. & Whitelock, P.A. (1987). Mira variables and the galactic bulge population. In Late Stages of Stellar Evolution, ed S. Kwok & S.R. Pottasch, pp 33-46. Dordrecht: Reidel.
Feast, M.W., Whitelock, P.A., Catchpole, R.M., Roberts, G. & Overbeek, M.D. (1984). Variable circumstellar obscuration of the carbon star R Fornacis. Mon. Not. R. astr. Soc., 211, no. 2, 331-7.
Feast, M.W., Woolley, R. & Yilmaz, N. (1972). The kinematics of semi-regular red variables in the solar neighbourhood. Mon. Not. R. astr. Soc. 158, no. 1, 23-46.
Forrest, W.J., Gillett, F.C. & Stein, W.A. (1975). Circumstellar grains and the intrinsic polarization of starlight. Astrophys. J., 195, no. 2, 423-40.

Fox, M.W. (1982). Photometry of red variables in 47 Tucanae. Mon. Not. R. astr. Soc., 199, no. 2, 715-23.
Fox, M.W., Wood, P.R. & Dopita, M.A. (1984). Shock waves in Mira variables. Astrophys. J., 286, no. 1, 337-49.
Frogel, J.A. (1983). The evolutionary state and pulsation characteristics of red variables in globular clusters. Astrophys. J., 272, no. 1, 167-74.
Frogel, J.A. & Elias, J.H. (1988). Red variables in globular clusters. Astrophys. J., 324, no. 2, 823-39.
Frogel, J.A., Persson, S.E. & Cohen, J.G. (1981). Infrared photometry of red giants in the globular cluster 47 Tucanae. Astrophys. J., 246, no. 3, 842-65.
Frogel, J.A., Persson, S.E. & Cohen, J.G. (1983). Infrared photometry, bolometric luminosities and effective temperatures for giant stars in 26 globular clusters. Astrophys. J. Suppl., 53, no. 3, 713-49.
Frogel, J.A. & Whitford, A.E. (1987). M giants in Baade's window. Astrophys. J., 320, no. 1, 199-237.
Gehrz, R.D. & Woolf, N.J. (1971). Mass loss from M stars. Astrophys. J., 165, no. 2, 285-94.
Gillet, D. (1988). The Balmer emission profiles in Mira stars. Astr. Astrophys., 192, no. 1, 206-20.
Glass, I.S. (1986). IRAS sources in the Sgr I window. Mon Not. R. astr. Soc., 221, no. 4, 879-85.
Glass, I.S., Catchpole, R.M., Feast, M.W., Whitelock, P.A. & Reid, I.N. (1987). The period-luminosity relation for Mira-like variables in the LMC. In Late Stages of Stellar Evolution, ed S. Kwok & S.R. Pottasch, pp 51-4. Dordrecht: Reidel.
Glass, I.S. & Feast, M.W. (1982). Infrared photometry of Mira variables in the Baade windows. Mon. Not. R. astr. Soc., 198, no. 1, 199-214.
Glass, I.S. & Lloyd Evans, T. (1981). A period-luminosity relation for Mira variables in the Large Magellanic Cloud. Nature, 291, no. 5813, 303-4.
Glass, I.S. & Reid, N. (1985). A survey for red variables in the LMC. Mon. Not. R. astr. Soc., 214, no. 3, 405-18.
Herbig, G.H. (1956). Identification of aluminium hydride as the emitter of bright lines observed in χ Cygni near minimum light. Publs. astr. Soc. Pacific, 68, no. 402, 204-10.
Hinkle, K.H., Hall, D.N.B. & Ridgway, S.T. (1982). Time series infrared spectroscopy of the Mira variable χ Cygni. Astrophys. J., 252, no. 2, 697-714.
Hinkle, K.H., Scharlach, W.W.G. & Hall, D.N.B. (1984). Astrophys. J. Suppl., 56, no. 1, 1-17.
Houk, N. (1963). V1280 Sagittarii and the other long-period variables with secondary periods. Astr. J., 68, no. 4, 253-7.
Hyland, A.R. (1974). Medium resolution stellar spectra in the two micron region. In Highlights of Astronomy 3, ed G. Contopoulos, pp 307-26. Dordrecht: Reidel.
Iben, I. & Renzini, A. (1983). Asymptotic giant branch evolution and beyond. Ann. Rev. Astr. Astrophys., 21, 271-342.
IRAS Science Team (1986). Atlas of low-resolution spectra. Astr. Astrophys. Suppl., 65, no. 4, 607-1065.

Joy, A.H. (1949). Spectra of the brighter variables in globular clusters. Astrophys. J., 110, no. 2, 105-16.
Jura, M. (1986). Mass loss from carbon stars. Astrophys. J., 303, no. 1, 327-32.
Jura, M. (1988). Mass loss from S stars. Astrophys. J. Suppl., 66, no. 1, 33-41.
Kahane, C., Maizels, C. & Jura, M. (1988). The bipolar outflow from the rotating carbon star, V Hydrae. Astrophys. J., 328, no. 1, L25-8.
Keenan, P.C., 1954. Classification of the S-type stars. Astrophys. J., 120, no. 3, 484-505.
Kholopov, P.N. et al. (1985). General Catalogue of Variable Stars. Moscow: Nauka.
Knapp, G.R. & Morris, M. (1985). Mass loss from evolved stars. Astrophys. J., 292, no. 2, 640-69.
Le Bertre, T. (1988). Optical and infrared observations of the carbon Mira R Fornacis. Astr. Astrophys., 190, no. 1, 79-86.
Little, S.J., Little-Marenin, I.R. & Bauer, W.H. (1987). Additional late-type stars with technetium. Astr. J.', 94, no. 4, 981-95.
Lloyd Evans, T. (1971a). Supergiant red variable stars of large amplitude in the Small Magellanic Cloud. Observatory, 91, no. 982, 118-20.
Lloyd Evans, T. (1971b). A search for red variable stars in the Magellanic Clouds. In The Magellanic Clouds, ed A.B. Muller, pp 74-78. Dordrecht: Reidel.
Lloyd Evans, T. (1974). Near-infrared photometry of globular clusters-II. The metal-rich cluster 47 Tucanae. Mon. Not. R. astr. Soc., 167, no. 2, 393-411.
Lloyd Evans, T. (1976). Red variables in the central bulge of the Galaxy. Mon. Not. R. astr. Soc., 174, no. 1, 169-84.
Lloyd Evans, T. (1980). Spectra of red supergiant variables in the SMC. Mon. Not. R. astr. Soc., 193, no. 2, 333-6.
Lloyd Evans, T. (1983a). Observations of red variable stars in globular clusters. Mon. Not. R. astr. Soc., 204, no. 3, 961-73.
Lloyd Evans, T. (1983b). Abundance sensitive parameters for red giants in globular clusters. Mon. Not. R. astr. Soc., 204, no. 3, 945-59.
Lloyd Evans, T. (1983c). S stars in ω Centauri. Mon. Not. R. astr. Soc., 204, no. 3, 975-84.
Lloyd Evans, T. (1984). Are there S stars in galactic globular clusters? Mon. Not. R. astr. Soc., 209, no. 4, 825-39.
Lloyd Evans, T. (1985a). Bright red variables of large amplitude in the Magellanic Clouds. Mon. Not. R. astr. Soc., 212, no. 4, 955-73.
Lloyd Evans, T. (1985b). Circumstellar material and the light variations of RV Tauri stars. Mon. Not. R astr. Soc., 217, no. 2, 493-506.
Lloyd Evans, T. (1987). Slow variability and circumstellar shells of red variable stars. In Circumstellar Matter, ed I. Appenzeller & C. Jordan, pp 541-2. (IAU Symp. 122). Dordrecht: Reidel.
Lloyd Evans, T. & Catchpole, R.M. (1988). The Westerlund-Olander sample of S stars in the southern Milky Way. Mon. Not. R. astr. Soc., submitted.

Lloyd Evans, T., Glass, I.S. & Catchpole, R.M. (1988). Long-period variables in the Small Magellanic Cloud. Mon. Not. R. astr. Soc., 231, no. 3, 773-81.

Lloyd Evans, T. & Menzies, J.W. (1973). Red variable stars in metal-rich globular clusters. In Variable Stars in Globular Clusters and in Related Systems, ed J.D. Fernie, pp 151-63. Dordrecht: Reidel.

Lloyd Evans, T. & Menzies, J.W. (1977). Near infrared photometry of globular clusters-III. The metal-rich clusters. Mon. Not. R. astr. Soc., 178, no. 1, 163-93.

Mayall, M.W. (1965). Variable star notes. J.R. astr. Soc. Can., 59, no. 5, 245-8.

Menzies, J.W. & Whitelock, P.A. (1985). A period-luminosity relation for Mira variables in globular clusters. Mon. Not. R. astr. Soc., 212, no. 4, 783-97.

Merrill, P.W. (1940). Spectra of Long Period Variables. Chicago: University of Chicago Press.

Merrill, P.W. (1960). Spectra of long-period variables. In Stellar Atmospheres, ed J.L. Greenstein, pp 509-29. (Stars and Stellar Systems, v.6). Chicago: The University of Chicago Press.

Merrill, P.W. & Greenstein, J.L. (1958). Double absorption lines in the spectrum of R Andromedae. Publs. astr. Soc. Pacific, 70, no. 412, 98-101.

Morris, S. & Wyller, A.A. (1967). Molecular dissociative equilibria in carbon stars. Astrophys. J., 150, no. 3, 877-907.

Neugebauer, G. & Leighton, R.B. (1969). Two-Micron Sky Survey. Washington, D.C.: NASA SP-3047.

Olnon, F.M., Baud, B., Habing, H.J., de Jong, T., Harris, S. & Pottasch, S.R. (1984). IRAS observations of OH/IR stars. Astrophys. J., 278, no. 1, L41-3.

Payne-Gaposchkin, C. (1951). The intrinsic variable stars. In Astrophysics, ed J.A. Hynek, pp 495-525. New York: McGraw-Hill.

Payne-Gaposchkin, C. (1954). The red variable stars. Ann. Harvard College Obs., 113, no. 4, 191-208.

Payne-Gaposchkin, C. (1971). The variable stars of the Large Magellanic Cloud. Smithsonian Contr. Astrophys., no. 13, 1-41.

Payne-Gaposchkin, C. & Gaposchkin, S. (1966). Variable Stars in the Small Magellanic Cloud. Smithsonian Contr. Astrophys., no. 9, 1-205.

Reid, N., Glass, I.S. & Catchpole, R.M. (1988). A survey for red variables in the LMC. Mon. Not. R. astr. Soc., 232, no. 1, 53-79.

Robertson, B.S.C. & Feast, M.W. (1981). The bolometric, infrared and visual absolute magnitudes of Mira variables. Mon. Not. R. astr. Soc., 196, no. 1, 111-20.

Rosino, L. (1951). The spectra of variables of the RV Tauri and yellow semiregular types. Astrophys. J., 113, no. 1, 60-71.

Rowan-Robinson, M., Lock, T.D., Walker, D.W. & Harris, S. (1986). Models for IRAS observations of circumstellar dust shells around late-type stars. Mon. Not. R. astr. Soc., 222, no. 2, 273-86.

Shapley, H. & Nail, V. McK. (1951). Magellanic Clouds II: Supergiant red variable stars in the Small Cloud. Proc. Nat. Acad. Sci., 37, no. 3, 138-45.

Smak, J.I. (1966). The long-period variable stars. Ann. Rev. Astr. Astrophys., 4, 19-34.

Thackeray, A.D. (1937). The excitation of emission lines in late-type variables. Astrophys. J., 86, no. 5, 499-508.
Tsuji, T. (1981a). Effective temperature scale of N-type carbon stars. J. Astrophys. Astr., 2, no. 1, 95-113.
Tsuji, T. (1981b). Spectra, colours and HR diagram of cool carbon stars. J. Astrophys. Astr., 2, no. 3, 253-76.
Tsuji, T., Unno, W., Kaifu, N., Izumiura, H., Ukita, N., Cho, S. & Koyama, K. (1988). V Hydrae: a carbon star in transformation to a bipolar nebula. Astrophys. J., 327, no. 1, L23-6.
Van der Veen, W.E.C.J. & Habing, H.J. (1988). The IRAS two-colour diagram as a tool for studying late stages of stellar evolution. Astr. Astrophys., 194, no. 1, 125-34.
Vardya, M.S., de Jong, T. & Willems, F.J. (1986). IRAS low-resolution spectrograph observations of silicate and molecular SiO emission in Mira variables. Astrophys. J., 304, no. 1, L29-32.
Wallerstein, G. (1985). Stellar stratigraphy. Publs. astr. Soc. Pacific, 97, no. 596, 994-1000.
Whitelock, P.A., Feast, M.W. & Catchpole, R.M. (1986). JHKL observations of IRAS sources-III. The galactic bulge. Mon. Not. R. astr. Soc., 222, no. 1, 1-9.
Whitelock, P.A., Pottasch, S.R. & Feast, M.W. (1987). Evidence for pulsationally driven mass-loss from Mira variables. In Late Stages of Stellar Evolution, ed S. Kwok & S.R. Pottasch, pp 269-72. Dordrecht: Reidel.
Whitford, A.E. & Rich, R.M., (1983). Metal content of K giants in the nuclear bulge of the Galaxy. Astrophys. J., 274, no. 2, 723-32.
Willems, F.J. (1986). IRAS observations of carbon stars. In Light on Dark Matter, ed F.P. Israel, pp 113-8. Dordrecht: Reidel.
Willson, L.A., Wallerstein, G. & Pilachowski, C.A. (1982). Atmospheric kinematics of high velocity long period variables. Mon. Not. R. astr. Soc., 198, no. 2, 483-516.
Wing, R.F. (1985). Photometric properties of peculiar red giants. In Cool Stars with Excesses of Heavy Elements, ed M. Jaschek & P.C. Keenan, pp 61-85. Dordrecht: Reidel.
Wood, P.R. (1975). Red variables. In Multiple Periodic Variable Stars, ed W.S. Fitch, pp 69-84. Dordrecht: Reidel.
Wood, P.R. & Bessell, M.S. (1983). Long-period variables in the galactic bulge. Astrophys. J., 265, no. 2, 748-59.
Wood, P.R., Bessell, M.S. & Fox, M.W. (1983). Long-period variables in the Magellanic Clouds. Astrophys. J., 272, no. 1, 99-115.
Wood, P.R., Bessell, M.S. & Paltoglou, G. (1985). Long-period variables in the Bar of the Large Magellanic Cloud. Astrophys. J., 290, no. 2, 477-86.
Woolf, N.J. & Ney, E.P. (1969). Circumstellar infrared emission from cool stars. Astrophys. J., 155, no. 3, L181-4.
Yorka, S.B. & Keenan, P.C. (1985). Spectral classification and the relations between peculiar giants. In Cool Stars with Excesses of Heavy Elements, ed M. Jaschek & P.C. Keenan, pp 15-8. Dordrecht: Reidel.

Irregular Red-Giant Variable Stars

M. Querci   F. Querci
Observatoire Midi-Pyrenees

1. Introduction

Aperiodic intrinsic variability becomes increasingly evident for a greater number of red giant and supergiant stars as observations become more technically refined and extended over a longer time. All red variables show an erratic temporal behavior to some degree. Even Miras, with their temporal changes in maximum brightness and irregular fluctuations in the cycle length, are not perfect clocks.

Apart from the Miras, the variable M, S, and C stars are either semiregulars (classified SRa, SRb and SRc) or irregulars (classified Lb and Lc) (e.g., F. Querci, 1986, p. 54). Their location among the various types of pulsational variables in the H-R diagram is shown on Fig. 1 on a Mbol/log Te diagram (Becker, 1987). As commented by Becker, evolutionary tracks of 1, 7, and 15 M$\odot$ models show that a red star may become: (1) an Lc or SRc type variable for the massive stars (M > 10 M$\odot$); (2) a Mira, SRa, SRb, Lb, or Lc and SRc for the intermediate-mass stars (10 M$\odot$ > M > 2.25 M$\odot$); (3) a Mira, Sra, Srb or Lb variable for the low-mass stars (M_{*} < 2.25 M$\odot$).

As presented by Bessell <u>et al.</u> (1988a,b) and Bessell (this Colloquium), an Australian group has undertaken a wide ranging program of theoretical and observational investigations of the spectra of red giants in order to obtain physical parameters – in particular, accurate temperatures – of Miras, semiregulars, and irregulars. Through IR color-color diagrams, they demonstrate that the semiregulars bridge the gap between the non-variable M0-M6 giants in the solar neighborhood and the Miras; this is due mainly to a temperature difference, the Miras being cooler than the SR variables. In addition to the luminosity and the total mass, the temperature therefore appears to be an important parameter in discriminating between the various pulsational behaviors of the PRG stars.

In fact, the primary question is to understand why a red giant is observed to be quasi-regular (Mira), or semiregular or fully irregular. Its intrinsic physical parameters, i.e., its location in the H-R diagram, play a sensitive role in the type of variability. This is demonstrated by the recent physical explanation by Buchler and Goupil (1988) involving a chaotic attractor which will be presented in Part 4, together with a discussion on other mechanisms such as multiperiodicity or randomness. Part 2 describes light curves of irregular variables, and Part 3 discusses radial velocity variations in some absorption lines and temporal changes in characteristic emission-line profiles well suited for depicting the presence of shock waves linked to the radial pulsation of these stars.

In this review we omit binaries and RCB stars as they are presented in other talks in this Colloquium.

2. Light curves

2.1 The variability types and their light curves

The initial classification of semiregulars and irregulars was based on their visual light curves. Generally speaking, the SR variables are characterized by a form of periodicity hidden by more marked irregular brightness fluctuations; the irregulars are characterized by a pronounced disordered variability. The following examples of light curves have been known for many years (cf. the review by Jacchia (1933) from which several of our examples are drawn). Today amateur astronomers, for example from the AAVSO group, endeavour to observe regularly some typical semiregulars and irregulars.

SRa variables are giants which present Mira-like behavior such as relatively constant periods and spectroscopic similarites. However, they have smaller light-curve amplitudes (<2.5 mag) and strong variations from one cycle to another in the amplitude and the shape of the light curve. Nevertheless, they have an appreciable range in periodicity ($35 < P$ (days) < 1200). Fig. 2 and 3 illustrate typical behavior of the light curves of SRa stars. For a time, S Aql shows double maxima; over other cycles, these are smoothed and the main minimum becomes broader. The carbon star RS Cyg likewise shows double maxima, as do some other Miras (Fig. 3); however, the total amplitude of the light curve is significantly larger in the Miras. Another example, the SRa variable V Hya, observed over 80 years (Mayall, 1965), shows unexpected deep minima (Fig. 4). These might reveal a binary nature of the star, which would be consistent with the transformation of V Hya to a bipolar nebula (Tsuji et al. 1988).

SRb semiregulars show quite individual light curves (Fig. 5) and poorly expressed periodicity, preventing the predictions of the epochs of minimum and maximum brightness. Periodic oscillations temporarily alternate with slow irregular variations or even with a constancy of the brightness (Fig. 6). These SRb variable stars are giants with mean periods of 20 to 2300 days.

SRc type variables are supergiants with an SRb behavior. Visual amplitudes are about one magnitude or less, as in μ Cep and α Ori, though a few stars present amplitudes up to 4 magnitudes, as in S Per and VX Sgr (Fig. 7). The number of such large amplitude variables is very limited, and if we suppose that the supergiants pass through both the small and large amplitude phase, the latter must be short. The SRc variables have periods of 30 to several thousand days.

Lb variables are irregularly variable giants without any trace of periodicity or with an occasional very weak periodicity (Fig. 8). W CMi has an irregular light curve, but during some intervals of time its light curve suggests a regular light period and during other periods constant brightness (Krempec, 1973). As the time interval between successive maxima may be long, a number of stars firstly classified Lb have been shifted to the semiregular types after longer time-series of observations were obtained. An example is VY Leo, classified Lb in the GCVS, whereas Maran et al. (1980) demonstrate it is a SRa variable with $P\sim1$ year, using a satellite for a long series of uninterrupted observations.

Lc irregulars are supergiants such as VY CMa (Fig. 9). Amplitude variations are generally small ($<$ 1.5 mag) and the time interval between two consecutive maxima is between several hundred and several thousand days. Such features imply that several years of observations are needed to decide whether a supergiant belongs to he SRc class or to the Lc class.

2.2 The various time-scales in light variations

Evidences of short-term, intermediate-term, and long-term light variations in the semiregular and irregular variables are stressed by the following examples.

(a) Short time-scale light variations are observed in the SRb stars R Crt (M8 II). Rapid variations with a time scale of about 1 hour in the DDO magnitudes and colors (over TiO bands and Ca I lines) have been detected by Livi and Bergmann (1982). Strong and rapid oscillations have also been observed in the U-B index between phase 0.80 and 0.90 (Bouchet, 1984; his Fig. 2) on the C SRb star, TW Hor; they are supposedly related to the rapid variations in the UV Fe II V1 emission lines (Bouchet et al., 1983) (see Part 3).

(b) Intermediate-term initially are depicted in the supergiant α Ori. As observed initially by Stebbins (1931), they are of several hundred days, superimposed on a main curve with cyclic variation of period about 5.8 years (Fig. 10).

(c) Long-term variations are observed in the supergiant SRc variables in which the light maxima decrease about 2 mag in one century (Maeder, 1980).

3. Spectroscopic observations

A Mira-like pulsating mechanism in semiregulars is indicated by some signatures.

3.1 S-shaped absorption-line radial-velocity curves

Regular temporal observations of lines in the SR and the irregular variables to yield radial velocities as a function of phase are still rarely available. However, S-shaped radial velocity curves are recognized in the following examples:

(a) through the CN absorption lines over the region from 6100 to 6700 Å observed by Sanford (1950) in the SR carbon stars RR Her (SRb), T CnC (SRa), and V Hya (SRa) (Fig. 11) (M. Querci, 1986);

(b) through IR CO lines in the M SRa variable X Oph (Fig. 12) (Hinkle et al., 1984);

(c) through IR atomic lines observed by Goldberg (1979) in the M supergiant SRc, α Ori (Fig. 13).

The curves have much lower amplitude, < 10 km/s, in the SR variables (it is 6 km/s in α Ori) than in the Miras, in which it is up to 30 km/s, indicating less available energy and/or larger damping.

The S-shaped curves are the signature of a shock-pulsation motion associated with a _radial_ global stellar pulsation driving acoustic waves. These waves are generated by turbulence at the top of the hydrogen convective zone and turn into shocks dissipating energy and heating the stellar layers as they propagate outward into layers of decreasing density.

3.2 Temporal variations in emission-line profiles

Emission lines also probe the presence of shock fronts in the atmosphere. Some lines are seen at some phases, or rather at some phases in some cycles.

Balmer emission data in the SR and L classes of M-type stars are summarized by Jennings and Dyck (1972), who consulted the literature over about 80 years. Only five M giants and supergiants are quoted as having shown Balmer emission. The presence of hydrogen emission is not likely to be regular from cycle to cycle or within a cycle. A Mira-like phase behavior is noted from observations by McLaughlin (1946) in the supergiant μ Cep in which Hβ, Hγ and Hδ emissions appear strongest just before maximum light, disappear as the star fades, and appear in absorption at minimum light. Sanford (1950) reports such behavior also in the SR carbon stars RR Her and V Hya, in which Hα emission is observed from 0.25 period before to 0.25 period after the maximum light. However, Hα in emission is not observed by this author in T CnC, which we have mentioned presents an S-shaped curve in the CN absorption lines. As for the C star WZ cas (SRa), a large temporal coverage shows episodic Hα or Hβ emission around the light maximum.

For some stars, such as α Ori, the hydrogen lines appear only in absorption and are never reported in emission. This might indicate that shocks are not strong enough in this star to excite observable hydrogen emission. In other stars, no hydrogen is detected either in emission or in absorption during even a long time internal -- an example being the C star TW Hor (Querci and Querci, 1985). That no hydrogen absorption is seen must mean that the chromospheric optical depth at line-center is small (Avrett and Johnson, 1984) or that the shock occurs only in very shallow layers.

On the other hand, ultraviolet Fe II emission lines around 3200 A and h and k Mg II emission are always present in giants and supergiants. They are variable in strength with time. In α Ori, where the Fe II lines are particularly well studied (see M. Querci, 1986, for a review of the literature), they are correlated with either an outflow on infall of material. In the C star TW Hor, they have been seen to appear and disappear on a time scale of 1 day (Fig. 14). This star also shows a particularly stochastic temporal behavior in its IUE spectra (Querci and Querci, 1985).

The Mg II flux in M giants is only slightly variable with time. In α Ori (Dupree et al., 1987) the Mg II h emission line is indicative of an outflow velocity at the epoch where the Fe II lines also indicate outflow from this star.

In the SR and L giants and supergiants the k line displays an asymmetry due to circumstellar absorption (e.g., Bohm-Vitense and Querci, 1987). Such CS absorption is particularly important in the irregular carbon star TX Psc (Eriksson et al., 1985), and it might alter any conclusion concerning the propagation of shocks into the upper atmosphere of these stars.

All the previous line observations are probing shocks that are due to radial pulsation modes. Generally speaking, these shocks appear to dissipate energy in higher layers than in the Miras since the hydrogen emission, so conspicuous around maximum light in Miras, is rarely observed in emission in the SR and L stars. However, even if the shock lacks enough energy or is too heavily damped to excite Balmer emission, it is sufficiently energetic to excite the Mg II and Fe II lines located in the upper atmosphere.

In fact, due to strong radiative damping, shock fronts may form, say, at the chromospheric temperature minimum in agreement with the model chromosphere of low-gravity stars by Schmitz and Ulmschneider (1981). These ideas may be linked to the short-period acoustic heating theory (e.g., see comments and examples in M. Querci, 1986). Indeed, the steep temperature gradients found necessary in chromospheric models to match observations of the carbon star TX Psc (Luttermoser et al., 1988) are suggestive of shocks at just this level. The recent suggestion of stochastically changing wave periods in the short period range (Cuntz 1987), giving interacting acoustic wave packets and generating a greater shock strength and a larger wave amplitude in the chromosphere, quite well account for stochastically variable lines, such as Fe II in TW Hor (Fig. 14).

Shock fronts progressing outward from the deeper layers, as happens generally for Miras, are obvious at times in irregular variables through hydrogen emission as is seen in some available examples of SR and in the supergiant μ Cep. For example, in μ Cep, this agrees with de Jager (1984), who shows that shock-wave dissipation starts deeper in the photosphere in supergiants which are near the upper luminosity limit in the H-R diagram.

The light curve, i.e. the brightness of the star at the moment of the spectroscopic observations, helps us to predict the line strength. For example, the Mira prototype, o Cet itself, did not show Balmer emission at all when its magnitude at maximum light was faint in June 1983; strong hydrogen emission is known to be linked to the brighter light maxima.

Though a basic common heating mechanism is likely at work in the atmospheres of Miras, semiregular and irregular giants and supergiants, a high shock efficiency occurs at a different atmospheric level from one star to another and changes with time in a given star.

Outflowing gas seen in the Fe II lines of, for example, α Ori and β Peg (Boesgaard, 1981) is also evidence of mass loss in SR and irregular stars. Models such as these by Cuntz (1987) or by Bowen (1988, and this Colloquium), which specifically apply to Miras, represent progress in the understanding of the pulsation mechanism associated with mass loss, which influences the course of the evolution of these stars.

4. Mechanisms of non-periodic phenomena

Since the IAU Colloquium on Non-Periodic Phenomena in Variable Stars in 1968, the physical explanations for irregular behavior of varying degree in semiregulars and irregulars have smoothly advanced without being fully satisfactory, if we except a recently proposed mechanism. Let us summarize our knowledge on the question, but refer the reader to IAU Colloquium 111 on Pulsating Stars (1988).

According to Whitney (1984), a semiregular may be either: (a) multiperiodic,, i.e., showing superimposed periods or beats which are the interaction of simultaneous modes of oscillation; or (b) truly irregular, i.e., explained either by randomness or by chaos.

To detect the presence and significance of periods in stars, periodograms are calculated. Various methods are available. The difficulty with the astrophysical data is that they are unevenly spaced in time and contain large amount of random noise, as noted by Horne and Baliunas (1986). These authors present an extremely valuable technique to predict periodicities. As an example, Karovska (1987) applied this technique, among others, to observations over 60 years of the SRc variable α Ori finding a multiperiodicity, of which a 1-year period is attributable to the fundamental mode of pulsation. Other examples of multiperiodicity in semiregulars are given in the reviews by Wood (1975) and by F. Querci (1986, p. 59). We shall not debate in this review the period ratios P_0/P_1 found in the literature. Theoretical values of such ratios and theoretical pulsation modes in SR and supergiants are discussed in Fox and Wood (1982).

As stressed and illustrated by Detre (1968), irregular stellar variability may be the observable effect of random succession of transitory events. Regarding chaos, Whitney (1984) defines it as changes that are not simply the summation of many small changes, but reflect a collective and cooperative behavior. The chaotic behavior is governed by non-linear equations. It is due to intrinsic forces and generally happens when the amplitude of the motion exceeds a critical value.

Discriminating between these various explanations of the aperiodic fluctuations of a variable has not always been fully satisfactory. An example is μ Cep, the irregular oscillation of which has been studied since 1848. The interpretation of its irregular behavior has shifted from random disturbances to a more satisfactory multiperiodicity (Fig.

15) as reviewed by Whitney (1984). However, this author remarks that the good fit over a sample of the light curve "does not prove that μ Cep has multi-periodic, because random or chaotic behavior can imitate multi-periodic behavior roughened by observational errors".

Another not totally satisfactory example might be the superimposition of sinusoidal variations (4 and 5 periods) to synthetize a light curve for α Ori (Fig. 10), though Karovska (1987) suggests a binary explanation to the excess of brightness around 1980-81 and 1985-86. In fact, Perdang (1985) emphasizes that interpreting stellar variability depends on the analytical tools used. The non-linear oscillation of a model may be strictly periodic or multi-periodic, but a refined analysis of the same model may establish that the oscillation is a stochastic (i.e. chaotic) motion. Today, it appears that chaos is a remarkably common process.

A decade ago, explaining disordered variability necessitated extrinsic stochastic mechanisms, as stressed by Perdang (1984), such as irregular surface features, spots, convective cells, or atmospheric veiling, that would have a larger effect on the light variation than have pulsational effects. It is now demonstrated that irregular aperiodic time behavior of a star may arise spontaneously as a result of the non-linear structure of the stellar hydrodynamics (Perdang, 1985).

In the spirit of non-linear (chaotic) dynamics, Buchler and Goupil (1988) propose a mechanism for the irregular variability shown by radially pulsating stars that are red giants. This mechanism is able to generate regularly modulated or erratic oscillations with period variations as well as intermittent oscillations (almost ceasing and restarting) in addition to steady oscillations. It is based on a mathematical model which involves the nonlinear interaction between the fundamental pulsation mode and one or more overtones in the special situation that the stellar model is near dynamical instability in the H-R diagram.

The reason for the dynamical instability is that the adiabatic index $\gamma = (\delta \ln P / \delta \ln T)$ becomes small ($\gamma < 4/3$) over the convective partial ionization regions (the hydrogen and first helium ionization zones in which the driving of the oscillation occurs) which becomes very extended (e.g., Tuchman et al. 1978). So, in the H-R diagram -- that is, on a luminosity/Te-plane -- a line can be found representing the boundary of the region of the dynamical instability for each stellar mass. Near this line the frequency of the fundamental mode, as well as its growth rate, become small with respect to the frequencies of the other modes (Fox and Wood, 1982).

Buchler and Goupil find a sufficient-dimensional dynamic to give rise to irregular behavior in considering the nonlinear interaction of such a dynamically marginally stable fundamental mode with the first overtone. Involving other overtones will only increase the complexity of the temporal behavior.

In consequence, the location of a red star in the H-R diagram -- that is, its physical properties such as luminosity, mass, temperature, density, chemical compositon -- sensitively determines the richness of its pulsational behavior. Our present limitation in applying a hydrodynamic mechanism to real stars is the observational situation, which has considerable difficulty determining clearly the actual number of dominant pulsation modes in these stars. Standard hydrodynamic techniques (fractal dimension test etc.) have to be applied. However, high quality photometric data spread over a long time are needed.

Let us note that other mechanisms involving an underlying chaotic dynamic to explain the arhythmic oscillations of the irregular stars have also been proposed, for example by Perdang and Blacher (1982), Perdang (1984), Regev and Buchler (1985), and Buchler and Kovacs (1987).

Nonradial oscillations might also have some role to play. Exploratory models for late-type giants and supergiants were made by Ando (1976). Spectroscopic evidence for nonradial behavior in the irregular red stars is still lacking. Maybe the presence of large-scale convective motions at the surface of supergiants such as α Ori might be a proof, if some irrefutable observation confirms their existence.

Finally, examples of stars undergoing a transition from irregular to regular pulsations should be given by the carbon stars observed with IRAS by Willems and de Jong (1988, and references therein), though the evolutionary scenarios are still very controversial (see papers by Kleinmann, Kwok, and Jura, this Colloquium).

5. Conclusion

The paucity of observations of semiregulars and irregulars, either photometric or spectroscopic, is evident. In particular, currently available photometric data are of insufficient accuracy for testing of pulsational modes and shock-wave travel and dissipation. Sustained observations, perhaps best performed from satellites for long series of uninterrupted data, are needed.

Consequently, we plead for more and better quality observations on these stars to test the theoretical results and contribute to the understanding of their irregular variability in the context of their evolution in the H-R diagram.

References

Ando, H. 1976, Publ. Astron. Soc. Japan **28**, 517.
Ashbrook, J., Ducombe, R.L., and Van Woerken, A.J.J. 1954, Astron. J. **59**, 12
Avrett, E.H., and Johnson, H.R. 1984, in Proc. Third Cambridge Workshop on Cool Stars, Stellar Systems, and the Sun, eds. S.L. Baliunas and L. Hartmann (Heidelberg: Springer-Verlag), p. 130.
Becker, S.A. 1987, in Proc. Conference on Stellar Pulsation, eds. A.N. Cox, W.M. Sparks, and S.G. Starrfield (Heidelberg: Springer-Verlag), p. 16.
Bessell, M.S., Brett, J.M., Scholz, M., and Wood, P.R. 1988a and 1988b preprints.
Boesgaard, A.M. 1981, Astrophys. J. **251**, 564.
Böhm-Vitense, E., and Querci, M. 1987, in Exploring the Iniverse with the IUE Satellite, eds. Y. Kondo, C. de Jager, and J. Linsky (Dordrecht: Reidel), p. 223.
Bouchet, P. 1984, Astron. Astrophys. **139**, 344.
Bouchet, P., Querci, M., and Querci, F. 1983, The Messenger, **31**, 7.
Bowen, G.H. 1988, Astrophys. J. **329**, 299.
Buchler, J.R., and Goupil, M.J. 1988, Astron. Astrophys. **190**, 137.
Buchler, J.R., and Kovacs, G. 1987, Astrophys. J. **320**, L57.
Coullet, P., and Spiegel, E.A. 1983, SIAM J. Appl. Math. **43**, 776.
Cuntz, M. 1987, Astron. Astrophys. **188**, L5.
De Jager, C. 1984, Astron. Astrophys. **138**, 246.
Detre, L. 1968, in Proc. IAU Colloq. on Non-Periodic Phenomena in Variable Stars, Budapest, p. 3.
Dupree, A.K., Baliunas, S.L., Guinan, E.F., Hartmann, L., Nassiopoulos, G.E., and Sonneborn, G. 1987, Astrophys. J. **317**, L85.
Ericksson, K., Gustafsson, B., Johnson, H.R., Querci, F., Querci, M., Baumert, J.H., Carlsson, M., and Olofsson, H. 1986, Astron. Astrophys. **161**, 305.
Fox, M.W., and Wood, P.R. 1982, Astrophys. J. **259**, 198.
Glasby, J.S. 1968, Journal AAVSO **5**, 67.
Goldberg, L. 1979, Quart. J., Roy. Astron. Soc. **20**, 361.
Hassenstein, W. 1938, Pub. Astr. Obs. Postdam **29**, Part 1.
Hinkle, K.H., Scharlach, W.W.G., and Hall, D.N.B. 1984, Astrophys. J. Supplement **56**, 1.
Horne, C.H., and Baliunas, S.L. 1986, Astrophys. J. **302**, 757.
Huggins, P.J. 1984, in "Mass Loss from Red Giants", ed. M. Morns and B. Zuckerman, (Reidel: Dordrecht.)
Jacchia, L. 1933, Le Stelle Variabili, Pub. Osservatorio Astronomico, Univ. di Bologna, Vol. II, n. 14.
Jennings, M.C., and Dyck, H.M. 1972, Astrophys. J. **177**, 427.
Karovska, M. 1987, in Proc. Conference on Stellar Pulsation, eds. A.N. Cox, W.M. Sparks, and S.G. Starrlield (Heidelberg: Springer-Verlag), p. 260.

Krempec, J. 1973, Studia Societatis Scientiarum Torunensis (Torun: Polonia) 5, n°2, Sectio F (Astronomia), p. 19.
Livi, S.H.B., and Bergmann, T.S. 1982, Astron. J. 87, 1783.
Luttermoser, D.G., Johnson, H.R., and Avrett, E.H. 1988, A Decade of UV Astronomy with the IUE Satellite, ESA, SP 287, p. 327.
Maeder, A. 1980, Astron. Astrophys. 90, 311.
Mantegazza, L. 1982, Astron. Astrophys. 111, 295.
Maran, S.P., Michalitsianos, A.G., Heinsheimer, T.F., and Stecker, T.L. 1980, in Proc. GSFC Conf. on Current Problems in Stellar Pulsations Instabilities, eds. D. Fischel et al., NASA TM-80625, p. 629.
Mayall, M.W. 1965, Roy. Astron. Soc. Canada J. 59, 245.
McLaughlin, D.B. 1946, Astrophys. J. 103, 35.
Perdang, J. 1984, in Proc. 25th Lirge Internat. Astrophys. Colloq. on Theoretical Problems in Stellar Stability and Oscillations (Univ. de Liege). p. 425.
Perdang, J. 1985, in Chaos in Astrophysics eds. J.R. Buchler, J.M. Perdang and E.A. Spiegel (Dordrecht: Reidel), p. 11.
Perdang, J., and Blacher, S. 1982, Astron. Astrophys. 112, 35.
Querci, F. 1986, in CNRS/NASA Monograph Series on Nonthermal Phenomena in Stellar Atmospheres, The M, S, and C Stars, eds. H.R. Johnson and F. Querci, chp. 1.
Querci, M. 1986, in CNRS/NASA Monograph Series on Nonthermal Phenomena in Stellar Atmospheres, The M, S, and C Stars, eds. H.R. Johnson and F. Querci, chp. 2.
Querci, M., and Querci, F. 1985, Astron. Astrophys. 147, 121.
Regev, O., and Buchler, J.R. 1985, in Chaos in Astrophysics, eds. J.R. Buchler, J.M. Perdang, and E.A. Spiegel (Dordrecht: Reidel), p. 285.
Sanford, R.F. 1950, Astrophys. J. 111, 270.
Schmitz, F., and Ulmschneider, P. 1981, Astron. Astrophys. 93, 178.
Smith, H.A. 1976, Journal AAVSO 5, 67.
Stebbins, J. 1931, Publ. Washburn Observatory, Univ. Wisconsin, 15, 177.
Tsuji, T., Unno, W., Kaifu, N., Izumiura H., Ukita, N., Cho, S., Koyama, K. 1988, Astrophys. J. 327, L23.
Tuchman, Y., Sack, N., and Barkat, Z. 1978, Astrophys. J. 219, 183.
Whitney, C.A. 1984, Bull. A.A.V.S.O. 13, 31.
Willems, F.J., and De Jong, T. 1988, Astron. Astrophys 196, 173.
Wood, P. 1975, in Proc. IAU Colloq. n°29 on Multiple Periodic Variable Stars, ed. W.S. Fitch (Dordrecht: Reidel), p. 69.

MODEL ATMOSPHERES WITH PERIODIC SHOCKS

G. H. Bowen
Astronomy Program, Physics Department
Iowa State University, Ames, Iowa 50011

Abstract. The pulsation of a long-period variable star generates shock waves which dramatically affect the structure of the star's atmosphere and produce conditions that lead to rapid mass loss. Numerical modeling of atmospheres with periodic shocks is being pursued to increase our understanding of the processes involved and of the evolutionary consequences for the stars. It is characteristic of these complex dynamical systems that most effects result from the interaction of various time-dependent processes. For example, rapid mass loss in the models is a joint consequence of the enormous extension of the atmosphere caused by shocks, and of radiation pressure on grains formed in the cool outer region; it is also affected by thermal relaxation processes, which determine the temperature distribution. The progress and significance of these modeling calculations will be reviewed.

1 INTRODUCTION

It has become clear in recent years that stars on the asymptotic giant branch (AGB) normally undergo mass loss so prodigious that it changes the entire course of their subsequent evolution. These stars end their lives as white dwarfs, most of relatively low mass (Weidemann 1987). The most rapid mass loss appears to be associated with evolutionary phases in which there is large-amplitude, long-period pulsation, notably in the Mira-class variables and in OH/IR sources. What changes occur in the outer structure and in the behavior of such stars because of their pulsation? How does mass loss result?

This paper is an attempt to summarize the results of modeling calculations which bear on those questions. The emphasis is necessarily on my own work, although a brief description of some other work will be given. A very great deal remains to be done. In order to make progress it has been necessary to make numerous approximations and to use estimated values for various input parameters that are uncertainly known at best. Nevertheless a picture has developed which seems to make sense, which increases our understanding of long period variable (LPV) stars, and which points to the areas most urgently needing further work.

One very important lesson learned from this modeling is an appreciation of the complexity of these stars' behavior. Simple pictures and simple solutions are appealing, but these are not simple systems. Many of the phenomena they exhibit can not be analyzed properly by studying one isolated process at a time. It would lead to very wrong conclusions, for example, to assume that the atmosphere of a Mira variable is normally in hydrostatic equilibrium and that when periodically disturbed by a shock wave it quickly returns to the original equilibrium conditions, which then prevail until the next shock passes through. Atmospheres in the dynamical models, and no doubt in the stars, are very different from that indeed. Their structure and behavior are determined by interactions, throughout the cycle, of hydrodynamic processes, including but not limited to shock waves; of thermodynamic processes, including but not limited to radiative transfer; of ionization and other chemical processes; of grain formation, growth, optics, and dynamics -- by a wide variety of time-dependent processes, in fact, many of which may be far from equilibrium in much of the extended atmosphere for much of the time. LPV atmospheres thus present a challenging problem -- difficult, rather messy, but very intriguing.

2 MODELING METHOD

The approach used in LPV atmosphere modeling calculations has been to consider the atmosphere separately from the interior, as follows. One constructs a model atmosphere whose inner boundary is placed at or somewhat inside the photosphere and studies the response of the model to periodic driving at the inner boundary which simulates the effects of a pulsating interior. Possible effects of the atmosphere on the interior have been neglected, which surely is a reasonable approximation for most purposes; the atmosphere may play a significant role in limiting the pulsation amplitude, however, as will be discussed later. In any case this approach has made possible much progress toward understanding the behavior of the atmosphere and the mechanism of mass loss.

Table 1. Assumed typical ranges for parameters of Mira variables.

Mass	1-2 $M_\odot$
Period	200-500 days
Radius	150-350 $R_\odot$
Effective temperature	2800-3000 K
Shock amplitude	25-35 km s^{-1}
Wind velocity	10 km s^{-1}
Mass loss rate	10^{-7} to 2×10^{-6} $M_\odot$ yr^{-1}

The objects most studied in modeling calculations of this kind are the Mira variables. These are cool giant LPVs of high luminosity, which commonly have considerable circumstellar dust and fairly large mass loss rates. Stellar properties assumed to be typical of Miras are shown in Table 1 (Willson 1982, 1988). Most LPV models have had parameters in or near these ranges.

No attempt will be made to give here a complete review of the relevant literature. For discussions of possible mass loss mechanisms, the excellent reviews by Castor (1981) and by Holzer & MacGregor (1985) are recommended. Modeling calculations which seem particularly significant with respect to the structure of LPV atmospheres have been carried out by Willson, by Hill, by Wood, and by Bowen; these will be described briefly.

Willson used analytic methods and a ballistic approximation for the motion of gas in the atmosphere to gain insight into the behavior resulting from shocks and to elucidate the conditions under which mass loss can occur (Willson & Hill 1979, Hill & Willson 1979). This work was elaborated and extended in later publications (e.g. Willson & Bowen 1985, 1986a, 1986b). Although it omits essential physical effects, it does provide a useful limiting case.

Numerical hydrodynamic calculations have been carried out by Hill (Willson & Hill 1979, Hill & Willson 1979), by Wood (1979), and by Bowen (1988a, 1988b). All assumed spherical symmetry but made no other assumptions about the form of the solution. Each then wrote the basic hydrodynamic equations, including artificial viscosity, and integrated these to obtain a description of the dynamical model. Significant differences in their work include the following. Hill used mostly a 5-$M_\odot$ model with a radius close to that now believed to correspond to the first overtone mode for the period employed; Wood used only a 1-$M_\odot$ overtone model; Bowen explored a grid of models that extended over a sizable range of masses and periods, including both fundamental and overtone modes, and was supplemented by systematic variation of other model parameters. Hill assumed that all thermodynamic processes, including shocks, were isothermal (except for one interesting case, not pursued further, in which an abrupt change to adiabatic behavior was made at large radii, where low density would be expected to give slow recombination of ionized hydrogen); Wood tried both isothermal and adiabatic conditions; Bowen used both completely isothermal and completely adiabatic conditions, but introduced the use in most models of density- and temperature-dependent thermal relaxation rates. Hill assumed a uniform, constant temperature throughout the model atmosphere at all times; Wood calculated the temperature as a function of radius (in his isothermal models) using a fictitious optical depth chosen to fit his interior calculations; Bowen used the Eddington approximation for a gray spherical atmosphere to calculate the radiative equilibrium temperature at each radius, assuming a uniform opacity estimated from the results of Alexander et al. (1983) to be suitable for the inner atmosphere. Both Wood and Bowen, but not Hill, made calculations for models that included radiation pressure on dust, with the assumed amount of dust (more precisely, its cross-section for radiation pressure) calculated in a simple, ad hoc way from the local radiative equilibrium temperature.

The results of all three of these investigators are similar, to the extent that their work overlaps. Shocks are formed in all cases, and the atmosphere becomes greatly extended. Completely isothermal models with no dust have very low mass loss rates, and completely adiabatic models have extremely high mass loss rates. Wood showed that addition of a suitably adjusted amount of dust to his isothermal model gave a realistic mass loss rate; Bowen confirmed this and explored the dust effects throughout his extended grid of models. In broad outline, at least, the picture seems solid and secure.

3 MODELING RESULTS

In order to illustrate typical results in some detail I shall use a specific model of my own. It is the 1.0-$M_\odot$, 320-day, fundamental mode model whose characteristics are listed in Table 2. Figures 1-4 show the behavior of this model, which will be discussed.

Table 2. Characteristics of Illustrative Model

Input parameters:
- M = 1.0 $M_\odot$
- P = 320 days
- Fundamental mode
- R = 240 $R_\odot$
- T_{eff} = 2960 K
- T_{eff} amplitude = 200 K
- Piston vel. ampl. = 3 km s^{-1}
- Dust: T_{cond} = 1400 K
  Radiation pressure
  cross section calc.
  to give a_{rad} = 0.95g

Calculated values:
Initial static model:
- L = 4000 $L_\odot$
- g = 0.48 cm s^{-2}
- v_{esc} = 40 km s^{-1}
- H = 0.025 $R_\star$
  (scale height)

Dynamical model:
- $\dot{m}$ = 2.5x10^{-7} $M_\odot$ yr^{-1}
- Max. shock ampl. = 33.5 km s^{-1}
- Wind speed = 11-12 km s^{-1}
- M_{bol} variation = 1.07

The input parameters and the calculations for this model were very similar to those described in Bowen (1988a). They differ in the use of more and finer zones in the model and of shorter time steps, to give better modeling; of a deeper inner model boundary (about 0.9 $R_\star$) to permit better study of conditions inside the photosphere; of slightly adjusted values of the dust condensation temperature (to 1400 K), of the assumed opacity (to 4x10^{-4} cm^2 gm^{-1}), and of the fraction of Lyman alpha radiation assumed to escape (to 0.1%); and of an imposed variation in the effective temperature (amplitude = 200 K), as part of the driving. Also included were the collisional transfer of energy from grains to gas atoms (Bowen 1988b).

A stable, reproducible dynamical model is generated by starting with a model in hydrostatic equilibrium, then increasing the velocity amplitude of the inner boundary (the "piston") slowly and smoothly from a tiny initial value (say 1 cm s^{-1}) to the desired final value. If piston oscillations are begun abruptly at full amplitude the first wave, moving outward in the steep density gradient of the static atmosphere, grows rapidly into an extremely strong shock which accelerates most of the model to speeds much greater than the escape velocity and destroys the model. When the piston amplitude is increased slowly, however, the model changes without disastrous transients to a very different density distribution which permits steady state behavior with periodic shocks. The adjustment of the model to a change in the driving amplitude is quite fast, in fact, requiring rather few pulsation cycles. Presumably stars can similarly adjust rather quickly and easily to changes in oscillation amplitude that might occur.

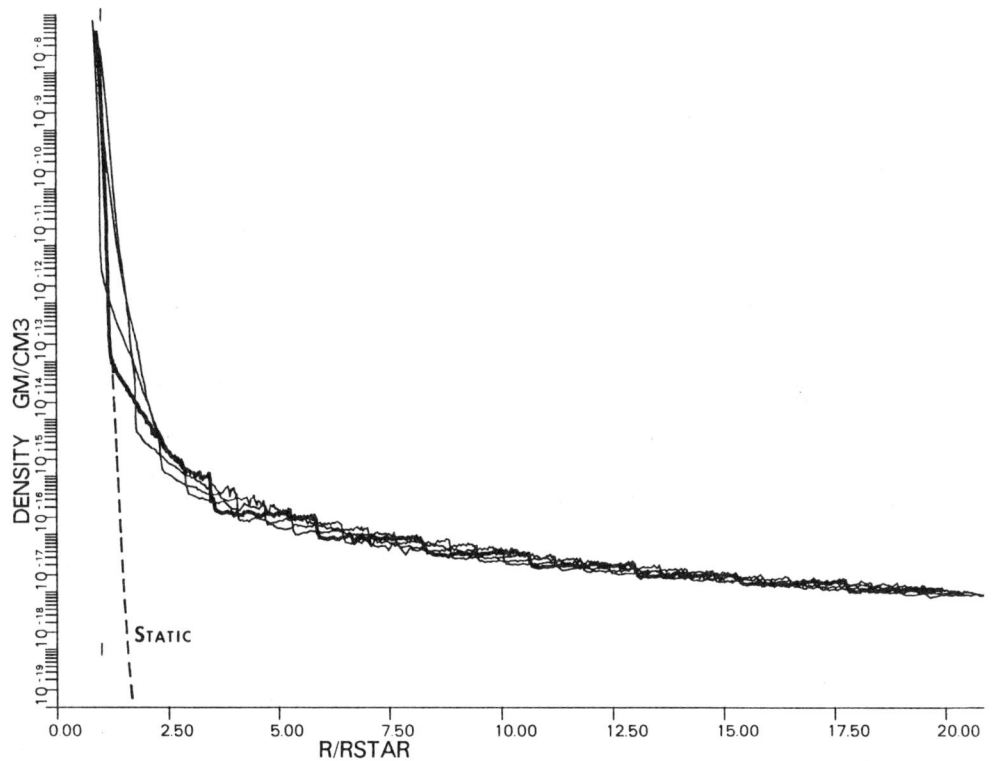

Figure 1. Density as a function of radius at phases 0.00, 0.25, 0.50, and 0.75. (Phase 0.00 shown bold.) The density distribution in the static model is shown for comparison.

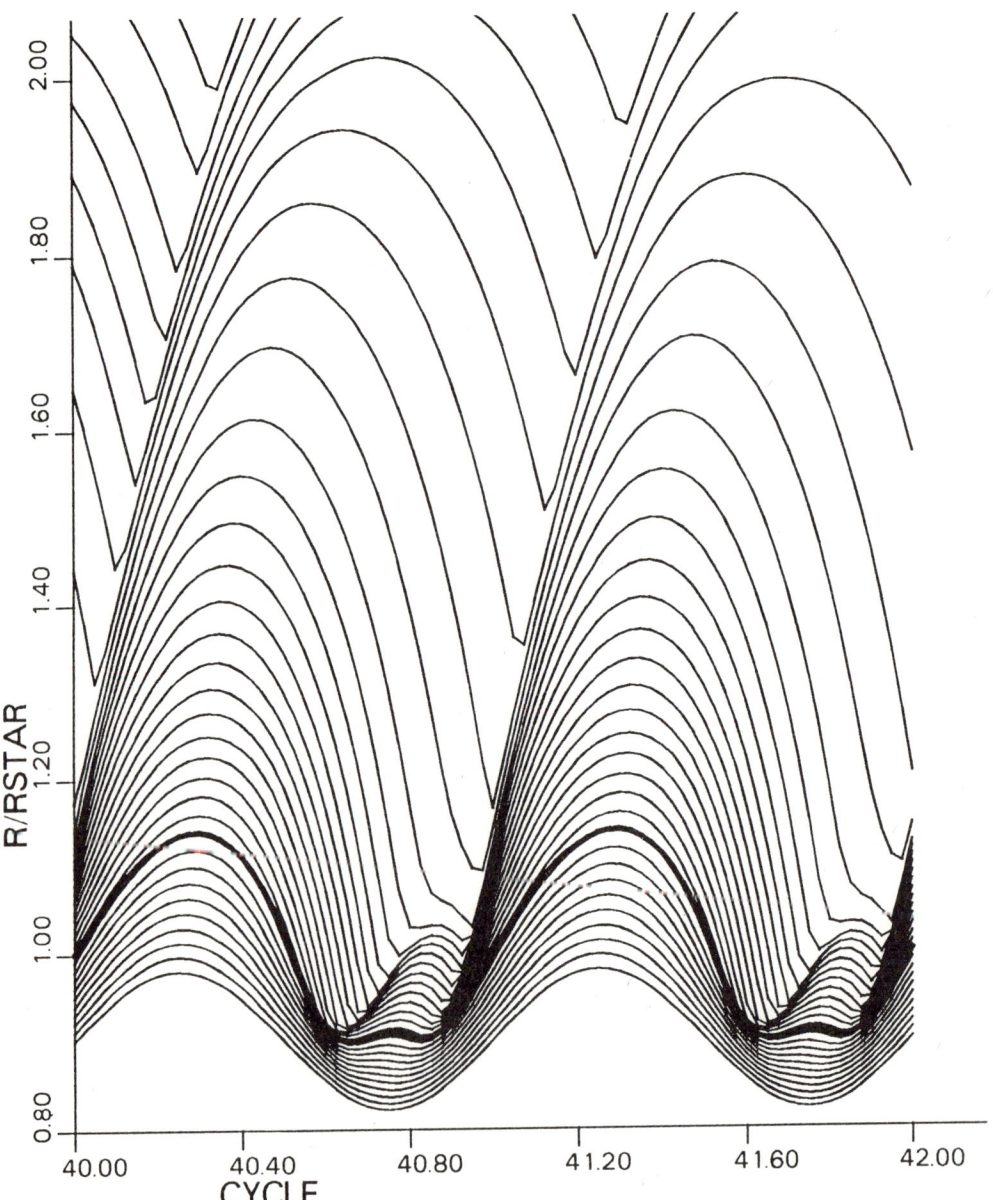

Figure 2. Radius of selected shells as a function of phase. The innermost line is the piston. The bold line is the photosphere.

Figure 1 shows the density distribution for the illustrative model at four phases. Apart from phase-dependent differences resulting from variations in the photosphere radius and from the formation and propagation of shocks, the curve can be described as consisting of two parts: inside the first shock, a steep exponential section like that of the static model; outside the first shock, a very gentle slope -- i.e. a very large scale height, which in fact increases with radius. The effect is to give a truly enormous extension of the atmosphere. The total mass in the outer region is not large, but it is this mass (and the low density gradient there) which makes possible the stable, steady state response of the model to periodic shocks, the formation of substantial amounts of circumstellar dust, and the development of an outflowing wind that gives rapid mass loss from the star.

If the piston amplitude is increased, the innermost shock is formed at slightly smaller radius, where the density and power dissipation are greater; beyond that radius the density curve runs parallel to the previous one, keeping a higher value at all radii. (The mass in the extended outer region is derived from the dense inner zones, of course, with negligible effect on them.) The effect is to increase the mass loss rate considerably -- at the cost of greater power input, almost all of which is dissipated at the inner shock and lost from there via radiation.

Figure 2 shows the radius as a function of phase for selected shells in the inner part of the model, where the density is relatively large and the mean outward velocity is small. Note the formation of a strong shock just outside the photosphere at about phase 0.0 of each cycle; these propagate smoothly outward, weakening as they go. Between encounters with these shocks, individual shells execute almost periodic motion along roughly ballistic trajectories. Note also the formation of a second shock near the photosphere at about phase 0.6; this does not propagate outward and appears rather limited and weak in the figure. In fact these second shocks dissipate a great deal of power because they occur in regions of relatively high density. To understand their formation, observe that motion inside the photosphere is essentially that of a standing wave whose amplitude increases with radius. At phase 0.0 shells just outside the photosphere start outward fast enough to follow semiballistic paths, but not fast enough to remain "levitated" for a full period, in the sense discussed by Willson (Hill & Willson 1979). They fall back into the material below and produce a shock. The propagating shock beginning at phase 1.0 is formed by the interaction of material whose postshock velocity at phase 0.0 was great enough to keep it levitated for one full period.

Figure 3 shows the same information as Figure 2 for a region that extends far enough outward to include dust formation (roughly 2-3 R_*). Note the rapid outward acceleration that occurs there and the transition beyond that to almost steady outflow. Because the mean outward velocity of the material in this lower density region is fairly large, a given shell encounters the (now weak) propagating shocks at time intervals much longer than one period.

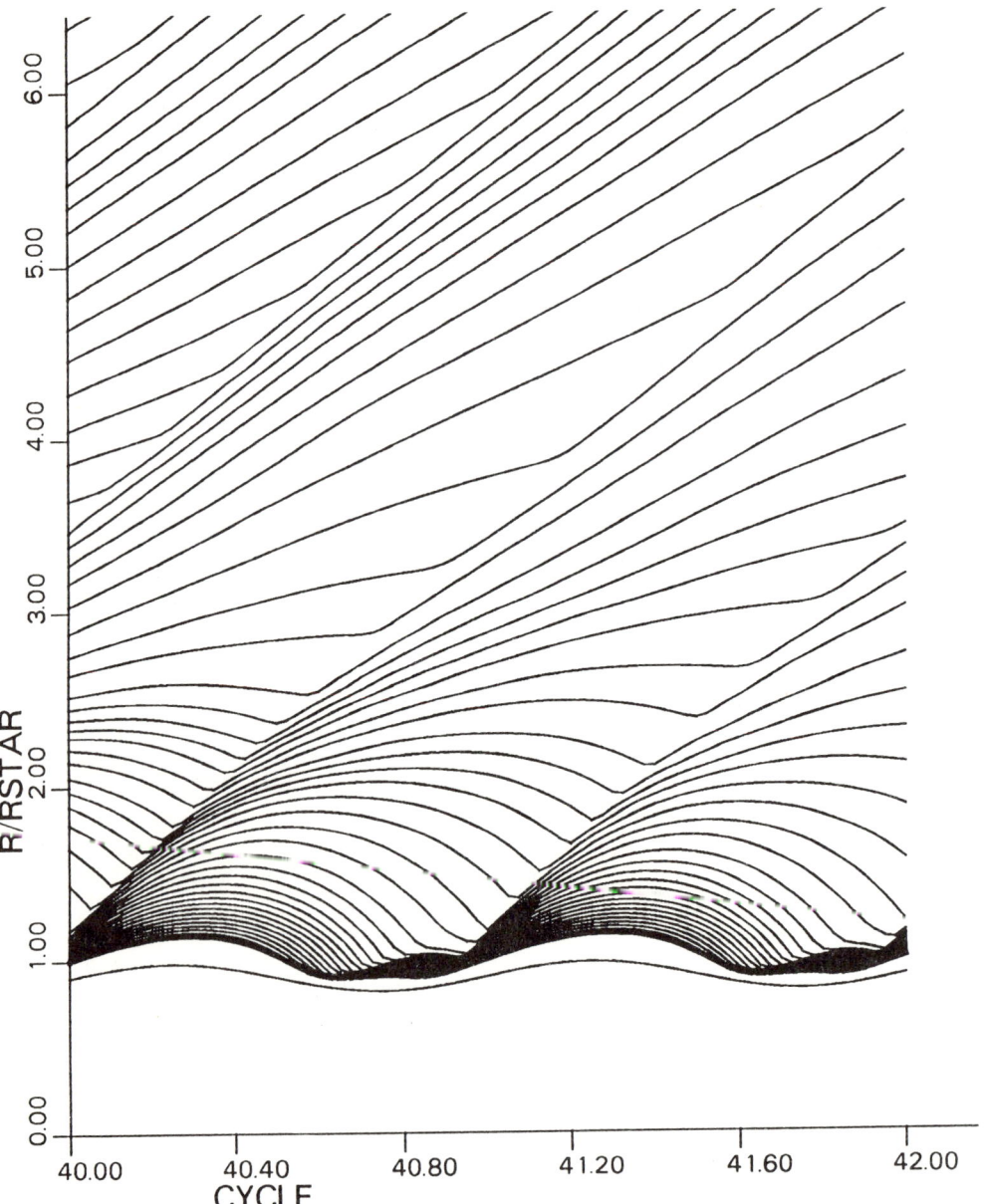

Figure 3. Radius of selected shells as a function of phase. (Same as Figure 2 except for the scale. Shells between the piston and the photosphere are omitted here for clarity.)

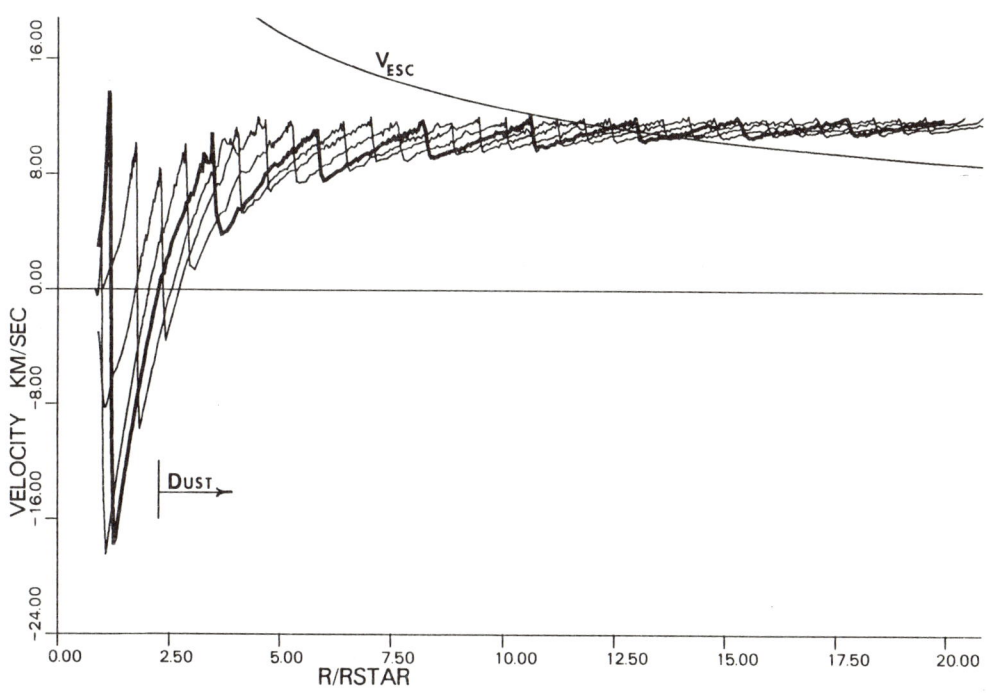

Figure 4. Radial velocity and gas kinetic temperature as functions of radius at phases 0.00, 0.25, 0.50, and 0.75. (Phase 0.0 shown bold.)

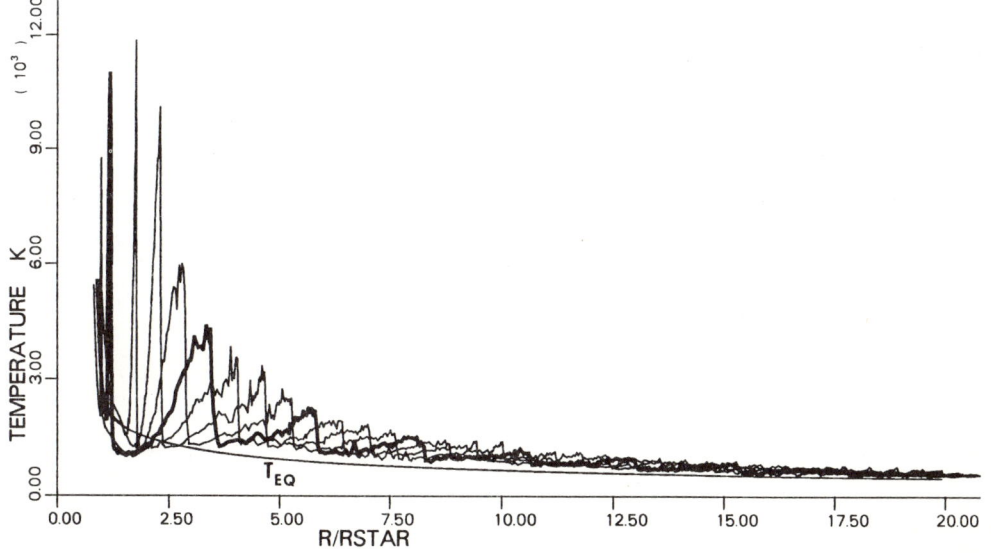

Figure 5. Same as Figure 4 except that dust was omitted from the model.

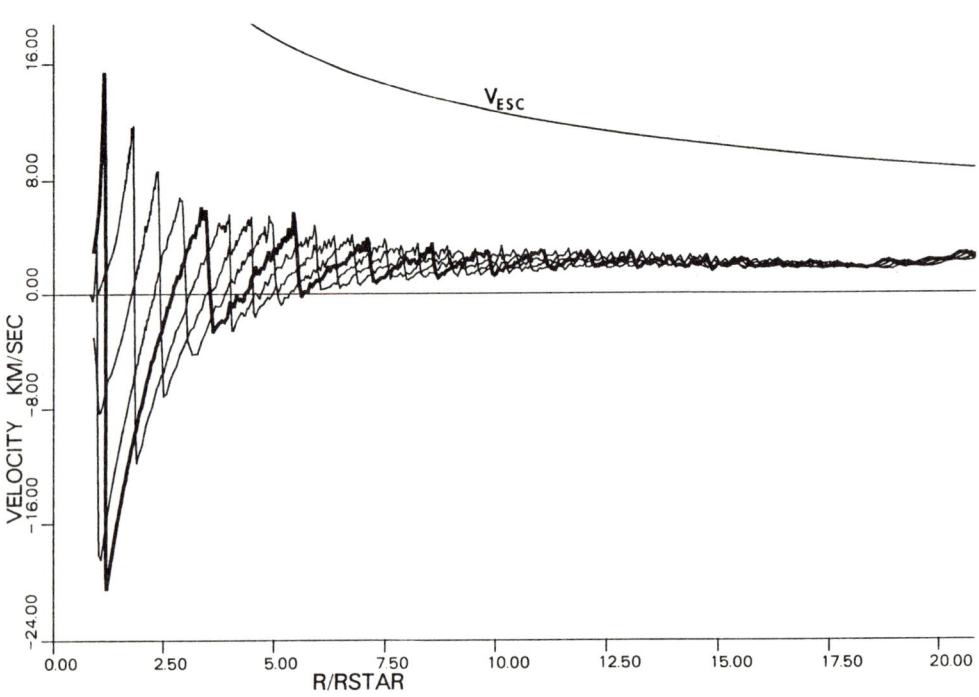

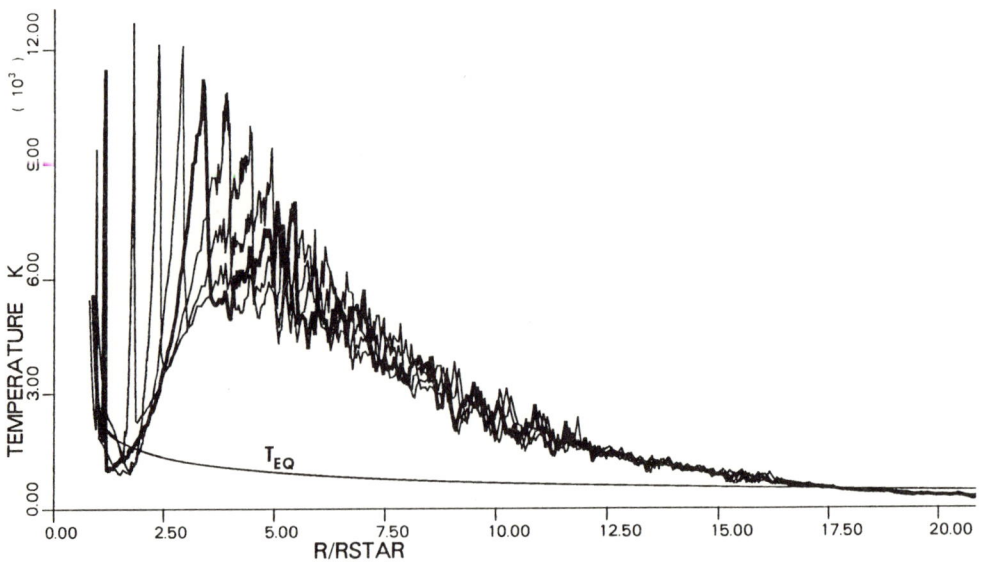

Figure 4 shows the radial velocity and the gas kinetic temperature as functions of radius for four phases. The velocity plot clearly shows the formation, propagation, and weakening of the shock waves; the rapid acceleration that occurs in the region where dust has formed; and the development of an almost steady wind with a speed exceeding the escape velocity. The temperature plot shows a sharp spike at the position of a strong shock in the relatively dense inner atmosphere, where the collision rates are high and collisional excitation is fast enough to give very rapid radiative cooling; processes there are effectively isothermal. At somewhat greater radii, where the density and temperature are much lower, radiative heating and cooling become so slow that processes there are effectively adiabatic. The dominant cooling mechanism is then adiabatic expansion, which occurs both in the regions between shocks and also at very large radii, because of the general outflow of material. That is sufficient to lower the temperature well below the radiative equilibrium temperature (T_{eq}) in some preshock regions and at all large radii; the wind is then quite cool (Bowen 1988a). The inclusion in the calculations of collisional transfer of energy as well as momentum to the gas by grains (Bowen 1988b) gives temperatures in the wind which are somewhat above T_{eq}, as seen here.

Figure 5 shows the quite different results obtained when dust is omitted from the calculations. The rapid acceleration seen in Figure 4 in the dust region does not occur. The gas kinetic temperature drops below T_{eq} in the region of very rapid, effectively adiabatic postshock expansion at roughly 1.5-2 R_*. Beyond that, however, expansion is too slow to give much cooling, and the temperature remains far above T_{eq} for a large distance. The temperature gradient does drive a very slow outflow at large r, and there is a small but nonzero mass loss rate. It should be added that off the scale at large r there is a sonic point, where acceleration of the gas to locally supersonic velocities occurs; but the density there is so low that the resulting mass loss rate is very small.

Table 3 summarizes in very schematic fashion the structure of the model of Table 2, with an indication of significant phenomena occurring in each region. A brief acounting of the power budget is also included. The results with other models would be similar. Note that the energy per unit time required to drive the mass loss (0.079 $L_\odot$) is an extremely small fraction of the star's luminosity, whereas the momentum required per unit time (140 $L_\odot/c$) is not small on that scale. Radiation pressure on grains appears to perform the essential function of coupling momentum from the star's radiation field to the gas, so as to make possible the observed outflow.

Note also in Table 3 that the maximum power delivered to the model by the piston is almost exactly equal to the star's mean luminosity. This is more or less fortuitous. (The piston velocity amplitude, 3 km s^{-1}, was chosen somewhat arbitrarily to be large enough to give well developed shocks which dissipate substantial power, but small enough to avoid supersonic piston velocities, odd resulting waveforms, and excessive power requirements. A range of driving amplitudes has been explored for this and many other models. There is nothing special about this particular choice beyond the considerations mentioned above.) It does suggest a conjecture, however. Interior models for stars of this type show very large oscillation growth rates (Ostlie & Cox 1986). It has never been clear, even with the best nonlinear models of the interior, what determines the limiting amplitude of pulsation. The large power requirement for mechanically driving the inner atmosphere, together with the power dissipation by shocks there, may play an important role in limiting the pulsation amplitude.

Table 3. Summary of the structure and power budget for the model of Figures 1-4. (Schematic)

Radius	Region	Phenomena	Power
>20 R_*	Wind	Wind speed approx. constant	**Wind**: (Net rate of energy change, from photosphere to large radius)
			Gravitational 0.069 $L_\odot$
			Kinetic 0.0025
		Cool wind	Thermal −0.0012
			Net: 0.070 $L_\odot$
15 R_*	Escape velocity exceeded		
		Heating/cooling by grains is small but significant	
		Heating/cooling by radiation is very slow	
		Shocks very weak	Shock power negligible
		Expansion --> cooling	
		Rapid outward acceleration	
2-3 R_*	Dust formation		
		Shocks propagate and weaken	**Postshock radiation**:
			Mean = 55 $L_\odot$
			Max. = 420 $L_\odot$
		(approx. ballistic behavior between shocks)	
R_*	Photosphere	Shocks form	**Continuum radiation**:
			Mean = 4300 $L_\odot$
		Weak traveling wave plus	**Wave transport**:
			Mean = 30 $L_\odot$
		Strong standing wave	Max. = 4300 $L_\odot$
0.9 R_*	Driving region	(Piston)	

Table 4 presents data selected from the grid of fundamental mode models described by Bowen (1988a). These show the mass loss rates for a range of masses and for a range of periods extending to values characteristic of OH/IR sources; all other model parameters were held constant to focus attention on the effects of mass and period. There are striking trends toward higher mass loss rates with either a decrease of model mass or an increase of pulsation period. The lower mass models with periods greater than about 500 days have very high mass loss rates and optically thick circumstellar dust -- properties associated with OH/IR sources. Such behavior would have important consequences for evolutionary tracks in mass-luminosity diagrams. An aging, mass losing AGB star, with increasing luminosity and core mass, would also increase in radius and pulsation period, increase its mass loss rate, and proceed more and more rapidly toward loss of all its envelope.

Table 4. Mass loss rates ($M_\odot$ yr^{-1}) for fundamental mode models including dust.

$M/M_\odot$	175 d	250 d	350 d	500 d	700 d	1000 d
2.0	....	9. x10^{-9}	5.8x10^{-8}	3.1x10^{-7}	1.2x10^{-6}	6.5x10^{-6}
1.6	....	3.2x10^{-8}	1.7x10^{-7}	6. x10^{-7}	2.3x10^{-6}	1.4x10^{-5}
1.2	2.2x10^{-8}	9. x10^{-8}	2.7x10^{-7}	9. x10^{-7}	7. x10^{-6}	4. x10^{-5}
1.0	3.4x10^{-8}	1. x10^{-7}	2.5x10^{-7}	2.1x10^{-6}	1.6x10^{-5}	1. x10^{-4}
0.8	5. x10^{-8}	3. x10^{-7}	7. x10^{-7}	8. x10^{-6}	4. x10^{-4}	....

4 FUTURE DIRECTIONS

Modeling results gained thus far have demonstrated that it is possible to construct reproducible, stable, steady state, spherically symmetric models which show apparently reasonable physical beahavior and have mass loss rates that are at least of the right order of magnitude. What now needs to be done? What are the most fruitful directions for further work?

There are several areas within the theoretical modeling which clearly need much further work. My own nominees for the most urgent of these, because each has potentially major effects on the large-scale structure and behavior of the models are the following.

1- Studies of the time-dependent, nonequilibrium chemistry of the atmosphere -- especially that of hydrogen, of course, since it is present in such large amounts, but extending also to other constituents which are known to be important radiative cooling agents, such as MgII. (By "chemistry" I mean not only conventional molecular reactions but also ionization/recombination.) The chemistry needs to be coupled directly to the hydrodynamic equations in order to treat properly the interactions between the various processes involved.

2- Improved treatment of the region inside and immediately outside the photosphere, and of the coupling between the atmosphere and interior. Energy transfer is very rapid there, and processes take place which probably shape the observable behavior of the star (e.g. light curves) and some aspects of the star's evolution (e.g. the limiting pulsation amplitude, which is a major determinant of the mass loss rate).

3- Studies of dust -- nucleation and growth of grains, their changing optical properties, their dynamics, their coupling to both the gas and to the star's radiation field. Dust plays a key role in these stars, but it has been treated only rather superficially so far.

Comparison of modeling results with observational results is urgently needed, on one hand to help interpret the observational data, and on the other to check on the validity of the models. This should also help to establish more accurately some of the rather uncertainly known parameters that enter the modeling calculations. The modeling has at last reached a level at which it should be rewarding to pursue such studies vigorously. Some work of this kind has been done (e.g. Beach et al. 1988, Brugel et al. 1988). Much more is needed.

And ultimately, of course, the most important goal of all, to me, is to use the understanding thus gained to learn more about the place in stellar evolution of these remarkable stars.

This work was supported in part by NASA grants NAG5-707 from the IUE program, and NAGW-1364 from the Astrophysics Theory Program.

REFERENCES

Alexander, D. R., Johnson, H. R., & Rypma, R. L. (1983). Ap. J., <u>272</u>, 773.
Beach, T. E., Willson, L. A., & Bowen, G. H. (1988). Ap. J., <u>329</u>, 241.
Brugel, E. W., Beach, T. E., Willson, L. A., & Bowen, G. H. (1988). in IAU Colloquium No. 103, The Symbiotic Phenomenon, ed. M. Friedjung. Dordrecht: Reidel. (In press)
Bowen, G. H. (1988a). Ap. J., <u>329</u>, 299.
Bowen, G. H. (1988b). in Pulsation and Mass Loss in Stars, ed. R. Stalio & L. A. Willson. Dordrecht: Reidel. p. 3.
Castor, J. I. (1981). in Physical Processes in Red Giants, ed. I. Iben Jr. Dordrecht: Reidel. p.275.
Hill, S. J. & Willson, L. A. (1979). Ap. J., <u>229</u>, 1029.
Holzer, T. E. & MacGregor, K. B. (1985). in Mass Loss from Red Giants, ed. M. Morrris & B. Zuckerman. Dordrecht: Reidel. p. 229.
Ostlie, D. A. & Cox, A. N. (1986). Ap. J., <u>311</u>: 864.
Weidemann, V. (1987). in Late Stages of Stellar Evolution, ed. S. Kwok & S. R. Pottasch. Dordrecht: Reidel. p. 347.

Willson, L. A. (1982). in Pulsations in Classical and Cataclysmic Variable Stars, ed. J. P. Cox & C. J. Hansen. Boulder: JILA. p. 269.
Willson, L. A. (1988). in The Use of Pulsating Stars in Fundamental Problems of Astronomy, ed. E. G. Schmidt. Cambridge: Cambridge University Press. (In press).
Willson, L. A. & Bowen, G. H. (1985). in Relations Between Chromospheric-Coronal Heating and Mass Loss in Stars, ed. R. Stalio & J. B. Zirker. Trieste: Osservatorio Astronomico di Trieste. p. 127.
Willson, L. A. & Bowen, G. H. (1986a). Irish Astr. J., 17, 249.
Willson, L. A. & Bowen, G. H. (1986b). in Lecture Notes in Physics, Vol. 254, Cool Stars, Stellar Systems, and the Sun, ed. M. Zeilik & D. M. Gibson. Berlin: Springer-Verlag. p. 385.
Willson, L. A. & Hill, S. J. (1979). Ap. J., 228, 854.
Wood, P. R. (1979). Ap. J., 190, 609.

Evolution of Oxygen-rich and Carbon Stars on the Asymptotic Giant Branch

Sun Kwok[1], Kevin M. Volk[2], and S. Josephine Chan[1]
[1]The University of Calgary, Calgary, Canada
[2]NASA Ames Research Center, Moffett Field, California, U.S.A.

I. Introduction

For many years, it has been commonly believed that oxygen-rich (M) stars evolve first to S stars and then to carbon (C) stars. However, the details of the transition are not understood. It is now accepted that the overabundance of carbon ([C/O]>1) in some asymptotic giant branch (AGB) stars is due to the dredge up of products of α capture and s-process elements after a number of thermal pulses (Iben 1975). Effects of convective overshooting and semiconvection in the dredge up process have also been considered (Castellani *et al.* 1985, Lattanzio 1986). The dredge up of carbon into the photosphere leads to the formation of carbon-based molecules, which absorption bands become the basis of spectral classification.

During the past decade, it is recognized that the carbon-richness of a star not only manifests itself in the photospheric spectrum, but also in the circumstellar environment as well. Probes of the circumstellar envelopes in the infrared and radio regions provide new means to characterize the chemical properties of the star. These methods are particularly useful in cases where the photosphere is heavily obscured by circumstellar absorption. Table 1 summarizes the photospheric and circumstellar spectral characteristics of M and C stars.

Table 1

O-rich stars	C-rich stars
Optical (photospheric)	
TiO, VO	C_2, CN
	3219 stars in GCCCS (Stephenson 1973).
Infrared (circumstellar)	
9.7 μm silicate dust feature	11.2 SiC dust feature
~2000 objects in *IRAS* LRS classes 21-29, 31-39.	538 objects in *IRAS* LRS classes 41-49.
Radio (circumstellar)	
1612 MHz OH maser	λ2.6mm rotational transition of CO
~400 detected as of 1987	~170 detected as of 1987, approximately half are from oxygen-rich stars.

While large numbers of stars have been classified optically as carbon stars (hereafter visual carbon stars) using, e.g. objective prism surveys, there have been an increasing number of carbon-rich objects which are discovered in infrared surveys, the most famous example being CW Leo (IRC+10°216).

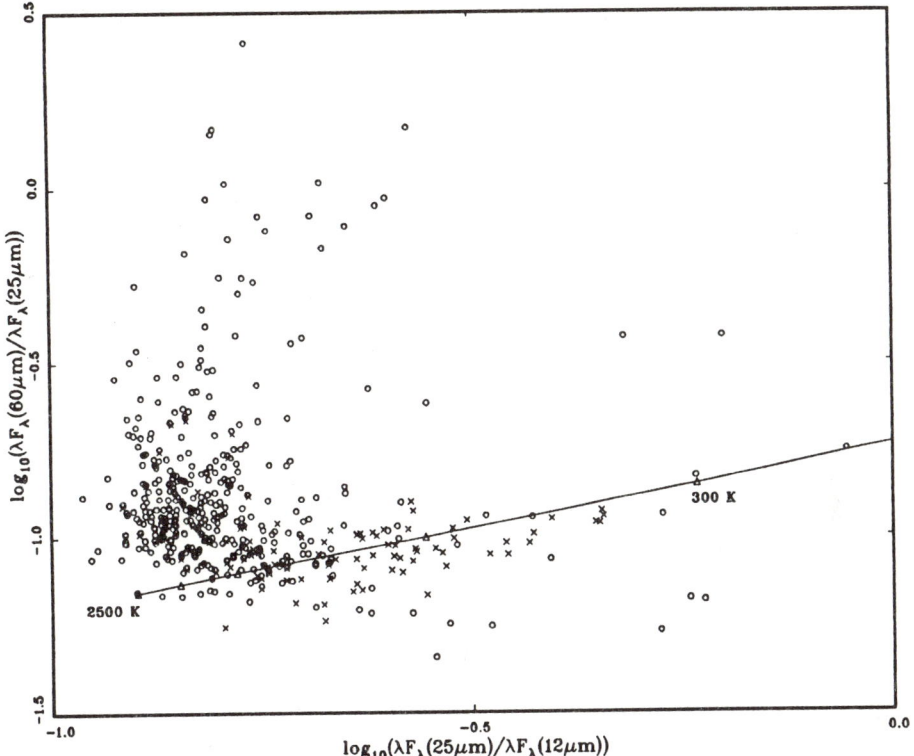

Figure 1. Color distributions of visual and infrared carbon stars. Circles: 386 visual carbon stars in GCCCS and MacConnell (1988) with good fluxes at 12, 25, and 60 μm *IRAS* bands; crosses: 111 stars in *IRAS* LRS classes 41-49 with good fluxes at all four *IRAS* bands.

CW Leo shows strong thermal emission line from the CO molecule, which suggests a carbon-rich nature. The Low Resolution Spectrometer (LRS) on *IRAS* has also discovered new objects which show the 11.2 μm silicon carbide feature, instead of the 9.7 μm silicate feature which is commonly observed in M stars. It is interesting to find that these infrared and radio carbon stars do not entirely overlap with the population of traditional visual carbon stars, and the inter-relationships between these three classes of carbon stars need to be clarified.

II. Color distributions of carbon stars

Many of the visual carbon stars were detected by *IRAS*. Figure 1 shows the distribution of 369 carbon stars in GCCCS and 17 southern carbon stars discovered by MacConnell (1988). Most of these objects have strong 60 μm excess, which was noted by Thronson et al. (1987). In contrast, radio carbon stars are found to cluster around the blackbody line (Zuckerman and Dyck 1986). Also plotted in Fig. 1 are 111 infrared carbon stars which are found to have the 11.2 μm SiC feature (Chan and Kwok 1988). Fig. 1 suggests that while visual, infrared, and radio carbon stars are all manifestation of carbon richness, they nevertheless occupy distinct parts of the color-color diagram.

III. Visual carbon stars as transition objects

The observed 60 μm excess and the detection of silicate feature in some visual carbon stars have led to the theory of Willems and deJong (1988) which proposes that visual carbon stars represent a transitionary evolutionary phase after an interruption of mass loss during the Mira variable phase. As the oxygen-rich circumstellar shell disperses into the interstellar medium, the carbon-rich photosphere will be observed as a carbon star while emission from the remnant envelope contributes

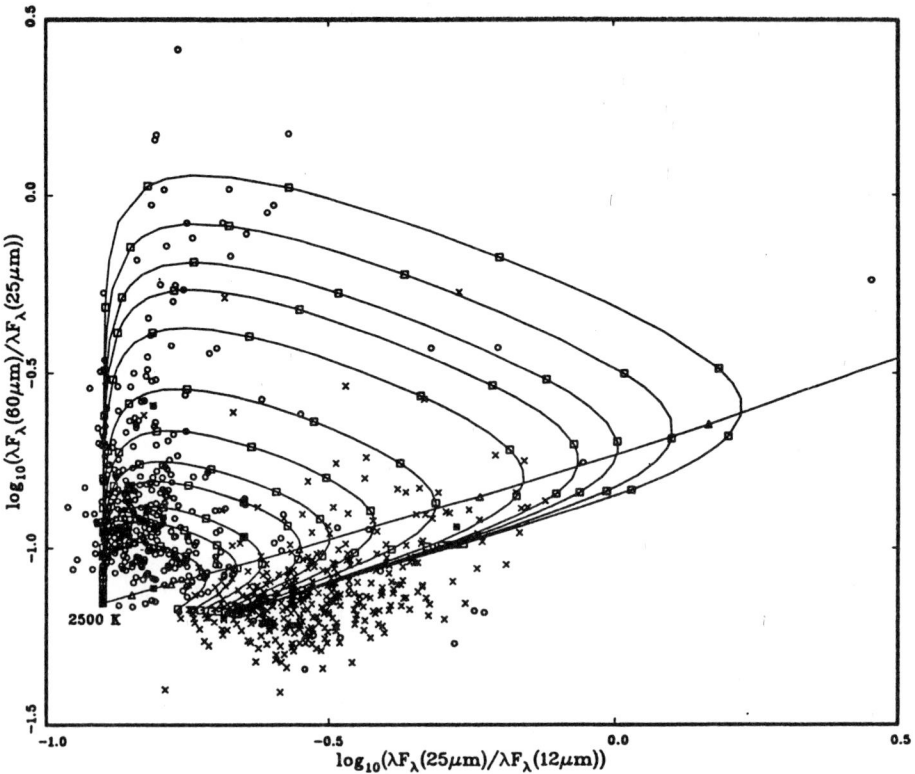

Figure 2. The evolutionary tracks for carbon stars on the color-color diagram. Also plotted are the visual carbon stars (circles) and M stars (crosses) showing the silicate feature in emission. The model tracks represent initial mass loss rates of 10^{-7} (inner most) to 10^{-5} $M_\odot$ yr^{-1} (outer most).

to the excess in far-infrared wavelengths. We have repeated the model calculations of Willems and deJong, while at the same time taking into account the effects of the silicate grain opacity function and temperature and density gradients in the circumstellar envelope. Model spectra were calculated for 12 initial mass loss rates (from 10^{-7} to 10^{-5} $M_\odot$ yr^{-1}) at 35 different epochs after the termination of mass loss. This resulted in a total of 435 spectra which were then convolved with the *IRAS* instrumental profiles to simulate the photometry measurements. Figure 2 shows the model evolutionary tracks on the color-color diagram. We can see that the tracks start in the area of the color-color diagram populated by M stars and describe loops of various sizes which pass through most of the visual carbon stars. After $\sim 10^3$ yr, the tracks begin to turn downward and after $\sim 5 \times 10^4$ yr, the shell has effectively totally dispersed and the stellar color resembles that of the photosphere.

Many visual carbon stars have ground-based photometry which can be used for detailed comparison with the models. A total of 123 stars were fitted by Chan and Kwok (1988). Figure 3 shows an example of such fits. The initial mass loss rate and the time since shell detachment can be determined from the fitting process. The excellent agreement between the models and observations gives us confidence in the correctness of the detached shell model.

As new SiC grains are formed, the optical depth of the circumstellar envelope will increase again and the 11.2 μm feature will become more prominent. Figure 4 shows the spectrum of the infrared carbon star T Dra (LRSC 45) fitted by a radiative transfer model using a r^{-2} density distribution. We can see that the dust continuum emission is already important at $\lambda \sim 5$ μm. We suggest that T Dra represents an intermediate object between visual carbon stars and radio carbon stars like CW Leo where the color temperature is only ~ 500 K. The very high mass loss rates of stars like CW Leo will quickly deplete the hydrogen envelope, leading to the formation of a carbon-rich planetary nebula.

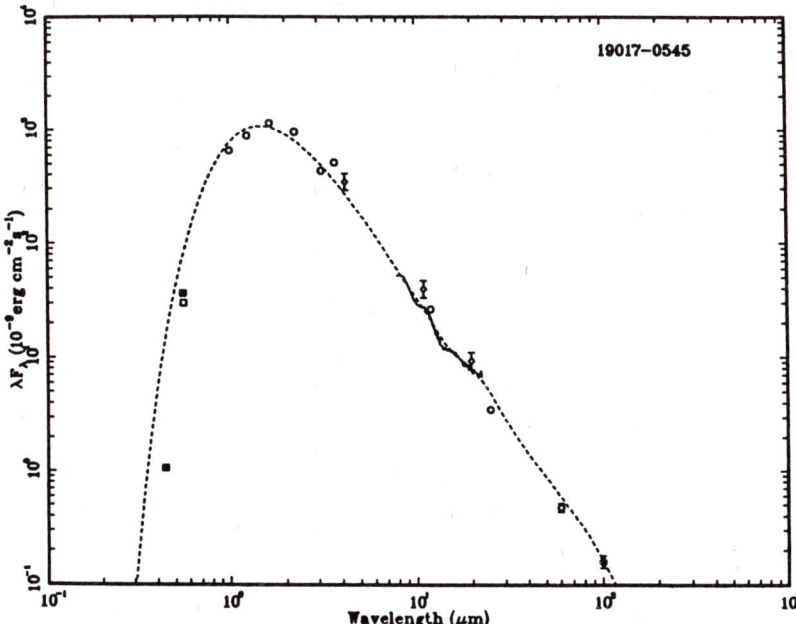

Figure 3. Model spectral fit to V Aql. The derived angular size of the inner radius is 0.7 arc sec and optical depth at 9.7 μm is 3.6(-3).

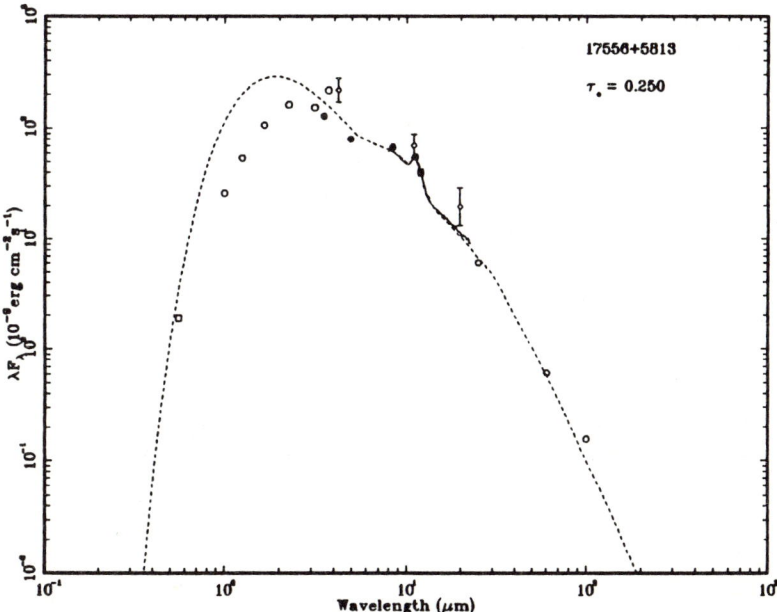

Figure 4. The spectrum of T Dra fitted by a radiative transfer model. The optical depth at 11.2 μm is 0.25.

IV. S stars

S stars show C/O abundance in between that of M and C stars and have often been suggested as transition objects between M and C stars. A cross-reference check of the General Catalogue of S Stars by Stephenson (1976) and the *IRAS* Point Source Catalog suggests that S stars belong to both LRS classes 21-29 (oxygen-rich) and 41-49 (carbon rich). S stars also do not possess unique colors but instead overlap extensively with both M and C stars in the color-color diagram. While S stars can be isolated through their photospheric spectra, we find no common infrared property which characterizes them as a separate group.

V. Radio observations of visual carbon stars

Recent CO observations have led to the detection of many visual carbon stars (Olofsson, Eriksson, and Gustafsson 1987, 1988). These detections confirm that visual carbon stars process extensive circumstellar envelopes as suggested by their 60 μm excess. Most interestingly, the carbon star S Sct was found to have a double-peaked CO profile, which is interpreted by Olofsson *et al* (1988) as due to a detached shell. They estimate that the outer radius of the shell to be > 140 arc sec. In comparison, model fitting of the infrared spectrum of S Sct by Chan and Kwok (1988) estimate an inner radius of 30 arc sec. An expansion velocity of 20 km s^{-1} implies that mass loss in S Sct terminated ~7000 (D/kpc) yr ago. The observation of the double-peaked profile therefore provides strong support that visual carbon stars are in between two mass-losing phases of evolution.

VI. Evolution of oxygen-rich stars on the AGB

While the above scenario may account for the evolution of carbon stars on the AGB, the observations of M stars with very deep silicate absorption features suggest that the optical depth of the circumstellar envelope must be very high ($\tau \sim 50$ at 9.7 μm). The corresponding mass loss rates have to be $> 10^{-5}$ M$_\odot$ yr^{-1}. With such high mass loss rates, the lifetime of such objects must be very short, probably $< 10^5$ yr. It is therefore unlikely that these objects will evolve to, or have evolved from, carbon stars. A reasonable hypothesis is that there is a branching on the AGB, where some stars evolve into carbon stars as described in §III, while others remain oxygen-rich.

The fact that OH/IR stars occupy a well-defined band on the color-color diagram was first noted by Olnon *et al.* (1984). Figure 5 shows a plot of the silicate emission (LRSC 25-29) and absorption (LRSC 31-39) objects on the color-color diagram. The difference in colors of these two groups of objects suggests that stars change from emission to absorption objects as the mass loss rate increases while they ascend the AGB. The band on the color-color diagram can therefore be interpreted as an evolutionary sequence (Bedijn 1987, Volk and Kwok 1988).

Since the remnant of the dust circumstellar envelope created during the AGB should still be present during the planetary nebula phase, planetary nebulae should be far infrared emitters (Kwok 1980). Approximately 1000 planetary nebulae were in fact detected by *IRAS* (Pottasch *et al.* 1984). The color distribution of 126 planetary nebulae is also shown in Fig. 5. These planetary nebulae are selected based on their high radio brightness temperature (T_b):

$$T_b(5GHz) = (c^2/2\pi k\nu^2)(F_\nu/\theta^2) \qquad (1)$$

where F_ν is the 5 GHz flux and θ is the angular radius of the nebula. T_b is a convenient parameter to use not only because it can be readily determined by observations, but also because it is distance independent. As the surface brightness of planetary nebulae is expected to decrease as a result of expansion, T_b is therefore also an age indicator. The observed range of T_b in planetary nebulae is between 0.1 K to 10,000 K, and the 126 nebulae selected here probably represent the younger members of planetary nebulae. We find that the infrared colors of these high surface brightness objects to be less affected by atomic line emission than more-evolved planetary nebulae. There are also indications in Figure 5 that the nebulae with $T_b > 1000$ K have higher color temperatures than the nebulae with T_b between 100 and 1000 K.

Also shown in Fig. 5 are evolutionary tracks for two stars with initial masses 1.5 and 8 M$_\odot$ on the color-color diagram based on the model calculation of Volk and Kwok (1988). The break points (point A) of the tracks represent the tip of the AGB, where mass loss is assumed to terminate. The ends of the tracks (point B) represent approximately 300 and 1300 yr after the end of AGB for the 1.5 and 8 M$_\odot$ cases respectively. Between points A and B, the circumstellar envelope is dispersing into the interstellar medium and physically detached from the star. These tracks successfully simulate the change in color of AGB stars, and also seem to be extending to the area of the color-color diagram

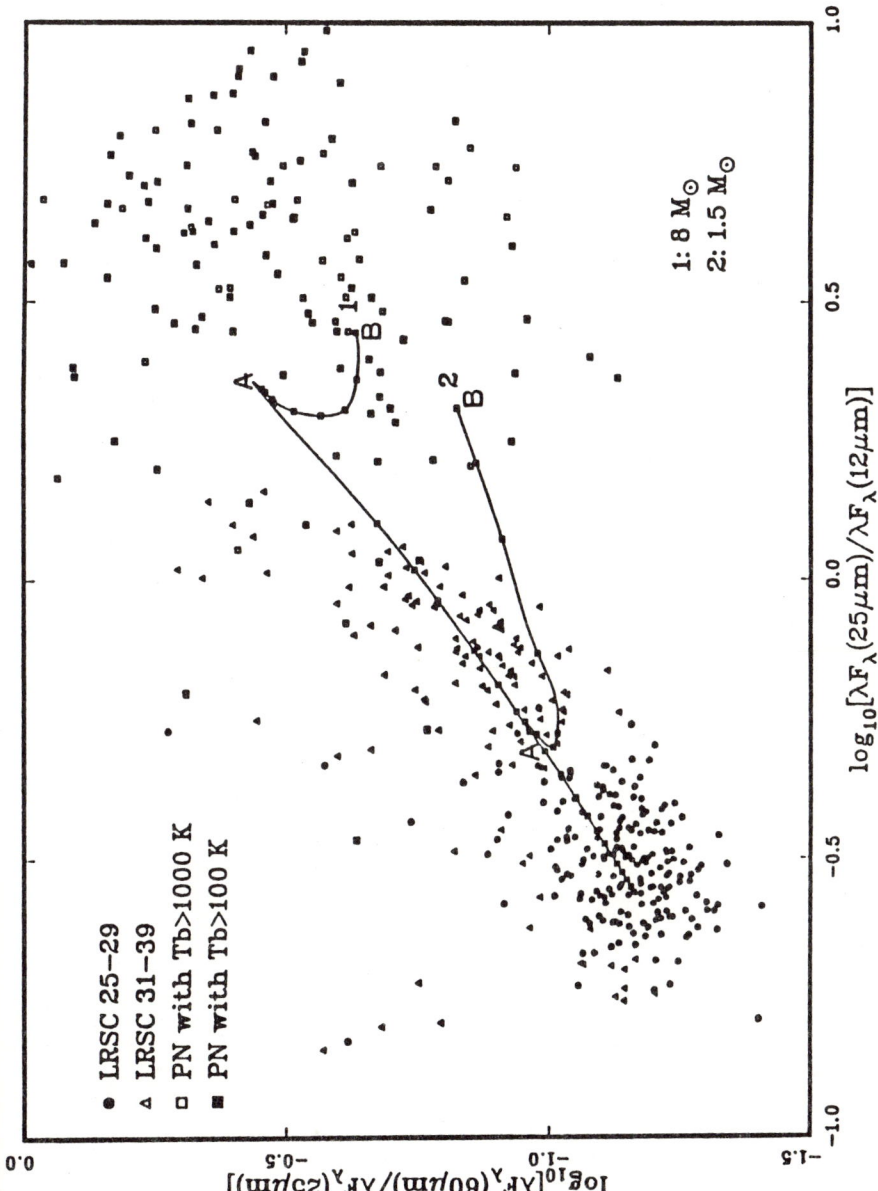

Figure 5. The evolutionary tracks for stars of masses 8 and 1.5 $M_\odot$ on the color-color diagram. The tracks originate from the area populated by AGB stars showing silicate emission features (circles), move upward through the area occupied by silicate absorption objects (triangles), and after the termination of mass loss (point A), shift to the right toward the area occupied by planetary nebulae (squares).

occupied by planetary nebulae.

VII. Proto-planetary nebulae

The separation of AGB stars and planetary nebulae in the color-color diagram suggests that transition objects between these two phases of evolutionary phases can be discovered by observing *IRAS* objects which occupy the area in the color-color diagram between these two groups. A number of objects showing double-peak energy distributions have been found (Parthasarathy and Pottasch 1987; Hrivnak, Kwok, and Volk 1988). Such energy distributions can be explained by the dispersal of the circumstellar envelope and the emergence of light from the central star as the optical depth of the envelope declines.

VIII. Conclusions

The concept of AGB stars evolving from M to S to C has been entrenched since the early days of spectral classifications. Recent photometric and spectroscopic observations from the *IRAS* satellite suggest that all AGB stars start as mass-losing M stars; some of which remain oxygen-rich while others become carbon rich. One possible parameter which may determine what branch a star will take is the inital mass. For a low mass star, the hydrogen envelope is small and the entire envelope may be lost before any dredge up occurs. For high mass stars, thermal pulse does not begin until the stellar luminosity is very high ($\sim 3 \times 10^4$ $L_\odot$ for a 6 $M_\odot$ star, Iben 1981). At such high luminosities, the mass loss rates will be very high and the hydrogen envelope may be totally depleted before enough carbon is dredged up to the surface. A summary of the suggested evolutionary scenario is given in Table 2.

Table 2

M-type Mira variables
mass loss rate $\sim 10^{-7}$-10^{-6} $M_\odot$ yr^{-1}

silicate feature in emission
LRSC 21-29

Intermediate mass stars:	High mass stars:	Low mass stars:
Carbon dredge up, [C/O]>1 oxygen atoms locked in CO, formation of silicate grains ceases.	Silicate feature in absorption LRSC 31-39 objects.	H-envelope depleted by mass loss before any dredge up occurs. Shell detaches, formation of oxygen-rich planetary nebulae.
Mass loss stops, circumstellar shell detaches. Excess at 60 μm, observed as visual carbon stars.	Mass loss rate increases to >10^{-3} $M_\odot$ yr^{-1} - OII/IR stars.	
Formation of carbon based grains (e.g. SiC), mass loss resumes - infrared carbon stars.	Depletion of H-envelope, shell detaches - proto-planetary nebulae (e.g. 18095+2704).	
Mass loss rate increases ($\dot{M} \sim 10^{-5}$ $M_\odot$ yr^{-1}) - radio carbon stars (e.g. CW Leo).	Central star ionizes circumstellar shell, formation of oxygen-rich planetary nebulae.	
Depletion of H-envelope, shell detaches, formation of carbon-rich planetary nebulae.		

This work is supported by the Natural Sciences and Engineering Research Council of Canada.

References

Chan, S.J., and Kwok, S. 1988, *Astrophys. J.*, in press.
Bedijn, P.J. 1987, *Astron. Astrophys.*, **186**, 136.
Castellani, V., Chieffi, A., Pulone, L., and Tornambe, A. 1985, *Astrophys. J.*, **296**, 204.
Hrivnak, B.J., Kwok, S., and Volk, K.M. 1988, *Astrophys. J.*, in press.
Iben, I. 1975, *Astrophys. J.*, **196**, 549.
_____1981, *Astrophys. J.*, **246**, 278.
Kwok, S. 1980, *Astrophys. J.*, **236**, 592.
Lattanzio, J.C. 1986, *Astrophys. J.*, **311**, 708.
MacConnell, D.J. 1988, *Astron. J.*, **96**, 354.
Olnon, F.M., Baud, B., Habing, H.J., deJong, T., Harris, S., and Pottasch, S.R. 1984, *Astrophys. J. (Letters)*, **278**, L41.
Olofsson, H., Eriksson, K., and Gustafsson, B. 1987, *Astron. Astrophys.*, **183**, L13.
_____ 1988, *Astron. Astrophys.*, **196**, L1.
Parthasarathy, M., and Pottasch, S.R. 1987, *Astron. Astrophys.*, **154**, L16.
Stephenson, C.B. 1983, *A General Catalogue of Cool Carbon Stars*, Publ. Warner and Swasey Obs., 1, No. 4 (GCCCS).
_____ 1976, *A General Catalogue of S Stars*, Publ. Warner and Swasey Obs., 2, No. 2.
Pottasch, S.R. *et al*. 1984, *Astron. Astrophys.*, **138**, 10.
van der Veen, W.E.C.J., and Habing, H. 1988, *Astron. Astrophys.*, **194**, 125.
Volk, K., and Kwok, S. 1988, *Astrophys. J.*, in press.
Willems, F., and deJong 1988, *Astron. Astrophys.*, **196**, 173.
Zuckerman, B., and Dyck, M. 1986, *Astrophys. J.*, **311**, 345.

IRAS LRS Spectral Class and Light Curve of M & S Miras

M.S. Vardya
Tata Institute of Fundamental Research
Homi Bhabha Road, Bombay 400 005, India

A large sample of 177 Miras, comprising 164 M and 13 S stars, has been examined to determine the dependence of 9.7 μm silicate emission, as revealed by their IRAS LRS Spectral class, on the visual light curve asymmetry factor, f. It is found that the silicate feature occurs not only in M (Vardya et al. 1986; Onaka & de Jong 1987) but in S Miras also only for $f \leq 0.45$. This, however, is only a necessary condition, as about one fifth of Miras with $f \leq 0.45$ do not show the 9.7 μm emission. This non-detection shows dependence on other parameters like the mean visual light amplitude. Non-detection is highest in the region $0.43 < f \leq 0.45$, as well as when mean amplitude is $\leq 5^m.0$. Though strong emission features in M Miras may occur for any value of f, very weak features are absent for small values of f, and the strongest feature tends to appear for large values of f. Infrared excess tends to increase with increase in the strength of the silicate emission and with decrease in the value of f.

Detection of silicate emission, viewed from the visual light curve classes (Ludendorff, 1928) is very high for α_1, α_2, and α_3 classes, decreases for α_4 and γ_1, and is negligible for β class. The strength of the silicate emission is highest for the α_1 class, decreases for (α_2, α_3, α_4)- classes, and is the lowest for the γ_1 class.

Coming to the S Miras, it is surprising that not a single S star shows a strong silicate feature, when even a C Mira, RV Cen, shows it. This may reflect a gradual change from M to S phase. This may also be due to silicate emission peak being somewhat shifted redward from 9.7 μm.

The above results can be understood qualitatively. However, a quantitative treatment of pulsation, shock waves, and condensation chemistry is essential for proper application.

References:

Lundendorff, H. 1928, Handbuch der Astrophysik, Springer Verlag: Berlin, vol. 6, chap. 2, p. 49.

Onaka, T. and de Jong, T. 1987, Late Stages of Stellar Evolution, eds. S. Kwok and S.R. Pottasch, D. Reidel, Dordrecht, p. 97.

Vardya, M.S., de Jong, T., and Willems, F.J. 1986, Astrophys. J. 304, L29.

Pulsational Periodicities in R CrB

J.D. Fernie
David Dunlap Observatory
University of Toronto

R CrB has been placed on the program of the Automatic Photometric Telescope Service (Genet et al. Pub. A.S.P. 99, 660, 1987) for continual monitoring while near maximum light UBV data are obtained on almost every clear night during the star's season and will be published in batches every few years.

The data for the seasons of 1985, 86, and 87 have been Fourier analyzed by Deeming's method and reveal only one pulsational period. The ephemeris JD 2446243.6 + 43.83 E fits all five epochs of maximum V light seen during these years with an rms deviation of 1.1 days, which is the typical uncertainty of an epoch. Moreover, these three seasons included one deep light minimum and one moderately deep minimum; the fact that the pulsational phase was preserved through these minima adds to the evidence tht the deep minima are probably due to a veiling effect and superficial to the star itself. However, while only one pulsational period seems present, there is strong amplitude modulation that gives the lightcurve a ragged appearance. A similar result has been found for GU Sgr by Lawson et al. (IBVS, No. 3178, 1988.) The physical cause of this amplitude modulation has not been found.

Photometry of R CrB in 1972/73 by Tempesti and De Santis (Mem. Soc. Ast. Italiana, 46, 443, 1975), however, reveal the presence of more than one period at that time. There is strong evidence for periods of 44 and 27 days, which might be interpreted as the fundamental and first harmonic modes of radial pulsation, and less strong evidence for a third period near 57 days, which is less easily interpreted. In any case, an attempt to synthesize the lightcurve from these three periods was unsuccessful, suggesting that once again strong amplitude modulation was present.

IRAS Infrared Fluxes of RV Tauri Stars

William P. Bidelman
Warner & Swasey Observatory, Case Western Reserve University
Cleveland, Ohio 44106

Data resulting from the IRAS survey relating to the stars of the RV Tauri class have been examined. Of a total of 78 objects definitely assigned to this class in the 4th edition of the GCVS, 33 are to be found in the Point-Source Catalogue, and for six of these spectral data are available in the low-resolution Spectral Atlas (to 22.6 µm) published by the IRAS Science Team (A & A Suppl. 65, 607, 1986). In the latter the pronounced difference in infrared emission between the carbon-rich star AC Her and most other RV Tauri stars, first noted by Gehrz in 1972, is striking, taking the form of a very marked continuation of that star's emission to at least 22 µm.

Further, this distinction is also nicely shown by the data of the Point-Source Catalogue, since for 29 of the 33 objects contained therein the 12-µm flux exceeds that at 25 µm, while for the four stars AD Aql, RU Cen, AC Her, and EP Lyr -- all carbon-rich -- the reverse is true. The star DY Ori, an uncertain member of the class, shares this characteristic, as do several other fairly bright mid-type high-luminosity variables: AI CMi, V1027 Cyg, AX Sgr, and V925 Sco.

Why only the carbon-rich RV Tauri stars exhibit this unusual infrared emission -- which, however, is also found in a great many even more exotic objects -- is a good question. The so-called "PAHs" seem to be prime suspects.

Another Look at the RV Tauri Period-Luminosity Relation

Glenn M. Wahlgren
Astronomy Programs, Computer Sciences Corporation
Space Telescope Science Institute

The Period-Luminosity (P-L) relation for the RV Tauri variables (DuPuy 1973) contradicts those of other variables in the instability strip as it predicts lower luminosity at longer pulsational periods. It is based primarily upon three globular cluster variables. DuPuy determined secular and statistical parallaxes from 23 field variables, defining a simple P-L relation with a slightly negative slope.

The importance of the P-L relation goes beyond the ability to predict luminosity. Takeuti and Petersen (1983) addressed the nature of the light curves as a resonance between the fundamental and first overtone modes. Their linear, adiabatic, radial pulsation models for the longer periods required an unreasonably small mass of $0.1M_{\odot}$ as a direct result of the P-L relation. Masses of the cluster RV Tauri stars are believed to be near $0.6M_{\odot}$ (Gingold 1976) from stellar evolution models.

As a test of the P-L relation's applicability to the field variables, Mv was determined for 19 RV Tauri stars using the results from synthetic spectrum fitting (Wahlgren 1986). From 2.5 Å resolution spectra, the fitting procedure determined the effective temperature, gravity, metallicity, and turbulent velocity. The stellar mass was assumed to be solar. For twelve variables Mv was found to be between -1 and -3. All but three variables had absolute magnitudes fainter than predicted by the P-L relation. Those variables brighter than predicted displayed hydrogen-line emission, signifying a possibly peculiar spectrum. Decreasing the mass would act to increase the value of Mv and the discrepancy with the P-L relation. The results do not support the P-L relation of DuPuy for the field RV Tauri variables. No P-L relation is evident, and it is suggested that Mv = -2 to -3 be used when a general knowledge of RV Tauri luminosity is required.

Coude spectra have been obtained for the Ca II H & K and Hα lines in RV Tauri and SRd variables. The data will be analyzed to determine whether luminosity can be estimated from the Wilson-Bappu effect and Hα profiles.

References

Dupuy, D.L. 1973, Ap. J. 185, 597.
Gingold, R.A. 1976, Ap. J. 204, 116.
Takeuti, M. and Petersen, J.O. 1983, Astron. Astrophys. 117, 352.
Wahlgren, G.M. 1986, Ph.D. Thesis, The Ohio State University.

Spectroscopic and Polarimetric Observations of AC Her and UU Her

M.A. Nook   J.A. Cardelli   K.H. Nordsieck
University of Wisconsin - Madison

We present preliminary results obtained between March and August of 1987 of a two-year spectropolarimetric study of 10 RV Tauri stars. The observations were made at the University of Wisconsin's Pine Bluff Observatory using the 36 inch cassegrain telescope with a Lyot polarimeter. The spectral resolution of the instrument was 6 A and covered the range between 4350 A and 7500 A. The polarimetry covered the same wavelength range but is binned in four broad bands at 4828 A, 5361 A, 6025 A, and 6877 A.

AC Her shows strong phase dependent variations in the hydrogen lines and metal lines. Period-to-period variations in the line strengths are interesting in that the light and color curves for AC Her are remarkably stable. Hα is seen in absorption during one period at phases 0.67 and 0.80. Mantegazza (1983) reports that AC Her is never observed with Hα in absorption, but it appears that Hα is partially filled in at all times, and this emission is generally attributed to residual emission from post-shock gas at most if not all phases (Cardelli and Howell, 1988). Polarimetrically, AC Her shows both phase dependent and wavelength dependent variations. The strongest wavelength dependence is seen following primary minimum, with a similar, though less strong, wavelength dependence near secondary minimum. The polarization near minima shows sharp increases towards the blue, suggesting the possible formation of very small dust grains and/or molecules within a few stellar radii of the surface.

The light curve for UU Her is much more erratic than that of AC Her and makes phase dependent discussions impossible. The best that can be done is to make comparisons based on the derivative of the light curve (i.e. increasing, decreasing, maximum, or minimum light) at the particular time of observation. Variations of the hydrogen and various metal lines are evident.

Time dependent polarization shows a clear increase during rising light, as is seen for AC Her, U Mon and R Sct. Wavelength dependence of the polarization was not observed for UU Her. The variations of the polarization in AC Her show a strong connection with the progapation of shocks through an extended atmosphere or circumstellar dust layer following minima, and a similar connection may exist for UU Her.

This work is supported by contract NAS5-26777 for the Wisconsin Ultraviolet Photo-Polarimeter Experiment.

Cardelli, J.A., Howell, S.B., 1988, submitted to Astron. J.
Mantegazza, L., 1983, Astron. Astrophys. Supp. Ser., 54, 379.

SHOCK-INDUCED BEHAVIOR OF ATOMIC SPECIES IN LPV ATMOSPHERES

J. N. Pierce (Mankato State University)

A dynamic model of the LPV atmosphere which calculates the time-dependent variations of density, temperature, velocity, and composition with radius has been used to predict changes in absorption line profiles with phase. General results based on density variations are shown in figure 1. Similar graphs, based on concentrations of individual species, are presented for six metals: Fe, Si, Mg, Ca, Cr, and Ti. Those species which remain essentially proportional to the density (e.g., most neutral metals) exhibit the same type of variation. Ionized metals show more individual variations, including some doubling at certain phases, caused by enhanced ionization in the regions traversed by the upper shock. The basic variation predicted by this model has been observed for CO lines in Chi Cygni. (Hinkle, Hall, and Ridgway 1982, Ap. J. **252**, 697)

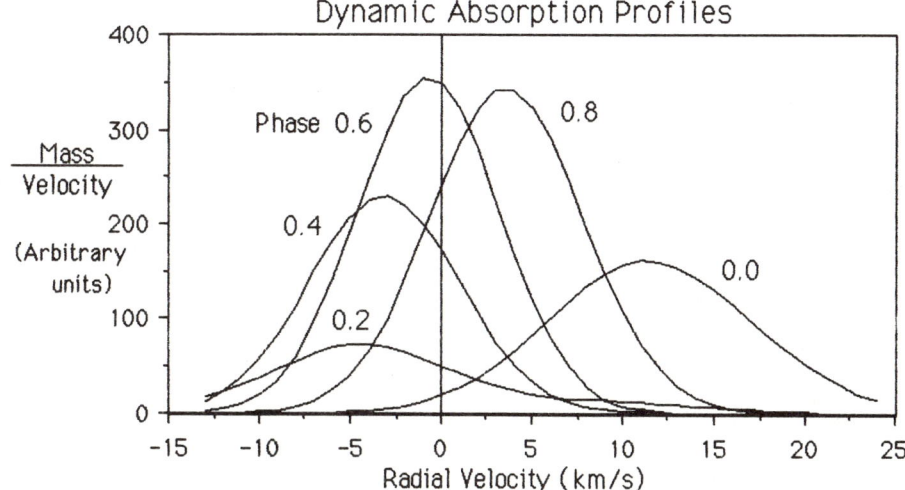

Figure 1. The ordinate is the column density (above the photosphere), binned by radial velocity and integrated over the photospheric disk. The vertical line marks the radial velocity of the star. The profiles are produced both by Doppler broadening and by the line-of-sight velocities induced by the passage of shock waves through the atmosphere. The redshifted profile at phase 0.0 (maximum light) originates in the high-density infalling gas above the lower shock. By phase 0.2, when the shock has become transparent, the original profile has faded and a blueshifted profile has arisen, formed by the high-density gas rising behind the lower shock. During the rest of the cycle this profile strengthens and shifts toward the red as the absorbing region begins to fall back in toward the photosphere.

Investigation of a Sample of Long-Period Variable Stars Possessing Maser Emission

I.L. Andronov   L.S. Kudashkina
Astronomical Observatory
Odessa State University
Odessa 270014, USSR

G.M. Rudnitskij
Sternberg Astronomical Institute
Moscow State University
Moscow V-234, 119899, USSR

An investigation of parameters of a sample of more than two hundred late-type long-period variable stars (LPVs) has been carried out. In more detail were studied 13 giants (Mira Ceti-type variables R Aql, RR Aql, RT Aql, R Leo, U Ori, U Her, R Cas, R Tau, Z Cyg, R Peg, U Aur and semiregular variables RT Vir and RX Boo) and 2 supergiants (S Per, PZ Cas). A considerable fraction of the sample stars (about one third) possess circumstellar maser emission in molecular spectral lines (OH, H_2O, SiO). Our aim was to elucidate the particularities of photometric characteristics of maser stars, such as period P, amplitude A, light curve asymmetry $f = (M-m)/P$, and, in prospect, to determine their status in course of their evolution on the asymptotic giant branch. An extensive comparison with non-maser LPVs was made.

New linear and parabolic light elements for six miras have been derived for the time interval 1957-1984. The study of photometric periods of PZ Cas, RT Vir, and RX Boo indicates the presence of multiperiodicity in these stars. The individual LPVs' light curves were also compared with the curves of variability of the flux density of maser radio emission in the 1.35-cm H_2O line. The correlation is noted between the H_2O-line flux rise and the brightness maximum height for the miras R Aql, R Leo, U Her, and U Ori. The mean cycle of appearance of a "high" optical maximum varies from 800 d to 1600 d for R Leo, U Her and U Ori, and ~600-800 d for R Aql. The "long" H_2O variability cycle has approximately the same length.

Dependences between P, A, f, and Sp have been analysed for the sample stars. The conclusion is confirmed that maser LPVs have smaller f values than non-maser ones; the lower the value of f, the higher the shock-wave velocity in the circumstellar shell. Thus, the presence of a maser is related to the strength of the shocks. the histogram of Sp-f shows that the fraction of maser stars increases gradually from M4 and reaches a maximum by M9, whereas the number of non-masers is maximum at M6 and decreases laterwards. It may be concluded that masers arise at a later evolutionary stage of an LPV.

Line Brightness Variations in Alpha Orionis - Phase Differences
and Radial Velocities

P. Joras
Institute of Theoretical Astrophysics
University of Oslo, P.O. Box 1029
N-0315 Oslo 3, Norway

Observations of the red, bright supergiant star α Orionis (Betelgeuse) during the past decade show significant variations in UV line fluxes and line shifts. Dupree et al. (1987) pointed out that the data from 1984-1987 indicates a periodicity of slightly more than one year in the B magnitude, UV continuum, and Mg II emission line fluxes. Our data shows that this conclusion applies to other strong and medium strong chromospheric lines as well. The variations seem to affect also the relative strength of lines within multiplets.

The UV line shift amplitudes are of the order 10-20 km/s and show the same general variability as the fluxes. Both fluxes and shifts show small phase differences from line to line. Moreover, there is a time lag of approximately 0.3 years in the chromospheric emission line fluxes relative to the photospheric radiation (B magnitude and UV conntinuum). These observations strongly suggest the existence of large scale velocity fields in the stellar atmosphere, possibly in the form of radially outwards propagating disturbances.

Reference

Dupree, A.K., Baliunas, S.L., Guinan, E.F., Hartman, L., Nassiopoulos, G.E. and Sonneborn, G. 1987, Ap. J., 317, L85-L89.

5. Chromospheres, Winds and Mass Loss

CHROMOSPHERES OF CHEMICALLY PECULIAR GIANT STARS

P.G. Judge,
Joint Institute for Laboratory Astrophysics,
University of Colorado, Boulder, Colorado 80309-0440, USA.

Abstract. A review is given of the chromospheres of evolved stars with peculiar chemical abundances, emphasizing the observed dependence of chromospheric properties upon the evolutionary status of the stars. Some old and new physical processes which are potentially important in determining the observed chromospheric features are discussed.

A sample of intermediate mass stars with useful observations in the radio, infrared and ultraviolet wavelength regions is compiled. After a résumé of "normal" M giants on their first or second ascent of the giant branch, attention is focussed on the MS, S and C stars, currently believed to be manifestations of the "third dredge up" phase during double shell burning on the asymptotic giant branch, and differential comparisons are made between the various groups. A systematic study of the "warm" $(T > T_{eff})$ and "cool" $(T \leq T_{eff})$ chromospheric components is made.

Several conclusions have been drawn: (i) for M giants the chromospheric heating has a weak dependence on the fundamental stellar parameters $(T_{eff}, log\ g)$, this is consistent with one heating mechanism (currently believed to be acoustic waves and/or remnant magnetic activity) which can account for the observed structure as a function of evolution, (ii) the heating drops sufficiently for the chromospheres to become progessively less ionized and more dusty as the star evolves, (iii) Carbon stars follow the same gradual trends of chromospheric heating with stellar parameters as the M stars, only when suitable corrections for chromospheric energy losses are applied, (iv) such corrections are inferred from spectral data (v) the few available MS and S stars have similar chromospheric activity (when accretion processes are unimportant) to M stars of similar effective temperature, (vi) the chromospheric spectra of the 3 S stars examined are consistent with photospheric studies suggesting that they are cool analogues of the Barium stars, (vii) it is not possible at present to place additional constraints on the evolutionary status of AGB stars on the basis of chromospheric data, (vii) questions concerning the nature of the temperature, density and velocity structure of MS, S and C stars and their chromospheric heating and mass loss processes must await answers from more sensitive observations, particularly in the UV, for instance with the *Hubble Space Telescope*.

1. INTRODUCTION

Detailed studies of the outer atmospheres of stars have been possible only over the last decade with the advent of data from the radio to ultraviolet (UV) wavelengths from telescopes such as the Very Large Array (VLA) and the International Ultraviolet Explorer satellite (IUE). IUE data have revealed information on the "warm" components (electron temperatures $T_e \geq T_{eff}$) of the outer atmospheres of stars across the HR diagram (Jordan and Linsky, 1987; Dupree and Reimers, 1987), and an overall picture has emerged concerning the nature of stellar outer atmospheres as a function of stellar parameters (e.g. $T_{eff}, Log\ g, v\ sin\ i$) and evolution. However, the majority of these studies focussed attention on stars which have not evolved beyond the K-giant phase of evolution. Other workers have studied very evolved stars using techniques which can detect cool material $(T \leq T_{eff})$ associated with the massive circumstellar outflows around such objects (see e.g. Olofsson, this volume, Mass Loss from Red Giants, Eds. Morris & Zuckerman, 1985, Late Stages of Stellar Evolution, Eds. Kwok & Pottasch, 1987).

Until fairly recently, few workers have studied outer atmospheres of late-M, MS, S, SC and C stars on the Asymptotic Giant Branch (AGB), because of the difficulty in detecting UV emission from these chromospherically "inactive" stars and because such stars do not (yet) possess the massive molecular envelopes associated with later stages of evolution. These stars are at a crucial evolutionary stage: theoretical models predict that essentially all stars with masses in the range $0.6M_\odot < M_* < 5M_\odot$ pass through this evolutionary phase (e.g. Lattanzio, 1988), where they undergo double shell burning and dredge-up with associated changes in surface abundances, substantial photospheric variability and mass loss which potentially can affect subsequent evolution. Helium flashes in the interiors of AGB stars are currently believed to be the primary source of "s-process" elements observed throughout the Galaxy (e.g. Audouze & Tinsley, 1976).

Currently there are two possible scenarios proposed to describe the observed distribution of peculiar red giants: traditional theory (Iben & Renzini, 1983) and analyses of photospheric abundances suggest that at least some stars follow the evolutionary sequence $M \to MS \to S \to SC \to C$ (e.g. Lambert, this volume). However, recent analyses of IRAS photometric and spectral data (Willems & de Jong, 1988) reveals that some S stars may not fit into this picture, and measurements of radial velocities in Barium stars and certain S stars (Jorissen & Mayor, 1988) suggest that Tc− deficient S stars may be "cool" counterparts of the K-giant Barium stars whose s−process abundances are currently believed to originate from mass transfer from a companion at earlier epochs (e.g. McClure, 1983; Smith & Lambert, 1988; Little-Marenin, this volume).

The definition of a "chromosphere" has recently required extension beyond the traditional models of solar–type structure, which are geometrically thin regions extending over several pressure heights with electron temperatures in the range $T_{min} \simeq 0.75 T_{eff} < T_e < 8000K$ (e.g. Linsky, 1980), heated by some non-thermal process such as the damping of mechanical, probably magnetic, waves. Even for stars as warm as the Sun, observations of molecular spectra (particularly CO in the infrared (IR)) imply a large chromospheric surface coverage of cooler regions ($T_e \leq T_{eff}$) (e.g. Heasley et al. 1978; Ayres & Testerman, 1981; Tsuji, 1988), perhaps associated with a runaway cooling effect due to the formation of molecules (Ayres, 1981; Kneer, 1983; Muchmore, 1986; Muchmore, Nuth & Stencel, 1987). Data from the IRAS satellite have revealed cool ($\leq$ a few hundred K) "dusty" regions around cool giants later than ∼M5 III (Stencel, Carpenter & Hagen, 1986), and radio and UV observations show evidence for geometrically extended warm regions ($T_e \leq 10^4 K$) in the coolest giants (Drake & Linsky 1986, Drake 1985). In the present paper I consider regions between the photosphere ($\tau(5000)$Å ~ 1) and the inner circumstellar envelope ($\sim$ several stellar radii), and thus include all of these observations. This picture of a "chromosphere" is a highly complex region in the near vicinity of the star, spanning a wide range of physical conditions (pressures, temperatures, compositions) where a wide variety of important, but as yet poorly understood, physical processes are occuring.

Understanding the chromospheric structure of AGB stars will provide vital constraints on the non-thermal heating, cooling, and deposition of momentum in the outer atmospheres of these important objects, on the mass loss and subsequent stellar evolution. Morris (1087) remarks, in the context of mass loss mechanisms: "The cause of the matter ejection is, however, understood only in broad outline because the near-stellar arena from which the matter is expelled is complex and usually cannot be resolved spatially". The chromosphere (as defined above) is the region where most of the non-thermal energy generated in the sub-photospheric convection zone is deposited in heating the circumstellar material and lifting it from the gravitational field of the star. Until we can adequately describe this region we cannot say that we understand the mass loss processes from cool, evolved stars.

Basic questions concerning the chromospheres of stars evolving up the AGB are presently unanswered– some are addressed below: how do chromospheric properties (densities, temperatures, dust/gas ratio, velocity fields, inhomogeneities, geometry) vary with the evolutionary changes in luminosity, effective temperature, chemical composition, photospheric variability? What can be inferred concerning mechanisms and physical processes responsible for the observed structures? What effect does binarity have on chromospheres of AGB and other chemically peculiar evolved stars? Recent reviews of relevant work are given by de la Reza (1987) and Johnson (1987).

The paper is organized as follows: Section 2 describes the stellar sample adopted. In Section 3 the chromospheres of the "non-variable" oxygen rich M giants are reviewed. In Section 4 the MS and

S stars are discussed, and Section 5 deals with the C stars. Section 6 discusses physical processes which may account for some of the observed properties, and Section 7 describes the stellar data in an evolutionary context.

2. STELLAR SAMPLE AND CHROMOSPHERIC DATA

Since typical M supergiant stars probably have masses in excess of the theoretical upper limit for helium shell flashes (e.g. Iben & Renzini, 1983), and since galactic N-type carbon stars have masses typically $\sim 1.4 M_\odot$ (Dean, 1976), I focus on M stars of luminosity class III to obtain a sample of AGB stars from early-M type through the carbon star phase which may represent an evolutionary sequence. The 42 stars examined are listed in Table 1, together with IRAS data (taken directly from the point-source catalogue and the LRS atlas). IUE and VLA data are listed in Table 2, together with sources listed individually.

The sample was chosen on the basis of stars having useful measurements from IUE, the VLA or both. H-α and the Ca II H and K lines and IR triplet can also provide useful constraints (Hagen, Stencel & Dickinson, 1983) These are not discussed in detail here since interpretation of these lines is less straightforward and there are observational problems with carbon stars (e.g. Johnson & Luttermoser 1987). Just 4 MS/S stars exist (all binaries) which satisfy these criteria, together with 13 carbon stars (3 R stars, 10 N stars). Note the observational bias of decreasing V-band flux (and hence distance, interstellar absorption, etc.) with more advanced phases of evolution. The discussion is therefore restricted to nearby stars and to long wavelength spectra obtained with IUE.

Photometric data are from the SIMBAD compilation of stellar data of the Centre du Données Stellaires, except where indicated. Angular diameters are from the Barnes-Evans V-R colour calibration (Barnes, Evans & Moffett, 1978) except other determinations are available. The UV data consist of observed ($F_\oplus$) and surface fluxes (F_*) ($erg\ cm^{-2}\ s^{-1}$) of Mg II (h+k), surface fluxes of the C II] $\lambda 2325$ Å multiplet and Al II $\lambda 2670$ Å, and the electron density N_e (cm^{-3}) derived from high-dispersion studies of the ratios within the C II] multiplet following early work by Stencel et al. (1981), modified using updated atomic data (Lennon et al., 1985)

Figure 1 identifies the stellar sample on the IRAS two–colour diagram of van der Veen and Habing (1988), revealing stars which have no detectable circumstellar (CS) dust (region I), "warm" dust associated with O-rich CS envelopes (regions II-IIIa), and warm (region VII) and cool (region VIa) C-rich CS envelopes. Only 6 stars have detectable silicate or silicon carbide (SiC) CS features in the LRS spectra (Table 1).

Figure 2 shows the fractional luminosity in the Mg II, C II] and Al II] chromospheric lines as a function of photometric colours and of the effective temperature (T_{eff}). Note the gradual decline in the fractional luminosity for the non-dusty or "clean" K and M stars, and the sharper decline for the dusty M and C stars.

3. THE "NON-VARIABLE" M GIANTS.

Strictly "non-variable" M giants do not exist (e.g. Querci, 1987); here I discuss M giants which are not regularly pulsating and whose variablity does not exceed a few tenths of a magnitude. Generally the chromospheric UV emission fluxes discussed below are relatively less variable than the underlying photosphere (Oznovich & Gibson, 1987), but sufficient data do not exist to discuss the variability of the radio and IR diagnostics.

"Warm" chromospheric components ($T_e > T_{eff}$)

"Warm" chromospheric regions around cool, low-gravity oxygen rich stars differ substantially from solar–like chromospheres (e.g. Judge 1988): Instead of an essentially hydrostatic chromosphere ($T_{eff} \leq T_e \leq 10^4 K$) overlaid by hotter material at transition region temperatures ($\sim 10^5 K$) and a hot ($\sim 10^6 K$) corona, the outer atmospheres of single K and M giants have a quite different structure in which no material hotter than $\sim 2 \times 10^4 K$ has been detected. Lines such as Mg II h and k therefore

Table 1
Basic data and IRAS fluxes for the program stars.

Name	spectrum	log g	T_{eff}	V	R	K	F_{12}	R_1	R_2	LRS	Notes
α Boo	K1 III	1.7	4250.0	−0.1	−1.0	−3.0	792.8	−1.73	−2.03		1
β UMi	K4 IIIBa			2.1	1.0	−1.2	160.3	−1.55	−2.08		
α Tau	K5 III SB	1.3	3850.0	0.9	−0.4	−2.8	699.4	−1.65	−1.95		2
β And	M0 III	1.6	3800.0	2.0	0.8	−1.8	286.4	−1.55	−2.03		4
μ UMa	M0 III SB	1.35	3850.0	3.0	1.8	−0.9	100.9	−1.50	−2.03	n/a	3
α Cet	M1.5 IIIa			2.5	1.2	−1.6	234.7	−1.55	−2.03		
β Peg	M2 II-III	1.2	3600.0	2.4	0.9	−2.2	387.3	−1.50	−2.00		4
π Aur	M3 III			4.3	2.6	−0.9	107.8	−1.47	−1.92		
σ Lib	M3 IIIa			3.3	1.7	−1.4	200.7	−1.68	−1.88		
μ Gem	M3 IIIab	1.0	3600.0	2.9	1.3	−1.9	304.5	−1.55	−2.08		7,8
γ Cru	M3.4 III			1.6	0.0		865.0	−1.47	−1.98		
δ^2 Lyr	M4 II		3385.0	4.3	2.5	−1.2	155.8	−1.38	−1.88		5
ρ Per	M4 II-III	0.8	3500.0	3.4	1.6	−1.9	308.6	−1.47	−1.92		6
β Gru	M5 III	0.3	3500.0	2.1	0.2		941.5	−1.50	−1.92	n/a	2
L^2 Pup	M5 IIIe			5.1	2.4	−1.8	2415.1	−1.15	−2.40	n/a	
W Cyg	M5 IIIe			5.4	3.0	−1.3	349.1	−0.97	−2.05	weak Si/C	
2 Cen	M5 III			4.2	2.1	−1.6	255.3	−1.60	−1.92		
R Lyr	M5 III			4.0	2.0	−2.1	370.8	−1.45	−1.85	n/a	
W Hya	M5 IIIe			7.0		−3.1	4200.0	−1.35	−2.03		
α Her A	M5 II	0.0	3220.0	3.5		−5.0	1515.1	−1.38	−1.77		5
g Her	M6 III	0.2	3250.0	5.0	2.5	−2.0	437.6	−1.17	−2.00		4
X Her	M6 III			5.7		−1.4	484.4	−0.75	−1.98	strong Si	
RZ Ari	M6 III	0.5	3295.0	5.9	3.5	−1.0	147.2	−1.47	−1.98		9
θ Aps	M7 III			5.5			733.8	−0.83	−2.00	Si	
R Dor	M8 IIIe			5.5	1.8	−3.8	5548.7	−1.25	−2.13	n/a	
4 o^1 Ori	M3⁻ S III SB	0.8	3450.0	4.7	3.5	−0.5	83.6	−1.47	−1.70		4,a
HR 363	S3+/2- SB	1.0	3600.0	6.4	5.1	1.6	11.6	−1.45	−1.68		4,a
HD 35155	S3,2 SB		3660.0	6.8	5.4	2.1	8.0	−1.50	−1.60		10
HR 1105	S3.5/2 SB	0.9	3550.0	5.1	3.5	0.2	41.0	−1.45	−2.10		6
CIT6	C4,3					1.1	3319.5	−1.10	−1.63	C	
S Cen	C4,5=R5			7.6	6.0		8.9	−1.42	−1.25	n/a	
UV Cam	C5,3=R8			7.9			11.1	−1.45	−1.63		
UU Aur	C5,3		2825.0	5.3	3.3	−0.7	232.1	−1.27	−1.47	C	7
TX Psc	C6,2		3030.0	5.0	3.2	−0.6	162.9	−1.53	−1.32	n/a	7
BL Ori	C6,2		2960.0	6.2	4.4	0.7	44.5	−1.25	−1.38		7
R Scl	C6,II		2550.0	5.8	3.7	0.0	162.1	−0.75	−0.45		7
TW Hor	C7,2			5.7	3.8	0.0	101.2	1.00	1.85		
NP Pup	C7,2			6.3	4.6	0.9	34.0	−1.05	−1.25		
U Hya	C7,3		2825.0	4.9	3.0	−0.7	205.5	−1.13	−1.58	n/a	7
T Ind	C7,3			6.0			47.2	−1.38	−1.15		
IRC+40540	C8,3.5					2.5	959.8	−0.78	−1.55		
IRC+10216	C9,5					1.3	47525.2	−0.78	−1.53	C	

Notes: Photometric data from SIMBAD except (a) from Eggen (1972). F_{12} is the IRAS 12μ flux, R_1 and R_2 the ratios $2.5 log(F_{25}/F_{12})$ and $2.5 log(F_{60}/F_{25})$ respectively. "Si" indicates that a silicate feature is present in the LRS spectrum, "C" a SiC feature. log g and T_{eff} are from the following sources: 1. See Judge (1986a) 2. See Judge (1986b) 3. Luck & Lambert (1982) 4. Smith & Lambert (1985). 5. Tsuji (1986) 6. Smith & Lambert (1986). 7. Lambert et al. (1986) 8. See Johnson & Luttermoser (1987) 9. Tsuji (1981) 10. See Johnson & Ake (1984).

Table 2
IUE fluxes, electron densities and VLA ionization fractions.

Name	$F(Mg\ II)_\oplus$	$F(Mg\ II)_*$	$F(C\ II])_*$	$F(Al\ II])_*$	$\log N_e$	F_{ion}	Notes
α Boo	9.0(−11)	1.7(5)	8.5(3)	2.7(3)	9.7	−0.5	1,2
β UMi	5.3(−11)	6.6(4)				−0.75	3,2
α Tau	1.7(−10)	7.0(4)	4.4(3)	2.2(3)	9.0	≤ −2.0	4,2
β And	4.2(−11)	3.7(4)	2.3(3)		9.0	≤ −1.5	5,2
μ UMa						≤ −0.3	2
α Cet						≤ −1.3	2
β Peg	≥ 5.8(−11)	≥ 3.3(4)			8.8	≤ −2.0	6,5,7,2
π Aur	≥ 1.6(−11)	≥ 5.0(4)				≤ −1.1	2
σ Lib						≤ −1.0	2
μ Gem	4.0(−11)	3.2(4)	3.2(3)			−1.42	8,2
γ Cru	1.6(−10)	1.6(4)	1.3(3)	6.4(2)	9.0		9,10
δ² Lyr	≥ 2.1(−11)	2.9(4)					11
ρ Per	3.4(−11)	2.1(4)	1.3(3)	6.6(2)	8.9	−2.39	12,2
β Gru	1.3(−10)	1.9(4)	1.9(3)	6.0(2)	8.5		4
L₂ Pup	5.1(−13)	9.2(1)		1.2(1)			11
W Cyg	2.4(−12)	1.7(3)					11
2 Cen	2.5(−11)	1.2(4)	1.2(3)	3.6(2)	8.0		12
R Lyr	2.8(−11)	1.3(4)	8.2(2)	4.1(2)	8.9		11
W Hya						≤ −4.1	13
α Her A						−1.95	2
g Her	6.5(−12)	2.0(3)	1.0(2)	8.0(1)	8.0		12
X Her	7.0(−13)						11
RZ Ari	4.1(−12)	3.9(3)	4.9(2)				8
θ Aps	1.6(−12)						8
R Dor	1.0(−12)	1.5(1)		2.4			11
4 o¹ Ori	2.5(−12)	3.6(4)	7.2(3)	≤ 2.3(3)			14
HR 363	1.6(−12)	5.9(4)	1.5(4)	≤ 3.7(3)			10
HD 35155	3.0(−12)	1.6(5)	1.3(4)	≤ 6.4(3)			15
HR 1105	5.0(−12)	2.2(4)	4.4(3)	≤ 1.1(3)			10
CIT6						≤ −3.3	13
S Cen	1.4(−13)	6.5(3)	6.5(3)	≤ 3.3(3)			8
UV Cam	8.4(−14)						8
UU Aur	2.2(−13)	4.4(2)	2.8(2)	8.8(1)			8
TX Psc	2.3(−13)	4.4(2)	2.8(2)	≤ 5.5(1)			8,16
BL Ori	8.0(−14)	5.7(2)	4.5(2)	≤ 2.9(2)			8
R Scl						≤ −3.2	13
TW Hor	2.6(−13)	9.9(2)	6.2(2)	2.0(2)			8,16
NP Pup	2.2(−13)	3.1(3)	1.6(3)	6.2(2)			8,13
U Hya	2.7(−13)	4.9(2)	2.0(2)	9.8(1)			8
T Ind	1.8(−13)						8
IRC+40540						≤ −3.3	13
IRC+10216						≤ −3.3	13

Notes: Fluxes given in erg cm^{-2} s^{-1}, N_e in cm^{-3}, and F_{ion} is the log of the ionization fraction from the VLA observations. References are from: 1. Judge (1986a) 2. VLA data from Drake & Linsky (1986) 3. Oranje (1986) 4. Judge (1986b) 5. Brown & Carpenter (1984) and Carpenter, Brown & Stencel (1985) 6. Stencel et al. (1980) 7. Byrne et al. (1988) 8. Johnson & Luttermoser (1987) 9. Carpenter, Pesce, Stencel, Brown, Johansson & Wing (1988, preprint) 10. IUE Archive 11. Stencel, Carpenter & Hagen (1986) 12. Eaton & Johnson (1988) 13. VLA data from Drake, Linsky & Elitzur (1987) 14. Ake & Johnson (1988) 15. Johnson & Ake (1984) 16. Variable: mean data taken

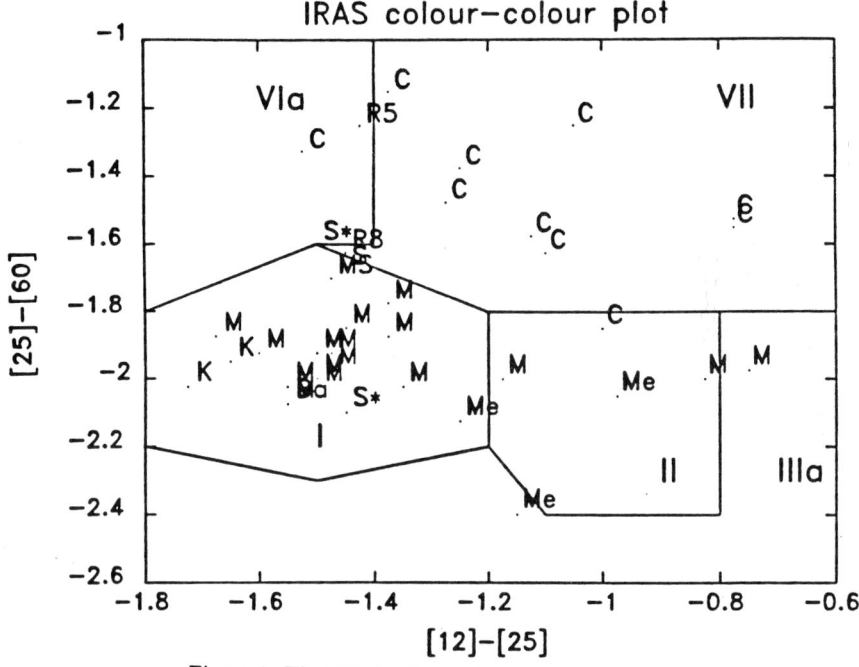

Figure 1. The IRAS colour-colour diagram of the stellar sample.

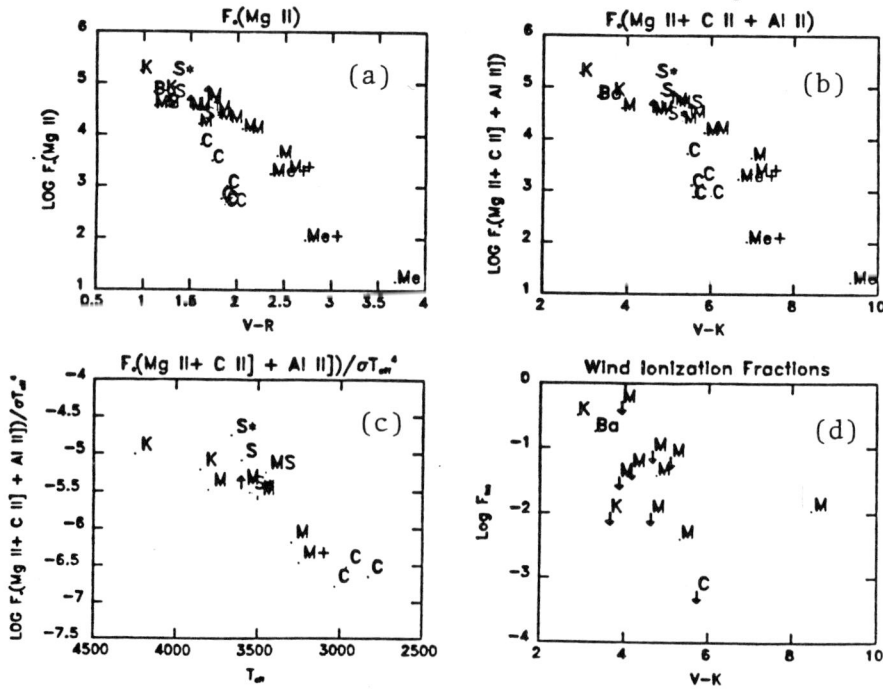

Figure 2. The behaviour of the observed chromospheric losses with photometric colours and T_{eff}. Binaries with evidence for accretion are marked with an asterisk, "dusty" M stars ($[25] - [12] > -1.2$) with a plus. All the C stars are "dusty" (see text).

remain optically thick relatively higher in the atmosphere where they show redward asymmetries indicative of substantial flows, which have been interpreted in terms of radially outflowing winds (Stencel and Mullan, 1980a,b; Drake and Linsky 1983; Drake 1985). Late-M giants show, in addition to the asymmetries, essentially permanent sharp circumstellar (CS) blue-shifted absorption features formed in the circumstellar envelope several stellar radii from the photosphere (e.g. Reimers, 1981).

In an evolutionary context the outer atmosphere becomes relatively "inactive" as a star spins down during its ascent of the giant branch through spectral types G-K III and it exhibits a "basal" level of activity perhaps related to purely non-magnetic waves (Schrijver, 1987) and /or a minimal level of dynamo activity associated with the reduced rotation rate (Rutten & Pylyser, 1988). Figure 2 shows quantitatively how the chromospheric heating falls off with later spectral type for M stars, analogous to a similar sample of stars studied by Steiman-Cameron, Johnson & Honeycutt (1985). This extends smoothly the earlier relation derived on the basis of G,K and warmer stars (e.g. Linsky & Ayres, 1978).

The chromospheric regions with $T_e \leq 2 \times 10^4$ become relatively more extended (in terms of $\Delta H/R_*$) in cool low gravity stars than in the dwarfs (Drake & Linsky, 1985; Judge 1986a,b), because (i) for a given chromospheric T_e and turbulent velocity field, the hydrostatic gas pressure scale height is inversely proportional to the stellar gravity g_* ($\Delta H_{scale} = P_g/g_*\rho$), i.e. proportional to the stellar radius *squared*; (ii) hydrodynamic "turbulence", as inferred by the observed widths of optically thin lines (e.g. Hartmann & Avrett 1984; Judge, 1986a,b), supports the chromospheric gas and momentum deposition drives a cool stellar wind, although the mechanisms remain poorly understood (e.g. Holzer, 1987). Judge (1988, in preparation) demonstrates that, even for α Ori (M2 Iab), which has a relatively massive stellar wind ($dM/dt \sim 10^{-6} M_\odot yr^{-1}$), hydrostatic equilibrium, when "turbulent" velocities are accounted for, dominates the momentum balance in the regions where UV photons are *created* owing to the $(density)^2$ dependence of the emissivity. The asymmetries in e.g. Mg II k are produced by scattering of photons at significantly lower column masses where the wind is well under way and where densities are so low that the contribution to the emissivity (or source function) is negligible.

Characteristics of "warm" chromospheric material are represented by IUE and VLA data. Several crucial types of data from observed emission lines are represented in Table 2. (i) For M stars Mg II integrated line fluxes represent a substantial fraction of the total non-thermal energy input above the photosphere (Linsky & Ayres, 1978, estimated the fraction to be $\sim 1/3$ from warmer stars, but Carpenter (1988) emphasizes that in the coolest M stars Fe II UV complexes contribute a comparable fraction to the total cooling). (ii) The electron density N_e yields a measure of the "mean" electron density in the "warm" chromosphere (proportional to the heating- see below). (iii) The ratios of line fluxes of C II] and Al II] with Mg II h & k yield constraints on the relative carbon abundance and on the effects of multiple scattering in the Mg II lines (the formation of Al II] and Mg II lines effectively differ only in their optical depths- the former are always optically thin in the chromosphere). The latter will be examined below in the context of the destruction of Mg II photons by absorption in "dusty" chromospheres, and depletion onto grains which may be formed at chromospheric heights.

The variation of the electron density N_e with $log\ g$ is plotted in Figure 3(a), extending the sample of Byrne et al. (1988) to later spectral types. Judge (1987) argues that these (and other) global chromospheric properties can be understood in terms of Ayres' (1979) explanation of the Wilson-Bappu effect, since these warm UV-line emitting regions are (i) close to turbulently-supported hydrostatic equilibrium (hence the behaviour with gravity g_*), and (ii) are heated by the same mechanism, Schrijver's (1987) "basal" level of chromospheric heating. Also plotted (Figure 3(b)) is $log\ N_e$ vs. $0.5[log\ g.A_{Fe}] + log\ T_{eff}^n$, where g and T_{eff} are stellar gravities and effective temperatures, and A_{Fe} are the metallicities relative to the Sun. According to Ayres' scaling law ($N_e \propto \tilde{F}g^{1/2}A_{Fe}^{1/2}T_{eff}^n$ where $F_*(Mg+..) \propto T_{eff}^{2n+1}$, here $n \simeq 3$ is estimated from the K and M stars of Figure 2), Figure 3(b) should yield ordinates proportional to the heating parameter $\tilde{F}$ ($erg\ g^{-1}\ s^{-1}$). Despite $\sim \pm 0.3 dex$ uncertainties in Figure 3(b), I conclude that the data imply a steady decline in $\tilde{F}$ from α Boo (K1 III) to the M stars. This potentially valuable method of examining the heating

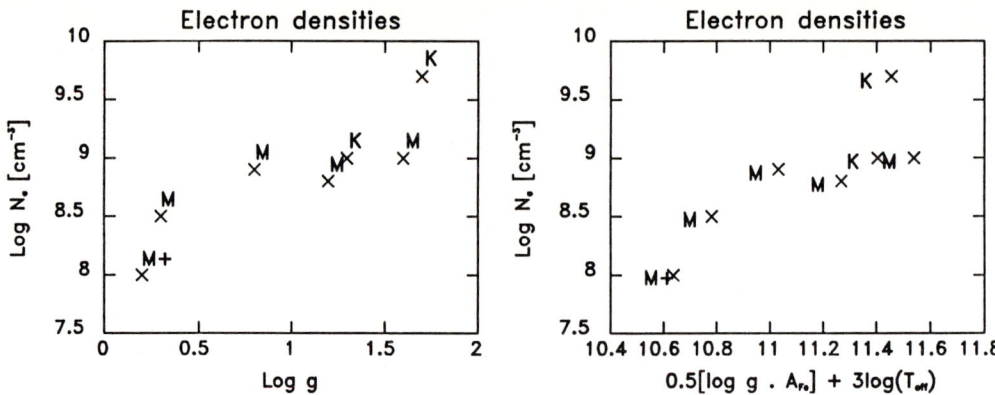

Figure 3. The variation of the observed chromospheric electron densities with (a) $\log g$, and (b) $0.5[\log g . A_{Fe}] + \log T_{eff}^n$, n=3 (see text).

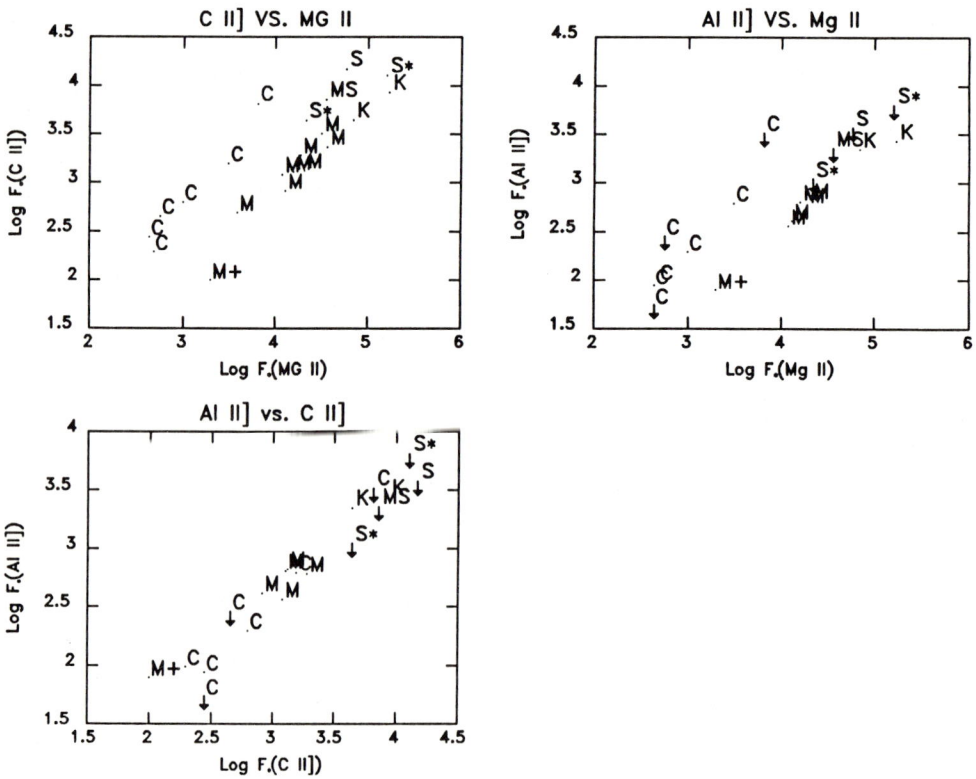

Figure 4. Flux-flux diagrams relating observed chromospheric losses in the lines of Mg II h & k, Al II] $\lambda 2670$ and C II] $\lambda 2325$.

as a function of spectral type, T_{eff} and evolutionary status should be pursued further.

Pioneering studies with the VLA also reveal valuable constraints on the warm chromospheric gas (Drake & Linsky, 1986, Drake, Simon & Linsky, 1987; Drake, Linsky & Elitzur, 1987). Table 2 lists the ionization fraction F_{ion} derived by these authors, which represents a first attempt to parametrize the fractional ionization of the emitting regions. In chromospheric models concerned with UV emission lines the fractional ionization can change by several orders of magnitude over the emitting regions (e.g. Judge, 1987). Nevertheless F_{ion} provides a useful starting point for comparing various stars, and it is clear that the ionization fraction of material within several stellar radii (the radio photospheric angular diameters are typically twice the optical photospheric diameters) declines with later spectral type. In an absolute sense, one problem remains: for some stars listed in the Table, F_{ion} has been derived from the general formula for the *total* mass loss rate of Reimers (e.g. 1981), which may or may not be appropriate for these stars (e.g. Drake, 1986).

"Cool" chromospheric components $(T < T_{eff})$

Three types of observations have been used to identify the presence of "cool" material above the photospheres of cool giants: (i) vibration-rotation IR molecular absorption bands, (ii) rotational molecular emission in the circumstellar envelopes (extending many stellar radii and the subject of Olofsson's review (this volume)), (iii) IR excesses due to circumstellar dust.

Tsuji (1987, 1988) provides evidence for a quasi-static, turbulent ($\xi_t \geq 5km/s$) new component in the outer atmospheres of M giants, from an analysis of equivalent widths, velocity profiles and variability of low-excitation CO lines compared with higher excitation lines. Tsuji argues that this new component could represent a transition zone between the warm chromosphere and cool wind. Although these spectra are certainly formed higher in the atmosphere than the photospheric lines, inspection of Tsuji's results reveal that, as expected by e.g. scaling from the work of Heasley et al. (1978), the absorbing layer has quite high column masses: for ρ Per (M4 II–III), a typical example, Tsuji's (1988) table 3 gives $log\ N_{CO} = 19.78 cm^{-2}$, yielding a column mass of $log\ m \simeq 0.1 g cm^{-2}$. The present author believes that, because of momentum and energy requirements, such large column masses imply that the absorbing regions lie close to the stellar photosphere, in disagreement with Tsuji's speculations, and that the absorbing regions co-exist with the warmer chromospheric regions with similar column masses inferred for the case of ρ Per e.g. from opacity sensitive line ratios of Fe II (Eaton & Johnson, 1988). This picture is consistent with Ayres' (1986) multi-component picture of α Boo (K1 III).

High quality IRAS data are available for $\simeq 5700$ oxygen rich giants and supergiants (Zuckerman, 1987; van der Veen & Habing, 1988). Analyses of data for individual stars have been made by Stencel, Carpenter and Hagen (1986), Stencel, Pesce and Hagen Bauer (1988) and recently by Kenyon, Fernandez-Castro and Stencel (1988) in studies of M–star components of symbiotic systems.

Kenyon et al. (1988) found that the onset of dust formation occurs near spectral type M5 III, as indicated by a [25/12] flux ratio > 0.25 (the Rayleigh- Jeans value for a photosphere near 2-3000 K) from the IRAS 25 and 12μ bands, i.e. essentially at the point where a star begins to pulsate as a long period variable (LPV), confirming earlier findings (e.g. Jura, 1986). From Figures 1 and 2 we find that *the M (and C) stars with dust have substantially reduced chromospheric UV emission* (in units of $F_*/\sigma T_{eff}^4$, but that the warm chromospheric components are not completely "quenched" in the late M giants, as discussed by Stencel et al. (1986). The crucial question of where the dust is formed is delayed until Section 6 where C stars will also be included.

4. MS AND S STARS.

Chromospheric data on MS and S stars are scarce, since the stars themselves are also rather scarce (Smith & Lambert, 1988, cite Stephenson, 1984, who lists 178 such stars with $V \leq 9.9m$). No SC stars have been observed in the UV, to the author's knowledge. These classes of stars include targets which should be examined in the UV with the Hubble Space Telescope, but some interesting results are available from IUE.

MS stars

4 o^1 Ori ($M3^-S\ III$), examined in detail by Ake & Johnson (1988) following earlier work by Peery (1986), is the only MS star for which valuable UV chromospheric data exist. Ake & Johnson argue that, although the SWP continuum spectrum (1200-2000 Å) reveals a white dwarf companion, the emission lines and LWP spectrum show no evidence for binary interactions through mass exchange and accretion. The chromospheric profiles of Mg II h & k are similar to those found for normal M giants, but the C II] $\lambda 2325$ multiplet is a factor of ~ 3 stronger relative to Mg II (and possibly Al II]) than in normal M giants (see Figures 2 and 4). These figures imply that 4 o^1 Ori has a slightly higher level of chromospheric activity (measured by e.g. $F_*(Mg\ II)$ or $F_*(Mg\ II)/\sigma T_{eff}^4$) than M stars of similar T_{eff}. This is probably "normal" chromospheric emission and not heating due to accretion since the flux of C IV $\lambda 1550$ has relatively a much lower upper limit.

Smith & Lambert (1985) have determined CNO, Fe and s-process element abundances from IR spectra, in a differential analysis with α Tau (K5 III). Although the C abundance is enhanced it is only by a factor of 1.5 relative to α Tau (K5 III), insufficient to account for the enhanced C II] flux. Possible interpretations will be discussed in Section 6.

Ake & Johnson's SWP spectrum shows (weakly) the usual lines of C I ($\lambda 1657$) and S I/Si II ($\lambda 1819$) again with normal M-star intensities (compared with spectra of e.g. Stickland & Sanner, 1981). However, the usually dominant O I ($\lambda 1304$) multiplet is clearly lower than in M stars of equivalent T_{eff}. A deep LWP high dispersion spectrum should be obtained which would potentially allow the measurement of the electron density, relative C II] abundance and chromospheric non-thermal line broadening parameters.

The IRAS data reveal that 4 o^1 Ori has a small 60μ excess which may indicate the presence of cool, relatively old dust perhaps from a previous mass-loss episode.

In a recent preprint, Smith & Lambert (1988) confirmed the presence of Tc in the photosphere of 4 o^1 Ori. They argue that MS and S stars with Tc are presently undergoing the thermally pulsing phase of the AGB, whereas the remaining MS and S stars ($\sim 37\%$ of their sample) probably acquired s-process elements from accretion of material from a companion AGB star at an earlier epoch ($\Delta t \geq 1.5 \times 10^6$ yr), which implies that the latter stars are cooler relatives of the Barium stars. Hence it appears that 4 o^1 Ori and other MS and S stars with Tc represent the first stage in the evolution from the M giant phase, following the "third dredge-up" (Smith & Lambert, 1986). Certainly, from the UV data for 4 o^1 Ori there is no evidence for significant mass transfer processes at the present time.

S stars

Tables 1 and 2 contain just three S stars which have been studied with IUE. The absence of Tc in HR 363 and HR 1105 (Smith & Lambert, 1985,1986) suggests that these stars are probably cool analogues of Barium stars, i.e. their s-process elements were accreted from their companions in an earlier period of mass transfer (Little, Little-Marenin & Hagen Bauer 1987; Smith & Lambert 1988). Except for HR 1105 these stars have excesses at 60μ indicative of cool dust perhaps associated with previous mass loss episodes of the companions, consistent with this picture.

HR 363, analysed by Ake, Johnson & Peery (1988), appears to be a non-accreting binary with an SWP spectrum similar to mid/late M giants, although Ake et al. note that their SWP image is flawed by a particle impact over the important Si II/S I blend near $\lambda 1820$. Further IUE observations by Ake examined from the archive reveal "normal" Mg II h & k profiles but, as can be seen from Figures 2 and 4, HR 363 has a relatively "active" chromosphere compared to normal M stars. The C II] lines are substantially enhanced, relative to the Mg II lines.

HR 1105 (Ake et al., 1988) and HD 35155 (Johnson & Ake, 1984; Peery, 1986) clearly have SWP spectra more closely resembling symbiotic stars than normal M giants since they show strong high temperature lines of N V $\lambda 1240$ and C IV $\lambda 1550$ associated with accretion processes and X-ray heating (in the case of N V).

The chromospheric surface fluxes of HD 35155 exceed those of any star in the sample, indicating that the "chromospheric" line fluxes also probably result from accretion onto the white dwarf companion. The Mg II line profiles reveal just red wings in emission indicating a substantial wind from the primary (Johnson & Ake, 1984).

HR 1105, although having evidence for accretion in lines formed near 10^5 K, has chromospheric fluxes similar to normal M stars suggesting that the dominant contribution is from the normal chromosphere rather than accretion. The C II] lines are enhanced by a factor ~ 2 compared with Mg II, relative to the M stars.

5. C STARS.

By comparison with the M (and S) stars, the C stars clearly have very different UV spectra (See Johnson & Luttermoser, 1987 and Figures 2 and 4) and dust properties (Figure 1). C stars also differ fundamentally in their degree of variability of chromospheric emission: studies by Querci & Querci (1985) and Johnson et al. (1986) revealed striking changes in UV emission lines of factors up to 5–10, greater than corresponding changes in the optical and UV continuum, on timescales as small as days (possibly hours). For some time it has been known that neither the H-α nor Ca II H & K traditional chromospheric features are seen in cool C stars (e.g. see the discussion by Johnson & Luttermoser, 1987). This is almost certainly due to low chromospheric densities and temperatures which (because of the NLTE effect of photon escape in Ly-α), yields very low chromospheric opacity in H-α compared with the M stars of higher gravity and effective temperature, and because the higher chromospheric opacities in resonance lines enhance escape in subordinate lines (e.g. the Ca II IR triplet lines) which may not be observed in emission against a strong background continuum. Querci & Querci (1985) have also argued that the regions where the K line is formed are probably cooled by H^-, leading to no observable emission in the K lines.

Warm C stars (R stars) have been studied in the UV by Eaton et al. (1985), cool C stars (N stars) by Johnson & Luttermoser (1987). TX Psc, the brightest C star showing detectable UV emission will be discussed in some detail below.

Dust Properties

The carbon-rich dust environment of our sample of C stars is clearly evident from Figure 1: all the C stars in the sample are "dusty". The 60μ excesses can be interpreted as evidence for cool dusty regions well separated from the star (and hence not embedded in the "chromosphere"), suggesting evolution following an earlier period of mass loss as implied by Willems & de Jong (1988) who included additional evidence from LRS spectra. Note that Zuckerman & Dyck (1986) interpreted the 60μ excesses in *very* dusty C stars (as opposed to M stars) in terms of different emissivities of grains in C- and O-rich environments.

Ultraviolet and Radio Properties

Figures 2 and 4 contain several important clues to the nature of C star chromospheres. For stars with several IUE observations a mean value for the fluxes was adopted. (Johnson et al. (1986) found that the C II] and certain Fe II lines approximately follow the Mg II line variability). The most striking features are the *apparent drop in the UV fluxes of chromospheric lines* relative to the M stars, and the *large relative flux of the C II] intersystem lines* (e.g. Johnson & O'Brien, 1983; Johnson, 1987). Figure 2(b) shows, for the first time, how the fractional energy losses in the measured lines of M and C star chromospheres varies with the effective temperature of the stars. The C II] line fluxes are much stronger than can be accounted for by the relative C abundance in M and C stars ($\simeq 0.4 dex$, Lambert et al. (1986)). Figure 4 reveals other spectroscopic clues to the nature of C star chromospheres: the optically thin Al II] and C II] line fluxes follow a trend consistent with the M stars, but the Mg II lines do not. Below, evidence is presented showing that the probable cause of these peculiar line ratios and fluxes is absorption, perhaps by dust embedded in the chromospheric regions as well as by overlying circumstellar gas.

Upper limit radio free-free data is only available for a handful of C stars (Table 2): these however yield very low values for F_{ion} and the chromospheric emission measures for these stars (Drake, Linsky & Elitzur, 1987) and indicate that the regions within $1-2R_*$ are essentially neutral, unlike early M and K giants.

TX Psc: The best-studied C star.

TX Psc is the only C star (to the author's knowledge) in which chromospheric emission has been detected at high dispersion with IUE: Eriksson et al. (1986) obtained Mg II h & k profiles (Figure 5) and some lines of Fe II. Eriksson et al. derive several important conclusions: (a) the lines are severely affected by CS absorption lines of Mn I and Fe I (possibly by an order of magnitude); (b) the Mg II line wings are marginally blue-shifted perhaps indicating outflow deep in the chromosphere (see Figure 5), but that (c) the interpretation of the line profiles is quite non-unique.

Point (a) is the only process by which the k/h ratio can be very much less than unity, as observed, implying that the observed radiative losses plotted in Figure 2 for TX Psc (and by implication other C stars, given Figure 4) *underestimate the true losses in Mg II by between factors of 2 to an order of magnitude*. Additional absorption in the interstellar medium over the M star sample is also expected since TX Psc, the brightest of the C stars in the UV, is significantly more distant ($d \simeq 300$pc). Eriksson et al. also point out the possibility that extinction by the observed CS grains could potentially lead to underestimates of the chromospheric cooling rates. This is discussed further below.

Recently Luttermoser et al. (1988) used the code PANDORA (Vernazza, Avrett & Loeser, 1973) to construct a one-component chromospheric model for TX Psc, adopting the Mg II h & k profiles, the C II] integrated flux and the overall appearance of the 2000-3200 Å, Mg I 2852 Å and H-α regions as observational constraints. This is the coolest model chromosphere ever made with very low densities, and it provides a needed starting point for future modelling. (A similar model for a late-M giant is urgently needed for comparison). The authors concluded that a strongly-constrained model could be found (within the limitations of their modelling procedure and diagnostics used), to account for the diagnostics. They derive, however, a very curious velocity field, (stationary high in the chromosphere, expanding at 40 km s^{-1} near the temperature minimum region) which may be an artifact of assuming a *smooth* velocity field (see Section 6). In addition, the PANDORA program takes no account of "Doppler drifting" of photons in frequency (Avrett, 1988, private communication) which dominates the transfer problem under conditions of low density and high opacity (Basri, 1980).

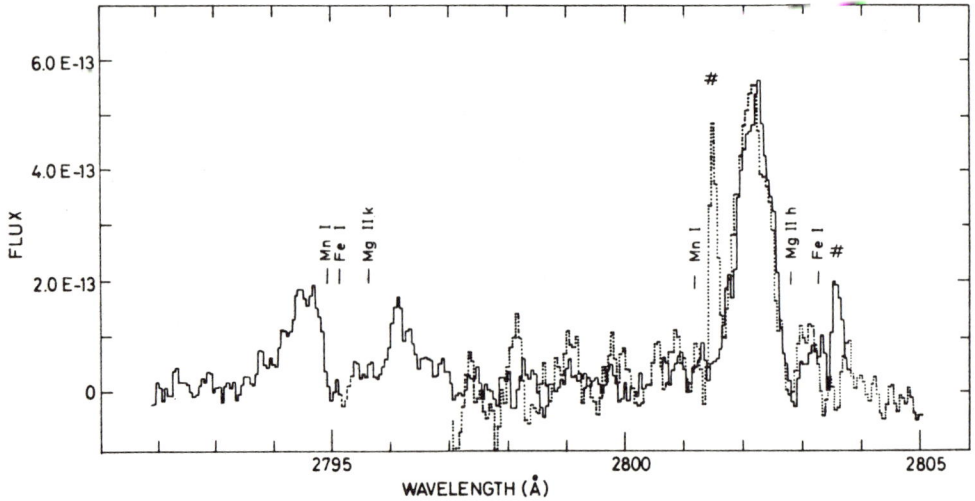

Figure 5. High dispersion spectra of Mg II h and k lines in TX Psc (Eriksson et al., 1986)

6. CHROMOSPHERIC HEATING AND RELATED PHYSICAL PROCESSES.

As noted in previous sections there are several crucial unanswered questions: What is the *actual* level of chromospheric heating in C stars? How can we derive this from the observations? How does this vary as the star evolves from M to C type? What is the underlying cause of the chromospheric variability of C stars? Do the chromospheric observations shed light on the proposed evolutionary sequences? Where is the dust formed in the outer atmospheres of cool giants and does this agree with current ideas of mass loss? Here attempts are made to answer some of these from the available data.

Chromospheric Energy Losses of M and C stars.

There are several problems concerning the use of Mg II (+ Al II]+C II]) fluxes to indicate chromospheric radiative losses, including (a) Absorption of Mg II k photons by overlying CS gas by e.g. Mn I, Fe I; (b) Extinction of UV photons by CS grains; (c) emission in e.g. Fe II complexes which e.g. may progressively increase with lower gravity; (d) Absorption in the ISM; (e) depletion onto grains of Mg (and Al) in O-rich environments (Snow et al. (1987)), (e) chemical fractionation caused by grain expulsion (Stencel, Pesce & MacGregor, this volume).

Here a method is proposed to overcome (a) and (d) by using the Al II] intersystem line flux as an indicator of the Mg II $h + k$ flux: model calculations for low gravity stars (Judge, 1988, in preparation) reveal that, unlike the Sun (Linsky & Ayres, 1978), the Mg II lines are effectively thin in these chromospheres: hence the emergent flux at the top of the chromosphere is proportional to the ion abundance and upward collisional rate for both Al II] and Mg II, integrated over the heated regions. Since Mg II and Al II] have very similar fractional ion abundances (e.g. Judge, 1986a) and excitation energies, their ratio remains approximately constant, over reasonable temperature ranges, in the absence of substantial depletions. This is evident for the dust-free M stars in Figure 4. Thus, from Figure 4, I suggest that the M stars form a calibration curve from which the C star Mg flux can be read off, given the Al II] flux. *This implies an increase in the chromospheric losses of between a factor of 5 and 10 for the C stars*, precisely the amount by which the C stars are lower than the extrapolated radiative losses for the M stars. Thus, in the absence of interstellar and CS extinction, the C stars do not differ greatly from the late-M stars in terms of chromospheric "activity". The Mg lines fluxes are reduced certainly by line absorption (observed in TX Psc) and probably by dust extinction (see below).

The C II] lines, although stronger than Al II], are not well suited because they have potentially different ionization fractions, excitation potentials (the ratio with Mg II is too temperature-dependent), abundances, chromospheric partial pressures and depletions onto grains (the latter two depending on the C/O ratio). The partial pressures of C I and C II will increase *non-linearly* with the C/O ratio such that a substantial enhancement of C I and C II ions may be expected as C/O becomes > 1. This may partially explain the enhancements in the C II fluxes seen in the MS, S and C stars.

Chromospheric cooling in a haze of e.g. Fe II lines, difficult to estimate from IUE spectra, implies that using the Mg II lines (with corrections mentioned above) underestimates the chromospheric heating. However, the relative cooling between M and C stars should still be represented by Mg II unless a systematic change occurs in the major cooling routes owing to different chromospheric conditions. Although fluorescent processes, important for exciting many Fe II multiplets (e.g. Carpenter, 1988), are enhanced relative to collisional ones in lower gravity stars (Judge, 1988), they serve merely to redistribute thermally created photons. Since thermal creation terms are relatively the same for Fe II and Mg II at the same N_e and T_e (with the exception of certain Fe II lines pumped by H Ly-α) and since Fe II does not explicitly remove photons from Mg II, then adopting Mg II lines in a comparative study should yield reliable results, provided the H Ly-α losses are smaller than Mg (which is certainly the case for β Gru (M5 III) - Judge (1986b)).

Strong Fe II emission (e.g. in TW Hor- Querci & Querci, 1985) can potentially result from pumping by H Ly-α photons generated by large shock heating. This idea is consistent both with the factor ~ 10 variability of these lines in TW Hor (Querci & Querci, 1985) and perhaps with the observation that strong Fe II emission persists even in dusty stars in which Ca II K emission is absent (Hagen, Stencel & Dickinson, 1983). Quantitative work is needed on the Fe II spectrum.

Co-existing Chromospheric Dust and Warm Gas?

Using a sample of M giants and supergiants Stencel et al. (1986) examined the relation between 'cool' dust and 'warm' chromospheric emission in the UV, and specifically tested Jennings' (1973) interpretation of earlier data that energy which would normally heat chromospheric gas and emit in e.g. Ca II H and K lines is re-directed into warming dust in stars which show dust emission. They concluded that not *all* of the available energy is so directed, and that significant alteration of radiation loss channels may occur. The present review confirms and extends the study of Stencel et al. to include a larger sample of C stars.

However, arguments presented here suggest that the dusty star chromospheres are heated to a similar degree as the M stars, but that the overlying opacity of CS lines (Eriksson et al., 1985) and possibly chromospheric dust obscures the Mg II lines. I suggest an additional interpretation of the data: the reduction of the observed chromospheric losses in dusty stars may not be due to the cooling instability of molecular formation in the chromospheres (Stencel et al.), but instead due to underestimates of the actual radiative losses from the UV observations (Section 5).

In addition to straightforward extinction of UV photons by e.g. CS dust, preferential destruction of resonance line photons may occur in the case where dust co-exists in regions where the lines are optically thick (Hummer & Kunasz, 1980). Resonance line photons (e.g. Mg II k) travel substantially further in space than optically thin lines (e.g. Al II]), thereby suffering greater extinction. An order of magnitude calculation using the scaling laws of Faurobert & Frisch (1985) reveals that this process, already observed in planetary nebulae (e.g. Barlow, 1983), may be important in the chromospheric conditions considered here, adopting a reasonable dust opacity at 2800 Å of $10^{1\pm1} cm^2 g^{-1}$. *This may be a useful diagnostic of the presence of dust in the regions close to the star where UV photons are created.* Note that the line profile may be altered since emergent photons from different parts of the profile travel different path lengths through the chromospheric regions: calculations extending those of Hummer & Kunasz should be performed. Chromospheric modelling using resonance line profiles may be affected.

Shocks: a possible diagnostic and a *caveat* concerning line profiles

Previous discussions of blue- (or red-) shifted emission components in chromospheric lines e.g. Mg II h & k (e.g. Eriksson et al. (1986), Luttermoser et al., 1988) have considered velocity fields which vary continuously and smoothly, over a mean photon path length, λ_p. With this assumption a blue-shifted emission core always implies outflow with the observed velocity shift (i.e. ~ 40 km s^{-1} for the lower chromosphere of TX Psc).

However, a very different situation occurs if the velocity field changes sharply over a smaller scale than λ_p (I. Hubeny, private communication). In the case of shock fronts accelerating outwards (such shocks surely must exist to provide the heating), emergent scattered line photons can be blue-shifted substantially from line centre, yielding a blue asymmetry in an opposite sense to the "smooth" regime where outward acceleration would yield a redward enhancement (e.g. Stencel & Mullan, 1980a). This is analogous to the Fermi acceleration process for particles crossing a shock front (Neufeld & McKee, 1988).

The rather unphysical *ad-hoc* velocity field of TX Psc derived by Luttermoser et al. (1988) assuming a smooth velocity field suggests instead that, in the regions where the Mg II h and k emission wings are formed, shocks of sufficiently large scale are propagating outwards and accelerating the gas, leading collectively to the observed profile. Indeed, TX Psc could be exhibiting a profile equivalent to those obtained recently by Bookbinder et al. (1988) in Miras. Calculations are planned by I. Hubeny which can correctly treat the complex transfer problem in the regime considered here. Such calculations are of crucial importance not only in stars with global shocks but also in the mesoturbulent chromosphere of the Sun (P. Goutebroze, paper presented at SMM workshop, Washington, July 1988). This may require a re-thinking of current ideas based on line profiles of optically thick lines.

An observational test will be provided by comparing profiles of optically thin and thick lines: the discrepancy in the velocities between Mg II and the less thick Fe II lines observed by Eriksson et al. already suggest that the shock model is more appropriate. Deeper observations should be obtained with IUE.

Heating and Mass Loss Mechanisms.

It is inappropriate here to discuss detailed heating processes, instead I briefly mention some mechanisms in the light of the present conclusions. Contrary to some previous authors, I suggest that for non-Mira M and C stars the heating mechanism varies smoothly with the effective temperature as the star evolves, pointing to a common heating mechanism, probably damped "short period" acoustic waves generated at the top of the convection zone, leading to emission in the UV lines observed (see e.g. Querci & Querci, 1985). These authors also conclude that for stars such as TW Hor (and TX Psc) with substantial variability of UV lines, global chromospheric changes (perhaps larger scale shocks?) associated with aperiodic acoustic and/or slow-mode magnetic waves (perhaps generated by the semi-regular photospheric pulsations), probably account for this time-dependent heating. Solar-like magnetic activity is not likely, but cannot be ruled out entirely.

The substantial mass loss observed from non-regular pulsating stars remains a problem (Holzer, 1987): periodic short period acoustic waves damp too soon to provide the energy to lift gas from the gravitational field of the star, and Alfvèn waves produce excessive wind velocities which are not observed. Recent exploratory work on stochastically-changing short period acoustic waves (Cuntz, 1987) concludes that merging strong shocks can lead to mass loss events. This work should be pursued further, since potentially this can account for many of the observations described above.

7. AN EVOLUTIONARY PICTURE.

Arguments have been presented above to allow important chromospheric parameters (heating, dust content) to be estimated from observations. For warmer stars and less evolved stars chromospheres have provided, via e.g. rotation-activity relations (See Jordan & Linsky's 1987 review), information on the evolutionary status of stars. Are currently available chromospheric conditions of M, MS, S and C stars consistent with current evolutionary pictures and can we learn from these conditions about AGB evolution?

No definite answer to these questions is currently possible. However, Figure 2 and earlier discussions suggest some interesting possibilities:

If the chromospheric losses drop precipitously as a star becomes dusty (i.e. if one adopts the 'raw' Figure 2 to represent the radiative losses) then this suggests strongly that the chromosphere has evolved from the relatively active M star phase to the C phase of evolution, either by spin-down following mass- and angular momentum loss (in the case where remnant magnetic activity is responsible for the M star emission (Rutten & Pylyser, 1988)), or by reduction of the acoustic flux (if short-period acoustic waves are responsible for the heating- Schrijver, 1987).

If the chromospheric losses are, as suggested in Section 6, smoothly varying with effective temperature, then the evolution of a single star up the AGB will barely affect the chromospheric emission. Instead, other processes such as the formation of a CS shell and associated dust and line absorption will simply dominate the observed spectrum. In this case dynamo activity is unlikely to be the source of chromospheric heating since the massive circumstellar shells observed will have removed substantial angular momentum and velocity from the star.

Data on the 3 S stars available for study in the UV regions so far suggest that these particular stars are cooler brethren of the Barium stars in which peculiar surface abundances have been acquired from accretion of a previous AGB star companion. The proposed sequence M–S–C cannot therefore be tested using present observations of chromospheres.

Subsequent evolution of the stars and enrichment of the ISM depends on the mass loss process:

future observations may be able to tell us whether chromospheric dust exists close enough to the stellar surface to produce the observed large mass loss rates with current theories.

FUTURE WORK.

Much of the work described here is of an exploratory nature requiring substantially improved data and analyses. It is hoped that readers may be inspired to obtain data with instruments such as the *Hubble Space Telescope* to examine some of the ideas put forward in this new, exciting field.

Acknowledgments. It is a pleasure to thank Bob Stencel for a great deal of help which has substantially enhanced the quality of this paper, Joe Pesce for providing measurements from the IUE archive at the RDAF, Boulder, and Don Luttermoser for results in advance of publication. This work has been supported by an SERC PDRA Research Assistantship at the University of Oxford and by grants NAG5-985 and NGL 06-003-057 from the National Aeronautics and Space Administration to the University of Colorado.

References.

Ake, T.B. & Johnson, H.R. (1988). Ap. J. <u>327</u>, 214-221.
Ake, T.B., Johnson, H.R. & Peery (1988). Paper presented at *A Decade of UV Astronomy with the IUE Satellite*, Goddard Space Flight Center, Greenbelt Maryland.
Audouze, J. & Tinsley, B.M. (1976). Ann. Rev. Astron. Astrophys. <u>14</u>, 43-79.
Ayres, T.R., (1979). Ap. J. <u>228</u>, 509-520.
Ayres, T.R., (1981). Ap. J. <u>244</u>, 1064-1071.
Ayres, T.R., (1986). Ap. J. <u>308</u>, 246-253.
Ayres, T.R., & Testerman, L. (1981). Ap. J. <u>245</u>, 1124-1140.
Barlow, M.J. (1983). <u>In</u> *Planetary Nebulae*, ed. D.R. Flower, pp. 105-128. Dordrecht, Holland. D Reidel Publ. Co.
Barnes, T.G., Evans, S & Moffett T.J. III (1978). M.N.R.A.S. <u>183</u>, 285-304.
Basri G.S. (1980). Ap. J. <u>242</u>, 1133-1143.
Bookbinder, J.A., Brugel, E.W. & Brown, A. (1988). Submitted to Ap. J. (Letts.)
Brown, A. & Carpenter, K.G. (1984). Ap. J. (Letts). <u>287</u>, L43-L46.
Byrne, P.B., Dufton, P.L., Kingston, A.E., Lennon, A.E. & Murphy H.M. (1988). Astron. Astrophys <u>197</u>, 205-208.
Carpenter, K.G. (1988). <u>In</u>, *Physics of Formation of Fe II lines Outside LTE*, pp. 95., eds. R. Viotti A. Vittone & M. Friedjung. Dordrecht, Holland: D. Reidel Publ. Co.
Carpenter, K.G., Brown, A. & Stencel R.E. (1985). Ap. J. <u>289</u>, 676-680.
Cuntz, M. (1987). Astron. Astrophys. <u>188</u>, L5-L8
Dean, C.A. (1976). Astron. J. <u>81</u>, 364-374.
Drake, S.A. (1985). <u>In</u> *Progress In Stellar Spectral Line Formation Theory*, eds. J.E. Beckman & L Crivellari, pp. 351-359. Dordrecht, Holland. D. Reidel Publ. Co.
Drake, S.A. (1986). <u>In</u> *Cool Stars, Stellar Systems and the Sun* (Proceedings Santa Fe, New Mexico) eds. M. Zeilik & D.M. Gibson, pp. 369-384. Berlin, Heidelberg: Springer-Verlag.
Drake, S.A. & Linsky, J.L. (1983). Ap. J. <u>273</u>, 299-308.
Drake, S.A. & Linsky, J.L. (1986). Astron. J. <u>93</u>, 602-620.
Drake, S.A., Linsky, J.L. & Elitzur, M. (1987). Astron. J. <u>94</u>, 1280-1290.

Drake, S.A., Simon, T. & Linsky, J.L. (1987). Astron. J. 92, 163-167.
Dupree, A. K. & Reimers, D. (1987). In *Exploring the Universe with the IUE Satellite*, ed Y. Kondo et al., pp. 321–353. Dordrecht, Holland: D. Reidel Publ. Co.
Eaton, J.A. & Johnson, H.R. (1988). Ap. J. 325, 355-371.
Eaton, J.A., Johnson, H.R., O'Brien G.T. & Baumert J.H. (1985). Ap. J. 290, 276-283.
Eggen, O.J. (1972). Ap. J. 177, 489-507.
Eriksson, K., Gustaffson, B., Johnson, H.R., Querci, F., Querci, M., Baumert, J.H., Carlsson, M. & Olofsson, H. (1986). Astron. Astrophys. 161, 305-313.
Faurobert, M. & Frisch, H. (1985). Astron. Astrophys. 149, 372-382.
Hagen, W., Stencel, R.E. & Dickinson D.F. (1983). Ap. J. 274, 286-301.
Hartmann, L. & Avrett E.H. (1984). Ap. J. 284, 238-249.
Heasley, J.N, Ridgway, D.F, Carbon, D.F., Milkey, R.W. & Hall, D.B.N (1978). Ap. J. 219, 970-978.
Holzer, T.R. (1987). In *Circumstellar Matter*, eds. I. Appenzeller & C. Jordan, pp. 289-305. Dordrecht, Holland: D. Reidel Publ. Co.
Hummer, D.G. & Kunasz, P.B. (1980). Ap. J. 236, 609-618.
Iben, I. & Renzini, A. (1983). Ann. Rev. Astron. Astrophys. 21, 271-342.
Jennings, M. (1973). Ap. J. 185, 197-208.
Johnson, H.R. (1987). In *Cool Stars, Stellar Systems and the Sun* (Proceedings Boulder, Colorado), eds. J.L. Linsky & R.E. Stencel, pp. 399-408. Berlin, Heidelberg: Springer-Verlag.
Johnson, H.R. & Ake T.B. (1984). In *Cool Stars, Stellar Systems and the Sun* (Proceedings Cambridge, Massachusetts), eds. S.L. Baliunas & L. Hartmann, pp. 362-364. Berlin, Heidelberg: Springer-Verlag.
Johnson, H.R., Baumert, J.H., Querci, F. & Querci M. (1986). Ap. J. 311, 960-968.
Johnson, H.R. & Luttermoser, D. G. (1987). Ap. J. 314, 329-340.
Johnson, H.R. & O'Brien, G.T. (1983). Ap. J. 265, 952-960.
Jordan, C. & Linsky, J.L. (1987). In *Exploring the Universe with the IUE Satellite*, ed Y. Kondo et al., pp. 259-293. Dordrecht, Holland: D. Reidel Publ. Co.
Jorissen, A. & Mayor, M. (1988). Astron. Astrophys. 198, 187-199
Judge, P.G. (1986a). M.N.R.A.S. 221, 119-153.
Judge, P.G. (1986b). M.N.R.A.S. 223, 239-268.
Judge, P.G. (1987). In *Cool Stars, Stellar Systems and the Sun* (Proceedings Boulder, Colorado), eds. J.L. Linsky & R.E. Stencel, pp. 294-308. Berlin, Heidelberg: Springer-Verlag.
Judge, P.G. (1988). M.N.R.A.S. 231, 419-444.
Jura, M. (1986). Irish Astron. J. 17, 322-330.
Kenyon, S.J., Fernandez-Castro, T. & Stencel, R.E. (1988). Astron. J. 95, 1817-1827.
Kneer F., (1983). Astron. Astrophys. 128, 311-317.
Kwok, S. & Pottasch, S.R. (eds.) (1986). *Late Stages of Stellar Evolution*, Dordrecht, Holland: D. Reidel Publ. Co.
Lattanzio, J.C. (1988). Preprint, to appear in the *Proceedings of the ACS Symposium on the Origin and Distribution of the Elements*.
Lambert, D.L., Gustafsson, B., Eriksson, K. & Hinkle, K.H. (1986). Ap. J. Suppl. Ser. 62, 373-425.
Lennon, D.J., Dufton, P.L., Hibbert A. & Kingston A.E. (1985). Ap. J. 294, 200-206.
Linsky, J.L. (1980). Ann. Rev. Astron. Astrophys. 18, 439-488.
Linsky, J.L. & Ayres, T.R. (1978). Ap. J. 220, 619-628.
Little, S.J., Little–Marenin, I.R. & Hagen Bauer, W. (1987). Astron. J. 94, 981-995.
Luck, R.E. & Lambert D.L. (1982). Ap. J. 256, 189-205.
Luttermoser, D.G., Johnson, H.R., Avrett, E.H. & Loeser R. (1988). Preprint.
McClure, R.D. (1983). Ap. J. 268, 264-273.
Morris, M. & Zuckerman, B. (eds.) (1985). *Mass Loss from Red Giants*. Dordrecht, Holland: D. Reidel Publ. Co.
Morris, M. (1987). P.A.S.P. 99, 1115-1122.
Muchmore D. (1986). Astron. Astrophys. 155, 172-174.

Muchmore, D., Nuth J. A. III & Stencel R.E. (1987). Ap. J. (Letts.) 315, L141-L146
Neufeld, D.A. & McKee, C.F. (1988). Preprint, to appear in Ap. J. (Letts).
Oranje, B.J. (1986). Astron. Astrophys. 154,185-196.
Oznovich, I. Gibson D.M. (1987). Ap. J. 319, 383-391.
Peery, B.F. (1986). In *New Insights in Astrophysics: 8 Years of UV Astrononmy with IUE*, ed. Rolf E., pp. 117-120. ESA SP-263.
Querci, F.R (1987). In *The M-type Stars*, eds. Johnson H.R. & Querci F., pp. 1-112. NASA SP–492.
Querci, M. & Querci, F. (1985). Astron. Astrophys. 147, 121-126.
Reimers, D. (1981). In *Physical Processes in Red Giants*, eds. I. Iben & A. Renzini, pp. 269-284 Dordrecht, Holland: D. Reidel Publ. Co.
de la Reza, R. (1987). In *The M-type Stars*, eds. Johnson H.R. & Querci F., pp. 373-408. NASA SP–492.
Rutten, R.G.M. & Pylyser, E. (1988). Astron. Astrophys. 191, 227-236.
Schrijver, C. J. (1987). Astron. Astrophys. 172, 111-123.
Smith, V.V. Lambert, D.L. (1985). Ap. J. 294, 326-328.
Smith, V.V. Lambert, D.L. (1986). Ap. J. 311, 843-863.
Smith, V.V. Lambert, D.L. (1988). Preprint.
Snow, T.P, Buss, R.H., Gilra, D.P. & Swings, J.P. (1987). Ap. J. 321, 921-936.
Steiman-Cameron, T.Y., Johnson, H.R. & Honeycutt, R.K. (1985). Ap. J. (Letts). 291, L51-L54.
Stencel, R.E., Carpenter, K.G. & Hagen, W. (1986). Ap. J. 308, 859-867.
Stencel, R.E., Linsky, J.L., Brown, A., Jordan, C., Carpenter, K.G., Wing, R.F. & Czyzak S. (1981) M.N.R.A.S. 196, 47P-53P.
Stencel, R.E. & Mullan, D.J. (1980a). Ap. J. 238, 221-228.
Stencel, R.E. & Mullan, D.J. (1980b). Ap. J. 240, 718.
Stencel, R.E., Mullan, D.J., Linsky, J.L., Basri, G.S. & Worden, S.P. (1980). Ap. J. Suppl. Ser. 4 383-402.
Stencel, R.E., Pesce, J.E & Hagen Bauer, W. (1988). Astron. J. 95, 141-151.
Stickland, D.J. Sanner, F. (1981). M.N.R.A.S. 197, 791-798.
Tsuji, T. (1981). Astron. Astrophys. 99, 48.
Tsuji, T. (1986). Astron. Astrophys. 156, 8-21.
Tsuji, T. (1987). In *Circumstellar Matter*, eds. I. Appenzeller & C. Jordan, pp. 377-378. Dordrecht Holland: D. Reidel Publ. Co.
Tsuji, T. (1988). Astron. Astrophys. 197, 185-199.
van der Veen, W.E.C.J. & Habing, H.J. (1988). Astron. Astrophys. 194, 125-134.
Vernazza, J.E., Avrett, E.H. & Loeser, R. (1973). Ap. J. 184, 605-631.
Willems, F.J. & de Jong, T. (1988). Astron. Astrophys. 196, 173-184.
Zuckerman, B. (1987). In *Cool Stars, Stellar Systems and the Sun* (Proceedings Boulder, Colorado eds. J.L. Linsky & R.E. Stencel, pp. 351-360. Berlin, Heidelberg: Springer-Verlag.
Zuckerman, B. & Dyck, H.M. (1986). Ap. J. 311, 345-359.

MOLECULAR RADIO LINE OBSERVATIONS OF MASS LOSS FROM RED GIANTS

H. Olofsson
Onsala Space Observatory, S–43900 Onsala, Sweden

Abstract. The number of molecules detected at radio wavelengths in envelopes around red giants stands presently at 36. Among these OH and CO have proven to be the most useful for the study of the physical characteristics of a circumstellar envelope. The mass loss rate of the central star can be relatively accurately estimated and it appears possible to trace its evolution with time. Also fascinating objects in transition from the red giant phase to the planetary nebula phase are becoming observationally accessible.

1. INTRODUCTION

Our knowledge of the asymptotic giant branch (AGB) and planetary nebulae (PNe) phases of late stellar evolution, as well as the transition stage, has increased substantially in recent years. The completion of new millimetre wave instruments and the data base provided by IRAS have made molecular radio lines an important, and in some cases unique, tool in this research. It is possible to follow the evolution of the "slow wind" mass loss as the star ascends the AGB, as well as its termination during the proto–PNe phase when the mass loss appears to change its character. The consequences for the chemical evolution of the Galaxy and the supernova rate are gradually becoming apparent. The circumstellar gas/dust envelopes (CSEs) formed by the mass loss are also the sites of a complex chemistry, whose effects beyond the red giant phase have yet to be established.

We will in this review concentrate on what has and can be learnt in this area through the use of OH and CO radio line observations. The large number of recent papers has made it impossible to properly reference all works, in particular the historical ones. We therefore in most cases mention only the most recent papers and presuppose that they contain proper references to previous works. A recent extensive review of radio and infrared observations of CSEs was given by Olofsson (1988).

2. MOLECULAR RADIO LINE OBSERVATIONS

Both the number of objects detected in circumstellar molecular emission and the number of circumstellar molecular species have increased substantially during the last few years. This has been possible due to the availability of the IRAS data base and the completion of new sensitive millimetre wave telescopes.

2.1 Circumstellar molecules

Table 1 summarizes the present situation for circumstellar molecules detected at radio wavelengths. The total number now amounts to 36 (two of the identifications are tentative). If we include those detectable only at infrared wavelengths (H_2, CH_4, C_2H_2, C_2H_4, SiH_4), the total number of circumstellar molecular species stands presently at 41. It is evident that the C–rich CSEs are apparently most productive, but the situation is improving also for the O–rich CSEs. However, it should be noted that for the C–rich, as well as for the O–rich, CSEs the scene is dominated by a single object, the extreme nearby carbon star IRC+10216 and the remarkable bipolar nebula OH231.8+4.2, respectively.

Table 1 Molecules Detected at Radio Wavelengths in Circumstellar Envelopes

Molecule	Number of Sources	
	C–rich	O–rich
CO	≈125	≈100
CN	3	
CS	6	3
SiO("thermal")	4	≈30
SiS	4	1
SiC_2	4	
SO		7
SO_2		16
H_2S		15
OCS		1
HCO^+	3	2
HCN	≈50	13
CH_3CN	3	
HNC	5	2
NH_3	2	1
HC_3N	6	
HC_5N	4	
HC_7N	2	
HC_9N	1	
$HC_{11}N$	1	
C_2H	4	
C_3H	2	
C_3H_2	3	
C_4H	4	
C_5H	1	
C_6H	1	
C_3N	5	
C_2S	1	
C_3S	1	
$HSiC_2$ (HSC_2) (?)	1	
NaCl	1	
AlCl	1	
KCl	1	
AlF (?)	1	
OH (maser)		>1000
H_2O (maser)	4	≈200
SiO (maser)		≈140

Armed with this wealth of molecular probes it is possible to study various effects of mass loss from evolved stars. From a statistical point of view it appears that the OH 1612 MHz [OH(1612)] maser line and the CO lines are the most useful probes of the CSEs. OH and CO observations of evolved objects will therefore be discussed at some length below.

2.2 OH and CO observations

2.2.1 OH.
It was the detection of OH maser emission from red giants in the late sixties that marked the beginning of radio astronomical research on evolved stars. The subsequent observations of OH(1612) maser emission from OH/IR–stars have substantially contributed to, and in many cases been crucial for, our understanding of the late stages of AGB evolution (see e.g., Herman and Habing, 1985a). Their strength and the absence of extinction at 18 cm make them observable throughout the Galaxy. Towards the galactic centre there are observational problems with extended absorption, but this can be avoided by using interferometers (Lindqvist et al., 1987). An OH/IR–star has also been detected in an external galaxy, the LMC (Wood et al., 1986). The fact that their distances can be fairly accurately estimated by combining linear size (from phase lag measurements) and angular size (interferometer observations) further strengthens their usefulness (Herman et al., 1985).

Red Giants and Supergiants. OH maser emission has now been detected towards O–rich red giants over a large range of mass loss rates: from the low mass loss rate ($<10^{-7}$ $M_\odot$ yr^{-1}), short–period Miras and semiregulars (Dickinson et al., 1986), through the long–period Miras and even longer–period OH/IR–stars up to the very top of the AGB ($>10^{-4}$ $M_\odot$ yr^{-1}) and perhaps beyond that (Nguyen–Q–Rieu et al., 1979; Herman and Habing, 1985b). When the ongoing surveys at Parkes and Arecibo of new OH/IR–sources, based on colour–selected IRAS sources (Lewis et al., 1985; Sivagnanam and Le Squeren, 1986), are completed the number of red giants detected in circumstellar OH(1612) emission will be well above a thousand (te Lintel Hekkert, 1987; Eder et al., 1988). A number of wellknown supergiants are also detected in OH maser emission: VY CMa, NML Cyg, S Per, VX Sgr, and IRC+10420 (this may be a proto–PNe) (Cohen et al., 1987; Diamond et al., 1987).

Proto–PNe and PNe. Habing et al. (1987) proposed that the non–variable OH/IR–stars with large mass loss rates are proto–PNe, i.e., they have or will shortly cease losing mass and increasingly hotter regions of the star will become uncovered. Modelling of the infrared radiation from these objects supports this suggestion (Bedijn, 1987). However, only upper limits to radio continuum emission were obtained until recently when Pottasch et al. (1987) managed to detect 2 stars at 2 and 6 cm. Both share, with the previously only known OH–emitting PNe Vy2–2 (Davis et al., 1979), the common characteristic that only a single OH(1612) feature is present. Recently, Payne et al. (1988) searched 79 PNe with the meagre result of only one detection, the very young PN NGC6302. Once again only one OH feature was present. To this list we should add M1–92 (Lepine and Nguyen–Q–Rieu, 1974), and possibly OH5.89–0.39 (Zijlstra and Pottasch, 1988). Vy2–2 also shows the rotationally excited OH ($^2\Pi_{3/2}$, J=5/2) 6035 MHz line in maser emission (Jewell et al., 1985). In NGC6302 the 6030 and 6035 MHz lines were seen in absorption (Payne et al., 1988). It seems that the statistics of OH–emitting (proto–) PNe can probably be improved substantially in the future by examining single–feature sources from the ongoing surveys of OH/IR–stars. There are at least three other OH–sources that appear to be in transition from the tip of the AGB and to the left in the HR–diagram, IRC+10420 (probably a supergiant; Lewis et al., 1986), OH231.8+4.2 (Morris et al., 1982), and IRAS16342–3814 (Likkel and Morris, 1988).

2.2.2 CO. The CO millimetre wave lines have become an increasingly important tool for the study of CSEs. They have the advantages of being fairly easily modelled, and being detectable in O–rich as well as C–rich CSEs. Their main disadvantages are the limited observational space (see §4.2), and the fact that they are easily excited and hence may mask more complicated density structures.

Red Giants and Supergiants. The number of CSEs detected in CO emission has increased dramatically during the last few years, and stands presently at well above 200, divided roughly equally between O–rich and C–rich objects. The mass loss rate range covered is $<10^{-7}$ M_θ yr^{-1} to $>10^{-4}$ M_θ yr^{-1} and seems to be independent of chemical composition (Knapp, 1987). Many surveys have been based on the IRAS point source catalogue, but also other criteria have been used. Pre–IRAS surveys were performed by Knapp and Morris (1985) and Wannier and Sahai (1986). Based on IRAS observations Zuckerman and Dyck (1986a,b, 1988), Zuckerman et al. (1986), Leahy et al. (1987), Likkel et al. (1987), and Nguyen–Q–Rieu et al. (1987), considerbly increased the number of detected CSEs. In particular, Nguyen–Q–Rieu et al. managed to detect fairly distant ones ($\approx$5 kpc) using the IRAM 30m telescope. Recently 13 OH/IR–stars were detected (Heske, priv. comm.). Many of the above mentioned surveys have concentrated on obscured objects, where little is known about the central stars. Olofsson et al. (1987, 1988a) approached this problem by observing bright carbon stars with relatively well defined photospheric characteristics. CO emission from supergiants is usually quite weak, but several of them have nevertheless been detected: α Ori (Huggins, 1987), VY CMa (Zuckerman and Dyck, 1986a), NML Cyg (Zuckerman, priv. comm.), IRC+10420 (Bachiller et al., 1988), and AFGL2343 (Zuckerman and Dyck, 1986b).

Proto–PNe and PNe. Up to recently only a few proto–PNe were detected in CO, and searches towards true PNe were mostly negative. However, also here the situation has improved, and the following proto–PNe are now detected: CRL618, M2–9 and HD44179 (Bachiller et al., 1988), CRL2688 (Kawabe et al., 1987), M1–92 (Knapp, 1986), and OH231.8+4.2 (Morris et al., 1987). The following objects are also potential proto–PNe objects: IRAS09371+1212 (Forveille et al., 1987), HD161796, SAO163075, and 89 Her (Likkel et al., 1987). IRC+10420 probably also belongs to this class (Bachiller et al., 1988). The PNe detected in CO are: NGC2346 (Huggins and Healy, 1986a; Healy and Huggins, 1988), NGC6302 (Zuckerman and Dyck, 1988), NGC6720 (Huggins and Healy, 1986a), NGC7027 (Masson et al., 1985), NGC7293 (Huggins and Healy, 1986b), and IRAS21282+5050 (Likkel et al., 1988). Also here the statistics are likely to improve considerably when new sensitive millimetre wave observations, guided by IRAS data, are performed.

3. CHARACTERISTICS OF THE CIRCUMSTELLAR ENVELOPES

We will in this section try to summarize our current knowledge of the characteristics of CSEs, e.g., the geometry, the kinematics, the density structure, the kinetic temperature, the lifetime, etc. The main emphasis is put on what has been learnt by using molecular radio line data.

3.1 The geometry

It has been known for some time that when observed with sufficient spatial resolution (using IR speckle interferometry or direct IR imaging) or with polarimetry the inner regions of many CSEs have a clearly non–spherical geometry (Dyck, 1987; Ridgway and Keady, 1988). In the few cases where the central region becomes visible in the optical, the structure is often bipolar, e.g., CRL618 (Schmidt and Cohen, 1981), CRL2688 (Ney et al., 1975), OH231.8+4.2 (Reipurth, 1987),

IRAS09371+1212 (Hodapp et al., 1988, Rouan et al., 1988), and M2–9 (Aspin et al., 1988). It has also been known that the presumed end–products of AGB evolution, PNe, have mainly non–spherical geometries (see e.g., Zuckerman and Aller, 1986; Balick, 1987). Zuckerman and Aller (1986) carefully examined the morphology of 108 PNe and concluded that ≈50% have a bipolar structure and ≈30% have an elliptical symmetry.

When it comes to the structure of the intermediate stage, the CSEs, the situation is much more unclear and hence unsatisfying. The distribution of H_2O masers associated with Mira variables are in many cases non–spherical (Lane et al., 1987). This is also the case for the OH main line and H_2O masers surrounding the supergiant VX Sgr (Chapman and Cohen, 1986). A number of OH/IR–stars (≈30) have been mapped in the OH(1612) line (Bowers et al., 1983; Diamond et al., 1985; Herman et al., 1985, Welty et al., 1987). It appears that the OH–emitting region is circularly symmetric (a thin shell, ≤300 AU, at a radius of ≈1500 AU) and by inference spherically symmetric, but there is considerable clumpiness. This symmetry agrees with the infrared speckle interferometry and imaging data on several OH/IR–stars obtained by Cobb and Fix (1987).

Single dish observations of thermal molecular line emission are usually not able to resolve any structure in the CSEs. Furthermore, millimetre wave interferometer data are sparse due to the limited number of instruments and the observational difficulties, and hence our knowledge is based on very scattered data. The CSE of the carbon star IRC+10216 has been mapped on several occasions. It appears spherical in the CO(J=1–0, 2–1) lines out to ≈40000 AU (Huggins et al., 1988), as well as in the HCN(J=1–0) line out to ≈5000 AU (Bieging et al., 1984). On the contrary, CO(J=1–0) maps of the carbon star V Hya show clear evidences for a bipolar outflow, as well as an extended symmetric CSE (Kahane et al., 1988b; Tsuji et al., 1988). The near–infrared spectrum provides evidence for high–velocity motions, ≈100 km s^{-1} (uncorrected for inclination) (Sahai and Wannier, 1988).

Observations of CSEs around objects believed to be in the transition phase often reveal complicated structures. The bipolar nebula OH231.8+4.2 appears to have a CSE similar in shape to that of V Hya. A CO(J=1–0) map reveals a bipolar outflow, aligned with the optical image, with an outflow velocity exceeding 140 km s^{-1} (Morris et al., 1987). The OH maser maps are extended along the same axis (Morris et al., 1982). However, the dominant CO emission, as well as that of most other molecules, originates from a core that is unresolved at a resolution of 21" (Morris et al., 1987). For a similar object, IRAS16342–3814, Likkel and Morris (1988) conclude from OH and H_2O observations that it is an evolved star with a bipolar outflow. The outflow velocity is about 130 km s^{-1}. CO(J=1–0) observations of another bipolar nebula, CRL2688, reveal expanding lobes around the optical nebulosity, as well as a large, cold CSE that is circularly symmetric (Heiligman et al., 1986; Kawabe et al., 1987). Nguyen–Q–Rieu et al. (1986) and Beiging and Nguyen–Q–Rieu (1988b) have found evidence for a disk– or toroidal–like density distribution lying perpendicular to the optical lobes from $NH_3(1,1)$ and HCN(J=1–0) observations, respectively. The CSE of CRL618, also a bipolar nebula, appears circularly symmetric in the CO(J=1–0, 2–1) lines, but the kinematics suggest that the CO gas takes part in a bipolar outflow (Bachiller et al., 1988). Finally, Masson et al. (1985) have mapped the CO(J=1–0) emission associated with the PN NGC7027. They conclude that the inner envelope is ellipsoidal in shape, and the ionized gas is expanding faster along the low density polar axis. However, at velocities close to the stellar velocity only 5–10% of the total flux appear in the interferometer map, suggesting an extended, resolved CSE. In the somewhat more developed PN

NGC2346, Bachiller et al. (1988) found remnant CO gas that participates in the bipolar motion seen in the visible. In NGC7293 the neutral gas is associated with the outer regions of the ring—like optical nebulosity, and it appears to be assembled in two expanding toroids (Huggins and Healy, 1986b).

3.2 The kinematics

The kinematics of a pulsating red giant atmosphere are extremely complicated with both rising and falling matter giving rise to shock waves (Hinkle et al., 1982). This is apparently also the case in the region between the pulsating atmosphere and the expanding CSE as inferred from SiO maser observations (Nyman and Olofsson, 1986). In fact, also the inner part of the CSE is expected to have a fairly complex velocity distribution, due to episodic mass ejection and acceleration (Keady et al., 1988).

On the other hand, the extended part of a CSE is expected to simply expand with a constant velocity since here no acceleration agents are effective. It now appears that the mechanism responsible for, at least, the final characteristics of the mass loss is radiation pressure on grains (the gas is dragged along) formed in the extended atmospheres of the stars (see Jura, these proceedings), i.e., the gas gains momentum from the grains. In the external parts of a CSE the grains drift supersonically through the gas with a velocity, v_d, obtained by equating the radiative force and the drag force (Kwok, 1975),

$$v_d = (\frac{v_e Q L}{c \dot{M}})^{0.5} , \qquad (1)$$

where v_e is the gas expansion velocity, Q is the ratio of the radiative momentum transfer cross section to the geometric cross section of the grains (averaged over frequency), L is the luminosity of the star, c is the speed of light, and $\dot{M}$ is the mass loss rate. Papoular and Pegourie (1986) have estimated that v_d can be as high as 20 km s^{-1}, but for high mass loss rate CSEs it is expected to be at most a few km s^{-1}. In the case of a negligible drift velocity the final expansion velocity of the gas is given by (Jura, 1984),

$$v_e = (\frac{\aleph_d L}{2 \pi r_c c})^{0.5} , \qquad (2)$$

where $\aleph_d$ is the dust opacity per unit mass (in fact, $\aleph_d L$ is weighted over all frequencies), and r_c is the dust condensation radius. It is expected from this that v_e should be only mildly dependent on the stellar luminosity since r_c is roughly proportional to $L^{0.5}$.

Most molecular line observations are consistent with a well defined terminal velocity for the gas that falls within a limited range. The compilation of Knapp (1987) shows that for 136 stars detected in CO emission, 90% lie within $5 \leq v_e \leq 25$ km s^{-1} and $<v_e> \approx 17$ and 18 km s^{-1} for O— and C—stars, respectively. The terminal velocities for OH/IR—stars fall in the same region (Eder et al., 1988), and are only mildly dependent on the luminosity (Jura, 1984). For CO— as well as OH/IR—stars the "large terminal velocity"—sources are concentrated to the galactic plane, and they are therefore presumably younger and hence more massive (Knapp, 1987; Eder et al., 1988; Zuckerman and Dyck, 1988). In only one case, VX Sgr, the SiO, H_2O, and OH interferometer maps can be interpreted in terms of an envelope that is still accelerating at ≈ 1000 AU (this corresponds to ≈ 50 stellar radii), possibly due to grain growth (Chapman and Cohen, 1986).

In the case of the transition objects it is clear from the discussion in §3.1 that the velocity structure once again becomes very complex, e.g., due to the onset of a fast stellar wind, expanding ionization fronts that are either ionization– or density–bounded, etc. In CRL2688 there is weak high–velocity CO emission ($\approx$40 km s^{-1}) from a small region near the centre of the optical image (Heiligman et al., 1986; Kawabe et al., 1987). In NGC7027 the spatial resolution is high enough to conclude that similar emission originates in a layer of neutral gas shocked by the expanding HII–region (Masson et al., 1985). The presence of bipolar structures has been inferred in a few cases: V Hya, OH231.8+4.2, IRAS16342–3814, CRL618, and NGC2346 (see §3.1). For the three former sources the outflow velocities exceed 100 km s^{-1}. The reason for the onset of such high–velocity flow is not known. The "normal state" mass loss mechanism, radiation pressure on grains, can be excluded in this case.

It should be mentioned that the kinematics may have an appreciable effect on the excitation of molecules. A velocity gradient allows infrared photons from the hot core to leak out into the envelope and enhance the excitation through infrared pumping. This may, in the external tenuous parts, have a considerable effect also on otherwise collisionally excited molecules like CO. In the same way it may affect the cooling of the gas (see §3.4).

3.3 The density structure and the total mass

The density at a radius r and in the direction (θ,φ) is intimately connected to the retarded mass loss rate, and the kinematics,

$$\dot{M}(t - \int_{r_i}^{r} \frac{dr}{v(r,\theta,\varphi)}, \theta, \varphi) = 4\pi r^2 \rho(r,\theta,\varphi) v(r,\theta,\varphi) \quad , \tag{3}$$

where

$$\rho = \mu m(H) n(H_2) = (1 + 4n(He)/n(H))(2n(H_2) + n(HI)) m(H) \tag{4}$$

It is therefore in principle possible to determine the spatial and temporal characteristics of the mass loss mechanism from the density structure and the kinematics. The discussions in §:s 3.1 and 3.2 indicate that these can be quite complicated. In the simplest case of spherical geometry, constant mass loss rate and expansion velocity, and $n(H_2) \gg n(HI)$ we obtain the following simple relation

$$n(H_2) \approx 10^6 \left(\frac{\dot{M}}{10^{-6} M_\odot \text{ yr}^{-1}}\right)\left(\frac{15 \text{ km s}^{-1}}{v_e}\right)\left(\frac{10^{15} \text{ cm}}{r}\right)^2 \text{ cm}^{-3}. \tag{5}$$

The evidences for a departure from the r^{-2} density law are sparse. Keady et al. (1988) conclude from a detailed study of infrared CO absorption lines towards IRC+10216 that acceleration occurs in steps within 14 stellar radii and for a constant mass loss rate this means an initial density gradient steeper than r^{-2}. Also, Bedijn (1987) in his IR continuum models requires steep initial density gradients for the low mass loss rate objects in order to fit the observed spectra. Grain growth and hence acceleration could be an explanation. A mass loss rate increasing with time would also lead to a steeper density gradient. According to Bedijn this will produce observable consequences in the IR spectra only for the highest mass loss rates, $>2\times10^{-4} M_\odot$ yr^{-1}, but this conclusion depends on the adopted $\dot{M}(t)$. Modelling of molecular line emission from the CSE of IRC+10216 requires a decreasing mass loss rate and hence a flatter density distribution (Sahai, 1987). Olofsson et al. (1988a) interpreted

CO(J=1–0) observations of the carbon star S Sct in terms of a detached CSE formed by a mass loss that decreased considerably ≈5000 yr ago. Another carbon star, U Ant, exhibits a remarkable CO(J=2–1) line profile that may be interpreted in the same way, but here the mass loss may have recommenced (Olofsson et al., 1988c).

An interesting question is whether the mass loss is continuous or occurs in the form of blobs of released material. A patchy density structure in the region where the acceleration starts is not unreasonable (see e.g., Bloemhof et al., 1985). Thermal instabilities may play a role here (Stencel et al., 1986; Muchmore et al., 1987). Alcock and Ross (1986a,b) have addressed this problem by examining the maps and spectra of SiO and OH(1612) masers. They suggest that the mass loss occurs in blobs of high density material that in the region of the OH(1612) masers have somehow developed into pancakes. High spectral resolution observations of OH(1612) masers are not inconsistent with this model (Cohen et al., 1987; Fix, 1987).

The total mass of a CSE within a radius R_e is approximately given by,

$$M(R_e) \approx \frac{\dot{M} R_e}{v_e} = 2 \times 10^{-3} \left(\frac{\dot{M}}{10^{-6} M_\odot \text{ yr}^{-1}}\right)\left(\frac{R_e}{10^{17} \text{ cm}}\right)\left(\frac{15 \text{ km s}^{-1}}{v_e}\right) M_\odot. \quad (6)$$

Thus, even a huge envelope like that of IRC+10216, where $\dot{M} \approx 2 \times 10^{-5}$ $M_\odot$ yr^{-1}, and $v_e \approx 15$ km s^{-1}, has a mass of only 0.5 $M_\odot$ within $R_e = 10^{18}$ cm. $M(R_e = "\infty")$ is of course an important parameter since it provides us with the total amount of mass lost during the AGB phase so far. However, to observationally directly estimate this quantity is not so simple. Adopting a certain probe will only yield the mass within a limited region, determined by chemical and excitation processes, unless uncertain extrapolations are being made. For most molecules the fractional abundance, $f(X,r) = n(X,r)/n(H_2,r)$, has a radial variation that is sufficiently complex to make them useless in this connection (Bieging and Nguyen–Q–Rieu, 1988a). In the case of the most stable species, CO, the extent of the detectable CSE is determined by photodissociation. Mamon et al. (1988) have carefully examined this problem by including self–, H_2–, and dust shielding. The extent of the CO envelope (defined as $f(CO,R_e) = f(CO)/2$, where $f(CO)$ is the freeze–out photospheric abundance) is a complicated function of $\dot{M}$ and v_e. An approximate relation for $\dot{M} > 10^{-7}$ $M_\odot$ yr^{-1}, $v_e = 15$ km s^{-1}, and $f(CO) = 4 \times 10^{-4}$, is given by,

$$R_e(CO) \approx 5 \times 10^{16} \left(\frac{\dot{M}}{10^{-6} M_\odot \text{ yr}^{-1}}\right)^{0.6} \text{ cm}. \quad (7)$$

This gives considerably lower values for the size of the CO envelope than does the model of Morris and Jura (1983). It is also consistent with modelling of circumstellar CO emission (Sopka et al., 1988). The dependence of $R_e(CO)$ on $f(CO)$ and v_e is approximately given by $(f(CO)/v_e)^{0.5}$. The corresponding time scale is given by,

$$t(CO) \approx \frac{R_e(CO)}{v_e} \approx 10^3 \left(\frac{\dot{M}}{10^{-6} M_\odot \text{ yr}^{-1}}\right)^{0.6} \text{ yr}, \quad (8)$$

for $v_e = 15$ km s^{-1}. Thus, we can only estimate the mass lost during the last ≈300 years for $\dot{M} = 10^{-7}$ $M_\odot$ yr^{-1} and ≈13000 years for $\dot{M} = 5 \times 10^{-5}$ $M_\odot$ yr^{-1}. The total detectable mass would be,

$$M(CO) \approx \dot{M} t(CO) \approx 1.2 \times 10^{-3} \left(\frac{\dot{M}}{10^{-6} M_\odot \text{ yr}^{-1}}\right)^{1.6} M_\odot. \quad (9)$$

This will exceed 1 M_θ only for $\dot{M} > 7 \times 10^{-5}$ M_θ yr^{-1}. $\dot{M}$ derived from CO millimetre wave data is for an unresolved envelope dependent on the distance, D, squared (see §4.2), and hence M(CO) is roughly dependent on D^3 and thus very uncertain.

Observations of material outside R_e(CO) can in principle be performed in the HI 21 cm line because sufficiently far out H_2 is also photodissociated (Glassgold and Huggins, 1983). However, these observations are not easily performed (see Knapp, 1985). Only recently were the first detections of HI in CSEs made. Bowers and Knapp (1987, 1988) detected HI in emission towards α Ori and omicron Ceti. In the case of α Ori the CSE is not representative since the star is so warm that hydrogen exists essentially in atomic form throughout the envelope (Glassgold and Huggins, 1983). omicron Ceti is cooler but still about 30% of the hydrogen is in atomic form when it leaves the star. Towards other CSEs only upper limits have been obtained (Knapp, 1985; Schneider et al., 1987). In the case of IRC+10216 Bowers and Knapp (1987) conclude that the CSE cannot be much larger than the size measured in CO. Detections of HI in PNe have been made: NGC6302 (Rodriguez et al., 1985), NGC6790 (Gathier et al., 1986), IC418 (Taylor and Pottasch, 1987), and IC4997 (Altschuler et al., 1986). Another potentially useful probe (and perhaps the best) for the total mass of a CSE is provided by submillimetre observations of dust emission (Sopka et al., 1985).

3.4 The temperature structure

The temperature of the grains is determined by the balance between the absorbed stellar radiation energy at "short" wavelengths and the reradiated energy at longer wavelengths. In the optically thin limit this results in a radial dependence for the grain temperature given by,

$$T_d(r) \propto r^{-2/(4+p)} , \qquad (10)$$

where p describes the wavelength dependence of the grain emissivity in terms of λ^{-p} (Sopka et al., 1985). For C–rich CSEs it appears that $p \approx 1$ over a wide wavelength range (Jura, 1986; Zuckerman and Dyck, 1986b), while for O–rich CSEs the dependence on wavelength is more complicated (Zuckerman and Dyck, 1986b). Close to the star the temperature may exceed that obtained from the optically thin approximation. Nevertheless eq.(10) combined with $T_d \approx 1000$ K at the dust condensation radius (Ridgway et al., 1986; Bedijn, 1987; Keady et al., 1988) should give a reasonable description of the radial dependence of the dust temperature.

The gas kinetic temperature is independent of the dust temperature. The heating of the gas is due to the supersonic streaming of the dust particles through the gas, i.e., gas–grain collisions (Goldreich and Scoville, 1976),

$$\frac{dq}{dt}\bigg|_{dust} = \frac{1}{2} n_d \sigma_d \rho v_d^3 , \qquad (11)$$

where n_d is the dust particle density, σ_d is the geometrical cross section of the grains, and ρ is the gas density. It is dependent on the mass loss rate (see eq.(1)) and may therefore vary considerably from star to star. The cooling is due to adiabatic expansion and radiative cooling via molecular rotational lines (mainly CO and H_2O in C–rich and O–rich CSEs, respectively; Goldreich and Scoville, 1976; Kwan and Linke, 1982), which in its turn is dependent on the dynamics through the escape of photons (Tielens, 1983). It therefore appears that the radial dependence of the kinetic temperature has to be obtained by solving the energy balance equation,

$$\frac{dT_k}{dr} = \frac{\gamma-1}{knv}\frac{dq_{dust}}{dt} - 2(\gamma-1)\frac{T_k}{r} - \frac{\gamma-1}{knv}\frac{dq_{rad}}{dt}, \quad (12)$$

for each individual CSE (γ is the ratio of specific heats, k is the Boltzmann constant, n is the gas particle density, and v is the gas expansion velocity). Huggins et al. (1988) recently did this for IRC+10216 by using CO(J=1–0, 2–1) data as a constraint. A reasonable fit to their result is given by (assumed distance 200 pc),

$$T_k(r) \approx 60 \, (\frac{r}{10^{16} \, cm})^{-0.7} \, K, \quad (13)$$

with a slightly weaker dependence on r for r>2×10^{17} cm. For the lower mass loss rate CSEs, with high drift velocities (see eq.(1)), we would expect T_k to be much higher than this, at least in the inner regions. In fact, it seems that eq.(13) has to be modified upwards also in the inner parts of the IRC+10216 envelope by as much as a factor of three (Sahai and Wannier, 1985; Sahai, 1987). On the other hand, the high mass loss rate CSEs could be significantly cooler. The presence of ice features at 3.1 μm in five CSEs are consistent with this (Rouan et al., 1988; van der Veen et al., 1988). Nevertheless eq.(13) has been widely used as a standard for CSEs of arbitrary mass loss rate and chemical composition. A second heating process may come into play in the external parts of a CSE, photoelectric heating due to the interstellar radiation field. Huggins et al. (1988) included this in the energy balance equation in an apparently successful attempt to explain the large size of the CO emitting region around IRC+10216. This may be an important process also for (proto–) PNe.

3.5 The post–AGB lifetime

One of the most fascinating stages of late stellar evolution is the transition from the tip of the AGB to the PN phase. Here we encounter the termination of the mass loss, the uncoverage of the bright stellar core, and the interaction between the ionized wind and the gradually dispersing CSE remnant of the red giant phase. However, molecular line observations of this phase give a somewhat confused picture. Non–variable OH/IR–stars are likely PNe–progenitors, but only two have been detected in radio continuum so far (see §2.2.1). The same number of OH–emitting PNe are known. CO emission appears to be a comparatively better diagnostic, but also here the results are contradictory. Some sources are strong and also rich in other molecular species, but most of them are exceedingly weak or undetectable. A case in point is CRL 618 where Bujarrabal et al. (1988) have found a chemistry very different from those of CRL2688 and IRC+10216. It appears that the molecular characteristics of these transition objects must be very sensitively dependent on the details of the evolution.

No doubt, part of the observational problem stems from the short time scales involved. The optical depth of a CSE, and hence the protection of the molecules, decreases rapidly after the termination of the mass loss as the following calculation shows. The column density of molecular hydrogen $N(H_2)=n(H_2,R_i)R_i$, where $R_i \approx v_e t$ is the inner radius of the CSE and t is the time since the mass loss ceased, is assumed to be related to the extinction via the relation $N(H_2)=0.94\times10^{21}A_v$ cm^{-2} mag^{-1} which seems applicable to interstellar clouds (Bohlin et al., 1978). Combined with the mass loss rate ,eq.(3), this gives,

$$A_v \approx 20 \, (\frac{\dot{M}}{10^{-6} \, M_\odot \, yr^{-1}})(\frac{yr}{t}) \, mag. \quad (14)$$

That is, for low mass loss rates the dispersion time is very short and even for

$\dot{M}=10^{-4}$ $M_\odot$ yr^{-1} it takes only ≈ 2000 years to reach $A_v=1$. The increasing penetration of the UV flux, both from the inside and from the outside, is bound to have profound effects on the chemistry. Sun and Kwok (1987) discussed the OH–emitting properties of a CSE during the transition stage.

3.6 Elemental abundances and isotope ratios

Elemental abundances and isotope ratios are two important time keepers of stellar evolution. However, studies of molecular line emission from the surrounding envelopes have made only marginal contributions to this area. In particular, elemental abundances are difficult to determine in this way with any confidence. For instance in the case of the molecular factories IRC+10216, CRL618, and CRL2688 it is in principle only possible to say that the objects are C–rich. There are clear differences in the molecular abundances between the sources but this can probably be attributed to different environmental conditions for the CSEs rather than differing elemental abundances of the central stars (Bujarrabal et al., 1988). That is, qualitative conclusions are in some cases possible, but never quantitative ones. This is because we have very little knowledge about how photospheric abundances are filtered into circumstellar abundances through a highly uncertain envelope chemistry involving, for instance, grains and photoinduced processes (Omont, 1987).

A case in point is the [HCN]/[CO] abundance ratio which, for an equilibrium stellar atmosphere chemistry, should be proportional to the carbon–excess, i.e., [C]/[O]−1, and hence could be used as a measure of this quantity for C–rich objects (from this point of view one would not expect HCN in O–rich objects). Olofsson et al. (1988b) have tested this by observing both species towards 22 stars with reasonably well determined photospheric [C]/[O]–ratios. The expected general trend is born out but there is also a substantial scatter. However, the detection of HCN also in O–rich objects has put some doubt on the usefulness of HCN as an indicator of C–richness. Hitherto 13 O–rich stars have been detected in HCN, several of them are normal Miras (Nercessian et al., 1988; Lindqvist et al., 1988). Lindqvist et al. even managed to detect CS in two cases, SiS in one case, and possibly HNC in one case. All three molecules are expected only in a C–rich environment. It appears that in an O–rich medium these molecular species are produced in a photoinduced circumstellar chemistry (Nercessian et al., 1988; Nejad and Millar, 1988).

Cases where the molecular line emission has aided in the classification of the objects are those where OH, H_2O, and/or SiO masers have been detected. In general it seems safe to conclude that if any of these masers are observed then the CSE is definitely O–rich. For instance, the recent detections of H_2O maser emission towards 4 carbon stars with silicate feature CSEs leave no doubt that the surrounding CSEs are O–rich (Benson and Little–Marenin, 1987; Nakada et al., 1987, 1988; Deguchi et al., 1988).

Isotope ratios are in general possible to estimate with higher accuracy. The circumstellar $^{12}C/^{13}C$–ratios fall in the ranges expected from stellar evolution, i.e., the ratio is on the average higher for C–rich CSEs than for O–rich CSEs (Knapp and Chang, 1985; Wannier and Sahai, 1987). The ratio, ≈ 40, obtained for IRC+10216 by Kahane et al. (1988a) using a number of optically thin lines seems secure. Jura et al. (1988) detected three ^{13}C–rich carbon stars ($^{12}C/^{13}C \approx 4$) in circumstellar $^{13}CO(J=1-0)$ emission. If anything, the circumstellar $^{12}C/^{13}C$–ratio is lower than that of the photosphere, and the ratio has been stable for at least 1000 years. It should be emphasized that $^{12}C/^{13}C$–ratios estimated from ^{12}CO and ^{13}CO millimetre wave observations are very uncertain since the excitation mechanisms and the spatial extents (if determined by self–shielding) are different. The $^{18}O/^{17}O$–ratio is significantly lower than the terrestrial value in three C–rich CSEs (Wannier and

Sahai, 1987). The $^{29}Si/^{28}Si-$, $^{30}Si/^{28}Si-$, $^{33}S/^{32}S-$, $^{34}S/^{32}S-$, and $^{37}Cl/^{35}Cl$–ratio towards IRC+10216 are all consistent with those in the solar system (Cernicharo and Guelin, 1987; Kahane et al., 1988a). A number of C–rich CSEs has been observed in the HC^{15}N(J=1–0) line (Andersson et al., 1987; Kahane et al., 1988a). In all cases ^{15}N is significantly underabundant with respect to the terrestrial value.

4. ESTIMATES OF THE MASS LOSS RATE

The mass loss rate (and its evolution with time) is the most important observational parameter for stellar evolution theories. Various methods, depending on the wavelength regime of the observational input, have been designed to estimate it (Goldberg, 1986). Often order–of–magnitude differences exist between different methods, showing the inherent difficulties involved. We will now discuss in some detail mass loss rate estimates based on OH and CO radio line observations.

4.1 OH 1612 MHz masers

The OH abundance at the inner boundary of the CSE is negligible, and the masing OH molecules are believed to be formed by photodissociation of H$_2$O molecules further out in the envelope (Huggins and Glassgold, 1982). When worked out in detail, taking into account the pumping of the masers by 35 μm photons (Elitzur et al., 1976), the OH(1612) masers are expected to lie in a thin shell (at a radius determined by the mass loss rate) surrounding the central star (Netzer and Knapp, 1987; Sun and Kwok, 1987). Observationally it is found that the shell tickness is ≈20% of the shell radius, which in the sample of OH/IR–stars observed by Herman et al. (1985) averages at ≈8×10^{16} cm.

Bowers et al. (1983) obtained an important empirical relation between $\dot{M}$ and R(OH), the radius of the OH(1612) emitting region,

$$\dot{M} = 10^{-6} \left(\frac{R(OH)}{10^{16} \text{ cm}}\right)^2 \, M_\odot \text{ yr}^{-1}. \tag{15}$$

This gives a distance independent measure of the mass loss rate in the case of phase–lag determined sizes. However, the absolute scale rests on estimates of $\dot{M}$ obtained using various other methods. Furthermore, R(OH) determined through phase–lag measurements will only be available for a very limited number of sources (say <100) due to the extremely demanding observations.

Baud and Habing (1983) derived a relation between $\dot{M}$ and L(OH), the luminosity of the OH(1612) maser (=SD^2, where S is the flux), based on an empirical relation between L(OH) and R(OH), the assumption of a saturated maser with a given column density, and an adopted OH abundance,

$$\dot{M} = 4\times10^{-14} \frac{S^{0.5}Dv_e}{f(OH)} \, M_\odot \text{ yr}^{-1}, \tag{16}$$

where S is given in Jy, D in pc, and v_e in km s^{-1}. This estimate is independent of other methods for estimating the mass loss rate, but it has the disadvantage of being dependent on D plus a number of assumptions.

Netzer and Knapp (1987) have produced a chemical model for circumstellar OH(1612) masers that gives the following relation between $\dot{M}$, R(OH), and v_e,

$$\dot{M} = 5 \times 10^{-6} \left(\frac{R(OH)}{10^{16} \text{ cm}}\right)^{1.4} \left(\frac{v_e}{15 \text{ km s}^{-1}}\right)^{0.6} M_\odot \text{ yr}^{-1}, \qquad (17)$$

provided that R(OH) coincides with the radius of the OH number density peak. The result is in reasonable agreement with the empirical relation obtained by Bowers et al. (1983), at least for the higher mass loss rates which is also the range on which eq.(15) is based. Relation (17) implies a mass loss rate close to 10^{-4} $M_\odot$ yr^{-1} for the average shell radius, 8×10^{16} cm, of OH/IR–stars obtained by Herman et al. (1985). Once again this is a distance independent estimate if phase–lag sizes are used, but it becomes proportional to $D^{1.4}$ if angular sizes are used. Eq.(17) may also be used to estimate the retarded time at which the OH(1612) maser samples the mass loss rate,

$$t(OH) \approx \frac{R(OH)}{v_e} \approx 10^2 \left(\frac{\dot{M}}{10^{-6} M_\odot \text{ yr}^{-1}}\right)^{0.7} \text{ yr}, \qquad (18)$$

for $v_e \approx 15$ km s^{-1}. Sun and Kwok (1987) have also constructed a model of circumstellar OH(1612) masers. They ignore the chemistry and assume an OH abundance. The location of the masers is constrained by the assumption of a minimum OH column density (which is related to the UV optical depth and hence the photoinduced chemistry) required for operating a saturated maser when pumped by 35 μm photons. These results give a theoretical explanation of the empirically derived relations between $\dot{M}$ and R(OH) and L(OH).

Even though the OH masers sample only a short interval of the mass loss history of an individual star, it is still possible to estimate an average time evolution by examining the OH luminosity function, provided that there is a relation between L(OH) and $\dot{M}$ (eq.(16)) and that L(OH) is not very dependent on the MS–mass. Baud and Habing (1983) used this method to derive,

$$\dot{M}(t) = \dot{M}_{min} \left(1 - \frac{t}{\tau(OH)}\right)^{-\alpha}, \qquad (19)$$

where τ(OH) is the lifetime of the OH emitting phase, and α was estimated to be ≈ 0.5. It was shown by Bedijn (1986) that such a mass loss evolution, with $\alpha \approx 1$, can be expected from the dependence of the scale height of the extended atmosphere on the decreasing stellar mass. In this scenario the Mira variables would correspond to the major portion of τ(OH), while the OH/IR–stars appear on the stage shortly before τ(OH).

4.2 CO

Millimetre wave lines of CO have the advantage, compared to OH, of being detectable towards O–rich as well as C–rich CSEs. The main disadvantage is the limited observational space (see below). In general, the mass loss rate estimated from observations of a CO line will be an average over a time interval given by eq.(8) or less depending on the excitation requirements. If we assume collisional excitation, i.e., the excitation temperature follows the kinetic temperature, eq.(13), the contribution to the intensity from the region r to r+dr is proportional to $r^{0.3}$ in the optically thick case and $r^{0.7}$ in the optically thin case (provided that the transition is effectively excited), i.e., more weight is given to the external parts. An estimate of the effectiveness of the excitation is obtained by putting $\exp(-E_u/kT_k(r))=1/2$, where $E_u=hBJ(J+1)$ is the energy of the upper level (B is the rotational constant). The J→J–1 transition is effectively collisionally excited within,

$$r_{1/2} \approx 2\times10^{17} \left(\frac{2}{J(J+1)}\right)^{1.4} \text{ cm}. \tag{20}$$

This results in $r_{1/2} \approx 4\times10^{16}$ and 2×10^{15} cm for the J=2–1 and 6–5 transitions, respectively. Radiative excitation as well as the inefficiency of collisional excitation at large radii will change this to some extent but the basic behaviour is likely to be correct.

Knapp and Morris (1985) modelled the CO(J=1–0) emission from CSEs with spherical geometry, constant mass loss rate, constant expansion velocity, and the kinetic temperature law given in eq.(13). They examined the dependence on the distance, the expansion velocity, the envelope size, and the CO abundance. For an optically thick line the following simple formula for the mass loss rate can be used,

$$\dot{M} = \frac{T_{mb} v_e^2 D^2}{A(\theta_b) f(CO)^{0.85}}, \tag{21}$$

where T_{mb} is the main beam brightness temperature of the CO(J=1–0) line, and $A(\theta_b)$ is a constant dependent on the size of the telescope. Eq.(21) is easily applied to data from an arbitrary telescope as long as the CSE is unresolved. For instance, Knapp and Morris derived $A(100") = 1.6\times10^{15}$ ($\dot{M}$ in $M_\odot$ yr^{-1}, v_e in km s^{-1}, and D in pc), which would scale to $A(22") = 3\times10^{16}$ for the IRAM 30m telescope. The major sources of error in the derivation of mass loss rates using eq.(21) are, apart from the obvious distance problem (which is more severe here than in the OH case), the uncertainty in f(CO), the possible spatial resolution of the CSE, and the use of the same kinetic temperature law for all CSEs.

In the optically thin case the situation becomes more complicated since radiative excitation plays an important role, as opposed to the optically thick case where the excitation is mainly collisional. This is because the infrared transitions become optically thin in the radial direction at the same time as the millimetre wave ones become optically thin in the tangential direction, and this allows infrared photons to flow out into the envelope and take part in the excitation (Schönberg, 1988). Therefore it is necessary to observationally obtain a S/N–ratio sufficiently high to be able to define the character of the emission. Nevertheless, van der Veen (1988) has shown that eq.(21) gives roughly the same result as a more sophisticated analysis even in the optically thin case.

We can use eq.(21) to obtain an estimate of the observational space,

$$D(CO) \approx \frac{f(CO)^{0.4}}{v_e} \left(\frac{A(\theta_b)\dot{M}}{T_{mb}}\right)^{0.5}. \tag{22}$$

For a detection limit of 0.1 K, and assuming $v_e = 15$ km s^{-1}, $f(CO) = 5\times10^{-4}$, and $\dot{M} = 10^{-5}$ $M_\odot$ yr^{-1}, this restricts us in distance to $D(CO) \approx 5$ kpc even for the IRAM 30m telescope. Individual objects more distant than this can of course be detected, in particular when using the J=2–1 line (there is a recent detection of an OH/IR–star in the Galactic Centre, Mauersberger et al., 1988), but high mass loss rate objects are rare and for survey work the amount of integration time spent per source is limited.

CO observations could also be used to sample the mass loss history but it requires considerable observational input (many lines that probe different regions, see eq.(20)) and careful modelling (Sahai, 1987). In some cases the CSEs are large enough to be mapped with sufficient spatial resolution to allow $\dot{M}(t)$ to be derived from the density distribution (Huggins et al., 1988; Olofsson et al., 1988a). The sample of sources detected in CO is far from complete and it is therefore not possible to derive the mass loss evolution from a CO luminosity function. It is also clear from the above discussion that in general at least CO(J=1–0) observations result in mass loss rates averaged over a time period given by eq.(8). Comparison of eqs (8) and (18) shows that t(CO) and t(OH) scale almost equally with $\dot{M}$ and that t(OH) is only about 10% of t(CO). In the case of a mass loss rate changing with time one can therefore expect substantial differences in the mass loss rate estimates based on the two species. If one assumes that the mass loss rate evolves with time according to eq.(19) with $\alpha=1$, and that the CO and OH(1612) data samples the mass loss rate at the retarded times given by eqs (8) and (18), respectively, one finds that at the end of the AGB $\dot{M}(CO)/\dot{M}(OH) \approx 0.2$, and in both cases the mass loss rate is underestimated. It appears that the peak mass loss rate is preferably obtained from IR continuum data. This has been discussed in more detail by van der Veen (1988), who also showed that OH, CO, and IR mass loss rates are in reasonable agreement except in the high mass loss rate range. In fact, for these objects the CO mass loss rate estimates are unexpectedly low, even taking into account the time evolution, suggesting a decrease in abundance or excitation.

5. CONCLUSIONS

It appears now that molecular radio lines can be used to study various aspects of the entire evolution from the early AGB to the PN phase. It is also evident that we have reached such a state of observational maturity that it will be extremely rewarding to analyze the data in considerably more detail with more sophisticated methods.

Acknowledgement. I am grateful to the Swedish Natural Science Research Council (NFR), Kungl. och Hvitfeldtska Stipendieinrättningen, and Wilhelm och Martina Lundgrens Vetenskapsfond for financial support.

References
Alcock, C., Ross, R.R. (1986a). Ap. J., 305, 837.
Alcock, C., Ross, R.R. (1986b). Ap. J., 310, 838.
Altschuler, D.R., Schneider, S.E., Giovanardi, C., Silverglate, P.R. (1986). Ap. J. (Letters), 305, L85.
Andersson, B.–G., Wannier, P.G., Olofsson, H. (1987). B.A.A.S., 19, 645.
Aspin, C., McLean, I.S., Smith, M.G. (1988). Astr. Astroph., 196, 227.
Bachiller, R., Gomez–Gonzalez, J., Bujarrabal, V., Martin–Pintado, J. (1988). Astr. Astroph., 196, L5
Balick, B. (1987). Astron. J., 94, 671.
Baud, B., Habing, H.J. (1983). Astr. Astroph., 127, 73.
Bedijn, P.J. (1986). In Light on Dark Matter, ed. F. Israel, p. 119, Dordrecht: Reidel.
Bedijn, P.J. (1987). Astr. Astroph., 186, 136.
Benson, P.J., Little–Marenin, I.R. (1987). Ap. J. (Letters), 316, L37.
Bieging, J.H., Chapman, B., Welch, W.J. (1984). Ap. J., 285, 656.
Bieging, J.H., Nguyen–Q–Rieu. (1988a). Ap. J. (Letters), 329, L107.
Bieging, J.H., Nguyen–Q–Rieu. (1988b). preprint.
Bloemhof, E.E., Danich, W.C., Townes, C.H. (1985). Ap. J. (Letters), 299, L37.
Bohlin, R.C., Savage, B.D., Drake, J.F. (1978). Ap. J., 224, 132.
Bowers, P.F., Johnston, K.J., Spencer, J.H. (1983). Ap. J., 274, 733.
Bowers, P.F., Knapp, G.R. (1987). Ap. J., 315, 305.

Bowers, P.F., Knapp, G.R. (1988). Ap. J., 332, 299.
Bujarrabal, V., Gomez–Gonzalez, J., Bachiller, R., Martin–Pintado, J. (1988). Astr. Astroph., 204, 242.
Cernicharo, J., Guelin, M. (1987). Astr. Astroph., 183, L10.
Chapman, J.M., Cohen, R.J. (1986). M.N.R.A.S., 220, 513.
Cobb, M.L., Fix, J.D. (1987). Ap. J., 315, 325.
Cohen, R.J., Downs, G., Emerson, R., Grimm, M., Gulkis, S., Stevens, G., Tarter, J. (1987). M.N.R.A.S., 225, 491.
Davis, L.E., Seaqvist, E.R., Purton, C.R. (1979). Ap. J., 230, 434.
Deguchi, S., Kawabe, R., Ukita, N., Nakada, Y., Onaka, T., Izumiura, H., Okamura, S. (1988), Ap. J., 325, 795.
Diamond, P.J., Johnston, K.J., Chapman, J.M., Lane, A.P., Bowers, P.F., Spencer, J.H., Booth, R.S. (1987). Astr. Astroph., 174, 95.
Diamond, P.J., Norris, R.P., Rowland, P.R., Booth, R.S., Nyman, L.–Å. (1985). M.N.R.A.S., 212, 1.
Dickinson, D.F., Turner, B.E., Jewell, P.R., Benson, P.J. (1986). Astron. J., 92, 627.
Dyck, H.M. (1987). In Late Stages of Stellar Evolution, eds S. Kwok and S.R. Pottasch, p. 19, Dordrecht: Reidel.
Eder, J., Lewis, B.M., Terzian, Y. (1988). Ap. J. Suppl. Ser., 66, 183
Elitzur, M., Goldreich, P., Scoville, N. (1976). Ap. J., 205, 384.
Fix, J.D. (1987). Astron. J., 92, 433.
Forveille, T., Morris, M., Omont, M., Likkel, L. (1987). Astr. Astroph., 176, L13.
Gathier, R., Pottasch, S.R., Goss, W.M. (1986). Astr. Astroph., 157, 191.
Glassgold, A.E., Huggins, P.J. (1983). M.N.R.A.S., 203, 517.
Goldberg, L. (1986). In The M–type Stars, eds H.R. Johnson and F.R. Querci, p. 245, Washington: NASA SP–492.
Goldreich, P., Scoville, N. (1976). Ap. J., 205, 144.
Habing, H., van der Veen, W., Geballe, T. (1987). In Late Stages of Stellar Evolution, eds S. Kwok and S.R. Pottasch, p. 91, Dordrecht: Reidel.
Healy, A.P., Huggins, P.J. (1988). Astron. J., 95, 866.
Heiligman, G.M., et al. (1986). Ap. J., 308, 306.
Herman, J., Baud, B., Habing, H.J., Winnberg, A. (1985). Astr. Astroph., 143, 122.
Herman, J., Habing, H.J. (1985a). Phys. Rep., 124, No.4, 255.
Herman, J., Habing, H.J. (1985b). Astr. Astroph. Suppl. Ser., 59, 523.
Hinkle, K.H., Hall, D.N.B., Ridgway, S.T. (1982). Ap. J., 252, 697.
Hodapp, K.–W., Sellgren, K., Nagata, T. (1988). Ap. J. (Letters), 326, L61.
Huggins, P.J. (1987). Ap. J., 313, 400.
Huggins, P.J., Glassgold, A.E. (1982). Astron. J., 87, 1828.
Huggins, P.J., Healy, A.P. (1986a), M.N.R.A.S., 220, 33P.
Huggins, P.J., Healy, A.P. (1986b), Ap. J. (Letters), 305, L29.
Huggins, P.J., Olofsson, H., Johansson, L.E.B. (1988). Ap. J., 332, 1009.
Jewell, P.R., Schenewerk, M.S., Snyder, L.E. (1985). Ap. J., 295, 183.
Jura, M. (1984). Ap. J., 282, 200.
Jura, M. (1986). Ap. J., 303, 327.
Jura, M., Kahane, C., Omont, A. (1988). Astr. Astroph., 201, 80.
Kahane, C., Gomez–Gonzalez, J., Cernicharo, J., Guelin, M. (1988a). Astr. Astroph., 190, 167.
Kahane, C., Maizels, C., Jura, M. (1988b), Ap. J. (Letters), 328, L25.
Kawabe, R., Ishiguro, M., Kasuga, T., Morita, K.–I., Ukita, N., Kobayashi, H., Okumura, S., Fomalont, E., Kaifu, N. (1987). Ap. J., 314, 322.
Keady, J.J., Hall, D.N.B., Ridgway, S.T. (1988). Ap. J., 326, 832.
Knapp, G.R. (1985). In Mass Loss from Red Giants, eds M. Morris and B. Zuckerman, p. 177, Dordrecht: Reidel.
Knapp, G.R. (1986). Ap. J., 311, 731.

Knapp, G.R. (1987). In Late Stages of Stellar Evolution, eds S. Kwok and
        S.R. Pottasch, p. 103, Dordrecht: Reidel.
Knapp, G.R., Chang, K.M. (1985). Ap. J., 293, 281.
Knapp, G.R., Morris, M. (1985). Ap. J., 292, 640.
Kwan, J., Linke, R.A. (1982). Ap. J., 254, 587.
Kwok, S. (1975). Ap. J., 198, 583.
Lane, A.P., Johnston, K.J., Bowers, P.F., Spencer, S.H., Diamond, P.J. (1987). Ap. J.,
        323, 756.
Leahy, D.A., Kwok, S., Arquilla, R.A. (1987). Ap. J., 320, 825.
Lepine, J.R.D., Nguyen–Q–Rieu. (1974). Astr. Astroph., 36, 469.
Lewis, B.M., Eder, J., Terzian, Y. (1985). Nature, 313, 200.
Lewis, B.M., Terzian, Y., Eder, J. (1986). Ap. J. (Letters), 302, L23.
Likkel, L., Forveille, T., Omont, A., Morris, M. (1988). Astr. Astroph., 198, L1.
Likkel, L., Morris, M. (1988). Ap. J., 329, 914.
Likkel, L., Omont, A., Morris, M., Forveille, T. (1987). Astr. Astroph., 173, L11.
Lindqvist, M., Nyman, L.–Å., Olofsson, H., Winnberg, A. (1988). Astr. Astroph.,
        205, L15.
Lindqvist, M., Winnberg, A., Matthews, H.E., Habing, H.J., Olnon, F.M. (1987). In
        Late Stages of Stellar Evolution, eds S. Kwok and S.R. Pottasch, p. 79,
        Dordrecht: Reidel.
Mamon, G.A., Glassgold, A.E., Huggins, P.J. (1988). Ap. J., 328, 797.
Masson, C.R., et al. (1985). Ap. J., 292, 464.
Mauersberger, R., Henkel, C., Wilson, T.L., Olano, C.A. (1988). Astr. Astroph.,
        206, L34.
Morris, M., Bowers, P.F., Turner, B.E. (1982). Ap. J., 259, 625.
Morris, M., Guilloteau, S., Lucas, R., Omont, A. (1987). Ap. J., 321, 888.
Morris, M., Jura, M. (1983). Ap. J., 264, 546.
Muchmore, D.L., Nuth, J.A., Stencel, R.E. (1987). Ap. J. (Letters), 315, L141.
Nakada, Y., Deguchi, S., Forster, J.R. (1988). Astr. Astroph., 193, L13.
Nakada, Y., Izumiura, H., Oraka,m T., Hashimoto, O., Ukita, N., Deguchi, S.,
        Tanabe, T. (1987). Ap. J. (Letters), 323, L77.
Nejad, L.A.M., Millar, T.J. (1988). M.N.R.A.S., 230, 79.
Nercessian, E., Guilloteau, S., Omont, A., Benayoun, J.J. (1988). Astr. Astroph.,
        in press.
Netzer, N., Knapp, G.R. (1987). Ap. J., 323, 734.
Ney, E.P., Merrill, K.M., Becklin, E.E., Neugebauer, G., Wynn–Williams, C.G.
        (1975). Ap. J. (Letters), 198, L129.
Nguyen–Q–Rieu, Epchtein, N., Truong–Bach, Cohen, M. (1987). Astr. Astroph.,
        180, 117.
Nguyen–Q–Rieu, Laury–Micolaut, C., Winnberg, A., Schultz, G.V. (1979).
        Astr. Astroph., 75, 351.
Nguyen–Q–Rieu, Winnberg, A., Bujarrabal, V. (1986). Astr. Astroph., 165, 204.
Nyman, L.–Å., Olofsson, H. (1986). Astr. Astroph., 158, 67.
Olofsson, H. (1988). Space Science Reviews, 47, 145.
Olofsson, H. Eriksson, K., Gustafsson, B. (1987). Astr. Astroph., 183, L13
Olofsson, H. Eriksson, K., Gustafsson, B. (1988a). Astr. Astroph., 196, L1
Olofsson, H. Eriksson, K., Gustafsson, B. (1988b). in prep.
Olofsson, H. Eriksson, K., Gustafsson, B. (1988c). in prep.
Omont, A. (1987). In IAU Symp. No. 120 Astrochemistry, eds M.S. Vardya and
        S.P. Tarafdar, p. 357, Dordrecht: Reidel.
Papoular, R., Pegourie, B. (1986). Astr. Astroph., 156, 199.
Payne, H.E., Phillips, J.A., Terzian, Y. (1988). Ap. J., 326, 368.
Pottasch, S.R., Bignelli, C., Zijlstra, A. (1987). Astr. Astroph., 177, L49.
Reipurth, B. (1987). Nature, 325, 787

Ridgway, S.T., Joyce, R.R., Connors, D., Pipher, J.L., Dainty, C. (1986). Ap. J., 302, 662.
Ridgway, S.T., Keady, J.J. (1988). Ap. J., 326, 843.
Rouan, D., Omont, A., Lacombe, F., Forveille, T. (1988). Astr. Astroph., 189, L3.
Rodriguez, L.F., Garcia–Barreto, J.A., Canto, J., Moreno, M.A., Torres–Peimbert, S., Costero, R., Serrano, A., Moran, J.M., Garay, G. (1985). M.N.R.A.S., 215, 353.
Sahai, R. (1987). Ap. J., 318, 809.
Sahai, R., Wannier, P.G. (1985). Ap. J., 299, 424.
Sahai, R., Wannier, P.G. (1988). Astr. Astroph., 201, L9
Schmidt, G.D., Cohen, M. (1981). Ap. J., 246, 444.
Schneider, S.E., Silverglate, P.R., Altschuler, D.R., Giovanardi, C. (1987). Ap. J., 314, 572.
Schönberg, K. (1988). Astr. Astroph., 195, 198.
Sivagnanam, P., LeSqueren, A.M. (1986). Astr. Astroph., 168, 374.
Sopka, R.J., Hildebrand, R., Jaffe, D.T., Gatley, I., Roellig, T., Werner, M., Jura, M., Zuckerman, B. (1985). Ap. J., 294, 242.
Sopka, R.J., Olofsson, H., Johansson, L.E.B., Nguyen–Q–Rieu, Zuckerman, B. (1988). Astr. Astroph., in press.
Stencel, R.E., Carpenter, K.G., Hagen, W. (1986). Ap. J., 308, 859.
Sun, J., Kwok, S. (1987). Astr. Astroph., 185, 258.
Taylor, A.R., Pottasch, S.R. (1987). Astr. Astroph., 176, L5.
teLintel Hekkert, P. (1987). In Late Stages of Stellar Evolution, eds S. Kwok and S.R. Pottasch, p. 83, Dordrecht: Reidel.
Tielens, A.G.G.M. (1983). Ap. J., 271, 702.
Tsuji, T., Unno, W., Kaifu, N., Izumiura, H., Ukita, N., Cho, S., Koyama, K. (1988). Ap. J. (Letters), 327, L23.
van der Veen, W.E.C.J. (1988). Thesis, University of Leiden.
van der Veen, W.E.C.J., Habing, H.J., Geballe, T.R. (1988). in prep.
Welty, A.D., Fix, J.D., Mutel, R.L. (1987). Ap. J., 318, 852.
Wannier, P.G., Sahai, R. (1986). Ap. J., 311, 335.
Wannier, P.G., Sahai, R. (1987). Ap. J., 319, 367.
Wood, P.R., Bessell, M.S., Whiteoak, J.B. (1986). Ap. J. (Letters), 306, L81.
Zijlstra, A.A., Pottasch, S.R. (1988), Astr. Astroph., 196, L9.
Zuckerman, B., Aller, L.H. (1986). Ap. J., 301, 772.
Zuckerman, B., Dyck, H.M. (1986a). Ap. J., 304, 394.
Zuckerman, B., Dyck, H.M. (1986b). Ap. J., 311, 345.
Zuckerman, B., Dyck, H.M. (1988). Astr. Astroph., in press
Zuckerman, B., Dyck, H.M., Claussen, M.J. (1986). Ap. J., 304, 401.

MASS-LOSING PECULIAR RED GIANTS: THE COMPARISON BETWEEN THEORY AND OBSERVATIONS

M. Jura
Department of Astronomy
University of California
Los Angeles CA 90024

ABSTRACT

The mass loss from evolved red giants is considered. It seems that red giants on the Asymptotic Giant Branch (AGB) are losing between 3 and $6\ 10^{-4}\ M_\odot\ kpc^{-2}\ yr^{-1}$ in the solar neighborhood. If all the main sequence stars between 1 and 5 $M_\odot$ ultimately evolve into white dwarfs with masses of 0.7 $M_\odot$, the predicted mass loss rate in the solar neighborhood from these stars is $8\ 10^{-4}\ M_\odot\ kpc^{-2}\ yr^{-1}$. Although there are still uncertainties, it appears that there is no strong disagreement between theory and observation. However, it could also be that we have not yet identified much of the source of the mass-loss from pre-white dwarf stars.

Approximately half the mass loss from AGB stars is from carbon-rich objects in agreement with previous estimates and the finding that about half of all planetary nebulae are carbon-rich. We conclude that an appreciable fraction, perhaps half, of all intemediate-mass main sequence stars spend $\geq 3\ 10^4$ years as carbon-rich stars.

1. INTRODUCTION

In the standard picture for their evolution, intermediate-mass stars that initially have between 1 and 5 $M_\odot$ (or perhaps as much as 8 $M_\odot$) on the main sequence eventually become white dwarfs with ~0.7 $M_\odot$ (Iben and Renzini 1983). It is also usually hypothesized that most of this mass loss occurs during the Asymptotic Giant Branch (AGB) phase of the star's evolution. Here, we critically review this hypothesis by using current data on mass loss from red giant stars.

In the standard scenario, just before a star becomes a white dwarf, it is the central star of a planetary nebula (see for example, Osterbrock 1974). Furthermore, it has long been thought that just before a star becomes a planetary nebula, it is a red giant undergoing extensive mass loss (Abell and Goldreich 1966; Kwok 1982). The usual hypothesis is that during the red giant phase, the star loses a large fraction, perhaps more than half, of its initial mass.

With improvements in millimeter and infrared technology, considerable progess has been made in understanding the mass-loss phenomena from red giants during the past 20 years (Zuckerman 1980; Olofsson 1985). Because AGB stars are intrinsically cool and because they have large amounts of circumstellar dust, they are primarily infrared sources; they are not very prominent, for example, in the Yale Bright Star Catalog (Hoffleit and Jaschek 1982). It is possible, however, to use the Two Micron Sky Survey (Neugebauer and Leighton 1969) and the IRAS data base to identify and study the local AGB stars. This has now been done systematically for carbon stars (Claussen et al. 1987), S Stars (Jura 1988), oxygen-rich stars (Kleinmann et al. 1988) and those stars which are losing large amounts of mass ($>10^{-6}$ $M_\odot$ yr^{-1}, Jura and Kleinmann 1988). About 1/3 of all the stars brighter than K = 3.0 magnitude in the sky are AGB stars. There also have been extensive surveys of the molecular emission from mass-losing AGB stars. The most prominently studied molecule is CO (see Knapp and Morris 1985, Zuckerman and Dyck 1986a,b, 1988, Zuckerman, Dyck and Claussen 1988, Olofsson, Eriksson and Gustafsson 1987, 1988), but there have been extensive surveys of other species such as HCN (Lucas, Guilloteau and Omont 1988) and OH (Engels 1979). With this extensive amount of data, it is now possible to make a much more quantitative test between the standard model for stellar evolution and the observed mass-losing AGB stars.

2. DISTANCES AND LUMINOSITIES

In order to determine the physical properties of these stars, it is necessary to infer their distances and luminosities. Unfortunately, there is no reliable way to determine the distances to AGB stars. In the Magellanic Clouds, carbon stars display a small dispersion in their apparent K-magnitudes (Frogel, Persson and Cohen 1980). Although the distances to the Clouds are uncertain (see, for example, Stothers 1988), the typical inferred luminosity is about 10^4 $L_\odot$, in agreement with theoretical predictions for AGB stars with core masses of 0.6 $M_\odot$ (Iben and Renzini 1983). However, at least for the oxygen-rich AGB stars there is probably a significant range in their luminosities. Jones, Hyland and Gatley (1983) use galactic rotation and the observed fluxes of OH/IR stars to infer that the luminosity of the star apparently increases with the outflow velocity of the gas (see also Jura 1984). Also, according to Glass and Lloyd-Evans (1980) there is a significant period-K magnitude relationship for oxygen-rich Miras in the Magellanic clouds. Therefore, the carbon stars may have an appreciably smaller variation in their intrinsic luminosities than do the oxygen-rich stars. However, Zuckerman and Dyck (1988) find that carbon stars with higher outflow velocities of their circumstellar gas lie closer to the galactic plane than do carbon stars with lower gas outflow velocities, and this suggests that the high-outflow velocity stars have higher luminosity. Such an effect is in agreement with our theoretical understanding of the mass-loss phenomena (Jura 1984). Also, on the basis of galactic rotation, Nguyen-Q-Rieu et al. (1987) argue that some carbon stars have luminosities appreciably greater than 10^4 $L_\odot$.

Even though the distances to AGB stars are somewhat uncertain, we can still hope to make reasonably accurate estimates of the integrated mass loss rate. If we assume that the local Milky Way is a very flat disk, and if our surveys penetrate far enough so that we detect essentially all the stars in an imaginary cylinder whose axis is perpendicular to the galactic disk, then the inferred mass-loss rate is not sensitive to the assumed luminosity of the mass-losing stars (Knapp and Morris 1985). This is because the inferred mass-loss rate per star varies as the distance to the star squared while the inferred surface density of stars varies inversely as the inferred distance squared. The product of the two quantities which gives the integrated mass-loss rate: the mass-loss per star and the number of stars per unit area is then independent of the inferred luminosity.

In conclusion, with an inferred luminosity of 10^4 $L_\odot$ and an insensitivity to the exact value of this inferred luminosity, we can hope to make a reasonable estimate of the integrated mass-loss rate from evolved stars in the solar neighborhood.

3. SURFACE DENSITIES OF AGB STARS IN THE LOCAL GALACTIC DISK

Because there is a probably smaller dispersion in their absolute luminosities, and because this conference is devoted to "peculiar" red giants, in this section we focus on the carbon stars and S-type stars. We assume a number density distribution of carbon stars, ρ, such that

$$\rho = \rho_0 \exp(-z/H)$$

where ρ_0 is the density of stars in the galactic plane, z is measured vertical to the plane and H is the exponential scale height of the stars. We may write for the surface density, σ, of stars in the solar neighborhood that:

$$\sigma = 2 \rho_0 H$$

Claussen et al. (1987) identified 81 carbon stars in the zone surveyed by the Two Micron Sky Survey within 1 kpc of the sun and they inferred that $\rho_0 = 100$ kpc^{-3}, $H = 200$ pc and $\sigma_0 = 40$ kpc^{-2}. If we include the 11 carbon stars listed by Jura and Kleinmann (1988) that lie within 1 kpc of the sun but are not within the Two Micron Sky Survey, the inferred volume and surface density and exponential scale height are not effectively changed. As discussed below, the inferred total mass-loss rate is substantially increased because the total mass return rate is dominated by the relatively rare stars which are faint at 2 μm because they are so obscured by their own circumstellar dust.

Jura, Joyce and Kleinmann (1988) have found that in the galactic anticenter direction, the surface density of high luminosity carbon stars does not change on a scale of ~ 4 kpc, although the total surface density of all stars almost certainly does decrease on this scale. This increase in the relative fraction of stars that are carbon red giants is probably a result of the lower average metallicity in the anticenter direction.

Jura (1988) has used the survey of Wing and Yorka (1988) to find that there are about 1/3 as many S stars as there are carbon stars in the neighborhood of the Sun.

Finally, Kleinmann et al. (1988) discuss the numbers of oxygen-rich AGB stars in the solar neighborhood. There are about an order of magnitude more oxygen-rich than carbon-rich mass-losing AGB stars in the solar neighborhood. However, when one restricts the analysis to only those stars losing large amounts of mass ($>10^{-6}$ $M_\odot$ yr^{-1}), then the numbers of oxygen-rich and carbon-rich stars are about equal (Jura and Kleinmann 1988).

4. MASS LOSS RATES

For most AGB stars, it is not possible to detect the molecular hydrogen which is thought to be the form for most of the mass that is lost (Glassgold and Huggins 1983). Therefore, it is necessary to use indirect means to measure the mass-loss rate. Currently, the two most common procedures are to use the intensity of the CO radio emission (Morris 1985, Knapp and Morris 1985), and the far infrared flux that is produced by the circumstellar shell (Sopka et al. 1985, Jura 1986a). There is reasonably good agreement between the two mass-loss rates as can be seen in Figure 1 where the mass-loss rates derived from the CO radio emission measured by Olofsson et al. (1987, 1988) divided by a factor of 2 are compared with the mass-loss rates derived by the formalism given in Jura (1986a). Since we do not know the ratio of [CO]/[H_2] or [dust]/[gas] in the outflows, and because there are uncertainties in the models, we find better agreement between the infrared and CO derived mass-loss rates if we apply a factor of 2 correction to one or the other. This factor of 2 should be taken as a measure of the systematic uncertainty associated with deriving the net mass-loss rates from AGB stars.

As reviewed by Jura (1986b), studies of the circumstellar envelopes around AGB stars are consistent with the view that the mass-loss process from these objects is a two step process. First, pulsations levitate the matter to a zone above the photosphere. Second, dust grains form and then radiation pressure expels the matter into the interstellar medium.

5. THE INTEGRATED MASS LOSS RATE FROM AGB STARS

Miller and Scalo (1979) have estimated the space densities, exponential scale heights and theoretical lifetimes for the main sequence stars in the neighborhood of the sun. If we assume that all the stars between 1 and 5 $M_\odot$ ultimately become white dwarfs of 0.7 $M_\odot$ (Liebert 1980), then we predict that the mass-loss rate from these stars is 8 10^{-4} $M_\odot$ kpc^{-2} yr^{-1} (see, for example, Jura 1987).

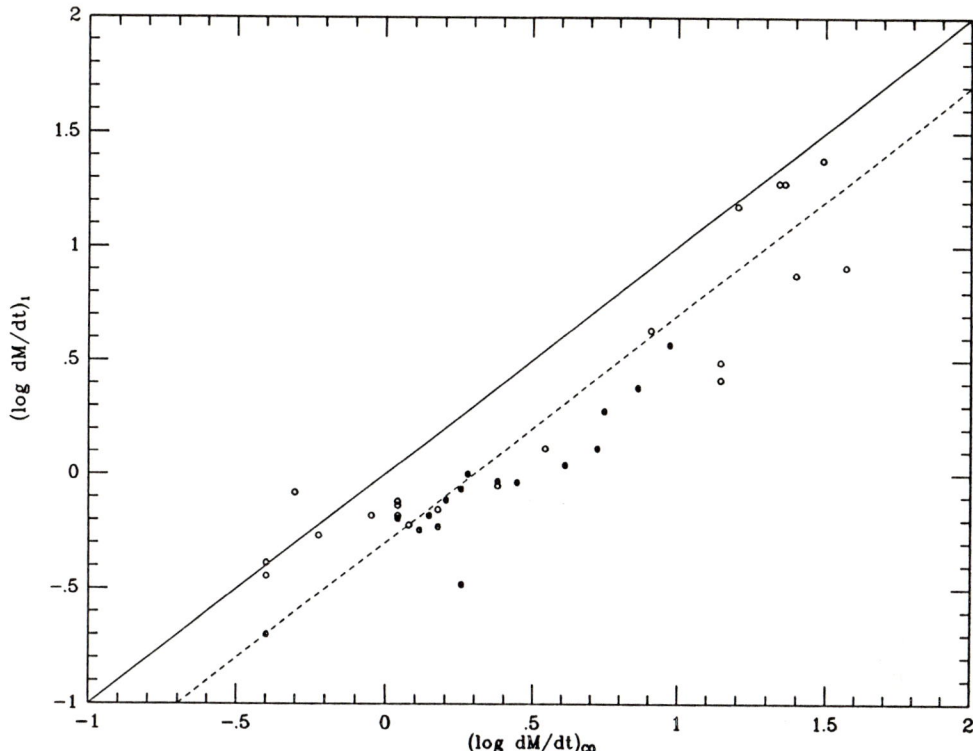

Fig. 1. Plot of the mass-loss rate derived from CO measurements, dM/dt_{CO}, vs. mass-loss rates derived from 60 μm fluxes and the formalism described by Jura (1986a). The units are 10^{-7} $M_\odot$ yr^{-1}. The filled circles are from CO data acquired at Onsala (Olofsson et al. 1987) while the open circles are from data acquired at SEST (Olofsson et al. 1988). The solid line is the fit if the two mass-loss rates equal each other; the dashed line is the fit if all the CO-inferred loss rates are reduced by a factor of 2. This figure is reproduced from Jura and Kleinmann (1988).

We now compare this theoretical rate with that inferred from the mass losing AGB stars. First, because the scale height of the carbon-rich AGB stars is about 200 pc, this implies that the progenitors of these objects usually have main sequence masses of 1.5 $M_\odot$, in agreement with the compilation of Miller and Scalo (1979).

We can use the infrared catalogs to identify all the mass-losing AGB stars in the solar neighborhood and then infer their mass-loss rates, to infer the total return rate by these stars into the interstellar medium. In Figure 2, we display a histogram of the mass loss rate for all the carbon stars in the neighborhood of the sun from the results from Claussen et al. (1987) and Jura and Kleinmann (1988). The "typical" carbon star is losing ~2 10^{-7} $M_\odot$ yr^{-1} (Claussen et al. 1987). However, a few carbon stars are losing more than 10^{-5} $M_\odot$ yr^{-1}. These stars, which are not very luminous at 2 μm because of circumstellar extinction, still amount to more than 20% of all the carbon stars, and therefore they dominate the return of matter into the interstellar medium by carbon stars. From the observed 60 μm flux and the procedures outlined by Jura (1986), we find that carbon stars, by themselves, are returning 1.6 10^{-4} $M_\odot$ kpc^{-2} yr^{-1} in the solar neighborhood. However, as discussed above, the calibration of the mass-loss rates by Olofsson et al. (1987, 1988) indicate that these mass-loss rates may be too low by a factor of 2. In this case, the return of mass into the interstellar medium by carbon stars is 3 10^{-4} $M_\odot$ kpc^{-2} yr^{-1}.

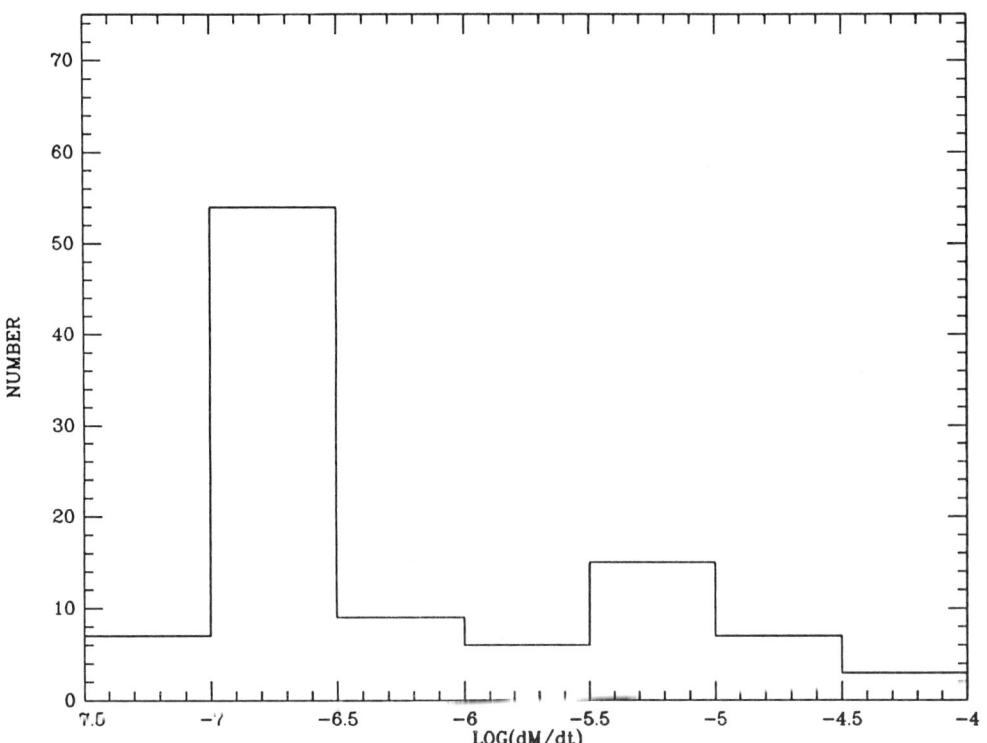

Fig. 2. Histogram of the mass-loss rates for all the carbon stars in the declination range -33°<δ<82° within 1 kpc from the Sun. The numbers are taken from Claussen et al. (1987) and Jura and Kleinmann (1988).

Another way to derive this integrated mass-loss rate is to use the surface density of very dusty carbon-rich stars of 12 kpc^{-2} (Jura and Kleinmann 1988) and an assumed mass-loss rate per star of between 1 and 2 10^{-5} $M_\odot$ yr^{-1} as expected if L = 10^4 $L_\odot$ and if the mass-loss rate is controlled by radiation pressure on the grains. The product of the surface density of stars and the mass-loss rate per star indicates that the carbon-rich objects could contribute between 1.2 and 2.4 $M_\odot$ 10^{-4} kpc^{-2} yr^{-1}, in agreement with the result derived above from summing the contribution from all the known carbon-rich stars.

Jura and Kleinmann (1988) have shown that very dusty oxygen-rich stars return about as much matter into the interstellar medium as do the very dusty carbon-rich stars. It also appears that the mass return rate by oxygen-rich stars into the interstellar medium is dominated by the relatively few objects that are losing large amounts of mass (Kleinmann et al. 1988). Therefore, the total return rate by both oxygen-rich and carbon-rich stars is about twice the total given above for the contribution simply from carbon-rich stars. Depending upon the calibration of the mass-loss rate from the infrared flux, this means that mass-losing AGB stars are returning between 3 and 6 10^{-4} $M_\odot$ kpc^{-2} yr^{-1}. Given the uncertainties, this result is in reasonable agreement with the predicted value 8 10^{-4} $M_\odot$ kpc^{-2} yr^{-1}.

A number of years ago, Zuckerman et al. (1976) argued that carbon-rich stars accounted for about half of all pre-planetary nebulae. More recently, Zuckerman and Aller (1986) have found that more than half of all planetary nebulae are carbon-rich. It appears that there is now strong evidence on the basis of this analysis of the mass losing stars that this picture is essentially correct: at least half of all main sequence stars enter into the carbon-star phase of evolution. Since there are about 12 carbon-rich high mass-loss rate carbon stars kpc^{-2} in the galactic disk in the neighborhood of the sun and since the death rate of all the main sequence stars between 1 and 5 $M_\odot$ in the solar neighborhood is 8 10^{-4} kpc^{-2} yr^{-1}, this phase of stellar evolution probably lasts for $\geq$ 3 10^4 years (Jura and Kleinmann 1988). During this carbon-star phase, which lasts 3 10^4 years, a 1.2 $M_\odot$ star loses between 1 and 2 10^{-5} $M_\odot$ yr^{-1} to shrink to a white dwarf of ~0.7 $M_\odot$. The data are therefore consistent with the standard model for the evolution of intermediate mass main sequence stars to white dwarfs.

6. CONCLUSIONS

1. We find that, given the uncertainties in the calibrations of the mass loss rates, there is reasonable agreement between the theoretical mass loss rate of 8 10^{-4} $M_\odot$ kpc^{-2} yr^{-1} for main sequence stars between 1 and 5 $M_\odot$ and the value inferred from the study of mass-losing red giants of 3-6 10^{-4} $M_\odot$ kpc^{-2} yr^{-1}.

2. It appears that about half of the mass that is lost is from carbon-rich stars. Therefore, roughly half of all intermediate-mass main sequence stars enter the carbon-rich phase for a duration of $\geq$3 10^4

years. We conclude that "peculiar" red giants are, in fact, a common
though short-lived phase of stellar evolution.

I thank Susan Kleinmann and Ben Zuckerman for their comments. This work
was partly supported by NASA.

7. REFERENCES

Abell, G. O., and Goldreich, P. 1966 P.A.S.P., 78, 232.
Claussen, M. J., Kleinmann, S. G., Joyce, R. R., and Jura, M.
   1987, Ap. J. Suppl., 65, 385.
Engels, D. 1979, Astr. Ap. Suppl., 36, 337.
Frogel, J. A., Persson, S. E., and Cohen, J. G. 1980, Ap. J., 239,
   495.
Glass, I. S., and Lloyd, Evans, T. 1981, Nature, 291, 303.
Glassgold, A. E., and Huggins, P. J. 1983, M.N.R.A.S., 203, 517.
Hoffleit, D., and Jaschek, C. 1982, The Bright Star Catalogue (4th
   rev. ed.; New Haven: Yale University Press).
Iben, I., and Renzini, A. 1983, Ann. Rev. Astr. Ap., 18, 263.
Jones, T. J., Hyland, A. R., and Gatley, I. 1983, Ap. J., 273, 660.
Jura, M. 1984, Ap. J., 282, 200.
Jura, M. 1986a, Ap. J., 303, 327.
Jura, M. 1986b, Irish Astr. J., 17, 322.
Jura, M. 1987, P.A.S.P., 99, 1123.
Jura, M. 1988, Ap. J. Suppl., 66, 33.
Jura, M., and Kleinmann, S. G. 1988, Ap. J., submitted.
Jura, M., Joyce, R. R., and Kleinmann, S. G. 1988, Ap. J., in press.
Kleinmann, S. G., Jura, M., Claussen, M. J., and Joyce, R. R. 1988,
   in preparation.
Knapp, G. R., and Morris, M. 1985, Ap. J., 292, 640.
Kwok, S. 1982, Ap. J., 258, 280.
Liebert, J. 1980, Ann. Rev. Astr. Ap., 18, 363.
Lucas, R., Guilloteau, S., and Omont, A. 1988, Astr. Ap., 194, 230.
Miller, G. E., and Scalo, J. M. 1979, Ap. J. Suppl., 41, 513.
Morris, M. 1985, in Mass Loss from Red Giants, eds. M. Morris and
   B. Zuckerman (Dordrecht: Reidel), p. 129.
Nguyen-Q-Rieu, Epchtein, N., Truong-Bach, and Cohen, M. 1987, Astr.
   Ap., 180, 117.
Neugebauer, G., and Leighton, R. B. 1969, Two Micron Sky Suyrvey
   (NASA SP-3047)
Olofsson, H. 1985, in Workshop on Submillimeter Astronomy, ed. P. A.
   Shaver and K. Kjar (Garching: ESO), p. 535.
Olofsson, H., Eriksson, K., and Gustafsson, B. 1987, Astr. Ap., 183,
   L13.
Olofsson, H., Eriksson, K., and Gustafsson, B. 1988, Astr. Ap., 196,
   L1.
Osterbrock, D. E. 1974, Astrophysics of Gaseous Nebulae (W. H.
   Freeman: San Francisco).

Sopka, R. J., Hildebrand, R., Jaffe, D. T., Gatley, I., Roellig, T., Werner, M., Jura, M., and Zuckerman, B. 1985, Ap. J., 294, 242.
Stothers, R. B. 1988, Ap. J., 329, 712.
Wing, R. F., and Yorka, S. B. 1977, M.N.R.A.S., 178, 383.
Zuckerman, B. 1980, Ann. Rev. Astr. Ap., 18, 263.
Zuckerman, B., and Aller, L. H. 1986, Ap. J., 301, 772.
Zuckerman, B., and Dyck, H. M. 1986a, Ap. J., 304, 394.
Zuckerman, B., and Dyck, H. M. 1986b, Ap. J., 311, 345.
Zuckerman, B., and Dyck, H. M. 1988, Astr. Ap., in press.
Zuckerman, B., Dyck, H. M., and Claussen, M. J. 1986, Ap. J., 304, 401.
Zuckerman, B., Palmer, P., Gilra, D. P., Turner, B. E., and Morris, M. 1978, Ap. J. (Letters), 220, L53.

Pre-Planetary Nebulae and R Corona Borealis Stars

Detlef Schönberner
Institut fur Theoretische Physik und Sternwarte

1. INTRODUCTION

With the advent of the IRAS satellite a completely new class of stellar objects became evident; viz., objects which are obviously in a rapid transition from the very tip of the asymptotic giant branch (AGB) into the planetary-nebulae region. These so-called pre-planetary nebulae (PPN) are characterized by stellar spectra indicative of supergiants mainly of F and G spectral types in conjunction with a large infrared excess due to a cool circumstellar dust shell. At the same time theoretical calculations through the AGB, with the inclusion of mass loss, and also through the following evolutionary stages down to the white-dwarf sequence, became available (Schönberner 1979, 1983; Wood and Faulkner 1986). These calculations predict evolutionary lifetimes of several 1000 years in the transition region between the tip of the AGB and the planetary-nebulae region. Thus a direct comparison between theory and observation now appears possible in the very early phase of this post-AGB evolution.

Another group of stars which occupy about the same region of the H-R diagram as PPNs, the R CrB stars (RCB), are well known to all astronomers, and their peculiarity has been known for more than 50 years. They are, however, one of the least studied groups of stars, and their evolution is not yet at all clear. As will be shown later in this review, they will most likely also evolve towards the final white-dwarf stage.

This review is organized as follows. After a short summary of post-AGB evolution, the theoretical predictions are compared with very recent observations of PPNs. Then follows a short account of our present knowledge about RCBs and a detailed comparison between the properties of both types of stars. Finally, a few statements about the evolution of RCBs are made.

2. TERMINATION OF THE AGB EVOLUTION

It is well established from evolutionary calculations that the luminosity of a giant at the Hayashi limit depends practically only on the mass, M(H), of the hydrogen-exhausted core (e.g. Paczynski 1970; Kippenhahn 1981). This statement holds also for stars on the AGB since it is mainly hydrogen burning that determines the overall course of evolution; i.e., L = L(H) for more than 80% of the evolutionary time. This luminosity, L(H), in turn determines the growth rate, $\dot{M}$(H), of the core according to

$$\dot{M}(H) = L(H)/(X \, E(H)) \;,$$

where E(H) (= 6.3×10^{18} erg/g) is the energy released per gram of hydrogen, and X is the hydrogen mass fraction of the stellar envelope. For a typical core mass M(H) = 0.6 M☉, $\dot{M}$(H) = 10^{-7} M☉/yr. This stellar core is already a very hot (pre) white dwarf, whose further evolution proceeds independently of the envelope as long as the latter contains sufficient mass (M(e) $\approx 10^{-4}$ M☉) to maintain the burning temperature in the hydrogen shell. The evolutionary tracks of AGB stars are determined by envelope expansion as the core M(H) grows by mass addition, and finally by contraction as the envelope M(e) drops below $\approx$ 0.05 M☉.

While the nuclear evolution of an AGB star is controlled by the core mass M(H), the total lifetime on the AGB is determined by mass loss from the surface of the star. This is a consequence of the observed stellar winds with mass-loss rates up $\dot{M}$(w) $\approx 10^{-4}$ M☉/yr for very luminous AGB stars (e.g. Knapp 1985; Kleinmann, this conference). Even if most stars do not reach such large rates, it is clear that always $\dot{M}$(w) >> $\dot{M}$(H) on the upper AGB (at the AGB limit, $\dot{M}$(H) = 6×10^{-7} M☉/yr). Thus, the lifetime at the tip of the AGB is very short (planetary-nebula formation!), and the subsequent post-AGB evolution is completely determined by the stellar structure at the AGB tip, which in turn is a function of the thermal-pulse cycle phase (cf. Iben 1984; Schonberner 1979, 1983).

Two evolutionary modes of an AGB star have to be considered -- the hydrogen-burning and helium-burning mode. Hydrogen burning makes up about 80% of the evolutionary time on the AGB and is characteristic of the majority of central stars (cf. Schonberner 1981). The star is in thermal equilibrium, and since there exists a one-to-one anticorrelation between residual envelope mass M(e) and effective temperature T(eff), the contraction time scale for the transit from the AGB tip to higher effective temperature is

$$t = -M(e)/\dot{M}(e) = M(e)/(\dot{M}(H) + \dot{M}(w)) \;.$$

For T(eff) < 10,000 K, M(e) < 0.001 M☉, and with $\dot{M}$(H) = const, the transition time scale is mainly determined by the wind term $\dot{M}$(w),

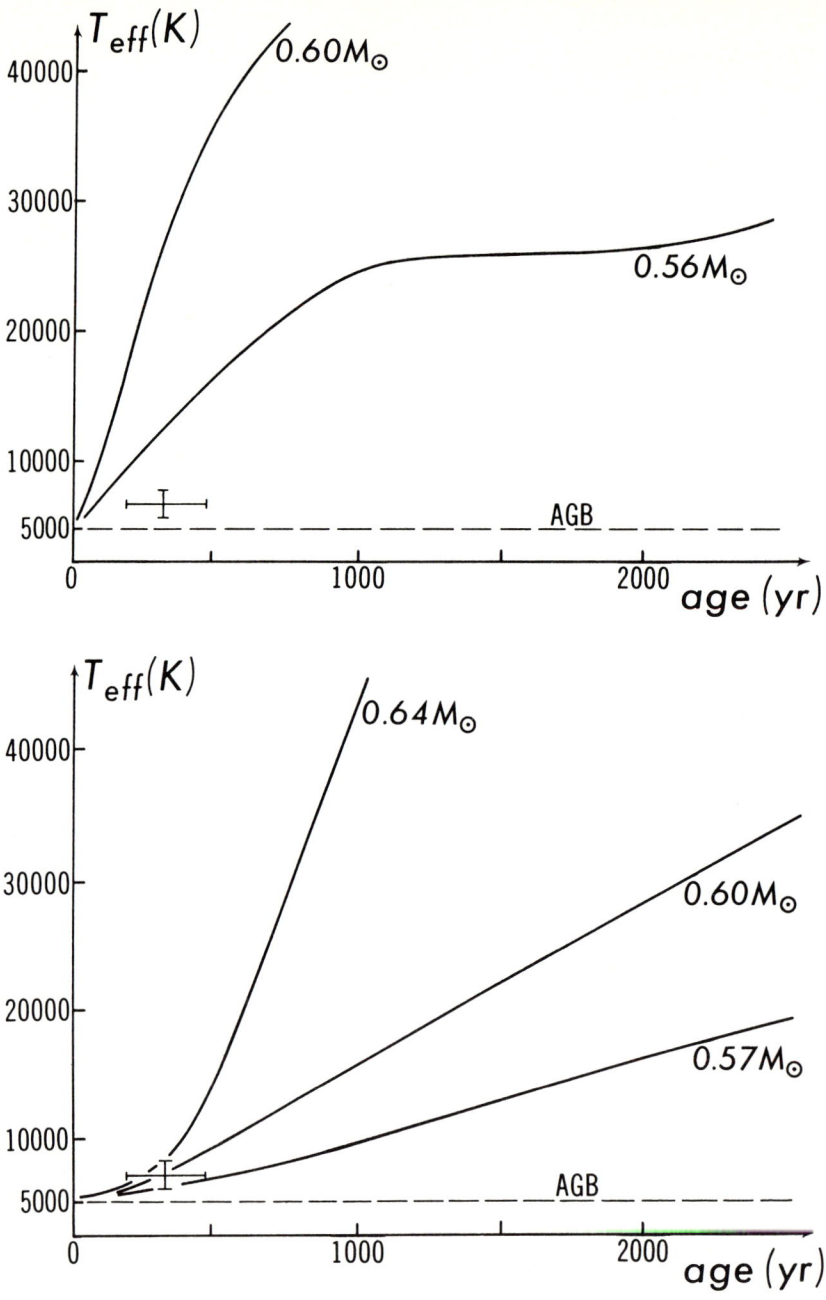

Figure 1. Post-AGB evolutionary tracks of different models displayed in a T(eff)-age diagram. Age zero corresponds to 5000 K. The upper part contains helium-burning models, taken from Wood and Faulkner (1986, 0.6 M☉) and Schönberner (unpublished, 0.56 M☉). The lower part shows the hydrogen-burning models of Schonberner (1979, 1983). In both parts the position of IRAS 18095+2704 is indicated (Hrivak et al. 1988).

especially in the vicinity of the AGB, where still $\dot{M}(w) > \dot{M}(H)$. Thus the evolutionary "speed" $\dot{T}(eff)$ of hydrogen-burning AGB remnants is practically determined by the size of post-AGB mass-loss rates and their variation with increasing effective temperature.

After a flash of the helium-burning shell, however, the star is completely out of thermal equilibrium and, should it be forced by mass loss to leave the AGB (planetary-nebulae formation), contracts according to the thermal time scale of the envelope. In this particular case, however, envelope means all matter above the helium-burning shell. It is difficult to estimate the transition time, but the computations show that it is much shorter than in the hydrogen-burning case (see Fig. 1).

Only two sets of evolutionary calculations with consideration of mass loss are available in the literature: Schonberner (1979, 1983) and Wood and Faulkner (1986). Schonberner (1983) assumed $\dot{M}(w) = 10^{-4}$ M☉/yr for $T(eff) \leq 5000$ K and $\dot{M}(w) = \dot{M}(w)$ (Reimers) otherwise, whereas Wood and Faulkner assumed $\dot{M}(w) = 10^{-5}$ M☉/yr for $T(eff) < 6300$ K and $\dot{M}(w) = 0$ otherwise. In both cases we have $\dot{M}(w) >> \dot{M}(H)$ in the vicinity of the AGB, and the mass loss affects a fast departure from the tip of the AGB. An illustration of the evolutionary speeds of different post-AGB models is presented in Fig. 1, where the effective temperatures vs. post-AGB ages are plotted. Age zero corresponds to 5000 K where the high AGB-mass-loss rate is assumed to cease. It is evident that a helium-burning model evolves much faster than a hydrogen-burning model of the same mass (cf. discussion above). For a typical remnant of 0.6 M☉, we find $\dot{T}(eff) \approx 40$ K/yr for the helium-burning and $\dot{T}(eff) \approx 10$ K/yr for the hydrogen-burning case. Please note that the curves for the hydrogen-burning models, taken from Schonberner (1979, 1983), contain the wind term $\dot{M}(w)$ according to Reimers (1975). This term is still important as long as $T(eff) < 10,000$ K (cf. Schonberner 1983).

3. PRE-PLANETARY NEBULAE

Stars close to the end of their AGB evolution are completely obscured by optically thick, relatively cool dust shells as the consequence of high mass-loss rates. When such a star leaves the AGB, the mass-loss rate will ultimately drop because the stellar surface is shrinking and warming. The dust shell continues to expand, but since there will not be sufficient replenishment of warm dust at the inner boundary of the shell, it becomes cooler and optically thinner until finally the stellar remnant may shine through (Bedijn 1987). By means of the IRAS satellite, several well-known supergiants which seem to fit this scheme have been observed. Mainly of spectral types F and G, these stars exhibit large IR excesses indicative of very cool ($T(d) \approx 100$ K) dust shells (Parthasarathy and Pottasch 1986; Likkel et al. 1987; Pottasch and Parthasarathy 1988). They are obviously low-mass remnants from the AGB instead of being massive young stars as one might judge from a first look at their spectra (cf. Luck and Bond 1984).

The IRAS data base also revealed the existence of non-variable OH/IR stars with properties just as described above: the peak of the IR emission is shifted further to longer wavelengths, and at $\lambda < 10$ μm the emission from the central remnant dominates the spectrum (Habing et al. 1989). These results suggest that pulsations are responsible for triggering high mass-loss rates. Recent theoretical calculations by Bowen (1988; also this conference) indicate the importance of periodic shocks caused by radial pulsations for increasing the extent of the stellar atmosphere and facilitating the formation of dust. Radiation pressure on dust grains then leads to high mass-loss rates. As a AGB giant becomes hotter by evolving off the AGB, pulsations will become less severe and will ultimately stop, leading to a decrease of the mass-loss rate.

Thus IRAS observations of non-pulsating OH/IR stars provides, for the first time, the opportunity to study the departure of stars from the AGB. By matching a dust-shell model to the observed IR flux one can estimate the inner dust-shell radius. Assuming a typical stellar luminosity of, say 6000 L☉, and with a typical expansion velocity of, say 15 km/s, one gets the time since the heavy mass loss ceased. The stellar temperature can be estimated by fitting the stellar component by a blackbody. Hrivnak et al. (1988) investigated the OH/IR source IRAS 18095+2704 this way and found a stellar component of T(eff) = 7000 K and an age of inner dust-shell boundary of $\approx$ 300 yr. The position of this object is shown in Fig. 1, and it appears that its position is consistent with hydrogen-burning model tracks of about 0.6 M☉.

It would be very useful to investigate more objects of this kind in order to improve our knowledge about the details of the transition from the AGB to central stars of planetary nebulae. Furthermore, the existing observations indicate that the space between the stellar surface and the inner dust shell is not empty. A small IR excess for $\lambda < 10$ μm hints at the existence of ongoing dust formation. A knowledge of this post-AGB mass-loss rate is very desirable since, as explained in the previous section, $\dot{M}(w)$ together with the nuclear term, $\dot{M}(H)$, determines the evolutionary speed.

A study of a larger sample of PPNs found by IRAS is presently underway (Van der Veen, private communication). Preliminary results are as follows: The central objects of these PPNs have effective temperatures ranging from 5,000 K to 20,000 K. Obviously the strong AGB mass loss continues until the remnant reaches about 5000 K. The distribution T(eff) vs. age corresponds to a mean evolutionary rate $\dot{T}(eff) \approx 10$ K/yr, in good agreement with the predictions of the hydrogen-burning post-AGB models of ≈ 0.6 M☉ by Schonberner (1979) (cf. Fig. 1). As already mentioned in the previous section, these computations considered a modest mass loss (Reimers 1975) with $\dot{M}(w) \gtrsim \dot{M}(H)$ ($\approx 10^{-7}$ M☉/yr). The consistency of Van der Veen's results with Schonberner's evolutionary calculations would then give the first hint that post-AGB mass-loss rates are of the order of the burning rates, i.e. $\dot{M}(w) \approx 10^{-7}$ M☉/yr.

4. THE R CORONA BOREALIS STARS

A comprehensive review on the evolutionary status and origin of RCB stars as a subgroup of extremely hydrogen-deficient stars is given in Schonberner (1986) and shall not be repeated here. Instead, in this review emphasis is put on certain aspects that have some bearing on PPNs. A few basic properties of RCBs, however, have to be mentioned first.

(1) RCBs are single, relatively cool supergiants with L/M $\approx 10^4$ L☉/M☉.

(2) They have an inert, electron-degenerate C/O core, growing by helium shell burning at a rate given by

$$\dot{M}(He) = L(He)/(Y\,E(He)) \, ,$$

where $E(He) = 6 \times 10^{17}$ erg/g and $Y = 1-Z$.

(3) Their extended envelope is virtually hydrogen-free, H/He 10^{-4}, and carbon is the most abundant element next to helium, C/He $\approx$ 0.003 ... 0.03 (number fractions: Schonberner 1975; Cottrell and Lambert 1982).

(4) From existing spectroscopic analyses and pulsational calculations, one can estimate a typical RCB mass of about 0.9 M☉ (Saio and Wheeler 1985; Weiss 1987a,b; Saio and Jeffery 1988).

(5) They certainly belong to an old stellar population because their galactic distribution has a scale height of $\approx$ 500 pc (M(bol) = -5 assumed).

Despite the fact that the history of RCBs is still not well understood (cf. section 6 below), the use of stellar models with the appropriate composition and with reasonable masses has proven somewhat successful (cf. Weiss 1987a). In fact, some of the above listed properties were derived from such models. Starting from a (hypothetical) helium main sequence, a helium-star model of about 1 M☉ evolves like a normal star, but with a larger luminosity, towards the Hayashi limit (lower branch). The model moves further upwards along the Hayashi line until contraction to the white-dwarf region sets in as the envelope mass falls below about 0.1 M☉ (upper branch).

An important piece of information is provided by the pulsational properties of RCBs, although neither the periods nor the effective temperatures are well known in most cases. For the best known cases, RY Sgr and R CrB, it appears that only on the upper, high-luminosity branch does the blue edge of the instability region extend far enough to higher effective temperatures to explain the pulsations in these two objects (Weiss 1987b). The case of RY Sgr is in particular interesting since its period is the best known, P = 38.6 d, but decreases according to $\dot{P}/P \approx -3 \times 10^{-4}$ yr^{-1} (Pugach 1977; Maracco and Milesi 1982; Kilkenny and Flanagan 1983). This period decrease indicates evolution towards the blue with a (pulsational) time scale of 3000 years. A stellar model of Weiss (1987a, 1987b) which closely matches the spectroscopically determined parameters of RY Sgr predicts P = 37 d and $\dot{P}/P = -5 \times 10^{-4}$ yr^{-1}. This model has M = 0.88 M$_\odot$ and L = 18,000 L$_\odot$ and is on the upper evolutionary branch. We conclude that RY Sgr evolves like an appropriate stellar model in thermal equilibrium (i.e., L = L(He)) towards higher effective temperatures with constant luminosity.

In the following we will assume that indeed only the high-luminosity (i.e. upper) branch is realized in nature and check whether the observations are consistent with this assumption (see also the discussion in secction 6). The known periods of RCBs range from 38.6 d (RY Sgr) up to $\approx$ 135 d (S Aps)*. The coolest models of Weiss at about 4000 K predict fundamental periods of about 400 d, far above 135 d, which corresponds to $\approx$ 5000 K. Thus one may conclude that either very cool RCBs with, say T(eff) below 5000 K, do not exist or that they are hidden behind optically thick dust shells. Support for the latter idea comes from non-adiabatic pulsational calculations by Saio and Wheeler (1985), which indicated that for M < 1.6 M$_\odot$ and T(eff) < 6000 K the pulsational amplitudes grow without bound. In reality one has, therefore, to expect substantial mass ejections in all these cases. However, a better knowledge of effective tempertures and pulsational periods of RCBs is important to further investigate the relation between periods and stellar temperatures. For example, Kilkenny and Whittet

*The fundamental mode seemed recently to have switched to the first overtone with $\approx$ 40 d, as reported by Kilkenny and Flanagan (1983).
(1984) estimated T(eff) = 4000 K for S Aps, a temperature which is incompatible with a period of 135 d.

As for R CrB itself, Gillett et al. (1986) detected by careful inspection of IRAS data a very cool (T(d) $\approx$ 30 K) and large (r(i) $\approx$ 0.7 pc) additional dust shell. With an expansion velocity of 20 km/s, this fossil dust shell is at least 30,000 yr old. Looking at appropriate models on the upper branch (Weiss 1987b, Table 7) one finds that the (contraction) time scale is about 40,000 yr at T(eff) = 4500 K, but only about 3000 yr at T(eff) = 7000 K (the present temperature of R CrB). Thus it appears that the fossil shell around R CrB was ejected when the star was much cooler, say T(eff) $\approx$ 4000 ... 5000 K. This interpretation would then be consistent with the above findings for RY Sgr. In this context it is interesting to note that fig. 1 of Walker (1985) clearly indicates very similar cool "fossil" shells around some other RCB stars. Because these cool shells appear to have some bearing on the evolutionary history of these objects, their thorough investigation is badly needed.

5. COMPARISON BETWEEN PROTO-PLANETARY NEBULAE AND R CORONA BOREALIS STARS

In this section we shall discuss similarities and dissimilarities between both groups of stars. Upon closer inspection of their properties it will become clear that they have little in common except that they both contain evolved stars with dusty circumstellar shells which populate similar regions in the H-R diagram. The typical properties are listed in Table 1.

Table 1: Typical properties of PPNs and RCBs

	PPNs	RCBs
envelope composition	normal	extremely hydrogen poor
stellar mass ($M_\odot$)	≈ 0.6	≈ 0.9
envelope mass ($M_\odot$)	$\lesssim 0.001$	$\lesssim 0.1$
temperature range (K)	5000 ... 10,000	5000 ... 7000 (16,000)*
dust temperature (K)	≈ 100	≈ 700
evolutionary rate (K/yr)	≈ 10	$\lesssim 1$
typical luminosity ($L_\odot$)	≈ 6500	$\approx 15,000$
core growth rate ($M_\odot$/yr)	$\approx 1 \times 10^{-7}$	$\approx 2 \times 10^{-6}$

* MV SGR (Jeffery et al. 1988)

Some comments seem to be in order. Firstly, the warm dust around RCBs indicates that their shells are closer to the stellar surface than is the case for PPNs. But this means only, of course, that the dust mass-loss rate is larger, and not necessarily the gas mass-loss rate! Note, for instance, that a RCB-like surface composition contains more refractory matter − namely carbon − than a solar-like composition. With a typical carbon-to-helium ratio as mentioned in section 4, one expects a gas-to-dust ratio of 10-100, the dust mainly being made out of amorphous or graphitic grains. This fact is important and should be considered when estimating total (i.e. dust plus gas) shell masses. For instance, solar-like compositions yield gas-to-dust ratios of about 250. Contrary to the PPNs, mass-loss rates for RCBs are not known. Feast (1986) estimates rates of about 10^{-6} $M_\odot$/yr, taking the peculiar photospheric composition into account and assuming that all carbon condenses. This rate is rather modest and can hardly compete with the burning rate (see Table 1). This would then explain why evolutionary calculations without mass loss correctly predict the observed evolutionary rates of RY Sgr (cf. section 4). Note that the PPNs have a burning rate lower by about a factor of ten, mainly due to the higher energy yield of hydrogen nuclear fuel, making their post-AGB evolution more dependent on the mass-loss rates.

Furthermore, the rather slow contraction of RCBs compared to PPNs explains the non-existence of helium-dominated planetaries. The total transition time from the Hayashi limit to the planetary-nebula region obviously exceeds the kinematical lifetime of any circumstellar shell. The so-called Wolf-Rayet central stars, which are believed to have hydrogen-free surfaces, possess nebular shells with solar composition. By contrast the RCBs will most likely turn into the so-called extremely hydrogen-deficient helium stars, none of which is known to possess a planetary, before they descend to the subdwarf and white-dwarf region (cf. Schonberner 1986).

6. COMMENTS ON THE ORIGIN OF R CORONA BOREALIS STARS

A detailed account about our understanding of the origin of RCBs has already been given elsewhere (Schonberner 1986); thus, only a few but nevertheless important points shall be emphasized again. There are two facts that exclude the possibilty that the RCBs are direct descendants of the AGB, contrary to earlier opinions. First of all, the evolutionary models (Schonberner 1977; Weiss 1987a) predict helium-envelope masses of the order of 0.1 $M_\odot$, which exceed the helium intershell mass in AGB stars with a comparable core mass by factors of 10 (cf. Paczynski 1975)! Secondly, deep envelope burning on the AGB, which would convert all the hydrogen into helium (Scalo et al. 1975), does not operate since the envelope composition of RCBs indicate a mixture of CNO-processed with triple-alpha-processed matter. The only hypothesis which seems to give at least a qualitative explanation for the existence of RCBs is that proposed by Webbink (1984) and further discussed by Iben and Tutukov (1985); namely, the merger of two close binary white dwarfs, one with a carbon-oxygen core, the other with a helium core. For more details, the reader is referred to the original papers cited above, or to Schonberner (1986).

It shall explicitly be emphasized here that the so-called "late-flash" scenario proposed by Iben et al. (1983) cannot explain the observed evolutionary lifetimes of RY Sgr and R CrB. Such a model is completely out of thermal equilibrium and has an evolutionary time scale in the vicinity of the AGB of about 100 year, as opposed to the observations which indicate time scales of several 1000 years for these two stars (cf. discussion of section 4). We think that this scenario, though it might be responsible for some exotic objects, will not work for the majority of RCBs.

Not much is known about the evolutionary status of a class of peculiar supergiants not mentioned at all in this review; viz., the extremely hydrogen-deficient carbon stars. Their spectra are very similar to those of RCB stars, yet they do not seem to possess any circumstellar dust shells (Feast and Glass 1973; Drilling et al. 1984; Walker 1985). Also, they do not pulsate, although Kilkenny (1988) seems to have detected low-amplitude, semi-regular variations in some cases. These may represent an extension of the RCB group to lower masses, and hence also to lower luminosity-to-mass ratios (Schönberner 1986).

In concluding this section, it must be stated that we are still far from an understanding of the origin of RCB stars.

The author gratefully acknowledges a travel grant from the DFG.

REFERENCES

Bedijn, P.J. 1987, Astron. Astrophys. 186, 136.
Bowen, G.H. 1988, Astrophys. J. 329, 299.
Cottrell, P.L., Lambert, D.L. 1982, Astrophys. J. 261, 595.
Drilling, J.S., Landolt, A.U., Schonberner, D. 1984, Astrophys. J. 279, 748.
Feast, M.W. 1986, Hydrogen Deficient Stars and Related Objects (Proc. IAU Coll. 87), ed. K. Hunger, D. Schonberner, N.K. Rao, (Dordrecht: Reidel), p. 151.
Feast, M.W., Glass, I.S. 1973, Mon. Not. R. Astron. Soc. 161, 293.
Gillett, F.C., Backman, D.E., Beichman, C., Neugebauer, G. 1986, Astrophys. J. 310, 842.
Habing, H.J., te Lintel Hekkert, P., van der Veen, W.E.C.J. 1989, Planetary Nebulae, (Proc. IAU Symp. No. 131), ed. J.B. Kaler, in press.
Hrivnak, B.J., Kwok, S., Volk, K.M. 1988, Astrophys. J., 331, 832.
Iben, I. Jr. 1984, Astrophys. J. 277, 333.
Iben, I. Jr., Kaler, J.B., Truran, J.W., Renzini, A. 1983, Astrophys. J. 264, 605.
Iben, I. Jr., Tutukov, A.V. 1985, Astrophys. J. Suppl. 58, 661.
Jeffery, C.S., Heber, U., Hill, P.W., Pollacco, D. 1988, Mon. Not. R. Astron. Soc. 231, 175.
Kilkenny, D. 1988, preprint.
Kilkenny, D., Flanagan, C. 1983, Mon. Not. R. Astron. Soc. 203, 19.
Kilkenny, D., Whittet, D.C.D. 1984, Mon. Not. R. Astron. Soc. 208, 25.
Kippenhahn, R. 1981, Astron. Astrophys. 102, 293.
Knapp, G. 185, Astrophys. J. 293, 273.
Likkel, L., Omont, A., Morris, M., Forveille, T. 1987, Astron. Astrophys. 173, L11.
Luck, R.E., Bond, H.E. 1984, Astrophys. J. 279, 729.
Marraco, H.G., Milesi, G.E. 1982, Astron. J. 87, 1775.
Paczynski, B. 1970, Acta Astron. 20, 47.
Paczynski, B. 1975, Astrophys. J. 202, 558.
Parthasarathy, M., Pottasch, S.R. 1986, Astron. Astrophys. 154, L16.
Pottasch, S.R., Partharasathy, M. 1988, Astron. Astrophys. 192, 182.
Pugach, A.F. 1977, Inf. Bull. Var. Stars No. 1277.
Reimers, D. 1975, Problems in Stellar Atmospheres and Envelopes, ed. B. Baschek, W.H. Kegel, G. Traving (Berlin: Springer), p. 229.
Saio, H., Jeffery, C.S. 1988, Astrophys. J. 328, 299.
Saio, H., Wheeler, J.C. 1985, Astrophys. J. 295, 38.

Scalo, J.M., Despain, K.H., Ulrich, R.K. 1975, *Astrophys. J.* **196**, 805.
Schönberner, D. 1975, *Astron. Astrophys.* **44**, 383.
Schönberner, D. 1977, *Astron. Astrophys.* **57**, 437.
Schönberner, D. 1979, *Astron. Astrophys.* **79**, 108.
Schönberner, D. 1981, *Astron. Astrophys.* **103**, 119.
Schönberner, D. 1983, *Astrophys. J.* **272**, 708.
Schönberner, D. 1986, *Hydrogen Deficient Stars and Related Objects* (Proc. IAU Coll. No. 87), ed. K. Hunger, D. Schonberner, N.K. Rao, (Dordrecht: Reidel), p. 471.
Walker, H.J. 1985, *Astron. Astrophys.* **158**, 58.
Webbink, R.F. 1984, *Astrophys. J.* **277**, 355.
Weiss, A. 1987a, *Astron. Astrophys.* **185**, 165.
Weiss, A. 1987b, *Astron. Astrophys.* **185**, 178.
Wood, P.R., Faulkner, D.J. 1986, *Astrophys. J.* **307**, 659.

Carbon Stars With Oxygen-Rich Circumstellar Chemistry

M.S. Vardya
Tata Institute of Fundamental Research

Abstract: Ten new carbon stars with oxygen-rich circumstellar shells have been discovered using IRAS LRS spectral class. The origin of these objects have been briefly discussed. It is likely that they may be a consequence of release of oxygen due to carbon condensation, though other scenarios cannot be ruled out.

Introduction

IRAS LRS spectra show 9.7 μm silicate emission in a large number of stars. The existence of this feature indicates oxygen-rich circumstellar (CS) chemistry. The presence of this feature is denoted in the LRS spectral classification (Olnon and Raimond 1986) by 2n and 6n, where n is a measure of the strength of the feature, and is given in the IRAS point source catalogue as well. Presence of this feature is expected in M stars but not in carbon stars. Therefore, it was a complete surprise when Little-Marenin (1986), Little-Marenin and Wilton (1986), and Willems and de Jong (1986) found ten carbon stars, including a Mira, with silicate emission. This discovery leads to several interesting possibilities about the existence and evolution of such objects.

The Sample

In order to increase the sample of such carbon stars, we have used the IRAS LRS spectral class along with the carbon star (Stephenson 1973) and variable star (Kholopov 1985) catalogues. In this way, we have discovered nine additional stars, all having LRS spectral class 2n. RT Oph, a Mira variable, can also be added to this list if its spectral classification (Kholopov 1985) of M7e(C) is confirmed. Note that the strength of the silicate emission, as denoted by n, is weak in a few cases, and may have resulted from noise in the spectra.

Results

The table gives the complete list of these carbon stars with oxygen-rich circumstellar shells discovered so far. The first ten

Carbon Stars With Oxygen Rich Circumstellar Shell

	R.A. (1950)	Decl. (1950)	CCS No.	Name	Sp. class	LRS	var. class	$\log\dfrac{S_{12}}{S_{25}}$	$\log\dfrac{S_{25}}{S_{60}}$	Source
1.	001714.9	+442553	11	VX And	C4,5J (N7)	24	SRa	0.534	0.586	V
2.	011321.1	+253019	63	Z Psc	C7,2 (N0)	22	SRb	0.473	0.541	V
3.	044822.5	+282634	254	TT Tau	C4,2-7,4 (N3)	22	SRb	0.492	0.218	V
4.	071755.2	+250540	716	BM Gem	C5,4J (Nb)	29	SRb	0.252	0.845	L,WD
5.	080017.0	+380330	1003	--	--	27	--	-0.042	0.568	L,WD
6.	085348.8	+200230	1344	T Cnc	C3,8-5,5 (R6-N6)	23	SRb	0.570	0.654	V
7.	085745.6	-603554	--	MC79-11	--	65	--	0.016	0.702	WD
8.	100906.5	-704925	1633	--	N,Mb	22	--	0.368	0.821	WD
9.	133421.7	-561321	2106	RV Cen	N3e	25	M	0.460	0.583	LW
10.	134417.7	-610930	2123	--	C::	27	--	0.135	0.710	WD
11.	175411.9	+111031	--	RT Oph	M7e(C)	22	M	0.339	0.787	V
12.	180037.1	-321310	--	FJF270	--	22	--	-0.051	0.409	WD
13.	185614.9	+141738	2684	UV Aql	C5,4-5 (N4)	22	SRa	0.536	0.456	V
14.	191355.3	+541206	--	NC#83	--	28	--	0.278	0.938	WD
15.	192322.7	+762746	2738	UX Dra	C7,3	23	SRa:	0.533	0.618	V
16.	192454.7	+232948	2733	--	C5,4 (N)	23	--	0.500	0.700	V
17.	194140.8	+342209	2783	--	C	22	--	0.494	0.775	V
18.	203504.0	+595456	2919	V778 Cyg	C4,5J (N)	29	Lb	0.224	0.967	L,WD
19.	231737.3	+465802	3184	EU And	C4,4 (R)	29	SR	0.099	0.961	WD
20.	235842.4	+600439	3214	WZ Cas	C9,2JLi (N1p)	23	SRb	0.509	0.550	V

Notes on the table:

1. $P = 369$ d; $\dot{M} > 1.6 \times 10^{-7}$ $M_\odot/yr$ (Olofsson et al. 1987); $T_{NIR} = 2400$ K (Willems 1987)
2. $P = 144$ d, $\dot{M} = 0.4 \times 10^{-7}$ $M_\odot/yr$; $T_{NIR} = 2800$ K
3. $P = 167$ d ($P = 364$ d ?); $T_{NIR} = 2550$ K
4. Periodic variation during some intervals, and shortly thereafter, not variable
6. $P = 482$ d; $f = 0.35$; $T_{NIR} = 2250$ K
7. (OH, H_2O)-maser
8. S Star?
9. $P = 446$ d, $f = 0.56$
10. H_2O maser
11. $P = 426$ d but variable; $f = 0.36$; (OH, H_2O)-maser; needs to be confirmed as a carbon star
12. Wrongly put as FJF 272
13. $P = 386$ d; $T_{NIR} = 2350$ K
15. $P = 168$ d, $f = 0.5$; $M = 1.8 \times 10^{-7}$ $M_\odot/yr$; light curve may correspond to type E; SB; $P \sim 340$ d
16. $T_{NIR} = 2200$ K
17. $N_{NIR} = 1200$ K
18. H_2O maser
19. H_2O maser
20. $P = 186$ d; $T_{NIR} = 2650$ K; VB A; light variations are sometimes taking place with double period; hydrogen emission not always detected; H_2 IR line detected

columns give serial number, right ascension and declination (epoch 1950) Stephenson's (1973) cool carbon star catalogue number, star name, spectral classification, IRAS LRS spectral class, type of variability, log S_{12}/S_{25}, and log S_{25}/S_{60}. Here S_λ is the flux density at wavelength λ in microns from the IRAS point source catalogue. The last column gives the source of discovery - L for Little-Marenin (1986), LW for Little-Marenin and Wilton (1986), V for this paper and WD for Willems and de Jong (1986). Notes at the end of the table give additional information on some of these stars. The spectral classification, especially the C subclass, is not available for about half the stars.

The figure gives a plot of log (S_{12}/S_{25}) vs log (S_{25}/S_{60}). Solid line is the blackbody curve. The number at the top of the star gives the LRS spectral class. The figure shows that most of the sources with weak or moderate strength of silicate emission lie near or below the blackbody line, as is true for normal carbon stars (cf. Willems 1987). However, all sources with spectral class 2n, with n > 7 are above the blackbody line, with smaller values of log (S_{12}/S_{25}) and larger values of log (S_{25}/S_{60}). A line above the blackbody line but slightly inclined to it demarcates the two regions. Note that class 6n may behave slightly differently. The dispersion of log (S_{12}/S_{25}) for weak silicate sources is small, barring a few exceptions, but is large for strong sources. In this sense also, weak silicate sources mimic normal carbon stars. The silicate emission strength appears to increase with increase in log (S_{12}/S_{25}) in the strong silicate sources.

Discussion

In an attempt to understand the causes of carbon stars showing oxygen-rich circumstellar dust, several possibilities have been advanced by Little-Marenin (1986) and by Willems and de Jong (1986).

1. An unusual chemical equilibrium exists in the CS matter, such that an oxygen-rich environment is produced out of carbon-rich material due to interaction of graphite and SiC grain formation with CO.

2. Due to rapid M → S → C star evolution, the star has become a C star since it (as an M star) ejected the oxygen-rich grains.

3. The system is a binary in which the C star is brighter, but the M star is the mass ejector.

4. This has been rapid, straight M → C star evolution, with oxygen-rich grains produced at the end of the M-star phase, which has now been ended by the conversion to a carbon star, as in 2 above.

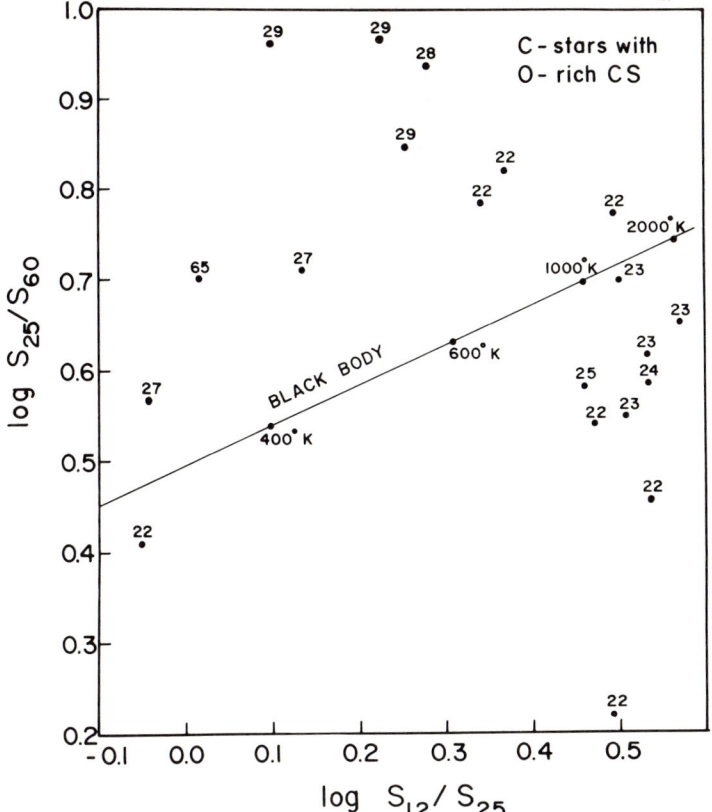

A color-color plot -- log (S_{25}/S_{60}) vs log (S_{12}/S_{25}) -- for carbon stars with oxygen-rich circumstellar shells. The number above each star gives the IRAS LRS spectral class. For comparison the solid line denotes a blackbody curve with various relevant temperatures marked.

Little-Marenin (1986) has discussed the first three possibilities but prefers the binary system hypothesis. The observation of water maser emission towards EU And (star no. 19) and the differences in velocity between photospheric absorption lines and the H_2O maser line have led Benson and Little-Marenin (1987) to propose it as the first radio spectroscopic binary, with the C star dominating the optical spectrum and the M star the far infrared; however, the binary nature is yet to be confirmed by observing periodic change in the radial velocity and the fact that the other component is an M star. Note that only two or three are known binary systems in our sample.

Willems and de Jong (1986) have also discussed the binary possibility but ruled it out; they prefer direct evolution from M to C star without passing through the intermediate S-star phase, especially for J-type carbon stars with ^{13}C overabundance. In the table only four stars are known to be J-type, two with strong silicate emission and two with silicate emission of moderate strength. The Mira variable RT Oph was classified earlier as M7e (Kukarkin et al. 1969) but is now classified as M7e (C) (Kholopov 1985); if this is indeed a carbon star, it may be pointing to the scenario suggested by Willem and de Jong (1986).

The above two possibilities, though not ruled out, may not be the answer for all such objects. But the solution via chemical equilibrium (no. 1), if it works, will have wide applicability. Tarafdar (1987) has carried out molecular equilibrium calculations with condensation for carbon-rich elemental composition; these show that condensation can increase the abundance of certain oxygen bearing compounds by several orders of magnitude, giving the appearance of oxygen-rich chemistry. Such calculations, but based on realistic conditions pertaining to circumstellar shells, may lead to a true picture. Such calculations may also hold the mystery of the presence of HCN in the oxygen rich circumstellar shells (cf. Deguchi and Goldsmith 1985).

Conclusion

Search for more carbon stars with oxygen-rich circumstellar shells should be made, and the known ones should be subjected to more detailed observations, besides theoretical calculations for condensation in a carbon-rich circumstellar conditions be undertaken. Then we may know the true nature of these objects.

References

Benson, P.J. and Little-Marenin, I.R. 1987, Astrophys. J. (Lettr.), 316, L37.
Deguchi, S., and Goldsmith, P.F. 1985, Nature, 317, 336.
Kholopov, P.N. 1985, General Catalogue of Variable Stars, 4th ed., (Moscow. Nauka, Publ. House), Vols. I-III.

Kukarkin, B.V. et al., <u>General Catalogue of Variable Stars</u>, 3rd ed. and supplements (Moscow: Astronomical Council of the Academy of Sciences of the USSR).
Little-Marenin, I.R. 1986, <u>Astrophys. J. (Lettr.)</u> **307**, L15.
Little-Marenin, I.R., and Wilton, C. 1986, in <u>Cool Stars, Stellar Systems, and the Sun</u>, ed. M. Zeilik and D.M. Gibson (Berlin: Springer Verlag), p. 420.
Olofsson, H., Eriksson, K., and Gustafsson, B. 1987, <u>Astron. Astrophys.</u> **183**, 413.
Olnon, F.M. and Raimond, E. 1986, <u>Astron. Astrophys. Suppl.</u> **65**, 607.
Stephenson, C.B. 1973, <u>A General Catalogue of Cool Carbon Stars</u>, Publ. Warner and Swasey Obs., Vol. 1, No. 4.
Tarafdar, S.P. 1987, <u>Astrochemistry</u>, eds. M.S. Vardya and S.P. Tarafdar, (Dordrecht: D. Reidel Publ. Co.), p. 589.
Willems, F.J. 1987, <u>Infrared Studies of Asymptotic Giant Branch Stars</u>, Ph.D. Thesis, University of Amsterdam.
Willems, F.J. and de Jong, T. 1986, <u>Astrophys. J. (Lettr.)</u> **309**, L39.

The Violet Flux Deficiency of Cool Carbon Stars

Donald G. Luttermoser   Hollis R. Johnson
Indiana University

A characteristic feature of carbon star spectra is the dramatic decrease in energy flux toward the violet. This flux decreases faster toward shorter wavelengths in N-type carbon stars than in either M giants or blackbodies of the same usual color temperatures, and an unknown violet opacity source in these carbon stars has consequently been the subject of speculation for the past 60 years. We investigate this longstanding problem of the violet flux falloff in cool carbon stars by directly testing, through synthetic spectra, the effects of many new and previously suggested opacity sources, based on currently available model atmospheres for carbon stars and M-giant stars. While several bound-free edges of neutral metals are important opacities, those of Na I at 2413 Å, Mg I at 2514 Å, and Ca I at 2940 Å are especially significant. Collectively, thousands of atomic lines, and in particular, the neutral metal resonance lines are the primary cause of the violet flux falloff. The severe violet flux decrease begins in cool carbon stars shortward of 4500 Å, roughly coincident with the redward wing of the extremely strong Ca I resonance line 4227 Å. The enormous line of Mg I at 2852 Å is one of the largest opacity sources in the near-ultraviolet region of these stars and influences the spectrum well into the visible. The pseudo-continuum of C_3 and the photodissociation continuum of CH both play noticeable but secondary roles. Other opacities that may affect the violet flux include the photodissociation of other hydrides, particularly NH and SiH, which have dissociation energies similar to CH. Synthetic spectra from the carbon-star models computed both with and without polyatomic molecules fit nicely to the collected observations of the well-observed carbon star TX Psc.

The severe violet flux falloff of these carbon stars is thus primarily a temperature effect. Carbon abundance plays an indirect role in determining the amount and depth of CO cooling as can be demonstrated by comparing synthetic spectra of a carbon-rich model and an oxygen-rich model with the same effective temperature and surface gravity -- the carbon-rich model is affected much more in the violet from the above mentioned opacity sources. If this scenario is correct, then the severity of the violet flux falloff as function of the "N" spectral type will be sensitive to a combination of both effective temperature and C/O ratio. This is indeed the case (as can be seen by comparing Table 1 and Table 4 of Lambert et al. 1986, Ap. J. Suppl. 62, 373).

Observations and Models for Red Giants with Unusual Dust

Ian Griffin   C.J. Skinner
Dept. Physics & Astronomy
University College London
Gower Street
London   WC1E 6BT

B.R. Whitmore
School Mathematical Science
Queen Mary College
Mile End Road
London   E1 4NS

We present near IR (H,K and L band) medium resolution ($\lambda/\Delta\lambda \approx 600$) spectra for a selection of 9 red giants which have previously been shown to exhibit anomalous dust emission as characterised by their IRAS LRS spectra. The objects observed (during UKIRT and AAT service time) include Carbon stars whose LRS spectra show the 9.7µm silicate feature and also M stars whose LRS spectra display an 11.3µm feature similar to that usually associated with emission from SiC dust grains.

Spectral classifications derived from our observations are, in all cases, in agreement with those derived for the same stars from optical spectra. This is shown to present some problems for a mooted binary model for these objects.

We present further evidence indicative that these objects are single stars by comparing the IRAS observations with models calculated using a comprehensive radiative transfer code that utilizes optical constants for silicate and SiC dust grains derived from the literature. Early models are suggestive that the amount of dust required to fit the observed silicate profiles does not necessarily produce enough optical depth at near IR wavelengths to obscure any reasonably luminous companion.

CO in OH/IR-Stars - on Excitation and Mass Loss

A. Heske, H.J. Habing, W.E.C.J. van der Veen   A. Omont, T. Forveille
        Sterrewacht Leiden, The Netherlands   Observatoire de Grenoble

Observations of CO in long period variables have been widely used to determine mass loss rates by applying models for CO line formation (e.g. Knapp and Morris, 1985) which use a simple method to take the impact from infrared radiation into account. Recent CO(2-1) and (1-0) observations of some more evolved OH/IR stars yielded much too low mass loss rates using these simple models, thus indicating that they cannot be extrapolated to far evolved AGB stars with optically thick circumstellar envelopes.

In order to investigate this effect in more detail we selected a sample of OH/IR stars with well known distances and IR and OH properties which should cover a wide range in optical depths of the circumstellar dust shells. It was observed in the CO(2-1) and CO(1-0) lines with the 30m-IRAM-telescope at Pico Veleta/Spain.

The following results were obtained:

1 - Mass loss rates determined from the CO(1-0) and CO(2-1) lines give too low values compared to those derived from the infrared continuum.

2 - The line ratio CO(2-1)/CO(1-0) increases with increasing IRAS colors when the 9.7 µm feature turns from emission into absorption.

3 - In some objects the expansion velocities derived from the CO(1-0) lines are larger than those derived from CO(2-1) or OH lines. The latter ones are comparable.

The main conclusion is that CO(1-0) and probably also CO(2-1) line intensities are suppressed if optically thick dust is present, i.e. the excitation temperature drops drastically when the optical depths of the dust in the 9.7 µm silicate feature.

Observations of the CO(3-2) line and comparison with present results should give more insight on excitation temperature and level population of CO.

Pulsation and its rôle for circumstellar features

A.Heske
Sterrewacht Leiden
P.O.Box 9513
2300 RA Leiden, The Netherlands

Circumstellar envelopes of cool giants and supergiants contain atomic and molecular gas, and dust. The charateristic spectral features of these different components can be observed at optical (atoms), at radio (molecules) and at infrared wavelengths (dust). Since the detection of circumstellar matter around giants and supergiants most studies concentrated on detemining mass loss rates from observations of a single component assuming steady mass loss during late stellar evolution. Nevertheless, the IRAS colour colour diagramme and evolutionary models rather point to a non steady evolution during the mass loss phase of the star.

Little is known how the structure of the circumstellar envelope and the spectral features evolve along the giant branch and by which stellar parameters they are determined.

To investigate these problems a sample of about 100 cool giants and supergiants with spectral types between K0 and M8, selected by their peculiar infrared colour (I–K), was observed at optical ($H\alpha$ and KI [potassium]), at infrared ($1 - 20\mu m$, including IRAS data) and at radio wavelengths (SiO at 43 GHz and CO at 115 GHz).

1 – A strong correlation was found between SiO masers, $H\alpha$ emission and multiple absorption components in the KI resonance absorption profile. It seems to hold for giants but *not* for supergiants. This suggests that shock waves act as a trigger for SiO maser emission and are responsible for multiple shell structure in envelopes of Mira variables.

2 – The asymmetry of the light curve and the period have significant influence on the strength of the $9.7\mu m$ emission feature, i.e. the optical depth of the dust.

3 – The expansion velocities derived from the thermal SiO(v=0) line are systematically about 2 km/s lower than those derived from the CO line.

This indicates that SiO emission originates in parts closer to the star where acceleration is still taking place.

These results lead to the conclusion that once pulsation becomes important during the evolution of cool giants the structure of circumstellar envelopes changes drastically.

IRAS09371+1212: A Unique Red Giant With Strong Emission in the 40-70 Micron Bands of Ice

A. Omont, S.H. Moseley, T. Forveille, P.M. Harvey, L. Likkel

IRAS 09371+1212 (sometimes called the Frosty Leo nebula) is up to now an absolutely unique object. Its stellar nature has been proved by the detection of circumstellar CO (J = 1-0) by Forveille et al. (1987), and by its stellar M giant spectral type, although its infrared colors and its visible appearance could have suggested a galaxy (Condon and Broderick 1986). It displays unique IRAS far infra-red colors which have been attributed by Forveille et al. (1987) to a strong emission in the 40-70 micron bands of ice. The presence of an extremely large quantity of ice is confirmed by extraordinary strength of its 3.1 micron absorption band (Rouan et al. 1988; Hodapp et al. 1988; Geballe et al. 1988).

The results of our February 1988 KAO observations (Omont et al. 1988) have confirmed its unusual nature: the most striking feature is the intensity of the far infrared bands of ice, by far the strongest observed up to now. Its dust temperature, confirmed to be smaller than 50 K by the KAO results, is by far the lowest known for a circumstellar envelope. This fact is certainly related to the anomalous abundance of ice and to its unusually low luminosity for a massive circumstellar envelope. Ice is clearly crystalline, and there is some indication from the results of our modeling that the emissivity index of the silicates is ~ 2, as in the model of Draine (1985). Its visible and near infrared appearance is also intriguing (Rouan et al. 1988; Clemens and Leach 1987; Hodapp et al. 1988). It clearly displays a bipolar nebular structure with a strong polarization and a thin dust disk seen edge on.

Clemens, D.P. and Leach, R.W. 1986, Optical Engineering.
Condon, J.J. and Broderick, J.J. 1986, Astron. J. 92, 94.
Draine, B.T. 1985, Astrophys. J. Supp. 57, 587.
Forveille, T., Morris, M., Omont, A., Likkel, L. 1987, Astron. Astrophys. 176, L13.
Geballe, T.R., Kim, Y.H., Knacke, R.F., Noll, K.S. 1988, Astrophys. J. Lett. 326, L65.
Hodapp, K.W., Sellgren, K., Nagata, T. 1988, Astrophys. J. Lett. 326, L61.
Omont, A., Moseley, S.H., Forveille, T., Harvey, P.M., Likkel, L. et al. 1988, in preparation.
Rouan, D., Omont, A., Lacombe, F., Forveille, T. 1988, Astron. Astrophys. 189, L3.

ER Del: A True Symbiotic S Star?

Hollis R. Johnson
Indiana University

Thomas B. Ake
Space Telescope Sci. Inst./CSC

ER Del = BD+8:4506 is a faint ($m \approx 10$) irregularly variable PRG star classified spectral type S5.5/2.5 (Ake 1979). Spectra show H alpha in emission and a strong UV continuum with emission lines of C IV, Si III], and C III] -- characteristic of symbiotic-like stars. Although a few other MS and S stars have hot companions, this is the first to show hydrogen emission lines in the optical region. It has been suggested (Smith & Lambert 1986; Little et al. 1987) that all Tc-deficient PRG stars are accidental; that is, they arose by mass transfer when the currently degenerate companion was itself an AGB star. Indeed, the Tc-deficient S stars HR 1105 (Ake, Johnson and Peery 1988) and HD 35155 (Ake and Johnson 1988) and the R8 star HD 59643 (Johnson et al. 1988) not only have hot, subluminous companions, but also are interacting systems - a sign the components are close enough for mass transfer to have occurred. An accretion disk is probably present.

In ER Del, the hot component is also variable, and the continuous flux at 1350 A is 2.5 times that of the companion in the brightest of these systems, HD 35155. Luminosities in C IV are: for HR 1105, 8×10^{30}; for HD 59643; 9×10^{30}; for HD 35155, 2.4×10^{31}; for ER Del, 1.5×10^{32} erg/s. Unlike these other stars, there is in ER Del no Mg II emission, and absorption features of Fe II are strong even at low dispersion. Thus the IUE flux is completely dominated by the secondary, whereas in the other systems the PRG is seen down to 2600 A. The H alpha emission is symmetric with a strong absorption core, and thus does not arise from a wind or stream from the primary; its presence in ER Del is no doubt due to the larger luminosity of the secondary. The brightness variability and range of excitation indicates we are observing stratified regions in a binary PRG + accretion disk system.

Ake, T.B. 1979, Astrophys. J., 234, 538.
Ake, T.B., Johnson, H.R. 1988, Astrophys. J., submitted.
Ake, T.B., Johnson, H.R., Peery, B.F. 1988, A Decade of UV Astronomy with the IUE Satellite, ed. E.J. Rolfe, ESA SP-281, in press.
Johnson, H.R., Eaton, J.A., Querci, F.R., Querci, M., Baumert, J.H. 1988, Astron. Astrophys., in press.
Little, S.J., Little-Marenin, I., Bauer, W. 1987, Astron. J. 94, 981.
Smith, V.V. and Lambert, D.L. 1986, Astrophys. J., 311, 843.

Fluorescence in the Outer Atmospheres of Red Giant Stars

K.G. Carpenter
CASA - University of Colorado @ NASA-GSFC

The outer atmosphere of a cool red giant star is an ideal locale for the operation of line fluorescence processes. Low plasma densities imply low rates of collisional de-excitation and thus allow radiative decay of levels populated by selective radiative pumping. There are many strong sources of line radiation (i.e. possible pumps) and numerous possible upward transitions from highly populated low-lying levels of abundant elements such as Fe II, thus providing many chance coincidences between potential pumps and lines to be pumped. These conditions ensure that many of the chromospheric emission features observed in the UV spectrum of such a star are affected by fluorescence. Many of the observed emission features originate from energy levels populated solely by radiative fluorescent excitation, including strong lines of S I, O I, CO, Ni II, Si I, Fe I and Fe II, as well as weaker lines from Cr II and Co II. Important pumps active in these atmospheres include hydrogen Lyman alpha, and individual lines of O I, C I, Si II, Fe II, and Mg II. In the case of Fe II, there are many additional features arising from upper levels whose populations, although primarily maintained by collisions, are also significantly affected by radiative fluorescent excitation. In fact, there may be virtually no level in Fe II not affected to one degree or another by direct decays or cascades down from levels populated by fluorescence, driven either by Lyman alpha or, in some cases, by lines of Fe II itself ("self-fluorescence").

Neither the spectrum itself nor the structure of the atmosphere can be properly understood without properly accounting for these fluorescent processes. Furthermore, detailed modeling of the processes has potential for providing critical information on the structure (density, temperature, velocity) of these outer atmospheres, including a determination of whether they consist of hot and cool layers or are multi-component atmospheres within which cool and hot plasmas co-exist at similar heights above the photosphere.

I summarize in this paper the processes active in these atmospheres. Grotrian diagrams which illustrate selected processes are presented; in particular, a grandiose Fe II diagram showing all levels of importance to the UV spectrum, and illustrating the impact of fluorescent excitation on virtually all of these levels is shown. A preliminary model of the Fe II ion, being prepared for use with MULTI and PANDORA is discussed.

Summary of Some Observations of Peculiar Red Giants With the IRAM 30M Radiotelescope

A. Omont, S. Guilloteau, R. Lucas, J.J. Benayoun, J. Cernicharo
E. Nercessian
    Observatoire de Grenoble, Universite Joseph Fourier
        M. Jura   M. Morris
                UCLA
    P.F. Goldsmith   D.C.Lis
    FCRAO University of Massachussets
    Nguyen-Q-Rieu - Observatoire de Paris-Meudon

    We present a brief description of the main results obtained on the following topics: (1) Sulfur-bearing molecules in 0-rich envelopes: SO_2 (104 and 160 GHz), H_2S (169 and 217 GHz) in 15 stars, SO (86 GHz) in 7 stars (all new detections except H_2S in OH231.8+4.2). Excitation temperatures and line profiles could suggest in some stars a main contribution from external layers for SO_2 and internal layers for H_2S.

(2) HCN in 7 0-rich envelopes. (3) HCO^+ in six envelopes (NGC 7027, CRL 618, IRAS 21282+5050, OH 231.8+4.2, CIT 6, VY CMa). (4) Observations of 7 new molecules (SO_2, SO, CS, HNC, NH_3, HCO^+, OCS) and of a rapid molecular outflow (v $\approx$ 100 m/s) in OH 231.1+4.2. (5) Discovery of the second strong millimeter-wave maser: HCN (0,2,0,J = 1-0) in CIT 6 (where it is strongly polarized) and in 5 other similar C-rich envelopes. (6) Measurements of large $^{13}CO/^{12}CO$ ratios in envelopes of 13 C-rich stars. In analysing these data, we show that the temperature profiles generally used for modeling the circumstellar envelopes can be quite wrong for small and very large mass-loss rates.

References

1. Lucas R., Omont A., Guilloteau, S., Nguyen-Q-Rieu 1986, Astron. Astrophys. Letters, 154, L12.
   Guilloteau, S., Lucas, R., Nguyen-Q-Rieu, Omont, A. 1986, Astron. Astrophys. Letters, 165, L1.
   Omont, A., Lucas, R., Guilloteau, S., Morris, M., Cernicharo, J. 1988, in preparation.
2. Nercessian, E., Guilloteau, S., Omont, A., Benayoun, J.J. 1988, Astron. Astrophys. in press.
3. Guilloteau, S., Omont, A., Lucas, R. 1988, in preparation.
4. Morris, M., Guilloteau, S., Lucas, R., Omont, A. 1987, Astrophys. J. 321, 888.
5. Guilloteau, S., Omont, A., Lucas, R. 1987, Astron. Astrophys. 176, L24.
   Lucas, R., Guilloteau, S., Omont, A. 1988, Astron. Astrophys. 194, 230.
   Goldsmith, P.F., Lis, D.C., Omont, A., Guilloteau, S., Lucas, R. 1988, Astrophys. J. in press.
6. Jura, M., Kahane, C., Omont, A. 1988, Astron. Astrophys. in press.

Grains of Meteorites, Originating in Cool Carbon Stars

Uffe G. Jørgensen

Niels Bohr Institute, Blegdamsvej 17, DK-2100 Copenhagen, Denmark

I. Introduction. The bulk of meteorites consist of material that condensed out of the primitive solar nebula, and their isotopic composition is therefore similar to that of the sun. Unfractionated meteorites (carbonaceous chondrites), nevertheless, contain a small fraction of inclusions that have a distinctly different isotopic composition of carbon, xenon, krypton and other elements (Zinner et al., 1987). These inclusions are therefore believed to be of pre-solar origin, and their composition and mineralogical structure carries information about the place where they were born.

Here I present the theory that inclusions of silicon carbide and diamonds in carbonaceous chondrites have their origin in the upper layers of cool carbon stars. The theory is in agreement with theories of late type stellar evolution, with high resolution stellar spectroscopy, as well as with available laboratory analysis on chemical, isotopic and mineralogical composition of carbonaceous chondrites. The SiC is calculated to be formed primarily in the early phases of carbon star evolution, whereas the diamonds are predicted to come from the later stages of evolution (Fig 2). The excess of heavy and light isotopes of xenon indicate that some of the diamonds have been in contact with a supernova of type I, that could have been triggered by mass flow from the same carbon star that created the diamonds.

II. Grains from carbon stars. Diamonds (C δ) and SiC (C β) with isotopic composition distinctly different from that of the sun, have been isolated from four meteorites (Lewis et al., 1987). Both grain types can form under the chemical and physical conditions that prevail in the upper atmosphere of cool carbon stars. SiC have been identified in the spectrum of several carbon stars.

Isotopic ratios of mainly $^{129}Xe/^{130}Xe$ and $^{12}C/^{13}C$ show that the grains didn't form neither in a supernova of type II nor in a local inhomogeneity of the pre-solar nebula. The atmosphere of a SNII progenitor has C/O < 1, and hence all the carbon bound in CO molecules, so that no carbon is left over for grain formation. The carbon rich interior zone of a SNII has $^{13}C/^{12}C = 0$ (which is not observed in grains from meteorites), and ^{129}I is there produced together with the Xe. If the grains were produced in chemical equilibrium in such matter, the highly reactive I would be included in the grains together with the Xe, and the ^{129}I would then later decay to ^{129}Xe which is not observed in excess in the meteoritic inclusions.

III. The silicon carbide. The SiC grains in meteorites are observed to have low $^{12}C/^{13}C$ compared to the sun. This is also normal in carbon stars with C/O near to unity (Fig 1). Accepting that the grains come from carbon stars, it is therefore concluded that SiC must come from the early phases of carbon star evolution. The Xe found trapped inside the SiC crystals is pure s-process Xe, which is the isotopic composition found in red giants (including the carbon stars).

IV. The diamonds. The diamonds have approximately solar $^{12}C/^{13}C$ which is usual in evolved carbon stars (higher C/O ratio). The diamonds, nevertheless, have an enhancement of heavy and light (Xe-HL) isotopes of Xe. These isotopes are expected to form only in supernova explosions, and the diamonds must therefore have been in contact with products of a supernova, after they were formed in a carbon star.

V. A binary system. Consider a close binary system, one of the components being a carbon star, the other a white dwarf. In the beginning the carbon star forms SiC grains, its radius is small and the mass flow (onto the white dwarf) is also small. Later the carbon star begins to form diamond grains. At this stage it is bigger and the mass flow onto the white dwarf is increasing. The accumulated mass on the white dwarf therefore grows fast now, and the chance for a supernova explosion increases dramatically. Therefore diamonds and not SiC grains will be contaminated with Xe-HL (as is found in carbonaceous chondrites). Reimers mass loss law predicts that $\dot{M} \alpha LRM^{-1}$ (where $\dot{M}$ = mass loss rate, L = luminosity, R = radius and M = mass of the star) for oxygen rich red giants, and there is observational indications that the mass loss increases considerably steeper (with luminosity) for evolved carbon stars.

VI. The model atmosphere prediction. High resolution spectroscopy of nearby carbon stars points at a relation between the $^{12}C/^{13}C$ and the C/O ratios, as do also stellar evolution theory, where pure ^{12}C (from 3α burning) is predicted to be mixed to the surface in AGB stars. Fig 1 shows this relation for the data presented by Lambert et al. (1986) for all of their stars with $^{12}C/^{13}C > 25$. Six stars that have considerably lower $^{12}C/^{13}C$, show no increase in $^{12}C/^{13}C$ with C/O and are assumed to be contaminated with e.g. CNO cycled material. Isotopic analysis of carbonaceous chondrites show that SiC have low $^{12}C/^{13}C$ and diamonds high $^{12}C/^{13}C$ (Lewis & Anders, 1983). Carbon rich model atmospheres (based on the code of Eriksson et al., 1984) predict that the molecular equilibrium favours (molecular progenitors of) diamond and graphite formation if C/O

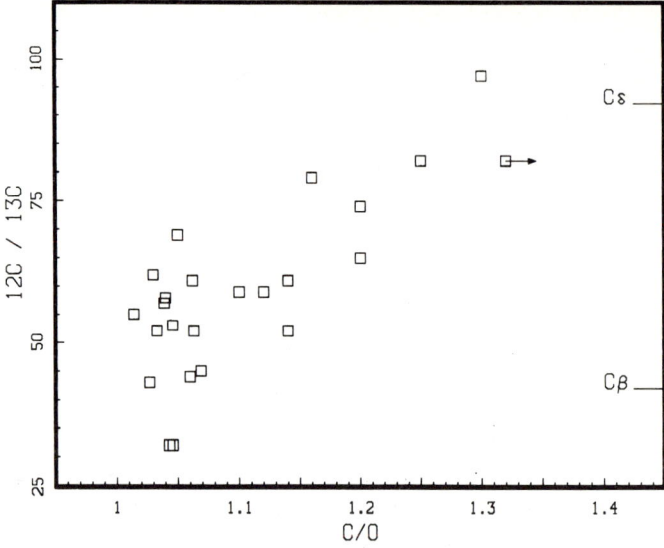

Fig 1. The $^{12}C/^{13}C$ and the C/O ratio of 24 of the brightest C stars in the sky. Also shown is the measured $^{12}C/^{13}C$ ratio of $C\beta$ and $C\delta$ grains in carbonaceous chondrites.

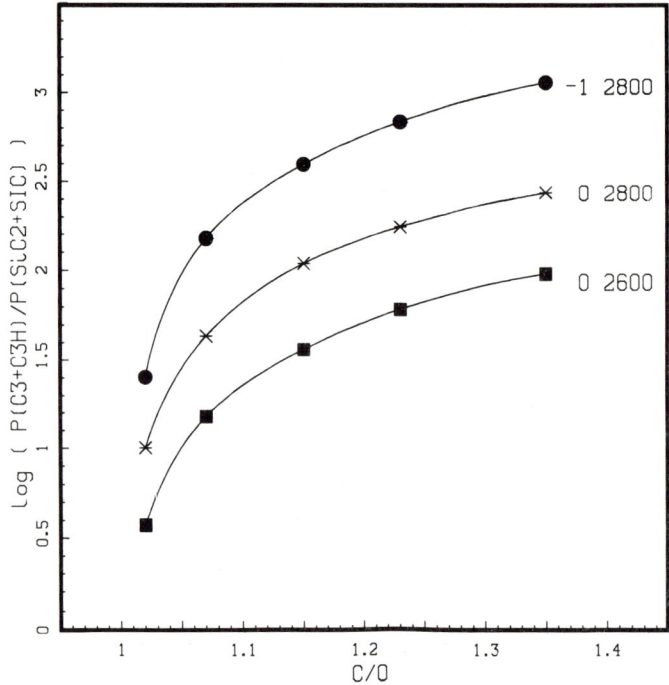

Fig 2. The ratio of expected molecular progenitors of diamonds and SiC, calculated as function of C/O in model atmospheres with $(\log(g), T_{\mathit{eff}}) = (-1, 2800), (0, 2800), (0, 2600)$, respectively.

(and hence $^{12}C/^{13}C$) is high, whereas SiC formation is favoured when C/O is close to unity (Fig 2). Model atmospheres combined with evolutionary theory (or high resolution spectroscopy) therefore predict diamonds with high $^{12}C/^{13}C$ and SiC with low $^{12}C/^{13}C$. This is also what is observed in carbonaceous chondrites.

VII. Carbon stars in the time of the solar system formation. Today the ratio of the number of red giants to the number of carbon stars, in our Galaxy, exceeds 100. Nevertheless, this ratio is strongly dependent on metallicity (Richer & Westerlund, 1983). The metallicity of the Galaxy was lower prior to the solar system formation, and carbon stars were therefore more common. The total amount of material transferred to the white dwarf SNI progenitors from carbon stars, is estimated to have been 9 times bigger than that transferred from oxygen AGB stars, at the time of the solar system formation (Jørgensen, 1988). Recent calculations (Gallino et al., 1988), furthermore, shows that the Kr isotopes found in the SiC grains can be predicted only from stars with metallicity lower than solar.

VIII. Conclusions. In conclusions it is found that the mineralogical, as well as the isotopic, composition that is found in small inclusions of carbonaceous chondrites, is in agreement with production of these grains in cool carbon star atmospheres, and that the diamonds enriched in Xe-HL were formed in a close binary system when a carbon star near its end of evolution were in the process of transferring material onto a companion white dwarf. This project has been supported by the Danish Natural Science Research Council.

Eriksson, K., Gustafsson, B., Jørgensen, U.G., Nordlund, Å. 1984, *Astron. Astrophys.* **132**, 37.
Gallino, R. et al. 1988, to be submitted to *Nature*.
Jørgensen, U.G. 1988, *Nature*, **332**, 702.
Lambert, D.L., Eriksson, K., Gustafsson, B., Hinkel, K. 1986, *Ap. J. Suppl.*, **62**, 373.
Lewis, R.S., Anders, E. 1983, *Scient. Am.*, **249**, 54.
Lewis, R.S., Ming, T., Wacker, J.F., Anders, E., Steel, E. 1987, *Nature*, **326**, 160.
Richer, H.B., Westerlund, B.E. 1983, *Ap. J.*, **264**, 114.
Zinner, E., Ming, T., Anders, E. 1987, *Nature*, **330**, 730.

He I λ10830 in the S Star HR 1105

Jeffery A. Brown   Hollis R. Johnson   Kenneth H. Hinkle
Astronomy Department, Indiana University   Kitt Peak National Obs.

Shcherbakov (1979) first reported that the S star HR 1105 shows a strong and variable He I λ10830 line. We confirm this report and present spectra of HR 1105 around 10830 A taken over approximately 0.3 orbit, from a program to obtain full orbital coverage for this 600-day system. Using Griffin's (1984) orbital solution, we have phased in the data presented by Shcherbakov with our own and found that for two coincidences of phase, the 10830 profiles at different epochs but same orbital phases are identical within the limitations of the data. The earlier observations were taken eight orbits (13 years) before ours.

Maximum strength of the He I feature occurs at conjunction (where the unseen secondary is in front of the primary), when the profile is pure absorption with an equivalent width of 2.2 A. For the next six months the profile is nearly a classic P Cygni profile. The equivalent width of the emission lobe is larger near quadrature (in May and July) than in February. Estimates of the strength of the absorption lobe are hampered by the presence of strong blends in the photospheric spectrum of the primary, but the total strength of the absorption lobe is roughly constant in the last three spectra.

The repetition of profile with phase suggests that the He I profile is modulated by the orbit, although extended monitoring of this long-period system would be needed to establish this. If this is the case, then HR 1105 could be labelled a marginally interacting binary system.

Some Condensation Calculations Using Chemical Equilibria

Christopher M. Sharp
Max-Planck-Institut für Physik und Astrophysik
Institut für Astrophysik
Karl-Schwarzschild-Straße 1
Federal Republic of Germany

A number of extensive chemical equilibrium calculations have recently been performed for temperatures below 2000 K, with mixtures containing over 200 gaseous species and the allowance for the formation of over 60 condensates. These calculations were based on the minimization of the Gibb's free energy.

For calculations performed with a solar mixture (C/O=0.55, but excluding the inert gases) at a total pressure of 500 dyn cm^{-2}, we found the most important condensed species to appear are $MgSiO_3$ and Mg_2SiO_4, with SiO_2 not appearing, which is in good agreement with Lattimer et al. (1978) and Tarafdar (1987). As expected, those gaseous species like SiO that contain at least one element that is also present in a condensed species, decrease sharply in abundance when the corresponding condensate appears. However, many flourine, chlorine and sulfur bearing species such as CS, increase in abundance relative to the case when condensation is neglected. This can be explained by these elements being released from other molecules containing an element that does condense out, such as SiS releasing sulfur when silicon condenses as silicates.

Calculations performed for a number of different C/O ratios show that for the carbon rich cases the condensation of graphite and SiC decrease the gas phase C/O ratio, and in the oxygen rich cases oxides and silicates increase the gas phase C/O ratio; in both cases this ratio moves closer to unity. Of particular interest are the weakly oxygen rich mixtures with C/O ratios greater than about 0.83, as sufficient oxygen is removed from the gas phase by oxides and silicates that graphite can condense out, and species like HCN and C_2H_2 which would otherwise have negligible abundances, can increase by several orders of magnitude.

Lattimer, J.M., Schramm, D.N., and Grossman, L. (1978), Ap.J. <u>219</u>, 230.

Tarafdar, S.P. (1987), in Astrochemistry, M.S. Vardya and S.P. Tarafdar, Eds., IAU Symposium No. 120, 559.

OH maser survey of very cool IRAS sources

P. te Lintel Hekkert and A. Heske
Sterrewacht Leiden
P.O. Box 9513, 2300 RA Leiden, The Netherlands

A.M. Le Squeren
Observatoire de Paris
Section de Meudon, DERADN, F-92190, Meudon, France

During our 1612 MHz survey of IRAS sources in Parkes, Australia, we found two objects with very large expansion velocities : IRAS15405–4945 and IRAS16342–3814 (P. te Lintel Hekkert et al. 1988, Astron. Astrophys. (Letters), in press.). These two sources and the well known source, OH231.8 + 4.2 , were discovered to have almost identical IRAS colours. The spectra of the three sources could be fitted with a 100 (±15) K black body curve. We than selected, using the IRAS colour–colour plot, a number of bright sources (F (60 μm) > 25. Jy) with similar colours to OH231.8 + 4.2 and IRAS16342–3814 . A small survey of about 150 sources was started in the last week of June and will be finished at the end of July 1988. Using the radio telescope in Nançay , we observed all four OH maser transitions, with the hunderd meter radio telescope at Effelsberg only 1612 MHz was observed, the sources found at Nançay were confirmed in Bonn and vice versa. From the preliminary analyses of the first 40 observations six sources were found of which three were not known before.

IRAS22036+5306 shows the characteristic two spike profile of OH/IR stars at the 1612 MHz and 1667 MHz transitions. This source has very red colours compared with "regular" OH/IR stars. Most likely this is an example of an OH/IR star which has stopped it's mass loss several hunderd years ago. Since than the original CSE has expanded and cooled. The presence of the 1667 MHz could also be interpret in this way (B.M. Lewis 1988, preprint) . It has been suggested by several authors (e.g. J. Sun and S. Kwok 1987, Astron. Astrophys. **185** , 285) that these cool OH/IR stars are PPN's.

The other two were found to have 1667 MHz emission only. IRAS17423–1755 showed a profile similar to the one of OH231.8 + 4.2 : a wide plateau of emission on top of which there is a broad spike. The profile of IRAS18491–0207 resembles the one of IRAS16342–3814 : two bunches of spikes, with a total velocity extent of 140 km · s^{-1}! The shape of the OH spectra suggest that the maser emission is unsaturated and pumped by photons from the high energy background (the hot star or nebula). The differences in shape of the profiles of the sources can be explained by assuming different orientation of the axis joining the two reflection nebulae. The high expansion velocity and the probable existance of the hot component in the system suggests that these are short lived stages of stellar evolution, most likely PPN's.

Molecular Cooling in the Outer Atmospheres of Red Giants

David Muchmore

Département de Physique, Université de Montréal
C.P. 6128, Succ. A, Montréal, H3C 3J7, Canada

In the outer atmospheres of cool giants and supergiants there is a competition between heating by shock waves which develop from noise lower down in the convection zone and radiative cooling. The post-shock cooling is most effective in keeping temperatures low and in damping shocks if molecules (especially CO) form. CO is important both because of its high dissociation energy (11 eV) and its large IR opacity.

If it is in LTE, CO can quickly cool atmospheric gases to $T \leq 1000$ K, permitting dust formation quite near a star's photosphere. However, in low density gas which is regularly shocked, chemical reactions may be too slow to keep molecular abundances in equilibrium at the momentary temperature.

Upper and lower bounds for plausible molecular formation rates for CO behind shocks are $d\ln N_{CO}/dt = kN_O$, where N represents particle density and k is a reaction rate constant with a value of 10^{-11} to 10^{-17}. For conditions appropriate for the region identified in empirical studies as the temperature minimum in Arcturus (K2 III), we find that the time scale for molecule formation is likely to be roughly (*very* roughly considering the broad range of plausible reaction rates) comparable with the time scales for shock repetition, which are a few hours for weak shocks and 1-2 days for stronger shocks (Cuntz and Muchmore, 1989).

Approximate solutions for the behavior of post shock temperatures show that CO is an effective coolant when the time scale for molecular formation is no more than about 10 times longer than the period between shocks.

Most significantly, we conclude that molecular abundances in the outer atmospheres of red giants will generally be out of equilibrium, lagging in time behind temperature and density fluctuations. At lower densities — farther from a star or in stars with lower g, i.e. the conditions expected in the majority of giant atmospheres — molecules would not have time to associate at all between shock heating events.

Cuntz, M. and Muchmore, D., 1988, *Astr. Ap.*, submitted

A Unified Formula For Mass Loss Rate of O to M Stars and its Effect on Stellar Evolution

S.P. Tarafdar
Tata Institute of Fundamental Research
Homi Bhabha Road, Bombay 400 005, India

A formula for stellar mass-loss rate has been derived using conservation equations of mass and momentum for coronal and continuous radiation driven wind. The derived mass-loss rate formula has been found to be consistent with the observed mass-loss rates for stars from O to M spectral type. Two constant parameters appearing in the mass-loss rate formula have been found to have values for special groups of stars like Be-stars and Wolf-Rayet stars different from each other and from the majority of stars. The effect of mass-loss according to the formula for the majority of stars on stellar evolution has been examined.

On the Nature of Radio Emission of Late-Type Giants

G.M. Rudnitskij
Sternberg Astronomical Institute
Moscow State University,
Moscow V-234, 119899, USSR

Long-period variable stars (LPV's) are considered. Pulsations of the surface of such a star result in formation in the stellar atmosphere of a shock wave in each cycle of the star's variability, ionizing the circumstellar gas which, recombining, gives rise to emissions in optical lines hydrogen and metals. I show the recombining layer behind the shock front to be optically thick in the free-free continuum at the radio wavelengths as short as 1 cm. At the gas temperatures behind the front $T = 20000$ K, the radio flux density at 1 cm from the front surface at a distance of a few hundred parsecs will be several or several tens of mjy. So far, the only positive result of searches for LPVS' radio continuum is the detection of radio flux from R Aql, obtained at different epochs by several authors. In October 1978 R Aql showed radio emission of 5.3 mjy at 14.9 GHz (Bowers and Kundu, 1979), and in August 1985 – 0.54 mjy at the same frequency (Drake et al., 1987); these figures are consistent with our model. Besides that, R Aql experienced stronger radio flares at longer wavelengths, up to a few hundred mjy (Woodsworth and Hughes, 1973; Estalella et al., 1983); these cannot be explained by thermal radio emission and require a nonthermal mechanism (synchrotron or cyclotron maser).

No continuum radio emission was found in other LPVS, possibly due to unfavorable observaional epochs. It is desirable to monitor systematically the nearest LPVS (R Cas, W Hya, O Cet, R Leo), especially at the variability phases when optical emissions are present.

LPVS' continuum radio emission may influence the intensity of their circumstellar molecular maser emission (OH, H_2O, SiO), provided the maser is initially unsaturated and the molecules are amplifying the stellar radio continuum. In this case, correlation must exist between the fluxes in the radio continuum and maser line (Rudnitskij, 1987).

References

Bowers, P.F. and Kundu, M.R. 1979, A.J., 84, 791.
Drake, S.A., Linsky, J.L., Elitzur, M. 1987, A.J., 94, 1280.
Estalella, R., Paredes, J.M., and Ruis, A. 1983, Astr. Ap., 124, 309.
Rudnitskij, G.M. 1987, in Circumstellar Matter. Proc. IAU Symp. 122
    (Dordrecht, Reidel), p. 267.
Woodsworth, A. and Hughes, V.A. 1973, Nature Phys. Sci., 246, 111.

POSSIBLE PROTO-PLANETARY NEBULAE

M. Parthasarathy
Indian Institute of Astrophysics

ABSTRACT: IRAS data for high-galactic-latitude F supergiants, a few luminous F-G stars, and a few peculiar (forbidden) emission-line stars reveal they have dust shells with characteristics similar to those observed in planetary nebulae. The objects described here appear to have experienced severe mass loss in the recent past on the AGB. These stars are most likely post-AGB stars or evolving from the tip of AGB towards the left in the H-R diagram and may be described as possible proto-planetary nebulae. From the IRAS data we find ten additional possible proto-planetary nebulae. From the IRAS data of HD 56126, $-53°$ 5072, HD 168625 and $-59°$ 6723 the luminosities, temperatures and masses of the dust shells are derived. HD 56126 (F5I), HD 168625 (B8Iae), and $-59°$ 6723 (B8Iae) appear to be in an evolutionary stage (post-AGB) similar to that of HD 161796 and HR 4049.

1. INTRODUCTION

The asymptotic giant branch (AGB) phase of evolution of intermediate and low mass stars is terminated by the ejection of the most of the hydrogen-rich outer envelope, resulting in a planetary nebula (Iben and Renzini 1983; Iben 1985). The superwind type of mass loss may lead to the formation of a planetary nebula and the termination of the AGB phase (Renzini, 1981). The Reimers mass-loss rate or low mass-loss rate will not be able to account for the formation of planetary nebulae. The formation of a planetary nebula may be preceded by a mass-loss rate of the order of 10^{-4} to 10^{-5} $M_\odot$ per year (Zuckerman 1978). Renzini (1981) proposed that superwind type of mass loss (10^{-4} to 10^{-5} $M_\odot$ per year) can take place for low and intermediate mass stars at the tip of the AGB phase. Iben and Renzini (1983) showed that once the star reaches a critical luminosity most of the residual hydrogen-rich envelope is ejected on a time scale very short compared with the previous AGB life time. Sabbadin et al., (1984) showed that the radius-expansion velocity diagram for planetary nebulae is accounted better by the superwind type of phenomenon. Recently several evolved stars and OH/IR stars have been found to show mass loss rate of the order of 10^{-4} to 10^{-5} $M_\odot$ per year (Knapp and Morris, 1985; Zuckerman and Dyck 1986a; Knapp 1987) which will become planetary nebulae within 10^5 years. Pottasch's (1984) definition of proto-planetary nebulae includes objects at the tip of AGB and also objects in a transition stage from the tip of AGB to the left in the HR diagram.

From IRAS data we find that the high galactic latitude F supergiants, a few luminous F-G stars, and a few peculiar (forbidden) emission-line stars to have dust shells with characteristics similar to that observed in planetary nebulae. These objects appear to have experienced severe mass loss in the recent past on the AGB. These objects may be proto-planetary nebulae. In this paper we present a brief report on these objects. From the IRAS data ten additional possible proto-planetary nebulae are detected and their properties are reported.

2. HIGH GALACTIC LATITUDE A AND F SUPERGIANTS

The presence of F-type supergiants at high galactic latitude was first noticed by Bidelman (1951). The evolutionary stage of these stars is not clear. HD 161796 and other stars in this group show small amplitude light and radial-velocity variations and also show switching of pulsation modes (Fernie 1986). The far infrared IRAS measurements of high galactic latitude F supergiants HD 161796 (F3Ib) and HD 187885 (SAO 163075) (F2I) and related stars were found to show strong far infrared excess due to large amounts of circumstellar dust (Parthasarathy and Pottasch 1986). The masses and sizes of the dust envelopes around HD 161796 and HD 187885 suggest that these stars suffered extensive mass loss in the recent past during the AGB stage of evolution. If the ratio of gas to dust mass is about 100, as it is in the interstellar medium, the total shell masses are of the order of 0.3 Mo. This is very similar to that found in some planetary nebulae. Lamers et al. (1986) and Waelkens et al. (1987) found the high-galactic-latitude A supergiants HR 4049 and HD 213985 also to have far infrared (IRAS) excess due to dust shells. The evolutionary stage of these two stars also appears to be similar to that of HD 161796 and HD 187885.

Recently we found ten additional luminous F-G stars to show far infrared (IRAS) exesss due to dust shells around them (Pottasch and Parthasarathy 1988). Their far infrared flux distributions and colours are similar to that of HD 161796 and planetary nebulae. The star HD 179821 (G5Ia) shows very large far infrared excess (12μm: 31.4 jy, 25μm: 647.7 jy; 60 μm: 515.8 jy; and 100μm: 166.8 jy). A low-dispersion spectrum of HD 179821 obtained with the image tube spectrograph and 1 meter telescope at the Kavalur Observatory shows that its spectrum is very similar to that of a G5Ia star. The MK - M(v) calibration yields an absolute magnitude M(v) = -8.0. The luminosity of the dust shell is found to be 2.5×10^4 L$_\odot$, and the mass of the dust shell is estimated to be 2.6×10^{-3} M$_\odot$. If the ratio of gas to dust mass is about 100, as it is in the interstellar medium, the total shell mass is of the order of 0.26 M$_\odot$. Zuckerman and Dyck (1986b) detected CO J = 1 $\rightarrow$ 0 emission from HD 179821. It has the largest radial velocity with respect to local standard of rest. Its CO outflow velocity is also large (34 km s^{-1}), which suggests that the star is of high luminosity. Zuckerman and Dyck (1986b) estimate the distance to be 6 kpc and suggest that it is probably an unusual oxygen-rich supergiant located far from the galactic plane. The optical region spectra of HD 161796, HD 187885, and HD 179821 clearly shows that they are high luminosity objects and the presence of detached cool dust shells suggests that they have experienced severe mass loss in the recent past. Detailed observational study of these stars may enable us to understand their evolutionary stage.

We have obtained short-wavelength SWP (1150Å to 2000Å) and long-wavelength LWP (1900Å to 3200Å) low resolution (6Å) spectra of HD 161796 and HD 188885 (only SWP spectrum) with the IUE satellite. The UV spectrum of HD 161796 shows no excess UV flux attributable to a hot degenerate companion. The ultraviolet spectrum of HD 161796 is consistent with it being a F3I star (Parthasarathy, Pottasch, Wamsteker 1988). The discontinuity at 1700Å due to SiI and the flux ratio log [f(λ)(5500Å)/f(λ)(1900Å)], with the calibration of Bohm-Vitense (1982), yields a temperature of 6300 K for HD 161796, which is in good agreement with the value detrmined by Fernie and Garrison (1984). There is no evidence for significant metal deficiency. From an analysis of the UV spectrum we find that the metal abundance is normal: [Fe/H] = 0.0. The 2000Å feature in the spectrum of HD 161796 shows no evidence for circumstellar or interstellar reddening. In spite of large IR (IRAS) excess and large amount of dust around the star, there is no evidence for reddening from 1600Å to 3.5 µm. This result suggests that the grain size in the dust shell is relatively larger causing wavelength independent extinction.

The blue-visual region spectrum of HD 187885 suggests that it is a F2I supergiant (Parthasarthy, Pottasch, Wamsteker 1988). However, the 1250Å to 1900Å low-resolution spectrum of HD 187885 is peculiar because of the presence of a strong emission feature around 1580Å and the possibility of a broad absorption feature in the 1657Å spectral region. In late A and early F stars the flux in the wavelength region λ <1700Å is very low. However, the UV spectrum of HD 187885 shows significant flux in the wavelength region 1400Å - 1700Å. The strong and broad (1560Å to 1610Å) emission at 1580Å may be due to SiI lines, since there is a crowding of SiI lines in this wavelength region. These lines can be in emission if the temperature at the surface drops less steeply than in normal A and F supergiants or there is an outward increase of temperature. This also makes the discontinuity at 1700Å due to SiI less steep, which seems to explain the observed less steep discontinuity at 1700Å. A broad absorption feature nearly 100Å wide centered around 1657Å appears to be present. This absorption feature may be due to CI or to quasimolecular absorption of H2 (Parthasarathy, Pottasch, Wamsteker 1988). High resolution UV spectrum of HD 187885 may enable us to understand the cause for the spectral peculiarities.

3. ADDITIONAL POSSIBLE PROTO-PLANETARY NEBULAE

A list of ten additional possible proto-planetary nebulae is given in Table 1. All these ten sources show far infrared flux distributions and colours similar to that observed in planetary nebulae. For some of the the optical counterpart is very faint. The flux ratio f(IR)/f(vis) is very high, which suggests that the optical counterpart is obscured by the dust envelope. The dust temperature T(d), the luminosity of the dust shell L(IR), and the mass M(d) of the dust shell around HD 56126, H 168625 and -59° 6723 are estimated and are given in Table 2.

HD 56126 is a F5I supergiant. The IRAS data shows a flux maximum around 25µm (table 1). The dust temperature is found to be 140 K. Van Genderen, Van Driel and Greidanus (1986) made VBLUW photometric observations and found it to show circumstellar reddening. The distance is estimated to be 5 kpc. The far infrared luminosity of the dust shell is found to be of the order of 1.4×10^4 L$_\odot$. A low dispersion spectrum obtained with the image-tube spectrograph at Kavalur Observatory with 1-

Table 1: Additional Possible Proto-Planetary Nebulae

	R.A (1950) DEC	V	Sp	IRAS fluxes (Jansky)			
	h m s ° ′ ″			12μm	25μm	60μm	100μm
IRAS	07 02 45.3 -79 34 23			22.8	82	41.6	13.2
HD 56126 SAO 96709	07 13 25.4 10 05 08	8.2	F5I	24.6	116.7	49.8	18.2
RAFGL4106	10 21 32.5 -59 16 53			200.8	1754.6	851.7	179.6
-53° 5072 SAO239853	12 17 33.3 -53 38 49	9.3	F2	1.0	20.7	7.4	2.4
-59° 6723 SAO243756	16 20 40.5 -59 56 38	9.8	B8Ia	0.4	11.1	12.1	4.7
IRAS	16 34 17.1 -38 14 18			16.3	199.8	290.1	138.2
IRAS	16 59 26.7 -46 56 14			45	297.3	130.5	33.5
RAFGL6815	17 15 04.6 -32 24 15			58	321.4	268.3	81.4
HD 168625 Hen 1681	18 18 25.5 -16 23 56	8.4	B8Ia	70.	325.6	116.6	
IRAS	20 02 48 39 10 03			41.8	210.8	142.6	45.9

Table 2: Luminosities, Temperatures and Masses of the Dust Envelopes

star	d(kpc)	F_{IR} (10^{-12} Wm^{-2})	L_{IR} ($L_\odot$)	T_d (K)	M_d ($M_\odot$)
HD 56126	5.2	16.5	1.4×10^4	140	3.8×10^{-4}
-53 5072		2.6			
HD 168625	6.2	43.9	5.3×10^4	140	1.4×10^{-3}
-59 6723	5.6	1.8	1.8×10^3	100	1.8×10^{-4}

telescope is very similar to that of a F4Ia supergiant. Spectra obtained in the red region show Hα in emission. The Hα emission strength and profile appears to be variable. Zuckerman, Dyck and Claussen (1986) find CO J = 1 → 0 emission (see also Likkel et al. 1987). The IRAS data (Table 1) suggests that HD 56126 is in an evolutionary stage similar to that of HD 161796.

HD 168625 is a B8Ia supergiant. The IRAS data shows a flux maximum at 25μm. It is also listed as an emission line object (Hen 1681). Near IR data also shows an infrared excess. The IRAS data yields a dust temperature of 140 K. The distance is estimated to be about 3 kpc. The luminosity of the dust shell is found to be 1.2×10^4 L$_\odot$, and the mass of the dust shell is found to be of the order of $M(d) = 3.3 \times 10^{-4}$ M$_\odot$.

$-59°$ 6723 (SAO 243756) is a luminous star in the southern milky way (Stephenson and Sanduleak, 1971). The MK spectral type is B8Iae. Klare and Neckel (1977) made UBV and polarization observations and find V = 9.79, B-V = 0.32, U-B = 0.26. The distance is estimated to be 5.6 kpc. The far infrared flux distribution and colours are similar to that observed in planetary nebulae. HD 168625 (B8Ia) and $-59°$ 6723 (B8Ia) appear to be similar to HR 4049 and HD 213985 (Waelkens et al. 1987).

Some of the objects described earlier in the text and some of the objects given in Table 1 were found to show CO J = 1 → 0 emission and also 1612 MHZ OH maser emission similar to that observed from evolved stars. The far-infrared characteristics of the circumstellar dust of the stars described here are very similar to those observed in planetary nebulae. These objects are not associated with any star forming regions. Some of the these objects are very far from the galactic plane. The most likely explanation for the presence of dust envelopes around these stars is that they experienced severe mass loss in the recent past on the AGB. They may be evolving from the tip of the AGB towards the left in the HR diagram and may be described as possible proto-planetary nebulae.

As proto-planetary nebulae evolve into early stages of planetary nebulae, part of the dust shell will begin to be ionized. We find few peculiar emission line (forbidden) stars to show far infrared (IRAS) characteristics similar to that observed in planetary nebulae. From the IRAS data we find peculiar emission line stars HD 51585, Hen 401, Hen 591, Hen 1013, He2-138, M2-9, Hen 1336, Hen 1357, Hen 1428, Tc 1, and M1-26 to show dust shells with characteristics similar to that observed in young compact planetary nebulae (Parthasarathy and Pottasch 1988). The central stars in these objects are rapidly evolving towards hotter spectral types. He2-138, Tc 1 and M1-26 are clearly far enough advanced and may be called planetary nebulae. Recently quite a number of sources have been alleged to be objects in the transition between the tip of AGB and planetary nebula stage. There are approximately less than 200 possible proto-planetary nebulae in the IRAS point source catalogue. The duration of the transition stage is relatively short and therefore we expect to see significant differences in the characteristics of the objects occupying the transition region. The transition region may include objects like non-variable OH/IR stars (Le Bertre 1987; Kwok et al. 1987) and also objects like CRL 618, IRC +10420, Vy 2-2, CRL 2688, and HD 161796, and on the other end of the transition there are young planetary nebulae like NGC 7027, NGC 6302, and IC 418. Detailed study of the objects in the transition region will enable us to understand their evolutionary stage and evolutionary sequence.

4. CONCLUSIONS

The far infrared (IRAS) flux distributions and colours of the stars described here suggests that the dust envelopes around these stars have characteristics similar to that observed in planetary nebulae and evolved stars. The characterics of CO and OH emission from these objects also suggests that they are evolved stars. Objects of the type described here are not found in young clusters, associations, or in star forming regions. The presence of dust envelopes with characteristics similar to those observed in planetary nebulae suggests that they suffered severe mass loss in the recent past on the AGB. The CNO abundances and $^{12}C/^{13}C$ isotope ratio (from CO J = 1 → 0 and 2 → 1 data) of these stars may enable us to understand their evolutionary stage. The relation and evolutionary sequence between the stars described here and stars such as OH/IR 17.7-2.0 (Le Bertre 1987) and OH/IR 349.2-0.2, OH/IR 0.9+1.3 (Pottasch et al. 1987) is not clear, however, they all appear to be in a transition phase between the tip of the AGB and young planetary nebula stage and may be described as possible proto-planetary nebulae.

REFERENCES

Bidelman, W.P. 1951, Astrophys. J., 113, 304.
Bohm-Vitense, E. 1982, Astrophys. J., 225, 192.
Fernie, J.D. 1986, Astrophys. J., 306, 642.
Fernie, J.D., Garrison, R.F. 1984, Astrophys. J., 285, 698.
Iben, I. 1985, Quart. J. Roy. Astron. Soc., 26, 1.
Iben, I., Renzini, A. 1983, Ann. Rev. Astron. Astrophys., 21, 271.
Klare, G., Neckel, Th. 1977, Astron. Astrophys. Suppl., 27, 215.
Knapp, G.R. 1987, in Late Stages of Stellar Evolution, ed. S. Kwok and S.R. Pottasch (Reidel:Dordrecht), p. 103.
Knapp, G.R., Morris, M. 1985, Astrophys. J., 292, 640.
Kwok, S. 1987, Astrophys. J., 321, 975.
Lamers, H.J.G.L.M., Waters, L.B.F.M., Garmany, C.D., Perez, M.R., Waelkens, C. 1986, Astron. Astrophys., 154, L20.
Le Bertre, T. 1987, Astron. Astrophys., 180, 160.
Likkel, L., Omont, A., Morris, M., Forveille, T. 1987, Astron. Astrophys. 173, L11.
Parthasarathy, M., Pottasch, S.R. 1986, Astron. Astrophys., 154, L16.
Parthasarathy, M., Pottasch, S.R., Wamsteker, W. 1988, Astron. Astrophys., 203, 117.
Parthasarathy, M., Pottasch, S.R. 1988, Astron. Astrophys., (submitted).
Pottasch, S.R. 1984, Planetary Nebulae, Astrophys. Space Science Library, Vol. 107.
Pottasch, S.R., Bignelli, C., Zijlstra, A. 1987, Astron. Astrophys., 177, L49.
Renzini, A. 1981, in Physical Processes in Red Giants ed. I. Iben, and A. Renzini (Dordrecht: Reidel), p. 165.
Sabbadin, F., Gratton, R.G., Bianchini, A., Ortolani, S. 1984, Astron. Astrophys. 136, 181.

Stephenson, C.B., Sanduleak, N. 1971, Publ. Warner and Swasey Obs. Vol. 1.
Van Genderen, A.M., Van Driel, W., Greidanus, H. 1986, Astron. Astrophys. 155, 72.
Waelkens, C., Waters, L.B.F.M., Ccassatella, A., Le Bertre, T., Lamers, H.J.G.L.M. 1987, Astron. Astrophys. 181, L5.
Zuckerman, B. 1978, in Planetary Nebulae, ed. Y. Terzian (Reidel: Dordrecht), p. 305.
Zuckerman, B., Dyck, H.M. 1986a, Astrophys. J. (Letters), 304, 394.
Zuckerman, B., Dyck, H.M. 1986b, Astrophys. J. 311, 345.
Zuckerman, B., Dyck, H.M., Claussen, M.J. 1986, Astrophys. J. 304, 401.

IRAS OBSERVATIONS OF SYMBIOTIC STARS

M. Parthasarathy   H.C. Bhatt
Indian Institute of Astrophysics
Bangalore - 560034, India

ABSTRACT. Of the 129 symbiotic stars in Allen's (1984) catalogue, 42 were found to be IRAS sources. Of these 42 IRAS sources, 22 are D-type (symbiotic Miras), 5 are D'-type (yellow symbiotics) and 15 are S-type. The separation of S, D and D' types into three distinct groups is clearer in the $\log[f\lambda(25\mu m)/f\lambda(12\mu m)]$ versus (H-K) diagram. The IRAS fluxes of S-type symbiotics are consistent with that observed from normal M giants. This result suggests that mass-loss rate from most of the S-type symbiotics is similar to that from normal M giants. The IRAS data of D-type symbiotics shows evidence for the presence of dust envelopes. The masses of the dust envelopes (10^{-6} to 10^{-7} Mo) around Miras in D-type symbiotics are similar to that observed in field Mira variables. These results suggest that mass-loss rates in symbiotic Miras are similar to those from field Mira variables and also that the mass loss from symbiotic Miras is pulsationally driven similar to that found in field Mira variables by Whitelock, Pottasch and Feast (1987). Analysis of IRAS data of yellow symbiotics M1-2, AS201, Cn1-1, Wray 157, and HD149427 suggests that they are young planetary nebulae containing a binary nucleus. M1-2, AS201 and Cn1-1 show evidence for the presence of evolved hot companions. The evolutionary stage of the late type (F-G) companions is not clear.

1. INTRODUCTION

The spectra of symbiotic stars show features commonly observed in late-type giants and high-excitation planetary nebulae. The late-type star in a symbiotic binary is typically a M-type giant or an AGB star or a Mira. The hot component of a symbiotic system is a hot white dwarf or probably the central star of a planetary nebula or an accretion disk surrounding a low mass main sequence star (Kenyon and Webbink, 1984). From the near-infrared observations symbiotic stars have been classified into three types (Webster and Allen, 1975; Allen, 1982). The S-type (stellar) symbiotics have stellar IR continua (Tcolour = 3000 K) corresponding to the associated M giants. The D-type (dusty: Tcolour = 800 K) symbiotics show the presence of circumstellar dust, and the cool component is a Mira variable. In the D'-type systems (yellow symbiotics) the dust is cooler (T < 500 K), and the cool component is of spectral type F-G.

Broad-band JHKL photometry has been obtained for nearly all symbiotics (Allen 1982; Kenyon and Gallagher 1983). However, 10μm and 20μm observations are available for only some bright symbiotics. The IRAS point source catalogue (Beichman *et al.* 1985) now provides

Table 1: IRAS OBSERVATIONS OF SYMBIOTIC STARS

	IR type	Observed Fluxes (Jansky)			
		12μm	25μm	60μm	100μm
EG and	S	3.22	0.89	<1.05	
M1-2	D'	1.79	2.59	1.53	
UV Aur	S	52.6	15.1	2.51	
Wray 157	D'	1.15	1.33	0.58	
RX Pup	D	204	101	12.1	6.9
AS 201	D'	1.22	2.84	2.14	1.08
He2-38	D	8.32	3.4	<3.72	
SY Mus	S	0.84	0.35		
BI Cru	D	18.9	13.3	<11.6	
SS 38	D	6.73	2.89	1.61	
RW Hya	S	0.88	<0.45	<0.4	
He2-104	D	9.35	8.62	5.98	
He2-106	D	33.9	21.0	4.61	29.3
He2-127	D	0.55	0.33	<0.68	
Cn1-1	D'	23.1	35.6	15.2	9.23
T CrB	S	0.61	0.24		
AG Dra	S	0.3	<0.75	<0.4	
He2-147	D	4.07	2.39	<0.4	
He2-171	D	7.36	3.54	0.61	
HD 149427	D'	2.65	8.16	1.45	
HE2-176	D	2.97	2.07		
Hen 1242	S	0.36	<0.25		
AS 210	D?	3.05	0.9	<0.4	
H2-5	D?	1.19	0.61		
H1-36	D	<18.2	20.8	4.35	
AS 245	D?	1.53	0.65		
Ap1-8	S?	0.85	0.53	3.14	34.4
SS 122	D?	1.45	0.84		
H2-38	D	3.35	1.69		
AS 296	S	0.88	0.35	<0.4	
He2-390	D	5.08	3.94	1.76	6.61
V3804 Sgr	S	0.51	<0.51	<0.44	
V443 Her	S	0.39	<0.25	<0.4	
CII Cyg	S	448	140	13.7	3.76
Hen 1761	S	0.42	<0.27	<0.4	
HM Sge	D	119	63.1	7.08	
CI Cyg	S	0.76	0.25	<0.45	
V1016 Cyg	D	45.1	26.9	3.03	
RR Tel	D	21.1	12.9	1.98	
AG Peg	S	1.49	0.47	<0.42	
Z And	S	0.58	0.21	<0.37	
R Aqr	D	1262	401	50.5	14.8

far-infrared fluxes from 12µm to 100µm for several symbiotic stars. The LRS 8-21µm data (Olnon and Raimond 1986) is available for few bright symbiotic stars. In this paper we report an analysis of the IRAS observations of symbiotic stars.

2. IRAS OBSERVATIONS

Allen's (1984) catalogue lists 129 symbiotic stars and 15 possible symbiotics. In Table 1 we list the IRAS observations of the symbiotics stars detected in the IRAS point source catlogue. The observed IRAS positions are in good agreement with their optical positions. The IRAS fluxes given in Table 1 have been colour corrected as described in the IRAS explanatory supplement. Only 8 symbiotic stars have LRS 8-21µm spectra (Figure 1).

3. ANALYSIS

Of the 129 symbiotic stars in the Allen's (1984) catalogue, 42 were found to be IRAS sources (Table 1); 22 are D-type, 5 are D'-type and 15 are S-type. Except D'-type symbiotics, the D-type and S-type symbiotics show decreasing IR flux with increasing wavelength. Most of the S-types do not show significant flux at 60µm. Only 3 D-type and 5 D'-type show flux at 100µm. He 2-106, Ap1-8, He2-390 show flux at 100µm. These three are close to the galactic plane, and the increased emission at 100µm may be due to IR cirrus. The flux ratio $f\lambda(12\mu m)/f\lambda(25\mu m)$ of S and D-types range from 1 to 3. This flux ratio from the data of D'-type symbiotics is in the range of 0.3 to 0.9. In figure 2 we show the flux ratios $\log[f\lambda(12\mu m)/f\lambda(2.2\mu m)]$ versus $\log[f\lambda(25\mu m)/f\lambda(12\mu m)]$. In Figure 2 the S-type symbiotics are in the same region as that defined by M giants, and D-types are in the same region as that defined by Mira variables. The D'-types appear to be well separated from S and D-types. H2-5, SS122, Ap1-8 and AS245 are in the border region of D-type symbiotic stars (Figure 2).

Figure 3 shows the (H-K) colour versus $\log[f\lambda(25\mu m)/f\lambda(12\mu m)]$; there S,D,D'-types are better separated than in Figure 2. Figure 3 clearly shows the large spread in the (H-K) colours and relatively less spread in the flux ratio $\log[f\lambda(25\mu m)/f\lambda(12\mu m)]$ of D-type symbiotics. From Figure 3 we clearly distinguish the five D'-types as a distinct group. The (H-K) colour and the flux ratio $\log[f\lambda(25\mu m)/f\lambda(12\mu m)]$ of SS122 suggests that it is a D-type symbiotic. SS122 appears to be similar to He2-176 and He2-147. He 2-147 was listed as a S-type symbiotic by Allen (1984). However, the position of He2-147 in Figures 2 and 3 suggests that it is most likely a symbiotic Mira. Ap1-8 and H2-5 are S-types; however, they appear to show excess flux at 12µm and 25µm compared to other S-types. AS245 also appears to show 12µm excess. It is listed as S-type, but its position in Figures 2 and 3 indicates that it may be a symbiotic Mira. The flux ratios of AS210 are similar to those of UV Aur and R Aqr. It is likely that AS210 may contain a carbon star or a symbiotic Mira. The IR types (S,D,D') based on Figures 2 and 3 and the far-infrared energy distributions of S type symbiotics are consistent with that expected from normal field M giants.

From IRAS data of S-type symbiotic stars we can conclude that they do not have dust envelopes similar to that observed in symbiotic Miras. These results suggest that the mass-loss rate in S-type symbiotics is

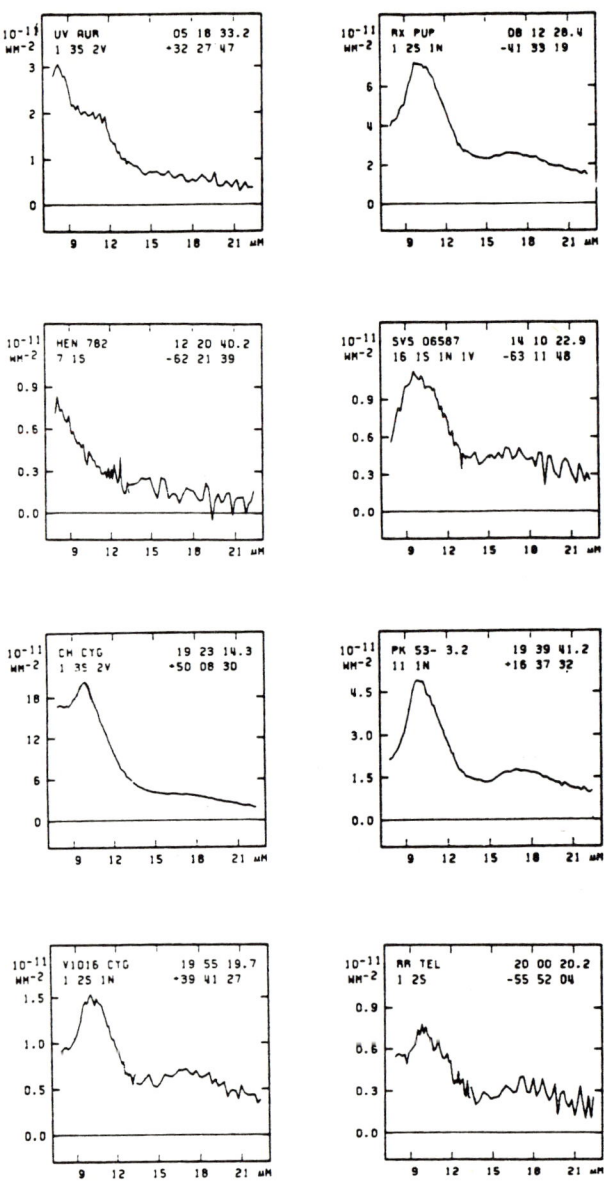

Figure 1. The low resolution 8 to 21 μm spectra of symbiotic stars.

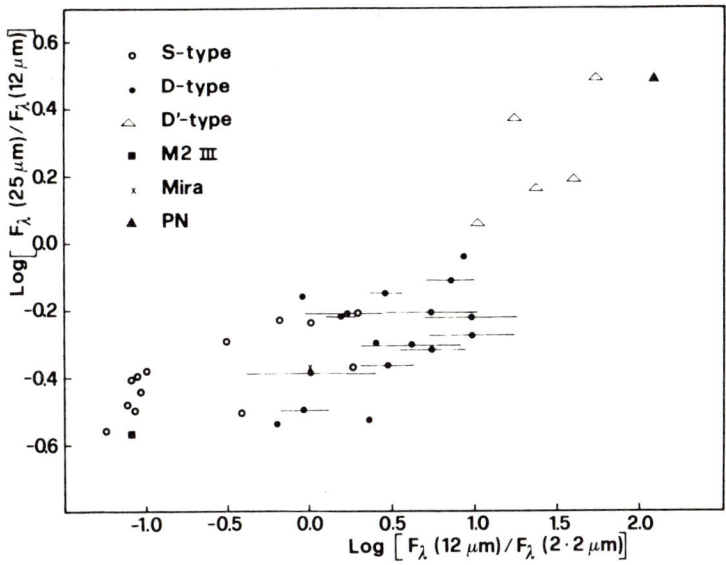

Figure 2. The colour-colour diagram for symbiotic stars. The solid lines indicate the amplitude of light variations of symbiotic Miras.

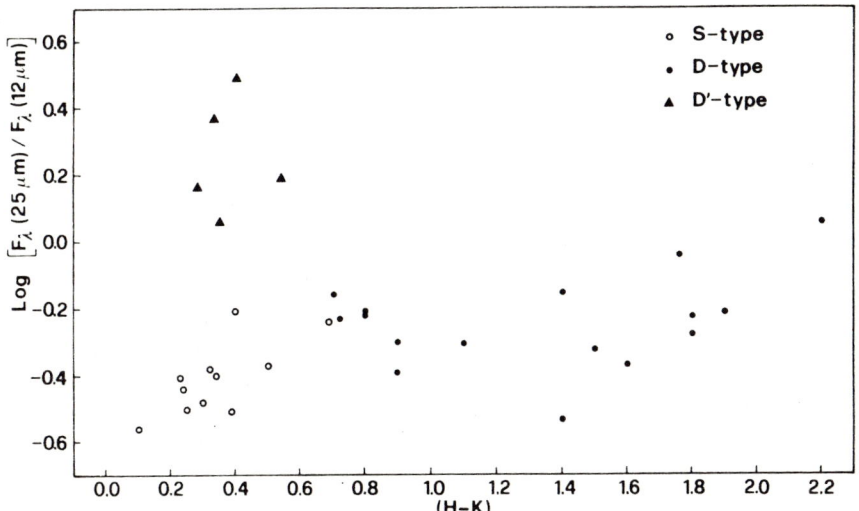

Figure 3. The colour-colour diagram for S, D and D'type symbiotics.

similar to that observed in normal M giants. However, a few S-type symbiotic stars like CH Cyg have dust envelopes and show relatively higher mass-loss rates. Kenyon, Fernandez-Castro and Stencel (1986) analyzed the IRAS pointed observations of few a S-type symbiotics and found similar results. The LRS spectra of the two S-type symbiotics (Figure 1) CH Cyg and UV Aur show evidence for 10-μm silicate emission feature. The IRAS fluxes of these two stars also indicate for the presence of dust shells. From an analysis of the data we find that the dust masses and dust temperatures to be of the order of 10^{-6} M$_\odot$ and 10^{-7} M$_\odot$ and 500 K and 500 K for CH Cyg and UV Aur respectively. In order to derive the characteristics of the dust shells around S-type symbiotics we need good estimates of distances and effective temperatures of the cool components.

3.1 SYMBIOTIC MIRAS

The IRAS data of all symbiotic Miras (D-type) (Table 1) show evidence for the presence of dust envelopes. Pulsation periods are known only for 10 symbiotic Miras (Whitelock, 1987). The period-luminosity relation found from the Mira variables is also found to be applicable to symbiotic Miras (Feast, 1984; Whitelock, 1987). The pulsation periods, K magnitudes and distances for 10 symbiotic Miras given in Table 2 are from Whitelock (1987). The M(bol) values are obtained from the relation M(bol) = 1.06-2.242 log(P) (Feast, 1984). The L values (pulsation amplitude in the L band) are estimated from the light curves and are given in Table 2. The far-infrared luminosities L(IR) given in Table 2 are derived from the IRAS fluxes (Table 1) and distances (Table 2). The dust temperatures given in Table 2 are derived from the IRAS fluxes. The average dust temperature in the dust envelopes of symbiotic Miras is found to 320 K. The masses of the dust envelopes M(d) are estimated from the equation (Hilderbrand, 1983)

$$M(d) = 4/3(a\rho d^2 f(\upsilon)/Q(\upsilon)B(\upsilon,T(d))), \qquad (1)$$

where d is the distance to the symbiotic Mira, $f(\upsilon)$ is the observed flux density (IRAS), $B(\upsilon,T(d))$ is the Planck function with dust temperature T(d), a is the radius of the dust particle, and ρ is the dust grain density. The quantity $a\rho/Q(\upsilon)$ is dependent on the dust properties and a value of $a\rho/Q(\lambda)(25\mu m) = 1.5\ 10^{-3}$ is adopted (Whitelock et al. 1987). The temperatures and masses of the dust envelopes of these ten symbiotic Miras (Table 2) are similar to that found for Mira variables by Whitelock et al. (1987). The dust envelopes areound BI Cru, He2-104 and H1-36 may have temperature gradients. We have not taken the temperature gradients into account. Whitelock et al. (1987) found evidence for pulsationally driven mass loss from Mira variables. They derived a relation between dust shell mass, pulsation period and pulsation amplitude (at L band).

$$\log M(d)\ (Mo) = 2.17 \log (P) +1.88\Delta L-13.38 \qquad (2)$$

We obtained the ΔL values for symbiotic Miras from their light curves (Whitelock, 1987), and from the pulsation periods we estimated the masses of the dust envelopes from the above equation. These values are also listed in Table 2 for comparison. From both these estimates of dust masses we conclude that the masses of the dust envelopes of symbiotic Miras are of the order of 10^{-6} to 10^{-7} M$_\odot$. These values are

Table 2 : SYMBIOTIC MIRAS

Luminosities Temperatures and Masses of the Dust Envelopes

	Period days	K mag	d kpc	M_{bol} mag	L mag	L_{IR} $L_\odot$	T_d K	M_d $M_\odot$	M_d (P& L) $M_\odot$
RX Pup	580	2.4	1.5	-5.1	0.88	1761	350	1.2×10^{-6}	1.9×10^{-6}
He2-38	433	4.4	3.0	-4.9	0.67	330	400	1.3×10^{-7}	4.0×10^{-7}
BI Cru	280	4.8	2.3	-4.4	0.1	547	300	6.9×10^{-7}	1.3×10^{-8}
He2-104	400	6.7	4.7	-4.8		1277	220	5.6×10^{-6}	
He2-106	450	4.7	2.8	-4.9	0.73	1164	250	3.0×10^{-6}	5.6×10^{-7}
H1-36	500	7.4	7.6	-5.0		6476	200	4.2×10^{-5}	
HM Sge	540	4.2	2.3	-5.1	0.7	2501	375	1.3×10^{-6}	7.3×10^{-7}
V1016Cyg	450	5.1	3.4	-4.9	0.5	2185	300	2.8×10^{-6}	2.1×10^{-7}
RR Tel	387	4.1	2.5	-4.7	0.55	565	300	7.2×10^{-7}	1.9×10^{-7}
R Aqr	387	-1.0	0.25	-4.7	0.8	241	500	4.0×10^{-8}	5.5×10^{-7}

Table 3:YELLOW SYMBIOTIC STARS

Luminosities,Temperatures and Masses of the Dust Envelopes

	Sp	K(mag)	d(kpc)	L_{IR} ($L_\odot$)	T_d (K)	M_d ($M_\odot$)
M1-2	G2Ib	9.8	2.0	61	220	2.6×10^{-7}
Wray157	G	9.4	2.5	51	260	1.1×10^{-7}
AS201	G2III	9.9	2.2	77	165	1.1×10^{-6}
Cn1-1	F5III	7.6	0.45	40	235	1.3×10^{-7}
HD149427	A-F	10.3	2.	150	250	4.0×10^{-7}

similar to those found for field Mira variables by Whitelock et al. (1987). These results suggest that the mass-loss rate from symbiotic Miras is of the same order of magnitude as that found from Mira variables.

The LRS 8-21μm spectra (Figure 1) of RX Pup, He2-106, HM Sge, V1016 Cyg and RR Tel show the 10μm silicate emission feature. However, the LRS spectrum of BI Cru (Figure 1) shows no 10μm silicate emission feature. The 8-21μm spectrum of BI Cru is in agreement with the hot grey continuum from dust without any additional silicate component. Roche et al. (1983) made a spectroscopic survey of 22 symbiotic stars in the wavelength region 2μm to 13μm. They found that most of the D-types (symbiotic Miras) show the 10μm silicate emission feature. Roche et al. (1983) also found a featureless smooth dust emission component. They found no silicate emission feature in BI Cru, AS210, He2-104, and He2-390. IRAS observations (Figure 1, Table 1) of these four symbiotic also shows that they do not have 10μm silicate emission feature. The 25μm and 60μm fluxes of He2-104 and He2-390 suggests temperature gradients in their dust envelopes. Roche et al. suggested iron based dust grains to explain the hot grey component in the 2 to 13μm spectra of symbiotic Miras. Whitelock (1987) suggests the presence of large dust grains in the dust envelopes of some symbiotic Miras, which may explain the featureless component in the LRS spectra. But very large particles are not required. According to Draine and Lee (1984) minimum sizes of between 3×10^{-5} and 10^{-4} cm will suffice and can produce neutral extinction.

3.2 YELLOW SYMBIOTIC STARS

M1-2, Wray 157, AS201, Cn1-1 and HD149427 show combination spectra (nebular emission lines + underlying stellar absorption spectrum) like the symbiotic stars. But the primary components of yellow symbiotics (D'-type) are of spectral type F-G. The dust temperature is also lower than that observed in symbiotic Miras. All these five objects show near-infrared excess and are also classified as planetary nebulae. In the flux-ratio diagrams (Figures 2 and 3) these five objects are well separated from symbiotics and symbiotic Miras. The far-infrared colours and flux distributions (Table 1) of these five yellow symbiotics are similar to those observed in planetary nebulae. In Figure 2 He2-104 and H1-36 are in the border region of yellow symbiotics. However, in Figure 3 they are clearly in the region of symbiotic Miras, and yellow symbiotics are well separated from the rest. Cn1-1 shows a F5III spectrum in the optical region with nebular emission lines superposed on it, and the UV (IUE) spectrum suggests the presence of a hot (O-B) companion (Lutz, 1984; Bhatt and Mallik, 1986). Cn1-1 shows strong emission lines characteristic of a high-density planetary nebula (Glass and Webster, 1973). The temperature of the hot (O-B) component is of the order of 70,000 K (Lutz, 1984; Bhatt and Mallik, 1986). The optical spectrum of AS201 shows characteristics of a G2III star (Kohoutek, 1977, 1987). Kohoutek (1977) considered AS201 as a possible photoplanetary nebula. Feibelman (1988) obtained (IUE) ultraviolet spectrum of AS201 and found evidence for a hot companion in strong CIII, CIV and HeII emissions and a fairly strong UV continuum. He argued that AS201 is a planetary nebula and estimated the temperature of the hot companion to be of the order of 60,000 K. Feibelman (1983) also studied the IUE ultraviolet spectrum of M1-2. In the optical region M1-2 shows the spectrum of a G2Ib star (O'Dell, 1966). Feibelman (1983) concludes that M1-2 is a young compact high-density planetary nebula of excitation class 5 or 6, and its nucleus is a binary consisting of a G2Ib star with a hot companion, probably a type O subdwarf. Recently Gutierrez-Moreno et al. (1987) studied the optical spectrum of HD149427 and concluded that it is a planetary nebula and not a symbiotic star.

The far infrared (IRAS) colours and flux distributions suggests that these five objects (Table 3) are most likely young compact high-density planetary nebulae. From IRAS data of these five objects we estimated the temperatures, luminosities and masses of the dust envelopes around them (Table 3). The masses of the dust envelopes are estimated from equation 1. Based on all the available observations we conclude that M1-2, Wray 157, AS201, Cn1-1, and HD149427 are young planetary nebulae containing a binary nucleus. The evolutionary stage of the late-type (F-G) components is not clear; they may be evolving leftward from the tip of the AGB in the H-R diagram, and the hot components may be subdwarfs or white dwarfs. Another possibility is that the late type (F-G) components are normal giants/supergiants and the dust envelopes are formed from the severe mass loss from the evolved companions which are now hot subdwarfs or white dwarfs. A CNO abundance study of the late-type (F-G) components of these five systems (Table 3) may enable us to understand their evolutionary stage. Kohoutek (1987) finds that the hot components of AS201, M1-2, and Cn1-1 are located on the Harman-Seaton sequence in the region of evolved central stars of planetary nebulae. Kohoutek estimates zanstra temperatures of 78000 K, 72000 K, and 85000 K for the hot components of AS201, Cn1-1 and M1-2 respectively.

4. CONCLUSIONS

Of the 129 symbiotic stars in the Allen's (1984) catalogue 42 were found to be IRAS sources. Only 13 D-type and 5 D'-type symbiotics show flux at 60μm. The separation of S,D and D'-types into three different groups is clearer in the $\log[f\lambda(25\mu m)/f\lambda(12\mu m)]$ versus (H-K) plot. The $f\lambda(25\mu m)/f\lambda(12\mu m)$ flux ratios and H-K colours of SS122, He2-147, AS245, and AS210 suggests that they may be symbiotic Miras. AS210 may contain a carbon star or a Mira. The IRAS fluxes and colours of S-type symbiotics stars are consistent with those observed from normal M giants. Excepting CH Cyg and UV Aur, most of the S-type symbiotic stars do not have dust envelopes. These results suggest that mass loss rate from the M giants in S-type symbiotics is similar to that from field normal M giants.

The IRAS data of D-type symbiotic stars (symbiotic Miras) show evidence for the presence of dust envelopes. The dust temperatures and masses of the dust envelopes are estimated for 10 symbiotic Miras with known pulsation periods. The dust temperatures are found to be in the range of 200 to 500 K. The masses of the dust envelopes are of the order of 10^{-6} to 10^{-7} $M_\odot$. The dust envelopes in few symbiotic Miras may have temperature gradients.

Analysis of the IRAS data of yellow symmbiotic stars (D'-type) M1-2, Wray 147, AS201, Cn1-1 and HD149427 suggests that they are young high density planetary nebulae containing a binary nucleus. They show evidence for the presence of a hot evolved companion. The evolutionary stage of the late type (F-G) companions in these systems is not clear; they may be evolving from the tip of AGB towards left in the H-R diagram.

REFERENCES

Allen, D.A. 1982, In The Nature of Symbiotic Stars IAU Colloquium No. 70, ed. M. Friedjung and R. Viotti (Dordrecht: Reidel), p. 27.
Allen, D.A. 1984, Proc. Astron. Soc. Australia, 5, 369.
Beichman, C.A., Neugebauer, G., Habing, H.J., Clegg, P.E., and Chester, T.J. 1985, IRAS Point Source Catalogue, JPL.
Bhatt, H.C., Mallik, D.C.V. 1986, Astron. Astrophys. 168, 248.
Draine, B.T., Lee, H.M. 1984, Astrophys. J. 285, 89.
Feast, M.W. 1984, Mon. Not. R. Astr. Soc. 211, 51p.
Feibelman, W.A. 1983, Astrophys. J. 275, 628.
Feibelman, W.A. 1988, in A Decade of UV Astronomy with IUE, ESA SP-281, p. 179.
Glass, I.S., Webster, B.L. 1973, Mon. Not. R. Astr. Soc. 165, 77.
Gutierrez-Moreno, A. et al. 1987, Preprint Dept. Astro. Univ. Chile.
Hilderbrand, R.H. 1983, Quart. J. Roy. Astron. Soc. 24, 267.
Kenyon, S.J., Fernandez-Castro, T., Stencel, R.F. 1986, Astron. J. 92, 1118.
Kenyon, S.J., Gallagher, J.S. 1983, Astron. J. 88, 666.
Kenyon, S.J., Webbink, R.F. 1984, Astrophys. J. 279, 252.
Kohoutek, L. 1977, in Planetary Nebulae (IAU Symp. 76) ed. Y. Terzian (Dordrecht: Reidel), p. 47.
Kohoutek, L. 1987, Astrophys. and Space Science 131, 781.
Lutz, J.H. 1984, Astrophys. J. 279, 714.
O'Dell, C.R. 1966, Astrophys. J. 145, 487.
Olnon, F.M., Raimond, E. 1986, Astron. Astrophys. Suppl. 65, 607.
Roche, P.F., Allen, D.A., Aitken, D.K. 1983, Mon. Not. R. Astron. Soc. 204, 1009.
Webster, B.L., Allen, D.A. 1975, Mon. Not. R. Astro. Soc. 171, 171.
Whitelock, P.A. 1987, Publ. Astro. Soc. Pac. 99, 573.
Whitelock, P.A., Pottasch, S.R., Feast, M.W. 1987, in Late Stages of Stellar Evolution, ed. S. Kwok and S.R. Pottasch (Dordrecht: Reidel), p. 269.

Formation of Helium Lines and Continua in a Late-Type Giant Star

Ramiro de La Reza   Celso Batalha
Observatorio Nacional   Observatorio Nacional & University of
                             California Berkeley

Due to the atomic structure and properties of neutral helium, its spectrum can be an excellent diagnostic of the hot components of the atmosphere of a late-type giant star. Having this in mind, we have studied the formation of the helium lines and continua in the K giant star β Cet. For that purpose different NLTE atmospheric models were constructed. We have also solved the NLTE problem for the formation of lines and continua for a model helium atom consisting of twelve bound levels and three ionized stages.

The theory is confronted with the observations of the He II 1640 Å and the 10830 Å lines of β Cet taken from the literature. We have also attempted to observe the D3 line of He I at 5876 Å (also in other red giant stars) but were successful only in setting upper limits of the order of 10 mÅ to the strength of this absorption line.

Because β Cet is an X-ray source, we have examined the influence of this radiation on the formation of the neutral helium spectrum. We find that the He II 1640 Å line is sensitive to the coronal radiation that also forms the continuum of He II at 228 Å. In fact, the layers of formation of both transitions are similar. The He II lines at 304 and 256 Å are formed at higher temperatures ($T \geq 10^5$ K) by collisional processes if the coronal fluxes are small and by a photoionization-recombination process if the coronal fluxes are large.

However, one important conclusion of our work is that the mechanism of formation of the neutral helium lines at 10830 and 5876 Å is not directly related to the coronal radiation but to the resonant He II radiation at 304 Å (especially by the emission at the wings of the line). In fact, this resonant radiative emission produces a strong contribution to the ionization of the proper He I continuum at 504 Å (and consequently influencing the 10830 Å line) by an overlapping radiation process.

In general, with the atmospheric models considered, our calculations produce lines somewhat smaller in intensity than those observed, with the exception of the residual D3 absorption line. A better adjustment to the observed absorption 10830 Å line could be obtained by means of non-homogeneous atmospheric models in which the overlapping between the 304 and 504 Å radiations could be more efficient.

THE 8-22μm EXCESS IN CARBON STARS FROM IRAS LRS SPECTRA

S. J. Little
Bentley College, Waltham, MA 02181, USA

I. R. Little-Marenin
Wellesley College, Wellesley, MA 02181, USA

We have measured the excess IR emission from carbon mira and SR variable stars from IRAS LRS spectra. The 8-22μm excess is defined as the ratio of flux above a 2500K energy distribution fit to the LRS spectrum at about 8μm. The carbon star LRS spectra show both emission and absorption features, which are incorporated into our 8-22μm excess. The most prominent feature in carbon stars is the 11.2μm SiC dust emission feature extending from 10μm to 13.8μm. We observe another emission feature of unknown origin which peaks between 8.4-8.7μm. The SiC emission feature is occasionally blended on the red side by an absorption feature (attributed to gaseous HCN + C_2H_2) which extends from about 12-16μm. Many of the spectra appear to turn down at the 8μm end due (?) to an HCN + C_2H_2 absorption feature located at 7.1μm. Carbon stars do not generally show as large an excess as the M mira variables do. The figure below shows our measured excesses for both carbon miras and carbon semi-regular variables. There appears to be little correlation of excess with period, however the mira variables show about twice the range of variation of excess that the semi-regular variables do. We find little correlation between our measured 8-22μm excess and the excesses of Jura (Ap. J., 303, 327, 1986) based on the ratio of 12μm flux to 2μm flux. Our data do support his conclusion that longer period variable stars show larger average excesses, but this is only true for mira variables in our analysis.

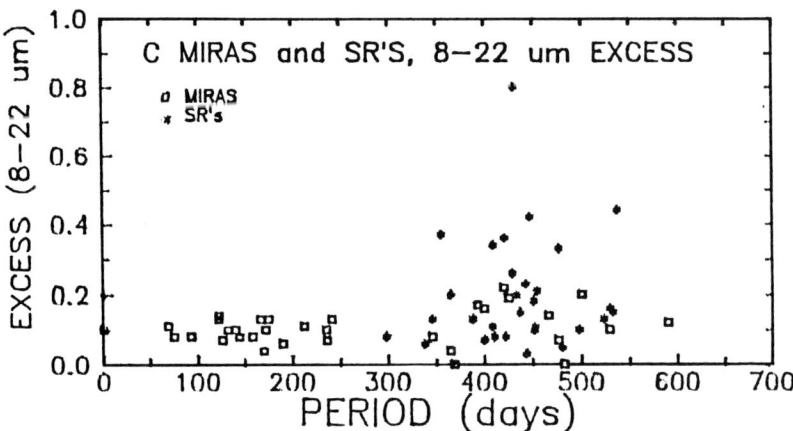

Absorption Spectrum of RCrB During the Light Minimum

N. Kameswara Rao   S. Giridhar   B.N. Ashoka
Indian Institute of Astrophysics

Observations of RCrB obtained during the 1986 and early part of 1963 light minima show the following changes relative to the spectrum obtained at maximum light.

The molecular band of C_2 (Swan System), CN bands at $\lambda 4215$ and low excitation metal lines are enhanced in absorption.

The strong singly ionized metal lines are weaker, e.g., Ti II, Sc II, consistent with the interpretation of filling in by the chromospheric emissions.

The equivalent width of absorption lines of CI either get weaker (i.e., 1986 minimum) or remain constant (1963 minimum) during the light minimum.

An absorption feature at $\lambda 4050.5$ Å probably due to C_3 occurs during the early decline and disappears by the end of the minimum.

THE VARIABILITY OF WATER MASER EMISSION ASSOCIATED WITH LONG PERIOD VARIABLE STARS

I.R. Little-Marenin, P.J. Benson and D. Goodwin
Wellesley College, Wellesley, MA 02181 USA

Since March 1988 we have monitored the 22.2 GHz H_2O maser line of 12 long period variable stars (LPV) with the 37 m telescope at the Haystack Observatory. We include the two Carbon stars V778 Cyg and EU And. The maser flux from V778 Cyg has varied by at least a factor of 5 from its detection level of 1.9 Jy on 1987 March with a possible period of a year. Figure 1 plots the 22GHz flux as a function of Julian date. We do not yet know whether the period will repeat. V778 Cyg also has shown variations in the intensity (by a factor of ~ 2) on a time scale as short as 15 hours. The water maser flux from EU And has decreased from its detection value of 8 Jy in 1986 November staying relatively weak at the 2-3 Jy level during the last year. We interpret both V778 Cyg and EU And as binaries, each with an M star component with a thick circumstellar shell and a C star component. The C star is brighter in the visual region where the system is classified whereas the M star is brighter at wavelengths > 5 μm. The circumstellar shell is the source of the strong silicate emission seen in the IRAS LRS spectra and of the water maser emission.

The first detected short period Miras, R Cet (P=166^d) and RZ Sco (P=157^d) show cyclic variations in the intensity of the maser emission. The maximum intensity for R Cet occurs progressively later at longer wavelengths, peaking in the infrared at phase 0.12 (in J), at 0.18 (in K) and between phase 0.20-0.25 at the radio (22.2 GHz) water maser line. However, the magnitude and the maximum of the maser flux appears to vary to a greater extent from cycle to cycle than does the visual lightcurve. Besides the above four stars we have also monitored the following eight stars: RX Boo (SRb, P=340^d), V CVn (SRa, P=192^d), AC Cyg (SRb, P=142^d), S Per (SRc, P=822), R Ser (Mira, P=356^d), T UMa (Mira, P=257^d), and T Vir (Mira, P=339^d). Preliminary indications are that many of the variables show maximum maser flux at phases between 0.2-0.3.

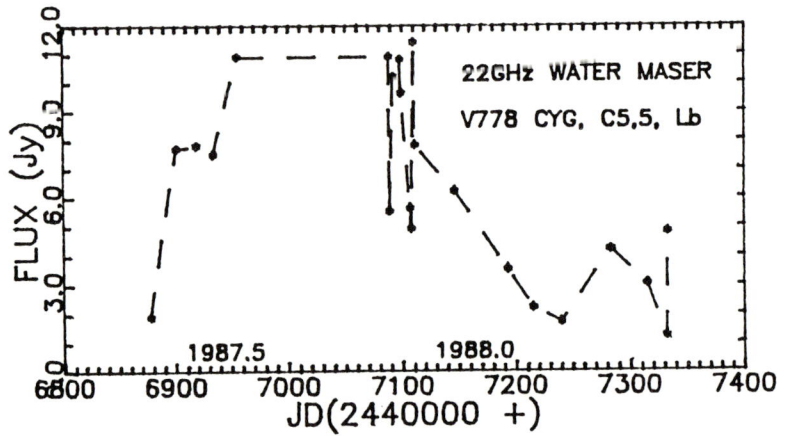

6. Outstanding Problems in Research on Peculiar Red Giant Stars

Panel Discussion

Johnson: Before beginning our closing panel discussion, we, the SOC, cordially thank our invited speakers, panelists, and all of you for your attendance and participation. It's been great to have you here! Now I call upon our panel and its chairperson, Al Glassgold (NYU).

Glassgold: Thanks. I believe this panel discussion will be a very useful way to summarize and end this excellent conference. Our panelists are, from the left, Bengt Gustafsson, Uppsala; George Wallerstein, Washington; Alvio Renzini, Bologna; Peter Wood, Mt. Stromlo; Ben Zuckerman, UCLA. We will begin with George Wallerstein.

Wallerstein: The commonly accepted doctrine (fortunately not yet dogma) is that cool stars evolve along the AGB through a sequence of types M-MS-S-SC-C. Along this sequence the abundance of ^{12}C increases by the mixing of helium-burning products to the stellar surface. At type SC the C/O ratio passes through unity. While this sort of sequence may well be followed by some stars I would like to point out some apparent inconsistencies as possible guidance for both discussion here and future research.

1. Among the carbon stars there are two groups that do not fit the sequence -- these are the early R stars (Dominy 1984) and the ^{13}C-rich carbon stars (often called type J) which constitute 13% of the sample of Lambert et al. (1986). The former do not fit because their luminosities are too low for the AGB, while the latter have converted ^{12}C to ^{13}C via hydrogen burning, presumably in a shell. Furthermore, the J stars often do not show s-process enhancements, thus indicating that they have not

passed through the S star phase but rather evolved directly from M to C.

2. The oxygen isotope ratios of the SC stars do not fit the sequence. Their $^{16}O/^{17}O$ ratios range from 75 to 1200 while the range for MS and S stars is 550 to 3000, and the range for the low-^{13}C carbon stars is 550 to 4000. Surprisingly, the $^{16}O/^{17}O$ range for Barium stars is 100 to 500, much like the SC stars. Furthermore the Ba stars have enhanced carbon, though with carbon still less than oxygen, and the coolest Ba star (HD 121447) has also been classified as the hottest SC star. Are the SC stars binaries that are about to dump on their companions? If so, we are observing them at a remarkable time because several SC stars show technetium, which indicates that they are still in the shell flashing stage. In any case, a search for binarism in SC stars would be worthwhile (Wallerstein 1988 and references therein).

3. A simple evolutionary sequence of M-MS-S-SC-C along the AGB should be reflected in the periods of the long period variables (LPV's) of those spectral types. However, if you eliminate the M-type LPV's of short period, high velocity, and small mass, there is no such correlation except for a small tendency for the C-type LPV's to have periods near 450 days. In fact, the discovery of the OH/IR stars shows that oxygen-rich LPV's are found with periods as great as 1000 days. They were not previously known because they are hidden in their own dust. Thus many AGB stars evolve to be stars of extreme radius without becoming carbon stars. Perhaps they are of higher mass than those which become carbon stars.

With these anomalies in mind, as well as others that have been discussed (or presented in poster papers) at this meeting, we should conclude that evolutionary sequences certainly depend on initial mass and are likely to be affected by both mass loss and mixing. While mass loss probably does not affect the evolution of the deeper layers of the star (until all the hydrogen is gone), mixing can affect the subsequent evolution so bifurcations or even trifurcations in evolutionary equences may take place due to convective events that are very difficult to model theoretically (Renzini 1987).

Wood: I think George partially answered his own question when he said there's a range of masses involved in the M, S, C stars, and that the luminosity or period where the transition from M to C star occurs is a function of the mass of the star. If you have a very massive star, it may never become a carbon star. The less massive it is, the lower its luminosity when it becomes a carbon star. In the Magellanic Clouds, where we also see this M-MS-S-SC-C sequence in a single cluster, we know that it does occur. How frequently it occurs is, I guess, another matter.

Little-Marenin: Is there a continuous sequence in period between the optical miras (which tend to have $P \lesssim 500$ days) and the OH/IR stars (many of which have much longer periods)? Is there a gap?

Wallerstein: I'm not sure. The dusty transition objects, like UX Cyg, WX Ser, and VX Sgr, which are rather obscured, were discovered among the OH sources but are not as heavily obscured as the pure OH-IR stars, which are usually discovered by the OH line, and which are found later in the IR and are virtually completely invisible. I think there's a continuity there.

Lloyd Evans: This regards the R stars, especially the early ones. Can one exclude the possibility that ^{13}C-rich stars evolve from early R → late R(J) → N(J) stars (i.e., evolution up the AGB from the red giant clump as a carbon star) on the basis of detailed abundances (e.g., N star abundance discussed by Lambert et al.)?

Lambert: I would like to go back to the start of this discussion and the problems with the M-S-C sequence. I think it's important to keep in mind what I would call normal stars in that sequence. Not all stars in that sequence may necessarily go through each step; it's possible they get so much carbon that they miss the S-star step, for example. Those are what we call normal, and I don't think there is any doubt it occurs roughly as we've outlined. Then there is a group of peculiar stars which we are discussing in this conference: the R stars, J-type N stars, cool hydrogen-deficient carbon stars, and R Cor Bor stars.

Let's say it takes X to produce an R-type carbon star. We don't know what X is, maybe a core flash. The R stars evolve, and they must account for some of the J-type N stars, so you would expect similarities in abundance between ^{13}C-rich cool carbon stars and the R stars. There are some; they are both ^{13}C-rich. The 4 or 5 N stars we looked at carefully are very ^{13}C-rich while the J-type R stars are moderately ^{13}C-rich. Neither group shows s-process enhancements. It's very hard to fit the J-type carbon stars into the M,S,C sequence because they're not s-process enriched. You can change carbon into nitrogen and fiddle with ^{13}C and the oxygen isotopes, but it's very hard to obliterate the s-process elements once you've made them.

The nitrogen abundances are of course a problem, and I pointed out in my talk that nitrogen abundances are a problem for the M-S-C sequence itself. Our carbon-star nitrogen abundances are not as enhanced as that sequence would lead one to expect. In our paper we acknowledged that and discussed ways in which it might be resolved; my hunch would be that we should push the effective temperatures up. I'd justify that, but then I'd have to write it down! (Laughter) The nitrogen doesn't agree with the K or M giant stars; it's a mystery. My gut feeling is that the nitrogen abundances we got for the carbon stars are systematically in error by some amount. That includes the J-type carbon stars, so I wouldn't refuse linking the J-type and early R stars on the basis of our nitrogen abundances. Also, there are undoubtedly objects that fit neither the M-S-C sequence or R-J-type sequence; maybe there's another sequence.

Stencel: George, you mentioned that the miras in 47 Tuc and other globular clusters share similar periods -- suggesting non-evolution of period. What is the evidence that mira periods do evolve?

Wallerstein: For miras in any rich globular cluster, the periods are similar. Of course, 47 Tuc is the best example, but there are others as well. If the cluster has two mira stars, the periods are similar.

Bowen: Concerning the question about whether there is a discontinuity between mira variables and OH/IR sources (at ~ 500 day periods); without wishing to claim too much accuracy for the theoretical envelope modeling, I simply point out that in the mira models there is a completely natural and continuous change from models showing the characteristics of a (somewhat dusty) optical mira to one with characteristics of an OH/IR source including very rapid mass loss and optically thick circumstellar dust shroud, for periods ~ 500^+ days. The result certainly suggests that this really is a continuous change, and that these objects are of the same kind, distinguished only by the natural changes in behavior that accompany the period increase.

Augason: We have observed ten bright, non-mira M and ten non-mira S stars. These are non-mira M and S stars with the same periods. The stars with longer periods are cooler. If luminosities are determined with the period-luminosity relationship, both the M and S stars form an AGB. No evolutionary sequence is apparent.

Feast: Regarding 47 Tuc. It's true that the miras in a cluster are concentrated around the same period, but one has got to remember that there are also low amplitude variables with shorter periods in that same cluster. So there is a sequence, and the large amplitude ones must be concentrated if one believes in evolution in the classical way. But the range is small, and that probably is consistent unless you can argue there are too many miras, which I don't think anyone has ever tried to do, and you would agree that there are not too many there.

Wood: It's about right. The number of stars in the cluster and the number of miras in the period range in which they exist is not inconsistent with evolution over a range from 90 to 320 days.

Feast: With the large amplitude stars, one is just isolating a very narrow region in the period luminosity relation.

Willson: The small range of periods of miras in globular clusters does not imply a lack of evolution in P for longer period, more massive stars. As is most clearly seen in the log M – log L ("Wood-Cahn") diagram, when a low mass star turns on as a mira, it already has a mass close to is final (WD) mass, so it does not need to change period very much. However, higher-mass stars must pass through a large range in P in going from where they turn on to their final (WD) masses.

Glassgold: We'll ask Alvio to make one or two short points.

Renzini: I have three points to touch on. (1) Is it worth recomputing synthetic AGB models (a la Iben & Truran 1978 or Renzini and Voli 1981)? (2) Why in the LMC and SMC are there no AGB stars brighter than Mbol ~ –6.5? Or, equivalently, what happens to the ~3 to ~8 M sun stars as they reach the thermally pulsing AGB stage? (3) What is the origin and evolution of R CrB stars? For this panel discussion I have singled out these three questions that to my taste are now particularly hot in the current theoretical research related to Peculiar Red Giants.

Concerning the first question, I have been frequently asked to update those calculations, taking into account the tremendous observational discoveries that have been made since earlier computations were completed. The old synthetic AGB models are in fact obsolete for a number of reasons, some of which have been mentioned by Lattanzio in his excellent review at this conference of the more recent theoretical work. I would now like to show you what I believe are the principal difficulties that for the time being tend to hamper rapid progress in this direction.

Back in 1978-1981 we used a very simple expression for the amount of mass dredged up after each thermal pulse. The whole dredge-up process was in fact described by only one very simple expression in which the dredge-up mass was given as a function of the core mass only -- what we used to call the "dredge-up law". The whole thing was contained in just one punched card! We now know that the real situation is enormously more complicated, and that besides its dependence on the core mass, (Mcore), the dredge-up law must depend also on the total stellar mass, (Mtot), on metallicity, (Z), on the pulse number, (Np), on the mixing-length parameter (α), and last but not least on the code and the physical assumptions concerning the boundaries of the convective regions. A "minimum" exploration of this parameter space may require computing evolutionary sequences for at least, say, 5 values of Mcore, 5 of Mtot, 3 of Z, 30 pulses, 3 values of α, and perhaps 3 "codes" for a total of 20,000 thermal pulses. Each thermal pulse requires the calculation of about 2,000 models, and each model takes about 10 seconds of CPU on a Cray computer, for a total of about 10 years of CPU time! (Laughter). So when people ask about another generation of synthetic AGBs, this is the kind of computational effort which is required.

One may argue that technology is advancing very fast and that perhaps I am a bit pessimistic in estimating the number of cases that should be computed, but certainly the calculation is very massive if one wants to do it properly. I have mixed feelings as to whether at this stage we should really embark in such a large computational effort. The reason is that the major uncertainty here comes from the way convective boundaries are handled. No doubt we now have an insufficient understanding of the physics of such boundaries, a situation that is hardly surprising, when one considers that we have to pick up the boundary between vanishing mixing and no mixing at all. It is worse than predicting the weather, as we don't have meteorological stations inside the stars! Therefore, I am skeptical that in the next few years there could be a decisive advancement of our physical understanding of convection boundaries. So, my impression is that for perhaps a long time we may better go the other way around, and use our ability of reproducing, e.g., the solar system distribution of s-process elements, to infer something about the dredge-up law.

The answer to the second question is perhaps easier to get. Magellanic Cloud studies have revealed that there are no AGB stars brighter than Mbol ~ -6.5 (whether none at all or just very few is presently an observationally unsettled question). This may imply that in AGB stars the core mass does not grow beyond ~0.85 $M_\odot$ (which corresponds to Mbol ~ -6.5), and the thermally pulsing (TP) AGB phase aborts. Why? Several possible explanations have been suggested, but no one has so far worked out a solution in convincing quantitative terms. A couple of years ago Wood and Faulkner reported convergence problems in their thermally pulsing AGB models with Mcore > ~0.85 $M_\odot$. (Iben found similar convergence problems in AGB models of core mass larger than those which led him to discover the third dredge-up process.) When reading the Wood &

Faulkner paper, I was immediately struck by the coincidence of the two reported values of the core mass: virtually no TP-AGB stars in the Clouds with Mcore > ~0.85 $M_\odot$ and bad convergence problems in TP-AGB models with Mcore > ~0.85 $M_\odot$. Perhaps there was link between the two aspects.

If the coincidence is not merely fortuitous, what may happen can be described (for short) as a dramatic increase of the radiation pressure at the base of the hydrogen-rich envelope, as the energy released by one helium-shell flash leaks out of the intershell region. Such an increase in radiation pressure inflates the envelope from below, and as the excess energy is radiated away from the star the envelope recollapses, shocks may be generated, and heavy mass ejection may result. If this scenario is correct, a few thermal pulses may suffice to eject the whole envelope. It is worth emphasizing that such a TP-driven envelope ejection would operate only in the more massive AGB stars, those in which Mcore grows above 0.85 $M_\odot$, or in which Mcore is already larger than this limit when stars initiate the TP-phase. In practice, this would apply in stars initially more massive than, say, 3 $M_\odot$, which represent a rather small minority.

This scenario can be tested both observationally and theoretically. From the observational point of view, a counterpart to these hypothetical objects could be searched among OH-IR sources with Mbol < ~ -6.5, with a very variable mass loss rate over timescales of order ~10,000 years being a possible signature. Such a variability could manifest itself as a complex density stratification of the circumstellar envelope. From the theoretical point of view, the switch from a quasi-static condition of the envelope to its fully hydrodynamical behavior could be followed in detail using an appropriate code, such as, e.g., the KEPLER code used by Woosley and Weaver to follow the quasi-static evolution of massive stars as well as their final supernova explosion. Progress in these directions may be considerably faster than for updating the dredge-up law.

Stencel: There is an interesting coincidence between the 0.85 $M_\odot$ core-mass limit for pulse-driven envelope ejection, and the estimated core mass for R CrB stars discussed earlier today. Could the R CrB stars be examples of initially "massive" stars which shed a lot of mass in an ejection? Gillett et al. (1986) estimated an upper limit of a few solar masses for the huge IR shell of R CrB itself.

Renzini: No, I disagree.

Wood: I would like to comment on the apparent lack of very luminous AGB stars in the Magellanic Clouds with $M_{bol} < -6$. Such AGB stars definitely exist (Wood, Bessell and Fox 1983, Ap. J., 272, 99; Hughes and Wood 1987, Proc. Astr. Soc. Australia, 7, 147) but appear to be fewer in number than expected given the number of Cepheids in similar areas of the Clouds (Wood, Bessell and Paltoglou 1985, Ap. J., 290, 477; Reid and Mould 1985, Ap. J., 299, 236). A possible explanation for the lack of such stars is a radiation-pressure ejection mechanism which comes into play during helium shell flashes (Wood and Faulkner 1986, Ap. J., 307, 659). Basically, at the peak of a helium shell flash, the luminosity escaping from the core of the star exceeds the Eddington limit there and the star has no option but to undergo a hydrodynamic envelope ejection. One way of overcoming this possibility is to let convection carry a significant fraction of the energy flux. A feature of stellar models of this type is that, the more massive the star, the larger is the fraction of the energy carried by convection. Hence, it may be possible to explain the AGB stars observed to have $M_{bol} < -6$ as the most massive of the AGB stars. It is worth noting that Richer, Olander and Westerlund (1979, Ap. J., 230, 724) found that the most luminous carbon stars in the LMC were J stars with high $^{13}C/^{12}C$ ratios, which indicates that envelope convection in these stars is penetrating right down to the H-burning shell during the interflash phase of AGB evolution. Such stars might be expected to carry a significant amount of convective flux at the bottom of their hydrogen-rich envelopes during helium shell flashes, and thereby avoid the radiation-pressure ejection mechanism (as well as undergoing dredge-up of carbon).

Glassgold: We will now hear from Bengt Gustafsson.

Gustafsson: I would like to comment on an old question. How good are classical models of photospheres, chromospheres, and envelopes? Could they be any better? By "classical" I mean that plane-parallel or spherical symmetry is assumed, together with static or stationary conditions and LTE. Furthermore, no back-reaction from upper layers (like the circumstellar envelope) or lower ones (like the chromosphere or photosphere) are considered, nor are magnetic fields. How good are these models? Are they internally consistent? Are they physically correct or at least reasonable? Are they in agreement with the observations? I claim "no", and I am not the first one to give that answer.

There have been numerous warnings and arguments from leading theorists. Also, the observational evidence is now rapidly growing, not the least for red-giant stars. This is not so astonishing - we knew long ago that they are more or less irregularly variable and quite tenuous. Signs of departures from LTE, of varying, non-spherical chromospheres, of giant prominences, of non-spherical and clumpy time-varying mass flows, of sometimes large-scale motions in shells, and of magnetic fields of significance for envelope dynamics are accumulating (see Gustafsson 1988, in Modeling the Stellar Environment: How and Why? (Proc. 4th IAP Meeting, in honor of J.C. Pecker, ed. P.J. Delache)).

At this meeting we have got more indications of interesting photospheric velocity variations in Arcturus (Irwin et al.), light variations in Betelgeuse (Joris), of non-steady, non-spherically-symmetric outflows from long-period variables and other red giants (Heske), of episodic mass ejections and clumpiness of outflows (Olofsson). We have heard about theoretical indications of non-periodic response to periodic shocks (Bowen) and noticed Muchmore's demonstration that molecular formation may be well out of equilibrium in shock treated regions, which could seriously affect the structures in very complicated ways (see also Sharp's constructive calculations that show how complicated and delicate the chemical equilibrium may actually be). There are indeed complex couplings between molecular cooling, dust formation, and dynamics and atmospheric structure in general, as has been suggested earlier by Ayres, Kneer, Muchmore, Stencel, and others. Some of this is reflected in the (admittedly only partly astrophysical) quite different double solutions found to the classical model-photosphere problem for M stars by Scholz, for carbon stars by our group, and even for the sun by Nordlund.

Those who think that these phenomena only matter for the outermost stellar layers should look closely at the recent films of the solar photosphere obtained by Scharmer. From all this one may well get the impression that stellar photospheres, chromospheres, and outer envelopes may be nothing but a bunch of overlapping, transient, erratic phenomena in beautiful interplay. Maybe the long-period variables are the simplest of all these objects because the large-scale radial pulsations force them to be regularly structured.

What could we do in this difficult situation? My main point is that we could do much better, for several reasons.

(1) We have seen an enormous increase in computing power and in the efficiency in numerical methods in recent years. More is to come. My impression is that the theorists in our field have not exploited these

resources to the full extent. Also, now is the time to start planning for
the use of future, and more efficient computers. Much more detailed
modelling of stellar atmospheres and circumstellar envelopes is already
possible - e.g. along some of lines beautifully sketched by Bessell and
Bowen in their talks. Also numerical model experiments, studying basic
physical processes and interaction between them, should be made.

(2) The rapid development of spatial interferometry, combined with
spectropolarimetry, opens up new and very promising possibilities to study
structures, such as departures from spherical symmetry. Also, the
development of CCD-like detectors for the infrared makes high-resolution
IR spectroscopy possible for many more stars than those studied so far.

(3) The semi-empirical modelling of the atmospheres and
circumstellar envelopes could be attempted much more systematically than
has been done until now. For example, as was shown by Tsuji, one could
use the high-resolution infrared FTS spectra already obtained for deriving
new and very interesting information on atmosphere and shell structures.

(4) The classical modelling is still important as a starting point
for any interpretation of observations. The hard work done by Alexander,
Augason, and Johnson, by Jorgensen, by the Australian group, and by others
to compile or calculate molecular line data, or to identify new opacity
sources, is of particular importance and will be of long-lasting value.

The question I posed at the start cannot be answered without
discussing _for what_ the models are to be used. For a general, overall
understanding? As a background for exploring specific physical processes?
For interpreting spectra, e.g., in terms of chemical abundances? As
regards the abundance analyses, the richness of lines in the spectra,
while being a nuisance in certain respects, often also allows the
possibility of finding many abundance criteria for a given chemical

element. One should exploit this and systematically search for those criteria least sensitive to the uncertainties in structure and velocity field. At least, one ought to combine any abundance analysis with a careful discussion of the effects of structural uncertainties (including thermal inhomogeneties) on the results obtained. Sometimes one might find that abundances, or abundance ratios, are astonishingly insensitive to the structural uncertainities; see, e.g., our discussion of CNO abundances in N stars (Lambert et al. 1986, Ap. J. Suppl. 62, 373).

The use of high resolution spectroscopy in abundance determinations is in my opinion still absolutely vital for high accuracy work - these results for suitable sets of apparently bright stars may then be used for calibrating lower-resolution criteria, suitable for exploration of more distant objects. Sometimes, this program is not possible to carry out, e.g., because the extragalactic stars are systematically different from the galactic ones. There, synthetic spectroscopy is helpful and necessary for a direct calibration of the lower-resolution criteria, but that should then be tested cautiously for the set of nearby stars with well known properties.

Obviously, many things can be done to improve our understanding of AGB star atmospheres and circumstellar envelopes. The major problem seems to be the lack of (wo)man power. How do we solve that?

Most important is not to make these stars less interesting, than they really are. They are very fascinating non-linear systems, where the complexity gradually grows - the number of degrees of freedom increases - from the deep photosphere to the interstellar medium. Yet, they are simple enough to be understood, accessible enough to be well observed, plentiful enough to be studied statistically.

The outer stellar layers are in fact appropriate testing grounds for current ideas about the growth of structures in non-equilibrium thermodynamics. I claim that this is of interest, not only to astrophysicists, but to scholars in many other fields, including cosmology. If cosmology has any bearing on the real world it must deal with the formation of complex structures in similar situations. "Black holes have no hair", but stars have a lot – and we should be proud of that and learn from them. In particular, don't shave them by misusing Occam's razor!

Glassgold: Peter Wood is now going to make a few comments.

Wood: I will make some comments on mass loss.

1. There seems to be some concern here about the use of the term "superwind" with regard to rapid mass loss. The term "superwind" was coined by Alvio Renzini, who noted that the average mass-loss rate required for the production of planetary nebulae ($\sim 2 \times 10^{-5}$ $M_\odot$ yr^{-1}), obtained by dividing the canonical mass for a planetary nebula by a typical planetary nebula lifetime, was much greater than that expected for red giants according to the Reimers mass loss formula $\dot{M} \propto LR/GM$. A question which arises is: does the "superwind" represent some type of mass-loss process different from that which occurs in, say, the optically visible Mira variables? I think the answer is probably "no", although schemes such as a switch in pulsation mode have been proposed in the past in an attempt to get a discontinuous increase in the mass loss rate (e.g. Jones et al. 1983, Ap. J., 273, 669). More recent estimates of mass-loss rates of pulsating red giants (Mira variables, IRC sources, OH/IR stars) obtained from observations of microwave emission from circumstellar CO, combined with luminosities derived from a period-luminosity relation, show that the mass loss rate experienced by red giant variables increases

dramatically with luminosity at the rate of a factor of 10 per ~0.3 mag increase in M_{bol}, at least until the radiation-pressure-driven wind limit is reached (Wood 1987, in Stellar Pulsation, ed. A.N. Cox, W.M. Sparks, and S.G. Starrfield (Springer-Verlag), p. 250). The increase in $\dot{M}$ with M_{bol} appears continuous, but it is much more rapid than predicted by the Reimers law. Thus there is probably no need to invoke some special mass loss mechanism in order to produce mass loss rates at "superwind" rates.

2. Another problem raised at this Colloquium is the observation of detached, cool shells around carbon stars indicating that a shell of matter was ejected typically a few thousand years ago. One possible way of doing this is with a helium shell flash. Detailed shell-flash calculations show that at a helium shell flash, the luminosity of an AGB star rises above the maximum interflash luminosity by ~0.5 mag for a few hundred years and then falls below the quiescent value by ~0.5 magnitude over a few thousand years. If the dependence of mass-loss rate on luminosity mentioned above for long-period variables applies, then the mass-loss rate during the shell-flash cycle might be expected to vary by a factor of 10 up and then down from the mass loss rate applying during interflash evolution. This may explain the detached shells observed by IRAS around many optically visible carbon stars. Given the typical maximum mass loss rates for red giants of ~3 x 10^{-5} $M_\odot$ yr^{-1} and that the duration of the luminosity peak at a helium shell flash is ~600 years (Wood and Zarro 1981, Ap. J., 247, 247), and that a much smaller mass loss rate exists outside the flash peak, a shell of mass ~0.02 $M_\odot$ would be ejected at a helium shell flash. This amount of shell mass is more than sufficient to be seen by IRAS according to the models presented by Sun Kwok presented at this symposium. These models predict typical shell lifetimes of ~3,000 years and, with a typical interflash time of ~50,000

years, we might expect roughly 2-10% of helium-shell-flashing stars to have detached circumstellar shells visible to IRAS at any given time.

Another explanation for the shells around carbon stars is that they are produced by some kind of sporadic ejection event. Such an event has been observed in the carbon long-period variable HV2379 in the LMC (Bessell and Wood 1983, MNRAS, 202, 31p). This star was observed to eject a shell of mass $M > 10^{-6}\ M_\odot$ in a manner very similar to that in R CrB stars.

Glassgold: I think we should discuss this particular point now. Ben, do you want to go?

Zuckerman: Yes. The question I want to ask Peter concerns the relative percentage of the carbon stars with 60 micron excesses. They are referred to in a paper by Van der Veen and Habing (A & A 194, 125 1988). When I looked at a preprint of their paper last year, I thought that possibly these 60-micron excess carbon stars could result from thermal pulsations as they suggested, and even though I don't know as much about as pulsating stars as you do, I could at least do the calculation on the time scales. With the numbers I published (1987, Proceedings of the Fifth Cambridge Workshop on Cool Stars, Stellar Systems, and the Sun, ed. R. Stencel and J.L. Linsky), I thought the percentages worked out very well. In other words, the percentage of carbon stars that show 60-micron excesses divided by the total number of carbon stars seems to fit in quite reasonably with the ratio of shell-flash to inter-pulse time. I personally feel that this is the explanation for the 60-micron excesses in carbon stars and the M stars too. Remember, Susan Kleinmann mentioned that more than 50% of the stars with 60-micron excesses in a certain box in the IRAS color-color plane that she showed were M stars. So there's no reason to suspect that the some thing won't be going on in them also.

Little-Marenin: Would you expect to see episodic ejections (i.e., multiple shells) around the two miras with decreasing P which are expected to come out of a helium-shell flashing episode (for example, R Hya)? Have these been searched for?

Wood: It really depends on where you are on that particular diagram. If you are down at 200 days, your mass loss rate has gone up from 10^{-7} to 10^{-6} $M_\odot$/yr; that's still not a terribly high mass loss rate that you might be getting to. If it's above 350 days, it should have a shell.

Willson: Over the past several years, it appears that our ideas and those of Wood have been converging -- and not only toward the Mira pulsation mode as the fundamental (F). Consider the thermal mass loss. In the original Wood/Cahn scheme, $\dot{M}$ increases gently until the F-mode onset leads to an ejection of a planetary nebula. In my orginal picture, I argued that the onset of the Mira F-mode pulsation brought about a large increase in $\dot{M}$. However, Ostlie and Cox pointed out that the growth rates become very large for small t_{KH}/P: up to 100%/cycle for $t_{KH}/P \sim 30$. This should be associated with large $\dot{M}$. On the log M - log L plot the line to $t_{KH}/P = 30$ falls near where Wood and Cahn had "PN ejection". So I would propose that $\dot{M}$ increases smoothly but also very rapidly for these stars with $t_{KH}/P \lesssim 30$. A rapid rate of increase of $\dot{M}$ at the end of the AGB is consistent with the interpretation of Bedijn, Baud and Habing, and their collaborators from interpreting IRAS spectra.

Kwok: I'd like to comment on two topics Peter mentioned. One is the slow mass-loss rate versus the so-called superwind, and the other is sudden ejection versus continuous ejection in changing the mass-loss rate. When Reimers proposed a formula based on late K and early M stars, we (at Univ. of Minn.) knew as early as 1970 that the mass-loss rate for many stars would be extremely high. For example, Merrill had demonstrated there is a continuous range of the optical depth of the silicate shell. As showed in my talk yesterday, the optical depth at 10 microns range over 3 orders of magnitude, so we knew there was a range of a factor of one thousand in mass-loss rate; in modern terms, $\dot{M}$ ranged between 10^{-7} and 10^{-4} $M_\odot$/yr. For years I have been skeptical about episodic ejection. The only discontinuity, as far as I see today, is still the onset of a total mass-loss mechanism because of the change in characteristics of the central star. It will go from a very extended star like an AGB star to a compact star like a white dwarf, and that will lead to a discontinuity as far as the mass-loss mechanism is concerned. Otherwise, as far as I can tell, and as supported by models, you can model it by a relatively smooth transition from 10^{-7} and 10^{-4} $M_\odot$/yr. I don't see any empirical evidence for any sudden ejection at any stage.

Glassgold: I think we better have a final word from Ben Zuckerman.

Zuckerman: I don't have a final word, but I want to raise a point about an apparent disagreement between observations and a long standing fundamental aspect of the theory of post-AGB star evolution. I think this dates all way back to Paczynski's work in the early 1970's. Let me show you what it is that is worrying me, and then maybe someone can very quickly clear this problem up.

We've had several discussions of evolution on the AGB and stellar pulsations in this meeting and many, many times over the last two decades. When a star leaves the AGB, according to theory, it runs at essentially constant luminosity toward very high temperatures (~10^5 K, see, e.g. the figures in Iben's article at this conference), and only then does the luminosity start to decline. The potential problem is that there are a variety of arguments from the optical, the radio, and the infrared that show for many stars there is a drop in luminosity at temperatures much lower than 10^5 K. If in fact there is an observational disagreement with this fundamental aspect of the theory that we've had now for over a decade, then I'm worried.

Let me show you a list of observations that I've put together. A couple in the optical are based on Peter Wood's papers (e.g. Ap. J. 307, 659, 1986) and private remarks to me at this meeting. Observations of low-temperature planetary nebulae with T = 20,000 K in both the Milky Way and the Magellanic Clouds seem to lie a factor of 3 or more below the AGB luminosity limit. Now Peter, wanting to avoid a contradiction with the theory, has explanations for the apparent discrepancy between the theory and observations for both the Milky Way and the Magellanic Clouds. The Milky Way discrepancy is explained away by an uncertain distance scale; observers just don't know how far away the planetaries are. In the Magellanic Clouds, Peter has suggested that dust absorption is the culprit. At any rate, he is clearly worried about this problem, and he has had to think of ways to place the burden on the observations because as far as I can tell, the theory is quite fundamental.

These are optical observations. Relevant observations in the infrared and radio have also accumulated over the last few years, including some by people in the audience. One of them has to do with an apparent decline in mass-loss rate which shows up in observations of post-AGB stars in a variety of ways. One way is the shape of the IRAS flux data. The typical circumstellar envelope around an AGB star with mass loss produces an IRAS spectrum with monotonically smaller fluxes as one goes from 12 to 100 μm. On the other hand, the spectrum of a star with a declining luminosity has a peak IRAS flux in either the 25 or 60 μm band.

Now a declining mass-loss rate could be due to one or more of three possible physical mechanisms. First is the cessation of pulsations, since as stars ascend the AGB, pulsations distend their atmospheres and promote mass loss according to what Dr. Bowen has told us. Another possibility is that dust grains stop forming as the temperature increases so that radiation pressure on these grains is insufficient to overcome gravity and lead to mass ejection. The third possibility, which is the most disturbing of the three, is that the luminosity of the central star declines. If radiation pressure drives mass loss, then a decline in the central star's luminosity will lead to declining mass loss. I say this is the most disturbing of the three possibilities, because this is the one that disagrees with what I think is pretty basic theoretical modelling.

Now we have three choices, so we could just say, well, for this observation of declining mass loss let's choose either option one or two as the explanation, but for the other observation that I would like to call to your attention there isn't any such choice. I think that declining luminosity is the only one that makes any sense. The problem arises when we take the ratio of the momentum in the wind deduced from CO to the bolometric luminosity, L, of the star. You can deduce $\dot{M}v_\infty$ momentum in the wind, from observations of CO rotational emission, as we've heard about a number of times through the meeting. You can deduce the star's radiative momentum by measuring the flux at all wavelengths, generally from IRAS observations. Now, if there are no serious errors in modelling the CO emissions, then this ratio is meaningful because both L and $\dot{M}$ depend on the distance to the star squared, and therefore the uncertain distance doesn't enter in. Eight objects detected by IRAS, several of which were studied by the French and their UCLA collaborators, have effective temperature less than 20,000 K and IRAS spectra consistent with declining mass-loss rates. For these 8 objects, $\dot{M}v_\infty c/L$ lies between 3 and 20. So the impression one has, if one believes radiation pressure drives these CO emitting envelopes, is that the luminosity has dropped by about a order of magnitude in many of these stars. I think there's a good chance that we're not making any stupid errors in our analysis since for NGC 7027, which is very hot, one would expect a drop in luminosity and indeed the momentum in its wind deduced using CO is about an order of magnitude lower than the momentum that is currently being carried by the radiation from its central star.

Glassgold: Let me just ask a technical question on the last way of making this argument. Are those CO measurements from bipolar outflows? Could there be some problems in the estimates of the rate in which the momentum is being sent out in the wind?

Zuckerman: I don't think so, because NGC 7027 gives a similar result. Also, all 8 of these objects have the unusual IRAS spectrum which is consistent with a declining mass-loss rate. You don't find these large ratios of $\dot{M}vc/L$ from CO emission for stars that don't also have the funny IRAS spectra that peak at 25 or 60 μm. In other words, if you model a star with an ordinary IRAS spectrum, you will find that in fact $\dot{M}v \leq L/c$. Finally, as Omont mentioned earlier in this meeting, if you model the OH-IR stars which have funny IRAS spectra, if anything, you get too <u>low</u> a mass-loss rate from CO. The error seems to be such that CO gives you a lower mass-loss rate than is really there, so I think that the chance of seriously overestimating a mass loss rate from CO is not very great.

Omont: I do agree with your point, of course. Nevertheless, I would like to put forth some caveats. The first one concerns the analysis of the CO data. I mentioned the case of OH-IR stars; those are very cold ones -- the most extreme ones. Probably the computed CO mass-loss rate is too small by an order of magnitude. But these are special objects which are not completely understood, and in that case, probably because you have some complex radiative transfer, you might achieve some much larger radiation pressure on grains and have some other consequences on heating the gas. So maybe it is not true that the CO mass is simply related to the real mass. Also, it is possible that the momentum could exceed the theoretically possible limit if the optical thickness is larger than 1; then $\dot{M}v > L/c$. Nevertheless, for some of these objects, especially for the few IRAS planetary nebulae, I don't see any way to avoid this decline in the luminosity before reaching 30,000 K.

Zuckerman: I think Alain is being very conservative. Let me just comment on one point that he raised -- on the large optical thickness in the infrared dust continuum. It's true in principle that if the envelope is very optically thick, you might get optical depth values of a few, but I don't think they'll be as high as 20. But, more important, some of the IRAS sources, which have ordinary spectra that decrease from 12 to 100 microns, are certainly very optically thick in the dust continuum and CO; they have enormous mass-loss rates -- just as big, I think, as the objects with funny IRAS spectra which seem to show $\dot{M}v > L/c$. It seems to me that this phenomenon, $\dot{M}v > L/c$, should show up in these objects which have normal IRAS spectra if it were simply a matter of large optical depths in the IR, but it never does. One always finds $\dot{M}v \leq L/c$ in objects with normal IRAS spectra, none of these order-of-magnitude discrepancies.

Glassgold: Let's accept the evidence and then ask what it means.

Zuckerman: Yes, but I want to make the argument as iron-clad as possible.

Renzini: I have several comments. The first is that evolution during the post-AGB phase takes place at constant luminosity insofar as hydrogen-burning models are concerned, while if the stars leave the AGB right at the time of a thermal pulse, then there is a decrease in luminosity during the transition from the AGB to the PN stage. This, however, is probably less than a factor of 10 (cf. Iben 1984, Ap. J. 277, 333). On the other hand, if I remember correctly, it is not infrequently the case that carbon stars as well as oxygen-rich stars exhibit a momentum flux in excess of that of the stellar luminosity. I remember Knapp et al. providing evidence for this, so right back on the AGB so there seems to be the same kind of situation you have mentioned.

Zuckerman: No, no! The discussion in that old (1982) paper by Knapp et al. has been superseded many times, and right now the situation is that with the best data and best analysis, the ordinary AGB stars do not have excess momentum; they lie within a factor of 2 of L/c. It is only these transition objects which seem to show us, time and time again, the much greater excess momentum.

Renzini: On the other hand, the ejection itself of the envelope in some cases might not necessarily be due to radiation pressure. The ejection might be due to processes which have nothing to do with your idea.

Zuckerman: Well, as Sun Kwok commented earlier, I think he believes that basically you start out with radiation pressure on dust grains and you end up with radiation pressure in resonance lines or something similar. This leads to these high velocity winds coming off the hot central stars, and there isn't any special thing that throws off a lot of mass. I personally agree with him. So unless you can show very clearly that something that does not have to do with radiation pressure is operating at just the critical moment here, I don't think it is a way out of the problem.

Renzini: There does seem to be a lot of confusion over the years about the semantics of sudden mass ejection and superwinds, a confusion that dies hard. (Laughter.) Anyway, for example, I would find it very difficult to distinguish operationally between a discontinuous transition from a "regular wind" to a "superwind" regime, and a continuous increase in mass loss rate by, say a factor of 10 which takes place while the luminosity increases by only 0.5 mag. Back to the momentum problem, I would like to mention the possibility that the final ejection may not necessarily be an individual, unique, dramatic event, but may well consist of a series of less dramatic events, each removing perhaps a small fraction of the envelope, until the whole ejection is accomplished. This kind of scenario would indeed apply to the thermal-pulse mechanism proposed by Wood for getting rid of the envelope in the more massive AGB stars. In such a mechanism mass ejection is due to processes which are not entirely radiative. So, I think we should leave open the possibility that (some) PNe may be ejected by processes other than radiation pressure, and Zuckerman's momentum argument can be interpreted in support of this possibility.

<u>Wallerstein</u>: I would like to comment on another aspect of mass ejection from massive stars which hasn't been mentioned here. It's suggested by Iben's evolutionary tracks. There exist quite a number of hot subdwarf stars with effective temperatures of 30-50,000 K. Some have hydrogen-rich envelopes, and some are helium stars. They don't have planetary nebulae around them at all!

<u>Iben</u>: They are merged white dwarfs!

<u>Parasarathy</u>: IRAS observations show no evidence for the presence of dust envelopes around subdwarf OB stars.

<u>Wallerstein</u>: I think that's correct. Nothing else is visible, and with a central star of 30-50,000 K you can very easily see emission lines.

Renzini: The question is that the time for the envelope to disperse may be far shorter than the time for the star to fade So we can have a bright "planetary nebula nucleus" without the "planetary nebula"!

Glassgold: Did you get the answer to your question, Ben?

Zuckerman: Well, I just want to make a comment about non-radiation pressure mechanisms in these objects. In IRAS 2128 +50, for example, the authors of a paper (Astr. & Ap. 198, L1, 1988) which includes two members of the audience state that their estimate of the distance to this cool PN corresponds to a luminosity that is less than the minimum luminosity a star can have and still be on the AGB according to the theoretical models. If you want to say there is no decline in luminosity, then you've got to increase the distance to IRAS 2128+50 and other bright transition objects. I feel that as we go through this list (and we have done this), you will find that it is very difficult to move all these stars to the distances required to get luminosities of a few times 10^4 $L_\odot$. Independent distance estimates place the star close by, and therefore it doesn't have the luminosity that it had on the AGB, independent of CO.

Iben: I think Alvio's point is that there may be no relationship whatsoever between $\dot{M}v$ and L/c. Any mass loss occurred at some point earlier than the observed L/c.

Zuckerman: No, what I'm saying is that this independent estimate of luminosity has nothing whatsoever to do with the CO. It is based solely on distance estimates.

Iben: The fact remains that your major argument is based on making estimates of $\dot{M}v$ and saying that this implies a value of L/c which is less than that anticipated on the AGB. The only direct evidence for luminosity is the thing you just now quoted.

Zuckerman: OK. If one goes through the objects in question, there will be a real problem in placing them at the appropriate distances. At any rate, one has to show there is some other mechanism that can throw off large amounts of molecular gas.

Iben: Tuchman, Sack, and Barkat (1979; Ap. J. 234, 217) a decade or so ago found a hydrodynamical ejection mechanism for planetary nebula. I don't know why Peter doesn't defend that because he did the same thing.

Wood: I don't believe it! (Laughter)

Zuckerman: I don't know if anyone believes these mechanisms anymore. There must be some reason why they've fallen into disfavor. One would have to look at it in detail.

Iben: It's tough to do!

Renzini: Yes. That's the only reason.

Glassgold: Probably we should stop this discussion. We're getting close to the 4:00 o'clock deadline. We have time for one brief comment.

Renzini: Let me just say a few words about my third point: R CrB stars. Just ten years ago I suggested a possible way for producing hydrogen deficient giants from single stars, and I still think that this scenario is in every respect far superior to the one reviewed this morning by Schonberner. In brief, when a final thermal pulse takes place in a post-AGB star where the hydrogen shell has already ceased burning, then the tiny residual hydrogen envelope can be ingested into the convective shell, at the base of which helium is burning. Hydrogen is then carried into the hot interior where it burns quite quickly, and the released energy may cause the expansion of the former intershell region, which is mostly composed of helium and carbon, which are indeed the dominant species at the surface of R CrB stars.

I am reasonably convinced that such a scenario can explain every major characteristic of R CrB stars -- such as dimensions, timescales, surface compositions, luminosities, and kinematics. Quite naturally the timescale of the R CrB "loop" comes out to be nearly 1/10 the "fading time" of the previous planetary nebula stage. (This follows from the identical ratio in the available hydrogen fuel in the two stages.) The lifetime of the R CrB stage is therefore extremely sensitive to the final mass of these stars, and may range from just a few years (as e.g. in V605 Aql) for $M_f = \sim 1\ M_\odot$, up to perhaps as much as 10,000 years for $M_f = \sim 0.55\ M_\odot$. Concerning surface abundances, beyond He and C, the presence of trace H, N, and occasionally s-process elements (as in U Aqr) is also a natural property of this model. In your place I would have no problem at all in choosing the best R CrB model! (Laughter)

Glassgold: I don't think we're going to be able to discuss this until the next meeting. Thanks to all the participants and to our hosts for an excellent time! (Loud applause.)

INDEX

abundances, elemental (see also chemical composition) 98, 99, 102-104, 109-128, 142, 148, 197, 322, 392, 393, 402
accretion 371
acoustic waves 263, 309-311, 317
Am stars, possible relation to Barium stars 60
AGB (see asymptotic giant branch)
angular diameter 82-87
asymptotic giant branch (AGB) evolution, termination of 349-351
asymptotic giant branch (AGB) stars (see also individual topics and star types)
    classification of 4-11
    CSE's of 349-357, 359-364, 367, 369, 370, 373, 383
    evolution of 58, 104-112, 161-173, 208-213, 225, 236, 284-290, 341, 342, 349-351, 412, 414
    in external galaxies 37-49, 104-106
    galactic distribution and space density of 15-24, 340, 348-351
    IRAS observations of 58, 306
    kinematics of 26-33
    luminosities of 37-49, 340, 341
    mass-loss rates from 229-231, 342-346, 349-354, 387, 424, 425
    photometry of 13-24, 35-41, 305-311, 367
atmospheric models (see model atmospheres)
Balmer lines 257, 262, 371, 369
barium stars 4-6, 31-33, 59, 60, 108-110, 139, 145, 150, 196-203, 223, 224, 317, 408
bifurcation in atmospheres (see inhomogenities)
binary stars 31-33, 108-110, 138, 139, 156, 196-203, 222, 232, 234, 303, 304, 362, 371, 375, 378
bipolar nebulae 325, 326
born-again AGB stars (see post-AGB stars)
brightness distribution (stellar surface) 81-84
C IV line 312, 371
^{13}C pocket (see mixing)
^{13}C/^{12}C ratio 111, 112-115, 140, 150, 331, 375, 407
carbon enhancement 112-117
carbon grains (see grains)
carbon monoxide (see CO)
carbon recombination opacity 164, 169-172, 180
carbon stars (see also AGB stars and individual topics)
    chemical composition of 112-117, 122, 136
    ^{13}C rich (J stars) 53, 140, 407-411, 416, 417
    CSE's of 278-284, 297, 303, 316, 359-364, 367
    in external galaxies 42-44, 414-417
    as extragalactic distance indicators 48, 49
    IRAS observations of 15, 16, 23, 24, 306, 402
    luminosities and space densities of 15-17, 21, 63, 340-342
    mass loss from 22, 367, 374, 375
    molecules in 52, 113, 116, 245, 246
    photometry of (see also IRAS) 37-44, 53, 244, 366
    production of 36, 44, 51, 104-106, 169-173, 224, 229-231, 284-287
    variability of 245, 246, 424
    violet and ultraviolet observations of 307, 313, 314, 366
Cepheids 242
CH 5, 7, 151
CH stars 31-33, 196-200, 202, 203

chemical composition 3, 109-128, 146-148, 211-213, 228, 304-318, 331, 332,
                      410, 411
chromospheres 252, 263, 303-318, 351, 366, 381, 401, 385
chromospheric heating and cooling 309-311, 315-317, 366, 381
circumstellar chemistry 359, 382
circumstellar (CS) dust 54, 55, 86, 248-250, 284, 287, 290, 292, 313, 351-355,
                         359-364, 367, 372, 377, 378, 408, 411
circumstellar envelopes (CSE)
    models of 327-331, 367, 369, 381, 422-423
    observations of 321-324, 370, 385-388, 404
circumstellar (CS) gas 232, 251, 304, 317, 321-324, 372
classification of PRG stars 4-11, 38-40, 54, 62, 63, 99, 104
CN 52, 63, 75, 113, 116, 152, 146, 402
C/N ratio 228
CNO 103, 104, 109-123, 152, 182, 183
CO
    photospheric (visual, infrared) 90-95, 102-104, 113, 146, 244, 282 302
    circumstellar (radio) 226, 232, 233, 302, 322-325, 333-335, 368, 370, 373,
                           410, 411
    cooling by 93-95, 316, 381
C/O ratio 61, 102-104, 247, 284, 366, 375, 379
colors 73, 74, 87-89
common envelope evolution 232, 234
condensation (see dust)
convection 93-97, 150, 152, 164, 169-172, 227, 233, 236, 416, 417
convective overshooting (see also convection) 164, 169-172
core breathing pulses 163, 164
core helium burning 162, 163, 416
core mass 158, 206, 211, 235, 349-351, 413-416
coronae, stellar 401
diamonds (see meteorites)
distances, stellar 37, 48, 49, 331, 332
double shell sources 152
dredge-up (third, on AGB) 110, 119-128, 134-135, 137, 161-173, 206, 227,
                           229-231, 245, 251, 413
dust (see also CS dust) 229-231, 272-279, 284, 290, 359-363, 367, 373-377, 379,
                         381
dust shell (see CS dust)
equilibrium 371, 381
evolution of PRG stars (see stellar evolution)
excitation, radiative 368, 372, 402
expansion velocity 369
fluorescence 372
Four color IR Sky Survey (AFGL) 16
galactic center (see nuclear bulge)
galactic evolution and PRG stars 36, 37
galaxies, PRG's in external 36-49, 46, 47
globular clusters 6, 7, 16, 242, 243, 322
grains (see also CS dust; dust)
    carbon 54, 355-357, 367
    silicates 284, 288-290
    silicon carbide 54, 367, 375
H_2 90, 327, 328
H_2O 68, 72-78, 153, 298, 404
HCN 65, 68, 155
helium core ignition (see core helium burning)

helium lines 378, 401
helium shell flash (see also thermal pulse) 349-352, 412, 413
hot bottom burning (envelope burning) 111, 112, 117, 152, 236, 356
H-R diagram 58, 266
Hubble constant 48, 49
hydrogen burning (see also thermal pulses) 213-220, 349-352, 356, 357
hydrogen-deficient stars 151, 353-357, 433, 434
hydrogen, neutral (H I) 151, 251, 262, 369
ice 370
incredibly thin hydrogen envelope scenario (see also post-AGB stars) 218-220
infrared excesses 288, 306, 307, 313, 316, 340, 345, 351, 385, 402, 423
infrared spectroscopy 54, 284
inhomogeneities, atmospheric 93-97, 316, 381
initial mass 390
initial mass function (see luminosity function)
interstellar extinction 15
IRAS observations 14, 24, 54, 234, 294, 305-307, 340, 351, 352, 359-361, 380, 384, 391-400, 428, 405, 409
IRAS color-color diagram 17, 54, 58, 233, 249, 250, 285-289, 299, 305-305, 363, 369, 395
IRAS low resolution spectra (LRS) 16, 24, 54, 292, 294, 359-361, 367, 394, 402, 429
IRAS sources (see individual stellar types)
isotopic ratios 117-122, 332, 333, 374, 375, 408
IUE observations 55, 262, 298, 299, 305-307, 371, 386
joy of stars 399-402
KAO observations 370
K giant stars 136, 145, 150, 198-202, 228, 234
kinematics
    of stars 5, 26-34, 56, 60, 240, 245
    of CSE (see also mass loss) 326, 327
line blanketing 68, 71-73, 87-89, 153, 157
lithium 150
Large Magellanic Cloud 10, 33, 45, 46, 51, 57, 238
long-period variables (see Mira variables)
LTE and departures from LTE 69, 89, 154, 372, 381, 401, 418
luminosities 28-30, 32, 46-49, 295, 331, 332
luminosity function, of PRG stars in external galaxies 44-49
Lyman α 315, 316
magnesium (Mg II) lines 234, 262, 307, 308, 313-315, 366
M-MS-S-SC-C sequence 14, 102, 107, 112-117, 134, 140, 197, 304, 407-411
M giant stars 17-24, 26-31, 84, 85, 111-114, 136, 141, 145, 147, 153, 288, 305-311, 323, 324
M supergiants (see also individual stars) 85-87, 136, 154, 242, 248, 260, 299, 323, 324
Magellanic Clouds (see also Large Magellanic cloud) 10, 15, 19, 29, 45, 46, 75, 104-107, 152, 197, 247, 248, 413-417, 426
main sequence stars 206, 223
masers 245, 251, 298, 326, 332, 333, 369, 380, 404
masses 27, 138, 353, 409
mass loss
    evolutionary effects of 169-172, 211-213, 224, 225, 226, 228, 229-231, 284, 349-351, 382, 390, 391, 403, 424
    mechanisms of 233, 234, 269-282, 317, 342, 382, 412
    observations of 14, 58, 226, 319, 323, 324, 352, 353, 421, 422
    rates of 14, 51, 211-213, 224, 233, 281, 275, 327-335, 342-346, 355-357, 368, 382, 384, 421, 422, 424, 425-430

mass transfer 60, 122, 139, 140, 199-203, 222, 223, 224, 234, 304
meteorites, origin of grains in 374, 375
microturbulence 91-93
microwave observations 321-324, 332-335
Mira variables (see also AGB stars and individual star types)
    infrared observations of 391-398, 402
    kinematics of 26-31, 50
    masers associated with 380, 404
    models of 72-78, 269-282, 297
    periods of 244-248, 409-412
    photometry of 244-248, 292, 298
    radio observations of 321-324, 383, 404
    space distribution and luminosities of 17-24, 340-342
    spectra and chemical composition of 61, 138-140
mixing 105, 114-120, 133, 134, 169-172, 190, 191, 409
mixing length 160, 395
model atmospheres:
    chromospheres 314, 366
    comparison with observations 73, 75-78, 81-97, 375, 377, 149, 151, 31
                                         366 399-402
    extended atmospheres 68-73, 157, 269-282
    photospheres 67-79, 141, 153, 236, 375, 399-402, 417-421
    with shocks 72-79, 269-282
molecules (see also individual molecules)
    molecular bands in infrared 113, 155, 284, 381
    in radio 284, 322-324, 373
    in visible 52, 61, 72-78, 90, 91, 95, 102-104, 113, 114, 284, 403
momentum loss 425-430, 432, 433
MS stars 133, 306-312
nova 218
nebula, reflection 55
neutron exposure 133, 176-179, 227
neutron source 109, 110, 127, 128, 133, 134, 176, 180-188, 227, 237
NGC 7027 410, 428
N/O ratio 235
nuclear bulge, red giants in, 10, 42, 246, 247
nucleosynthesis 227
number of PRG stars 27, 42-44, 49, 51, 377, 423
objective prism survey 63
OH 95, 322-324, 332, 333, 380, 428, 429
OH/IR stars 30, 352, 368, 409, 415
opacities 68, 153, 155, 157, 180, 236, 351, 366
oxygen stars (see K giants, M giants)
oxygen 102-104, 113-123, 229-231, 408
PAH molecules 233, 294
period-luminosity relation
    of Mira variables 28-31, 138, 241, 289
    of RV Tauri stars 295
photodissociation (see excitation)
photometry 35-41, 53, 57, 59, 73-77, 244, 259, 260, 298, 366, 367, 391-39
photospheres (see model atmospheres)
planetary nebula 113, 114, 208, 209, 226, 232, 233, 235, 314, 315, 388, 3
                      425-427, 433
planetary nebula nuclei 211-220, 235, 351, 352, 425-427, 432
polarization 296
population II stars 141, 154

post-AGB stars 55, 146-148, 196, 211-220, 228, 330, 331, 348-353, 356, 357, 384-389, 425-430
pre-planetary nebulae (PPN) 207-210, 228, 232, 233, 290, 323, 324, 351-352, 354-357, 380, 384-389
proto-planetary nebulae (see pre-planetary nebulae)
pulsation 228, 269-282, 293, 353-356, 369, 383, 424
radiative excitation (see excitation; fluorescence)
radial velocities 56, 60, 139, 144, 219, 299
radio observations 284, 288, 311, 313, 321-324, 373, 383
R Cor Bor (R CrB) stars 57, 215, 293, 348, 353-357, 387, 403, 416, 423, 433
red-giant branch 206, 223, 228, 237
Roche lobe 222, 234
r-process 183-185, 190
R stars 10, 53, 197, 203, 407-410
semiconvection (see convection)
shell flashes (see thermal pulses)
shocks, effect on atmosphere 297, 381
shocks, models of 263, 269-282, 316, 317
shocks, observations of 251, 263, 269-282, 297, 316
silicate grains (see also dust; circumstellar (CS) dust) 284, 288-290
silicon carbide (SiC) 54, 284-286, 290, 367, 375
spectral classification (see classification)
spectral energy distribution 74-79, 82-84, 151, 153, 155
spectral lines (see also individual atoms and molecules)
　　in infrared 61, 261, 378, 401
　　in radio 321-327, 369
　　in ultraviolet 55, 305-317, 366, 369, 372
　　in visible 4, 5, 8, 55, 89, 107, 261, 296, 369
spectral type 4-11
spectrophotometry 54
spectroscopic binaries 222, 223
s-process elements 60, 108, 109, 111, 127, 128, 131-141, 169, 170, 187, 190, 191, 237
s-process enhancement 104, 110, 111, 123-128, 131-141, 154, 156, 176, 183-188, 190, 191, 206
S stars 16-19, 26-31, 104, 108, 109, 114, 117, 124-126, 136, 197, 245, 288, 306-313, 317, 371, 378
spectroscopy 53, 261, 262
star formation 36, 47
stars, groups of
　　ζ Aur 234
　　barium (see barium stars)
　　carbon (see carbon stars)
　　ZZ Ceti 218-220
　　CH (see CH stars)
　　CH-weak (weak G-band stars) 7, 9
　　Eta carinae 30
　　globular cluster (see globular clusters)
　　M31 46
　　M giants (see M giants)
　　M supergiants (see M supergiants)
　　N55 46
　　N300 46
　　NGC 2403 47
　　nuclear bulge (see nuclear bulge)
　　Omega centauri 30
　　R stars (see R stars)
　　S (see S stars)
　　SC stars 27, 245
　　strong CN stars 4
　　R CrB stars (see R Cor Bor stars)
　　subgiant CH stars 196, 197
　　RV Tau 154, 242, 294-296
　　47 Tuc 242, 243, 411, 412
stars, individual
　　α Tau 84, 85
　　Arcturus 82, 83, 94, 144, 306, 307, 311
　　Betelguese 85-87, 261, 262, 264, 299

μ Cep 264
μ Gem 85
μ Leo 2
BD -21.3873 156
HD 35155 156
HD 36598 33
HD 39853 150
HD 49500 7
HD 130255 4-6
HD 182040 151
NT Tel 30, 243
HR 1105 378
HR 3126 55
IRC+10216 15, 326, 328
FG Sge 146-148
SN 1987A 238
TX Psc 155, 306, 307, 314

5 Cet 234
IRAS 212824+5050
TW Hor 256
o Cet 111, 257
R CrB (see R Cor Bor stars)
30 g Her 91-93
AC Her 290
UV Her 290
ER Del 156, 371
V CrB 32
χ Cyg 61, 111, 251, 261
RV Tau 288-290, 294, 295

stars, joy of research on 399-402
stellar evolution
    descriptions of 36, 37, 55, 58, 160, 104-106, 124, 125, 131, 152,
        169-172, 221, 234, 290-292, 308, 339, 349-354, 356, 357, 412
    with mass loss 225, 349-351, 362
    modeling of 169-172, 211-220, 225, 238, 394-396; 413, 414
    observational tests of 104-106, 110-128, 164, 290, 317, 318, 397, 404, 407-410
stellar wind (see also mass loss; mass transfer) 222, 378
subdwarfs 431
supergiants (see also M supergiants 242, 314, 315, 348, 353-357, 430, 431
supergiants of spectral types F, G as post-AGB stars (see also post-AGB stars)
    146-148, 348, 351, 384-388
superwind (see also mass loss) 211-213, 217, 228, 384, 421, 422, 424, 425
surface brightness distribution 84-87
symbiotic stars 156, 371, 378, 391-400
technetium (Tc) 29, 102, 107-109, 128, 131-141, 197, 227, 251, 304
temperatures, effective 70, 73, 83, 86, 349-352, 366, 418
thermal pulses 30, 105-107, 109-128, 130, 161, 164, 182, 183, 206-208, 213-218,
        225, 236, 349-352, 415, 422, 423
TiO 56, 68-78, 73, 103, 104, 146, 284
Two Micron Sky Survey (TMSS) 13, 14, 15-24, 24, 340
variable stars
    irregular 136, 241, 258-266;
    mira (see also Mira stars) 19, 20, 136, 241-252, 258, 292, 298, 424
    models of 263-276, 291, 297
    periods of 260, 261, 264, 293
    semi-regular 19, 20, 136, 241-252, 258, 261, 262
    types and light curves 254, 260, 298
velocity dispersion, RG stars 26-29, 32
VLA observations 298, 302
water vapor (see H_2O)
white dwarfs 208-220, 223, 356, 357, 431
x-rays 401

OCT 2 4 1989